Lecture Notes in Computer Science 1125
Edited by G. Goos, J. Hartmanis and J. van Leeuwen

Advisory Board: W. Brauer D. Gries J. Stoer

Springer
*Berlin
Heidelberg
New York
Barcelona
Budapest
Hong Kong
London
Milan
Paris
Santa Clara
Singapore
Tokyo*

J. von Wright   J. Grundy   J. Harrison   (Eds.)

Theorem Proving in Higher Order Logics

9th International Conference, TPHOLs'96
Turku, Finland, August 26-30, 1996
Proceedings

Springer

Series Editors

Gerhard Goos, Karlsruhe University, Germany

Juris Hartmanis, Cornell University, NY, USA

Jan van Leeuwen, Utrecht University, The Netherlands

Volume Editors

Joakim von Wright
Jim Grundy
John Harrison
Åbo Akademi University, Department of Computer Science
Lemminkäinengatan 14A, 20520 Turku, Finland

Cataloging-in-Publication data applied for

Die Deutsche Bibliothek - CIP-Einheitsaufnahme

Theorem proving in higher order logics : 9th international conference ; proceedings / TPHOL '96, Turku, Finland, August 26 - 30, 1996 / J. von Wright ... (ed.). - Berlin ; Heidelberg ; New York ; Barcelona ; Budapest ; Hong Kong ; London ; Milan ; Paris ; Santa Clara ; Singapore ; Tokyo : Springer, 1996
  (Lecture notes in computer science ; Vol. 1125)
  ISBN 3-540-61587-3
NE: Wright, Joakim von [Hrsg.]; TPHOL <9, 1996, Turku>; GT

CR Subject Classification (1991): B.6.3, D.2.4, F.3.1, F.4.1, I.2.3

ISSN 0302-9743
ISBN 3-540-61587-3 Springer-Verlag Berlin Heidelberg New York

This work is subject to copyright. All rights are reserved, whether the whole or part of the material is concerned, specifically the rights of translation, reprinting, re-use of illustrations, recitation, broadcasting, reproduction on microfilms or in any other way, and storage in data banks. Duplication of this publication or parts thereof is permitted only under the provisions of the German Copyright Law of September 9, 1965, in its current version, and permission for use must always be obtained from Springer-Verlag. Violations are liable for prosecution under the German Copyright Law.

© Springer-Verlag Berlin Heidelberg 1996
Printed in Germany

Typesetting: Camera-ready by author
SPIN 10513526   06/3142 – 5 4 3 2 1 0   Printed on acid-free paper

Preface

This volume contains the proceedings of the *The 9th International Conference on Theorem Proving in Higher Order Logics* (TPHOLs'96). The previous meetings in the series were known initially as HOL Users Meetings, and later as Workshops on Higher Order Logic Theorem Proving and its Applications. The new name for the series reflects a broadening in scope of the conferences, which now encompass work related to all aspects of theorem proving in higher order logics, particularly when based on a secure mechanization of logic. As a sign of the broad scope of the conference, these proceedings contain papers describing work using the Alf, Coq, HOL, Isabelle, LAMBDA, LEGO, NuPrl, and PVS theorem provers.

The forty-six papers submitted to TPHOLs'96 were generally of high standard. All submissions were fully refereed, each paper being read by at least three reviewers appointed by the programme committee. Twenty-seven papers were selected for presentation as full research contributions. These are the papers contained in this volume. The conference also continued the tradition of its predecessors of providing an open venue for the discussion and sharing of preliminary results. Thus the programme included an informal poster session where twenty researchers were invited to present their work. The poster papers are available in a supplementary proceedings produced as a General Publication of the Turku Centre for Computer Science (TUCS).

The organizers are pleased that Mike Gordon and Andrzej Trybulec accepted invitations to be guest speakers at the conference. In addition to the two invited lectures, the conference also included two tutorials, by Paul Jackson and Christine Paulin-Mohring.

The conference was sponsored by the Turku Centre for Computer Science, the Research Institute of the Foundation of Åbo Akademi, and the Academy of Finland. Their financial support is gratefully acknowledged. We also want to thank Christel Engblom, Sirpa Nummila, and Gundel Westerholm who assisted in matters of local organization.

August 1996

Joakim von Wright
Jim Grundy
John Harrison

Conference Organization

Conference Chair:

Joakim von Wright (Åbo Akademi)

Programme Committee:

Flemming Andersen (Tele Danmark)
Albert Camilleri (Hewlett-Packard)
Tony Cant (DSTO)
Elsa Gunter (AT&T)
Joshua Guttman (MITRE)
John Herbert (SRI)
Paul Jackson (U. Edinburgh)
Ramayya Kumar (FZI Karlsruhe)
Tim Leonard (DEC)
Paul Loewenstein (Sun)
Tom Melham (U. Glasgow)
Tobias Nipkow (TU München)
Christine Paulin (ENS Lyon)
Larry Paulson (U. Cambridge)
Tom Schubert (Portland State U.)
David Shepherd (SGS-THOMSON)
Phil Windley (BYU)
Joakim von Wright (Åbo Akademi)

Organizing Committee:

Jim Grundy (Åbo Akademi)
John Harrison (Åbo Akademi)
Joakim von Wright (Åbo Akademi)

Invited Speakers:

Mike Gordon (U. Cambridge)
Andrzej Trybulec (U. Warsaw, Białystok)

Tutorial Speakers:

Paul Jackson (U. Edinburgh)
Christine Paulin (ENS Lyon)

Additional Referees:

David Basin
Paul E. Black
Rosina Bignall
Christian Blumenröhr
Annette Bunker
Roy L. Crole
Anthony Dekker
Katherine Eastaughffe
Dirk Eisenbiegler
Jens Chr. Godskesen
Andrew Gordon
Jim Grundy
Kelly Hall
John Harrison
Michael Jones
Trent Larson
Thomas Långbacka
Brendan Mahony
Michael Norrish
Chris Owens
Maris Ozols
Kim Dam Petersen
Jimmi S. Pettersson
Christian Prehofer
Emil Sekerinski
Kaisa Sere
Donald Syme
Marina Waldén

Contents

Translating Specifications in VDM-SL to PVS 1
S. Agerholm

A Comparison of HOL and ALF Formalizations of a
Categorical Coherence Theorem 17
S. Agerholm, I. Beylin, P. Dybjer

Modeling a Hardware Synthesis Methodology in Isabelle 33
D. Basin, S. Friedrich

Inference Rules for Programming Languages with
Side Effects in Expressions 51
P. E. Black, P. J. Windley

Deciding Cryptographic Protocol Adequacy with HOL:
The Implementation 61
S. H. Brackin

Proving Liveness of Fair Transition Systems 77
H. Busch

Program Derivation Using the Refinement Calculator 93
M. Butler, T. Långbacka

A Proof Tool for Reasoning About Functional Programs 109
G. Collins

Coq and Hardware Verification: A Case Study 125
S. Coupet-Grimal, L. Jakubiec

Elements of Mathematical Analysis in PVS 141
B. Dutertre

Implementation Issues About the Embedding of
Existing High Level Synthesis Algorithms in HOL 157
D. Eisenbiegler, C. Blumenröhr, R. Kumar

Five Axioms of Alpha-Conversion 173
A. D. Gordon, T. Melham

Set Theory, Higher Order Logic or Both? 191
M. Gordon

A Mizar Mode for HOL 203
J. Harrison

Stålmarck's Algorithm as a HOL Derived Rule 221
J. Harrison

Towards Applying the Composition Principle to Verify a
Microkernel Operating System 235
M. R. Heckman, C. Zhang, B. R. Becker, D. Peticolas, K. N. Levitt,
R. A. Olsson

A Modular Coding of Unity in Coq 251
B. Heyd, P. Crégut

Importing Mathematics from HOL into Nuprl 267
D. J. Howe

A Structure Preserving Encoding of Z in Isabelle/HOL 283
Kolyang, T. Santen, B. Wolff

Improving the Result of High-Level Synthesis
Using Interactive Transformational Design 299
M. Larsson

Using Lattice Theory in Higher Order Logic 315
L. Laibinis

Formal Verification of Algorithm $\mathcal{W}$: The Monomorphic Case 331
D. Nazareth, T. Nipkow

Verification of Compiler Correctness for the WAM 347
C. Pusch

Synthetic Domain Theory in Type Theory:
Another Logic of Computable Functions 363
B. Reus

Function Definition in Higher Order Logic 381
K. Slind

Higher Order Annotated Terms for Proof Search 399
A. Smaill, I. Green

A Comparison of MDG and HOL for Hardware Verification 415
S. Tahar, P. Curzon

A Mechanisation of Computability Theory in HOL 431
V. Zammit

AUTHOR INDEX 447

Translating Specifications in VDM-SL to PVS

Sten Agerholm

The Institute of Applied Computer Science (IFAD), Forskerparken 10,
DK-5230 Odense M, Denmark, E-mail: sten@ifad.dk

Abstract. This paper presents a method for translating a subset of VDM-SL to higher order logic, more specifically the PVS specification language. This method has been used in an experiment where we have taken three existing, relatively large specifications written in VDM-SL, hand-translated these to PVS and then tried to type check the results. This is not as simple as it may sound since the specifications make extensive use of subtypes, via type invariants and pre- and postconditions, and therefore type checking necessarily involves some theorem proving. In trying to prove some of these type checking conditions, a worrying number of errors were identified in the specifications.

1 Introduction

In a research project entitled "Towards industrially applicable proof support for VDM-SL", we aim at developing tool support for proving theorems about specifications written in the VDM Specification Language (VDM-SL) [6]. We would like to base our work on available theorem proving technology.

The goal of our first experiments is to investigate what the benefits and limitations of existing theorem provers are in reasoning about "realistic" specifications. To accomplish this, we have taken a number of existing non-trivial specifications written in VDM-SL and hand-translated these to the specification language of the theorem prover PVS [13, 8]; some initial experiments were also made with the HOL system [9]. We then tried to reason about the translated specifications, focusing on proving selected type checking conditions. Such conditions arise since the use of type invariants and pre- and postconditions in VDM-SL specifications is translated to type predicates (subtypes) in PVS.

The main contribution of this paper is a method for translating a subset of VDM-SL to the higher order logic of PVS. This can be viewed as a shallow embedding (see [4]) of a subset of VDM-SL in PVS. The "embedding strategy" has been employed on three existing, realistic specifications, and some subtle errors have been found in the specifications as a result. The example specifications, which have all been developed by others, range from an approximately 200 line confidential industrial specification, a complex 500 line specification, parts of which have been published in a journal, and a 1000 line yet unpublished specification (which has nevertheless recently been submitted for publication in a case studies book on formal methods).

An important common aspect of the three specifications is that they have all been developed using the IFAD VDM-SL Toolbox [7, 10]. This supports syntax checking, type checking for possible type correctness (type checking is undecidable for VDM-SL), and pretty-printing facilities that allow ordinary text and pictures to be mixed with formal specification in LaTeX documents. Since these facilities were used in developing the examples, we found no trivial errors, such as syntax or simple (static) type errors. Other features of the Toolbox such as its support for interpreting, testing and debugging specifications were not used, except in the industrial case study. In fact, we did not find any errors in this example, which had been tested extensively.

The paper is organized as follows. We start out with a relatively long presentation of the translation strategy in Section 2. The reader may wish to postpone this section for a later more in depth study. In Section 3 we present some of the errors we found in the example specifications. Section 4 concludes.

2 From VDM-SL to the Higher Order Logic of PVS

A large subset of VDM-SL can be translated into higher order logic. This section describes such a translation into the PVS specification language in enough detail to support manual translations, but might be too informal to support an implementation of the translation process. The translation method may be viewed as a (very) shallow embedding of VDM-SL in PVS. The syntax of VDM-SL constructs is not embedded, we work with the "semantics" of the constructs directly. Therefore it can be argued that the translation is not safe (in a logical sense). Roughly speaking this is a disadvantage of shallow embeddings in general, though some shallow embeddings are more safe than others, e.g. if they define constants very close to the constructs of the original language. The translation of VDM-SL to PVS is relatively safe in this sense since PVS and VDM-SL share many concepts and constructs.

We more or less ignore the fact that a VDM-SL expression can be undefined; there is one common undefined element to all types, which can be viewed as sets. For example, an expression is undefined if a divisor is zero, if a finite sequence is applied to an index outside its indices, or if a pattern match in a let expression fails. Most of such situations are handled implicitly by restrictions on the translation of various constructs, in combination with type checking in PVS. For instance, PVS itself generates type checking conditions to ensure that we do not specify division by zero. Moreover, the translation does not support partial functions. All VDM-SL functions are translated to PVS functions and PVS generates an obligation stating that a given function must be total.

Below we discuss how various constructs of VDM-SL specifications can be translated. We show VDM-SL expressions in the ASCII notation supported by the IFAD VDM-SL Toolbox.

2.1 Basic Types, the Product Type and Type Invariants

Boolean and number types can be viewed as sets and translated directly to the corresponding types in PVS. It is not necessary to edit arithmetic expressions and logical connectives, they have the same symbols and names in PVS. Specially, the connectives **and** and **or** are used as "and-also" and "or-else" in both the Toolbox version of VDM-SL and PVS. Hence, it is allowed to write "`f(x) and x/x = 1`" if one can prove the automatically generated condition "`f(x) IMPLIES x /= 0`".

Restricted quantifications are translated to quantifications over PVS subtypes. Hence, a universal quantification like `forall x in set s & P[x]` can be translated to `forall (x:(s)): P[x]`. Note that the `&` is replaced with `:` and we insert brackets around the binding.

In VDM-SL it is possible to specify a subtype of a type by writing a predicate that the elements of the subtype must always satisfy. This predicate is called an invariant. As an example consider the definition of a type of positive reals:

```
Realp = real
inv r == r >= 0
```

This introduces a new constant `inv_Realp` implicitly, which equals the invariant predicate. The definition is translated to PVS as follows:

```
inv_Realp(r:real) : bool = r >= 0
Realp: TYPE = (inv_Realp)
```

The type `Realp` is defined as a subtype of the built-in type `real`.

The product type of VDM-SL can be translated directly to the product type of PVS. VDM-SL tuples are written using a constant `mk_`, e.g. `mk_(1,2,3)`. This constant could be ignored in the translation, but instead we have introduced and usually use a dummy constant `mk_`, defined as the identity function. The only way to split a tuple in VDM-SL is using pattern matching on `mk_`, e.g. in a let expression like `let mk_(x,y,z) = t in e`. The translation of pattern matching is discussed in more detail in Section 2.6.

2.2 Record Types

VDM-SL records are translated directly to records in PVS, there is only a slight difference in syntax. Moreover, PVS does not automatically introduce a constructor function `mk_A` as in VDM-SL for a record definition called `A`, but the function is easy to define. Below, we first give an example of how to translate a standard record type definition and then give a more complicated example where an invariant is also specified.

Standard Records We consider a simple example, which defines a record of points in the two-dimensional positive real plane. The VDM-SL definition of the record type is (`Realp` was introduced above):

```
Point:: x: Realp
        y: Realp
```

This is translated to PVS as follows:

```
Point: TYPE = [# x: Realp, y: Realp #]
```
In both cases, field selectors `x` and `y` are introduced implicitly. Unfortunately, PVS does not support the VDM-SL notation for field selection, which is the standard one using a dot and the field name, e.g. `p.x` for some point p. Instead fields are viewed as functions that can be applied to records. Thus the VDM-SL expression `p.x` translates to the PVS expression `x(p)`. The PVS notation is less convenient, e.g. a nested field selection as in `r.a.b.c` translates to `c(b(a(r)))`.

In VDM-SL each new record definition generates a new "make" function for building elements of the record. In order to allow a direct translation of such expressions to the same expressions in PVS, the constructor function is defined along with a record definition. The constructor for points is:
```
mk_Point(a:Realp,b:Realp) : Point = (# x:= a, y:= b #)
```
In VDM-SL such make constructors are also used for pattern matching in function definitions and in let expressions but this is not possible in PVS. Instead we use Hilbert's choice operator (epsilon), as described in Section 2.6.

VDM-SL also provides an `is_` test function for records, which is sometimes used to test where elements of union types come from. Since we do not support union types properly (see Section 2.4) we shall ignore this constant.

Records with Invariants Sometimes a VDM-SL record definition is written with an invariant. For instance, an equivalent way of defining the point record above would be the following definition
```
Point:: x: real
        y: real
   inv mk_Point(x,y) == x >= 0 and y >= 0
```
where an invariant is used to specify that we are only interested in the positive part of the two-dimensional real plane. We translate this as follows:
```
inv_Point(x:real,y:real) : bool = x >= 0 and y >= 0

Point: TYPE = {p: [# x:real, y:real #] | inv_Point(x(p),y(p))}

mk_Point(z:(inv_Point)) : Point = (# x:= x(z), y:= y(z) #)
```
Note that we restrict the arguments of `mk_Point`. For instance, the following definition does not work
```
mk_Point(a:real,b:real) : Point = (# x:= a, y:= b #)
```
since we cannot prove that this yields a point for all valid arguments.

There is one small semantic difference between VDM-SL records and the above representation in PVS. In VDM-SL it is allowed to write `mk_Point(1,-1)` though this will not have type `Point`. The only use of such a (kind of) junk term would be in `inv_Point(mk_Point(1,-1))`, which would equal false. In the PVS representation, this invariant expression is written as `inv_Point(1,-1)`.

2.3 Sequences, Sets and Maps

In this section, we briefly consider the translation of finite sequences, finite sets and finite maps. Invariants on these types are treated much like in the previous section on records.

Finite Sets A type of finite sets is provided in the PVS `finite_sets` library. The type is defined as a subtype of the set type, which represents sets as predicates. Most operations on sets exist, or can be defined easily. One annoying factor is that for instance set membership, set union and set intersection are all prefix operations in PVS. E.g. one must write `member(x,s)` for the VDM-SL expression "`x in set s`". Moreover, user-defined constants must be prefix and one cannot define new symbols. PVS supports only a simple and restricted syntax of expressions.

Finite Sequences VDM-SL finite sequences can be represented as finite sequences or as finite lists in PVS. The difference is that finite sequences are represented as functions and finite lists as an abstract datatype. There is more support for finite lists, so we have usually chosen this type as the representation. An advantage of sequences is that indexing is just function application. However, with both representations one must be careful since indexing of sequences starts from one in VDM-SL and from zero in PVS. The safest thing to do is therefore to define new indexing operations in PVS and use these for the translation.

Finite Maps PVS does not support finite maps, so an appropriate theory must be derived from scratch. In doing this, one could probably benefit from the paper on finite maps in HOL by Collins and Syme [5], who have implemented their work in a HOL library. However, many operations on finite maps are not supported in this library, so an extended theory of finite maps must be worked out.

As a start one could just axiomatize maps in PVS, e.g. by introducing maps as an uninterpreted subtype of the function type, with a few appropriate definitions (and axioms). In fact, for the examples very little support was needed.

A representation of maps using functions has advantages. Map application will just be function application and map modification can be translated to PVS `with` expressions. For example, the VDM-SL map modification `m ++ { 1 |-> 2, 2 |-> 3 }`, where a map m from numbers to numbers is modified to send 1 to 2 and 2 to 3, translates to `m with [ 1 |-> 2, 2 |-> 3 ]`.

2.4 Union Types

In VDM-SL, the union of two or more types corresponds to the set union of the types. Thus, the union type is a non-disjoint union, if two types have a common element this will be just one element in the union type. Higher order logics do not support non-disjoint unions, but support disjoint sums (unions) or abstract datatypes as in PVS. In general, a VDM-SL union type cannot be translated easily to a PVS datatype. However, if the component types of the union are disjoint then this is partly possible. The translation is only satisfactory when the component types are quote types; these correspond to singleton sets.

Union of Disjoint Types The union of disjoint types can be represented as a new datatype with constructor names for the different types. This representation is not perfect, the component types does not become subtypes of the union type as in VDM-SL. For example this means that the operators defined on the individual types are not inherited as in VDM-SL, where the dynamic type checking ensures that arguments of operators have the right types. In the special case where all components of the union type are new types, it might be possible to define the union type first and then define each of the component type as subtypes of this. Such tricks would not be easy to employ in an automatic translation.

Enumerated Types An enumerated type is a union of quote types, which is written using the following ASCII syntax in VDM-SL: ABC = <A>|<B>|<C>. This can be translated almost directly to PVS, with some minor syntax changes: ABC: TYPE = {A,B,C}. PVS does not support identifiers enclosed in < and >.

2.5 Function Definitions

Functions are translated directly to PVS functions. As mentioned above, we do not consider partial functions, though PVS supports partial functions via subtypes and other ways could be formalized (see e.g. [2, 1]). Polymorphic functions are not considered at the moment.

Standard explicit function definitions, which are function definitions that do not have postconditions, can be translated directly to PVS, if they are not recursive. A precondition will be translated to a subtype predicate. If functions are recursive we must justify they are total functions in PVS. It is up to the translator to specify an appropriate measure, which is decreased in each recursive call, for the termination proof. Moreover, VDM-SL supports mutual recursive function definitions which would not be easy to translate. The example specifications used only few recursive definitions and these were very simple.

Implicit function definitions, which are specified using pre- and postconditions only and have no function body, can be represented using the choice operator. Almost equivalently, one can also use function specification, which is a way of defining partially specified functions; it is only specified on a subset of a type how a function behaves, and this is specified by an "underdetermined" relation, not an equation [14, 11].

Implicit Definition Let us first consider the following semi-abstract example of an implicit function definition in VDM-SL:

```
f(x:real,y:real) z:Point
pre p[x,y]
post q[x,y,z]
```

where the variables in square brackets may occur free in the precondition p and
the postcondition q. This translates to the following PVS definitions:

```
pre_f(x:real,y:real) : bool = p[x,y]

post_f(t:(pre_f))(z:Point) : bool = let (x,y) = t in q[x,y,z]

f: FUNCTION[t:(pre_f) -> (post_f(t))]
```

The precondition is translated to a predicate on the arguments of the function
and the postcondition is translated to a binary relation on the arguments and
the result. The function itself is defined as an uninterpreted constant using a
dependent function type: given arguments t satisfying the precondition it returns
a result satisfying the postcondition, or more precisely, a result related to t by
the postcondition. This relation may be underdetermined, i.e. it may specify a
range of possible values for a given input, but the function will always return
a fixed value in this range. If the precondition is not satisfied the result is an
arbitrary value.

As a result of an uninterpreted constant definition, the PVS type checker
generates an existence condition, which says that we must prove there exists a
value in the specified type. Hence, above we must prove there exists a function
from the precondition to the postcondition. In general, proving this condition
can be non-trivial, since one must usually provide a witness, i.e. a function of
the specified form. (For instance, it would be difficult to prove that there exists
a square root function.)

Explicit Definition Explicit definitions of recursive functions can be problematic for automatic translation since a translator must insert a well-founded measure for proofs of termination. This is easy enough when the recursion is simple, which it is for primitive recursive functions over numbers and abstract datatypes, but for more general recursive functions this can be hard. PVS has some strategies for proving termination in simple cases.

An explicit function definition has no postcondition but instead a direct
definition, and perhaps a precondition. Let us consider a standard example of a
primitive recursive function on the natural numbers:

```
fac: nat -> nat
fac(n) == if n = 0 then 1 else n * fac(n-1)
pre 0<=n
```

This translates to the following PVS definitions:

```
pre_fac(n:nat) : bool = n >= 0

fac(n:(pre_fac)) : recursive nat =
  if n = 0 then 1 else n * fac(n-1) endif
  measure (lambda (n:(pre_fac)): n)
```

Note that we have inserted "recursive" and "measure", which are part of the syntax for recursive function definitions in PVS. The measure is used to generate conditions for termination of recursive calls. (The precondition is redundant above, it is included for illustration.)

2.6 Pattern Matching

Pattern matching plays an important role in VDM-SL specifications. It is used frequently to get access to the values at fields of a record, and it is the only way to get access to the values of the components of a tuple. We can represent a successful pattern matching but not a failing one, since we do not represent undefined expressions. However, either undefined expressions are avoided due to type checking, or they are represented by arbitrary values, i.e. values of a certain type that we do not know anything about.

Pattern Matching in Let Expressions Here are some examples which use a record type A with three fields a, b and c. Assuming x, y and z are variables, the following VDM-SL let expression

```
let mk_(x,y,z) = e1 in e2[x,y,z]
```

can be translated to exactly the same term in PVS, except that the tuple constructor mk_ must be omitted for the expression to parse. The following VDM-SL let expression with a pattern match on the record type A

```
let mk_A(x,y,z) = e1 in e2[x,y,z]
```

can be translated to the following PVS term:

```
let x = a(e1), y = b(e1), z = c(e1) in e2[x,y,z]
```

The field selector functions are used to destruct the expression. This corresponds to the way that PVS itself represents pattern matching on tuples (using project functions). If one of the variables in the VDM-SL expression was the don't care pattern, written as an underscore _, then we could just replace this with a new variable. We do not allow constants in patterns in let expressions, since they do not make much sense (they are however allowed in VDM-SL).

The following VDM-SL "let-be-such-that" expression

```
let mk_A(x,y,z) in set s be st b[x,y,z] in e[x,y,z]
```

can be translated to

```
let v = (epsilon! (w:(s)):
    let x = a(w), y = b(w), z = c(w) in b[x,y,z]) in
let x = a(v), y = b(v), z = c(v) in e[x,y,z]
```

where we use the choice operator to represent the looseness in the VDM-SL specification. Don't care patterns are translated as suggested above, by introducing new variables. We allow constants and other values in let-be-st expressions. For instance, we can translate

```
let mk_A(x,y,0) in set s be st b[x,y] in e[x,y]
```

into the PVS term

```
let v = (epsilon! (w:(s)):
    let x = a(w), y = b(w), n = c(w) in n = 0 and b[x,y]) in
let x = a(v), y = b(v) in e[x,y]
```

where we include a test in the body of the epsilon.

Pattern Matching in Cases Expressions The following VDM-SL cases expression

```
cases e:
    mk_A(0,-,z) -> e1,
    mk_A(x,1,z) -> e2,
    others      -> e3
end
```

can be translated to the following conditional expression in PVS:

```
cond
    a(e) = 0 -> let z = c(e) in e1,
    b(e) = 1 -> let x = a(e), z = c(e) in e2,
    else     -> e3
endcond
```

PVS's built-in cases expression only works on abstract datatypes.

Pattern Matching in Function Definitions Pattern matching can be used on arguments in a function definition, where the patterns are typically variables (or don't care patterns which are translated to new variables). We can treat this by inventing a new variable using the function definition and then extending the body with a let expression to represent the pattern match. This approach is also used in the formal semantics of VDM-SL.

2.7 State and Operations

A VDM-SL specification may contain a state definition, which specifies a number of variable names for elements of the state space. The state space is known to operations and nowhere else. The state definition is essentially a record definition and is therefore represented as a record type in PVS. Operations are represented as state transformations, i.e. functions which, in addition to the operation's input values, take the initial state as an argument and return the output state as a result (and possibly an explicit result value). Hence, operation definitions can be translated in a similar way as functions.

The body of operation definitions may contain assignments and sequential compositions. Assignments are translated to PVS with expressions and sequential compositions are represented using let expressions. In this paper we do not consider conditions (which should be easy) and while loops (which probably could be translated to recursive functions). More exotic features such as exception handling are also excluded from consideration.

Assume we have the following state definition in VDM-SL:

```
state ST of
      x: real
      y: real
      z: real
end
```

This can be translated to:

```
ST: TYPE = [# x: real, y: real, z: real #]
mk_ST(x:real,y:real,z:real): ST = (# x:=x, y:=y, z:=z #)
```

Now assume we have the sequence:

```
x:=5; y:=3; z:=1
```

This can be translated to

```
lambda (s:ST):
  let s1 = s  with [x:=5],
      s2 = s1 with [y:=3],
      s3 = s2 with [z:=1]
  in s3
```

or simply to

```
lambda (s:st): s with [x:=5, y:=3, z:=1]
```

since the assignments are independent in this example.

3 Errors in Example Specifications

The method presented in the previous section for translating VDM-SL specifications to PVS was used on three realistic specifications. A number of errors were identified in the specifications by simply trying to prove the type checking conditions, which were generated automatically by PVS. Some of these errors are listed below. The third industrial specification had been extensively tested using the IFAD VDM-SL Toolbox, and is also smaller and less complex than the other two. No errors were found in this specification.

The errors themselves are not a great contribution of the paper and should perhaps mainly be read as small and funny, but also worrying, examples of the errors that people make in writing formal specifications (and programs).

However, what the errors document is that in specification debugging one can benefit from working with specifications in a formal way, e.g. as suggested in this paper. However, other alternatives for validation such as testing could have found some of the errors as well.

We have divided the errors in three categories: (1) those that were pointed out directly during the proof of a type checking condition, (2) those that probably could have been found easily by testing specifications, (3) other errors, some of which are quite subtle, e.g. due to parentheses problems.

3.1 Directly by Type Checking Condition

Below we present two errors which were displayed clearly in a proof attempt to prove a type checking condition.

First we consider a function which is wrong (or unsafe) because it may perform a division by zero:

```
foo1: real * real -> real
foo1(a1,a2,a3) ==
    ... complex body containing  expr/(...a1...a2...a3...)  ...
pre g(a1,a3) = g(a2,a3)
```

Due to the division, PVS generates a proof obligation to ensure that the divisor is not zero. It is not possible to prove this from the precondition. In fact, the function is later used with arguments which we cannot exclude could lead to division by zero.

Next consider the function definition of foo2 below, which contains a call to the previous function foo1 in its body:

```
foo2: real * real * real * real * real * real -> real
foo2(a1,a2,a3,a4,a5,a6) ==
    ... complex body containing  foo1(a3,a4,a1)  ...
pre g(a2,a1) = g(a4,a1) and g(a5,a2) = g(a6,a2) and ...
```

Since the first function foo1 has a precondition which is translated to a subtype in PVS, the body of foo2 will generate a type checking condition which says that we must prove "g(a3,a1) = g(a4,a1)" (the precondition of foo1) assuming the the precondition of foo2. This is not possible, since pre_foo2 says "g(a2,a1) = g(a4,a1)". The problem is that pre_foo2 uses a2 where it should use a3.

3.2 Lack of Testing

The following error could probably have been found by testing. Apparently it was made because the developers did not fully understand how the function floor actually works. The intension is that the following function must divide a real time line into discrete points like $\ldots, -p, 0, p, \ldots$, where p is some constant:

```
foo3 : real -> real
foo3(r) ==
  if r >= 0 then floor(r / p) * p
  else floor((r / p) - 1) * p
```

Hence, it is wrong to subtract one in the argument of `floor` in the `else` branch since, assuming p is five, then for instance `foo3(-7.5)` yields `floor(-1.5-1)*p`, which equals -15 and not the desired -10. Probably the developers thought that for instance `floor(-2.5)` equals -2, in fact it equals -3. The right specification of `foo3` is therefore:

```
foo3 : real -> real
foo3(r) == floor(r / p) * p
```

3.3 Subtle Errors

In this section we discuss some quite subtle errors. It would probably be difficult to find these by testing; whether they were found would of course depend on how extensive the testing was.

In an exists expression with a variable binding list that covers several lines due to type information, implication was used wrongly instead of conjunction:

```
exists ... & p => q
```

This should have been:

```
exists ... & p and q
```

In writing the long binding list (ended by &) the developers may have forgotten that they were writing an existential and not a universal quantification.

The next error we consider was found in a ten line invariant, which contained a list of conjunctions:

```
(a1 /\ a2 => c1) and
(b1 => b2) and
(exists ... & ...) and etc
```

Quite a typical shape for an invariant, but after a careful reading, also of the documenting text, it turned out that the right parenthesis after c1 was wrong: the assumptions a1 and a2 should hold of all conjuncts. Hence, the invariant expression should have been:

```
( a1 /\ a2 =>
    c1 and
    (b1 => b2) and
    (exists ... & ...) and etc )
```

There were more problems with parentheses. Consider the following term which in its original form was five lines long:

```
x <= y and x = y => p
```

We realized by careful reading that the statement did not make much sense with the right parsing of the expression, where parentheses would be put as follows: `(x <= y and x = y) => p`. Hence, the intended parsing must have been

```
x <= y and (x = y => p).
```

A similar example:

```
(exists ... & ..............) =>
...long .... expression....    and
(exists ... & ..............) =>
...long .... expression....    and     etc
```

Should be:

```
((exists ... & ..............) =>
...long .... expression....)   and
((exists ... & ..............) =>
...long .... expression....)   and     etc
```

In one of the specifications the author had more or less misunderstood scoping rules. Hence, parentheses were often omitted around existential and universal quantifications, but not always. Instead scoping rules were often indicated using indentation! This type checks but the resulting statements are strange.

4 Conclusion

The experiment presented in this paper used the theorem prover PVS. We showed how it is possible to translate a subset of VDM-SL to the PVS specification language. The fact that we were able to employ the translation method on three realistic specifications give some indication that the chosen subset is big enough for many real-life specifications. We used the PVS system to reason about the translated specifications, in particular, PVS automatically generated the type checking conditions we had to prove. It was in trying to prove these conditions that we found the errors discussed in Section 3.

The general experience from the present experiment is that PVS has a very nice specification language (though it is not possible to define new infix constants) and useful theorem proving facilities. The hand translation of the example VDM-SL specifications was feasible because the PVS specification language and the chosen subset of VDM-SL are quite close syntactically, and the VDM-SL specifications were available in files. The PVS support for records and subtypes was particularly useful, though its syntax for record field selection by prefixing is less convenient than VDM-SL's postfix syntax. The PVS type checking of subtypes was exploited for precondition and invariant checking in VDM-SL, and to generate implementability conditions for function definitions.

PVS facilitates proofs at a fairly non-tedious level, due to the integrated decision procedures and rewriting techniques. Low level proof hacking using for instance associativity and commutation properties of arithmetic operations is usually not necessary. However, the real difficult side of theorem proving is still difficult, for instance, understanding the application (and formalizing it right), inventing proofs, and generating suitable lemmas.

The size and complexity of the chosen examples turned out to be a challenge to PVS. Type checking one specification could take 10-15 minutes[1]. Moreover, the built-in strategy for proving type checking conditions was not feasible for many of the conditions. Often it would run for hours without terminating, probably because it rewrote too much using definitions. The same conditions could sometimes be proved manually in a few minutes. One specification generated 160 type checking conditions, so it was not feasible to apply the strategy to each one individually. This was nevertheless what we had to do (since it did not terminate on some conditions). A timer facility would have been useful to force the strategy to give up after a certain period of time.

One may like or dislike the PVS Emacs interface. Though we are used to Emacs, we did not like it, for instance because we felt that the way in which buffers popped up and destroyed existing Emacs windows was confusing and irritating. We also felt that the quite frequent switching between buffers that we had to do became somewhat of a bottleneck. Moreover, the interface was unreliable and it was often necessary to restart PVS when Emacs ended up in a state where you could not execute important PVS commands (see Footnote 1).

Initially, we also tried to use the HOL system for the experiment. However, the logic of the HOL system was not close enough to VDM-SL to make manual translation feasible. In particular, HOL does not directly support subtypes and records, which were used extensively in the examples, and specifically type checking conditions therefore had to be generated manually during the translation. It is almost impossible to do that well (and remember all of them) and the size of the translation tended to explode (one line of VDM-SL could easily require many lines of HOL). Thus we quite quickly gave up on using HOL, but if the translation process was automatic the immediate infeasibility of using HOL would disappear. Note however that such an automatic translation would be considerably more complicated than an automatic translation to PVS, e.g. it would have to support the generation of type checking conditions. We do not think it would be a very difficult task to automate the translation to either system, but it would be a relatively time consuming task.

The outcome of our experiments has been that we have dropped using both HOL and PVS as the main theorem proving component of our proof support tool. Instead we intend to use Isabelle [12] to formalize the proof theory for VDM-SL [3, 11], and perhaps support access to the facilities of HOL and PVS via Isabelle oracles. One reason for not using PVS is that we think PVS may be too difficult to tailor for our particular task since it is a closed system and

---

[1] We used the version of PVS available in February 1996, the version available in June 1996 is faster and more stable.

does not have a meta language. The benefits of PVS type checking we intend to obtain via a minor extension of the type checker of the IFAD VDM-SL Toolbox, with which the proof support tool will be tightly integrated. A main task will be to build a graphical user interface for the proof support tool, since neither HOL, Isabelle nor PVS supports one that we feel non-expert theorem proving persons would be able to use efficiently.

It is an interesting question what the level of rigor should be in formal methods. The example specifications treated in this paper have shown that it is fairly easy to find serious errors in formal specifications, even though the specifications have been both syntax checked, (lightly) type checked, and nicely documented using pretty-printing facilities. Moreover, it is interesting to note that some of these errors probably could have been found by testing specifications, as supported in the IFAD VDM-SL Toolbox, but other errors seem more subtle. The investigations presented in this paper suggest also the use of theorem proving facilities for specification debugging.

Acknowledgments

I would like to thank Peter Gorm Larsen for his support and interest in the work reported here. The work was funded by the Danish Research Councils.

References

1. S. Agerholm. *A HOL Basis for Reasoning about Functional Programs*. PhD thesis, BRICS, Department of Computer Science, University of Aarhus, December 1994. Available as Technical Report RS-94-44.
2. S. Agerholm. LCF examples in HOL. *The Computer Journal*, 38(2), 1995.
3. J. Bicarregui, J. Fitzgerald, P. Lindsay, R. Moore, and B. Ritchie. *Proof in VDM: A Practitioner's Guide*. FACIT. Springer-Verlag, 1994. ISBN 3-540-19813-X.
4. R. J. Boulton, A. D. Gordon, M. J. C. Gordon, J. R. Harrison, J. M. J. Herbert, and J. Van Tassel. Experience with embedding hardware description languages in HOL. In V. Stavridou, T. F. Melham, and R. T. Boute, editors, *Theorem Provers in Circuit Design: Theory, Practice and Experience: Proceedings of the IFIP TC10/WG 10.2 International Conference*, IFIP Transactions A-10, pages 129–156. North-Holland, June 1992.
5. G. Collins and D. Syme. A theory of finite maps. In E. T. Schubert, P. J. Windley, and J. Alves-Foss, editors, *Proceedings of the 8th International Workshop on Higher Order Logic Theorem Proving and its Applications*. Springer-Verlag, September 1995. LNCS 971.
6. John Dawes. *The VDM-SL Reference Guide*. Pitman, 1991. ISBN 0-273-03151-1.
7. R. Elmstrøm, P. G. Larsen, and P. B. Lassen. The IFAD VDM-SL Toolbox: A practical approach to formal specifications. *ACM Sigplan Notices*, 29(9):77–80, September 1994.
8. M. Gordon. Notes on PVS from a HOL perspective. University of Cambridge Computer Laboratory, see http://www.cl.cam.ac.uk/users/mjcg/PVS.html, August 1995.

9. M. J. C. Gordon and T. F. Melham, editors. *Introduction to HOL: A Theorem-proving Environment for Higher-Order Logic*. Cambridge University Press, 1993.
10. IFAD World Wide Web page. http://www.ifad.dk.
11. P. G. Larsen. *Towards Proof Rules for VDM-SL*. PhD thesis, Technical University of Denmark, Department of Computer Science, March 1995. ID-TR:1995-160.
12. L. C. Paulson. *Isabelle: A Generic Theorem Prover*, volume 828 of *Lecture Notes in Computer Science*. Springer-Verlag, 1994.
13. PVS World Wide Web page. http://www.csl.sri.com/pvs/overview.html.
14. H. Søndergaard and P. Sestoft. Non-determinism in functional languages. *The Computer Journal*, 35(5):514–523, October 1992.

A Comparison of HOL and ALF Formalizations of a Categorical Coherence Theorem

Sten Agerholm[1], Ilya Beylin[2] and Peter Dybjer[2]

[1] The Institute of Applied Computer Science (IFAD), Forskerparken 10, DK-5230 Odense M, Denmark
[2] Department of Computer Science, Chalmers University of Technology, S-41296 Gothenburg, Sweden

Abstract. We compare formalizations of an example from elementary category theory in the systems HOL (an implementation of Church's classical simple type theory) and ALF (an implementation of Martin-Löf's intuitionistic type theory). The example is a proof of coherence for monoidal categories which was extracted from a proof of normalization for monoids. It makes essential use of the identification of proofs and programs which is fundamental to intuitionistic type theory. This aspect is naturally highlighted in the ALF formalization. However, it was possible to develop a similar formalization of the proof in HOL. An interesting aspect of the developments concerned the implementation of diagram chasing. The HOL development was greatly facilitated by an implementation of support for such equational reasoning in Standard ML.

1 Introduction

We compare the two proof assistants ALF and HOL by using them for implementing a proof in elementary category theory[3]. This proof was presented by Beylin and Dybjer [3] and shows how a Curry-Howard interpretation of a formal proof of normalization for monoids almost directly yields a coherence proof for monoidal categories. It is an interesting example of an application of intuitionistic type theory and can be viewed as part of the larger enterprise of "constructive category theory" in the sense of Huet and Saibi [10].

The paper shows how the ALF formalization makes essential use of the Curry-Howard interpretation while the quite parallel HOL development does not. In this case study we make a systematic comparison of the two systems, both with respect to their very different logic bases (ALF is based on Martin-Löf type theory and HOL on Church's simple type theory) and with respect to more pragmatic aspects. In particular, we show how the HOL development benefitted in an essential way from having available Standard ML as a metalanguage for writing tool support for reasoning about congruence relations. This was very

---

[3] A basic reference is Mac Lane [11]. Two recent books oriented towards computer scientists are Barr and Wells [2] and Pierce [19].

useful for proof by "diagram chasing" – the category theorists way of presenting equational reasoning.

Throughout the paper we show the HOL and ALF proofs side by side. Excerpts from the HOL development are shown in `tt`-style, whereas excerpts from the ALF development are displayed as they appear on the screen (we have used the print-facility provided by ALF's window interface). We use these contrasting styles to assist the reader to keep the interleaved developments apart.

Only some key parts of the developments are shown. The full proofs can be retrieved by ftp from `//ftp.cs.chalmers.se/pub/users/ilya/FMC` and from `//ftp.ifad.dk/pub/users/sten` respectively. We would also like to refer to Beylin and Dybjer [3] for a more complete presentation of the proof and for more background and motivation. For a more complete presentation of the tool support for diagram chasing, see Agerholm [1].

The paper is organized as follows. In Section 2 we give a brief introduction to the systems. In the remaining sections we compare the two developments. Section 3 discusses the implementation of the normalization algorithm for monoids and Section 4 discusses the implementation of the free monoidal category and the proof of the coherence theorem. Section 5 concludes.

2 The Systems and their Underlying Logics

The main objective of this section is to introduce Martin-Löf type theory and the ALF system. We also give a brief overview of the HOL system.

2.1 Higher Order Logic

The HOL system [8] is a mechanized proof-assistant for proving theorems in higher order logic, a version of Church's simple type theory extended with ML-style polymorphism. HOL is implemented in an interactive programming environment which supports the functional programming language Standard ML. The HOL logic and all theorem proving support are implemented in SML and a user can also program new proof strategies (tactics) in SML. Roughly speaking, a proof is an application of SML functions which yields a theorem as a result. The type system of SML ensures that theorems can only be created as consequences of the five axioms and the eight primitive inference rules. This makes HOL a safe system, also with respect to user extension. In this paper we only very briefly introduce the HOL logic in order to present syntax used later. The reader may wish to consult the HOL book [8] or Gordon's original paper [7].

The terms of the HOL logic can be variables, constants, λ-abstractions (written `\x.t`) and applications (written `t1 t2`). The usual logical connectives, such as conjunction `/\`, disjunction `\/`, implication `==>`, and negation `~`, are represented as constants. Types can be atomic types (like `bool` for the boolean truth values), type variables (like `*`, `**` or `*o`), compound types (like `*#**` for the product type), and function types like `*->**`, which denote functions from the domain type * to the range type **. Type variables range over and can be instantiated to any type. All terms must be well-typed in the usual sense.

The HOL logic is a higher order logic, so it is possible to quantify over variables of any type. Universal and existential quantification are written !x. t and ?x. t respectively, where t may contain x. Sets of elements of any type are represented as predicates, and the set notation {x | p[x]} represents the set of elements (of some fixed type) satisfying the predicate p.

2.2 Martin-Löf Type Theory and ALF

The ALF ("A Logical Framework") System is a proof assistant which supports proof in Martin-Löf type theory. This theory was introduced by Martin-Löf in the beginning of the seventies [13] and exists in several versions [14, 15]. ALF implements the most recent version presented by Martin-Löf in 1986. This version is monomorphic and intensional, see Nordström, Petersson, and Smith [17]. ALF also supports a rich class of inductive definitions, see Dybjer [6], and definition by pattern matching, see Coquand [5].

Martin-Löf type theory is intended to be a full-scale framework for constructive mathematics and at the same time a programming language. The core of the theory is a λ-calculus with dependent types. Intuitionistic predicate logic is obtained by the Curry-Howard identification: propositions are types, and proofs of a proposition are programs of the corresponding type. Therefore, Martin-Löf type theory is a suitable framework for program extraction from constructive proof. The research that this paper is based on is an example of this: it shows how a proof of coherence for monoidal categories can be extracted from a proof of normalization for monoids in a certain sense.

We have used Window ALF, a version of ALF which was implemented by Lena Magnusson [12]. Because of the Curry-Howard analogy, to prove a theorem in ALF is the same as writing a program "witnessing" the truth of the theorem. This is a fundamental difference between ALF and HOL (and many other proof-assistants), where the proof instead is presented as a sequence of tactics.

Both the theorem and the program (proof) is interactively synthesized by the user. At each stage the user can inspect the type of goals (placeholders in the proof), and their possible completions. During this process the user benefits from ALF's window interface. Several windows are maintained including a "scratch area", where the current incomplete proof is displayed, and a "theory area", with relevant definitions and theorems from earlier developments. The proof can be built by pointing and clicking in the windows or from the menus. Seeing the proof term gives a high degree of control of the proof and may sometimes make it easier to find concise proofs. One can contrast the philosophy behind ALF, which emphasizes the proofs themselves as objects of independent interest, to the more traditional view that truth is the central notion and proofs are only the tool for finding truths.

There are three principal ways of building types in type theory. Firstly, there are dependent function types, written $(x \in X)Y$ in ALF, where the type Y may depend on the variable x. A non-dependent function type, where Y does not depend on x, is often written $(X)Y$. Moreover, a repeated dependent function type, such as $(x \in X)(y \in Y)Z$ is written $(x \in X; y \in Y)Z$, and $(x \in X)(y \in X)Y$ is written $(x, y \in X)Y$. Secondly, there is the type **Set** of sets (which is also used as the type of propositions). Thirdly, each set is a type.

Sets are inductively defined by listing their constructors with types. These inductive definitions look similar to recursive datatype definitions, but are more general, since also families of sets (dependent sets) can be defined under this general scheme. This mechanism is very powerful and subsumes for example all basic data structures as well as logical primitives. So there are no built-in constructions in ALF like tuples or Booleans.

If X and Y are types which represent propositions, then $(X)Y$ means X implies Y. If Y represents a proposition depending on the variable $x \in X$, then $(x \in X)Y$ represents the universal quantification: Y for all $x \in X$.

ALF implements a *monomorphic* version of type theory. As a consequence it maintains a lot of type information which is redundant or uninteresting, and the full size of the proofs can in some cases be very large. However, the user can instruct ALF to suppress unwanted type information during display.

Moreover, ALF implements *intensional* type theory. In this theory proofs of equalities have to be manipulated explicitly and the primitive notion of equality of functions is not extensional. Below, we shall discuss how the use of intensional equality affects the proof of coherence.

To sum up, programming (and proving!) in ALF feels much like programming in a standard functional language, but with the extra expressiveness of dependent types added. It is clear that the abstract syntax of ALF is similar to that of ordinary functional programming languages, but there are some differences in concrete syntax. For example, parentheses are used in the Pascal, rather than the ML, fashion.

3 Normalization for Monoids

We first show how to formalize a proof of normalization for monoids.

We begin by defining the set of binary words (binary trees) over a given set. Then we introduce the least congruence relation which identifies such words up to associativity and unit laws. The free monoid is then defined as the set of congruence classes of binary words with respect to this congruence.

We also define a normalization algorithm which maps a binary word to a congruent binary word in normal form. This normalization algorithm yields a decision algorithm for congruence of binary words (equality in the free monoid) by testing syntactic equality of normal forms.

3.1 Binary Words

Our first task is to define the set of "binary words", that is, the least set closed under the following rules:

– a variable in a given set X is a binary word,

– the product of two binary words is a binary word,
– the empty word is a binary word.

To define this set in HOL we use the datatype definition package [9], which supports Standard ML style notation to specify new recursive types. Binary words can thus be defined as a datatype bw:

   bw = e | Var of X | Ox of bw => bw .

The notation Ox of bw => bw means that binary word product Ox takes two arguments, both of type bw. Moreover, product is specified to be infix and associates to the right; hence $a\, Ox\, b\, Ox\, c$ is the same as $a\, Ox\, (b\, Ox\, c)$.

In ALF we use the basic inductive definition mechanism. We give all constructors with their appropriate types to the system in the following way:

$$BW \in \mathbf{Set}$$
$$e \in BW$$
$$Var \in (x \in X)\, BW$$
$$\otimes \in (A, B \in BW)\, BW$$

The product is a prefix operator in ALF, which does not support infix notation.

Our next task is to define the least congruence relation on binary words generated by the axioms stating associativity of word product and that the empty word is a unit with respect to product. In HOL we represent this as a relation cbw:bw->bw->bool, which is defined using the inductive definition package (see [16, 4]). The inductive clauses are the standard equivalence relation rules (reflexivity, transitivity and symmetry):

```
                 a cbw b   b cbw c           a cbw b
   a cbw a       -----------------           ---------
                     a cbw c                 b cbw a
```

the congruence law for product:

```
     a cbw b    c cbw d
     -------------------
     a Ox c cbw b Ox d
```

and finally, the unit and associativity laws (which have no hypotheses):

   e Ox a cbw a   a Ox e cbw a   a Ox (b Ox c) cbw (a Ox b) Ox c.

The inductive definitions package represents the rules as implications.

In ALF this binary relation is represented as an inductively defined binary family of sets

$$cbw \in (A, B \in BW)\, \mathbf{Set}$$

As for binary words we list the constructors for each of the inductive clauses. Note that these constructors build proof objects witnessing that two elements are related. The standard equivalence relation rules are represented by the following constructors for cbw:

$$\iota \in (A \in BW) \, cbw(A,A)$$
$$o \in (g \in cbw(B,C); f \in cbw(A,B)) \, cbw(A,C)$$
$$sym \in (A,B \in BW; f \in cbw(A,B)) \, cbw(B,A)$$

the congruence law for product:

$$\otimes \in (f \in cbw(A,A'); g \in cbw(B,B')) \, cbw(\otimes(A,B), \otimes(A',B'))$$

and the unit and associativity laws:

$$\lambda \in (A \in BW) \, cbw(\otimes(e,A), A)$$
$$\rho \in (A \in BW) \, cbw(\otimes(A,e), A)$$
$$\alpha \in (A,B,C \in BW) \, cbw(\otimes(A, \otimes(B,C)), \otimes(\otimes(A,B), C))$$

The elements of $cbw(A, B)$, that is, the proofs that the words A and B are congruent, will later be used as arrows of the free monoidal category with one modification: we use an equivalent definition of congruent binary words, where symmetry is not a primitive but a derived rule. In this alternative definition, we remove symmetry, and instead add symmetric versions of the associativity and unit laws as inductive clauses:

$$\lambda' \in (A \in BW) \, cbw(A, \otimes(e,A))$$
$$\rho' \in (A \in BW) \, cbw(A, \otimes(A,e))$$
$$\alpha' \in (A,B,C \in BW) \, cbw(\otimes(\otimes(A,B), C), \otimes(A, \otimes(B,C)))$$

The point is that this definition is more closely related to the definition of monoidal category. (Note that this is not relevant to the HOL development, which does not use these proofs of congruence as arrows in the free monoidal category.)

3.2 The Normalization Theorem

The normalization theorem states that there is a function mapping a binary word to a normal binary word, such that two binary words are congruent iff they are mapped into equal binary words.

To formalize this in HOL we construct a normalization function `Nf:bw->bw` and show that

```
|- !a b. a cbw b = (Nf a = Nf b) .
```

In ALF equivalence is defined as bi-implication. So we wish to construct a function $Nf \in (A \in BW)BW$ and proofs of

$$nf \in (A,B \in BW; f \in cbw(A,B)) =(Nf(A), Nf(B))$$
$$nf' \in (A,B \in BW; h \in =(Nf(A), Nf(B))) \, cbw(A,B)$$

where we have stated each part of the implication separately. Here we see how proving these two implications in ALF means defining two functions: namely nf, which maps a proof of congruence of binary words to a proof of equality of normal forms, and nf', doing the converse.

The second property will be proved as a corollary of the property that each binary word is congruent to its normal form:

$$\forall \in (A \in BW) \; cbw(A, Nf(A))$$

The normalization function Nf is defined in terms of an auxiliary "meaning" function which takes a binary word as an argument and returns a function on binary words[4]. In HOL this function Mean:bw->bw->bw is introduced by a structural recursive definition on binary words:

```
Mean e n = n
Mean(a Ox b)n = Mean b(Mean a n)
Mean(Var x)n = n Ox (Var x)
```

using the data type definition package. Then we define Nf a = Mean a e.

The corresponding structural recursive definition in ALF uses the pattern matching mechanism. Given an inductively defined set (or family of sets) we can instruct ALF to define a function by case analysis on the different constructors[5].

$$\text{Mean} \in (n \in BW; A \in BW) \; BW$$
$$\text{Mean}(n, e) \equiv n$$
$$\text{Mean}(n, \otimes(A_1, B)) \equiv \text{Mean}(\text{Mean}(n, A_1), B)$$
$$\text{Mean}(n, \text{Var}(x)) \equiv \otimes(n, \text{Var}(x))$$

$$Nf(A) \equiv \text{Mean}(e, A)$$

The property that congruent binary words are mapped into equal normal forms is a corollary of the lemma that congruent binary words are mapped into equal meanings. In HOL we prove the following theorem

```
|- !a b. a cbw b ==> (!n. Mean a n = Mean b n)
```

by rule induction on cbw, supported by the inductive definition package. This is a one line proof:

```
cbw_RULE_INDUCT_TAC THEN ASM_REWRITE_TAC[Mean] .
```

The tactic performs a rule induction on cbw and then rewrites each of the seven cases with the corresponding definition of Mean and the assumptions.

---

[4] In fact, this function is best understood as a meaning homomorphism from the free monoid (obtained as the set of binary words up to congruence of binary words) into the monoid of functions on binary words. See [3] for more information.

[5] This function has the arguments in a different (and more unnatural) order than the corresponding HOL function Mean above.

In ALF we see how a proof by induction is represented by a function defined by recursion, here on the proof f that two binary words are congruent:

mean $\in$ $(A, B \in$ BW; $n \in$ BW; $f \in$ cbw$(A, B))$ =(Mean(n, A), Mean(n, B))
    mean$(\_, B, n, \iota(\_)) \equiv \; =_{id}$(Mean$(n, B)$)
    mean$(A, B, n, o(g, f_1)) \equiv \; =_{trans}$(mean$(A, B_1, n, f_1)$, mean$(B_1, B, n, g)$)
    mean$(\_, \_, n, \otimes(f_1, g)) \equiv \; =_{trans}$(mean$(B_1, B'$, Mean$(n, A_1), g)$, conorm(mean$(A_1, A', n, f_1), B'$))
    mean$(\_, \_, n, \alpha(A_1, B_1, C)) \equiv \; =_{id}$(Mean$(n, \otimes(A_1, \otimes(B_1, C))))$
    mean$(\_, B, n, \rho(\_)) \equiv \; =_{id}$(Mean$(n, \otimes(B, e)))$
    mean$(\_, B, n, \lambda(\_)) \equiv \; =_{id}$(Mean$(n, \otimes(e, B)))$
    mean$(\_, \_, n, \alpha'(A_1, B_1, C)) \equiv \; =_{id}$(Mean$(n, \otimes(\otimes(A_1, B_1), C)))$
    mean$(A, \_, n, \rho'(\_)) \equiv \; =_{id}$(Mean$(n, A))$
    mean$(A, \_, n, \lambda'(\_)) \equiv \; =_{id}$(Mean$(n, A))$

Note how the right hand sides of the recursion equations represent proofs of congruence. For example, $=_{id}$ builds a proof of reflexivity and $=_{trans}$ combines two proofs by transitivity.

The property that a binary word is mapped into a congruent binary word is a corollary to the lemma that the meaning function is congruent to the product of binary words. This theorem is stated as follows in HOL:

|- !a n. n Ox a cbw Mean a n

and proved by induction on bw supported by the datatype definition package.

In ALF the corresponding proof by induction is represented as the following definition of the function ν using an auxiliary function ξ:

$\xi \in (n \in$ BW; $AA \in$ BW$)$ cbw$(\otimes(n, AA)$, Mean$(n, AA))$
    $\xi(n, e) \equiv \rho(n)$
    $\xi(n, \otimes(A, B)) \equiv o(o(\xi($Mean$(n, A), B), \otimes(\xi(n, A), \iota(B))), \alpha(n, A, B))$
    $\xi(n, Var(x)) \equiv \iota(\otimes(n, Var(x)))$
$\nu \in (A \in$ BW$)$ cbw$(A, Nf(A))$
    $\nu \equiv [A]o(\xi(e, A), \lambda'(A))$

4 Coherence for Monoidal Categories

The point of Beylin and Dybjer [3] is to show how a proof of coherence for monoidal categories (see e.g. Mac Lane [11], Chapter VII) can be extracted with only a little extra work from the proof of normalization for monoids. The idea is to first use the Curry-Howard isomorphism and then add a little extra structure expressing what it means for two proof objects to be equal. In this way we get a free monoidal category from the free monoid, a normalization functor from the normalization function, and a proof of coherence from the proof of normalization. The reader is referred to [3] for full details. Here we shall concentrate on how this piece of category theory is represented in HOL and in ALF respectively.

4.1 Objects and Arrows of the Free Monoidal Category

A category has a set of objects and a set of arrows (morphisms) where each arrow has a domain and a codomain that are objects. A category must have an associative composition of arrows, and an identity arrow for each object.

An object of the free monoidal category will be a binary word:

$$\text{Obj} \in \textbf{Set}$$
$$\text{Obj} \equiv \text{BW}$$

An arrow will be a proof that two binary words are congruent. In ALF we can therefore use our definition of (proof-objects for) congruence of binary words directly as our definition of arrow in a monoidal category (see Section 3.1):

$$\text{Hom} \in (A, B \in \text{Obj}) \ \textbf{Set}$$
$$\text{Hom} \equiv \text{cbw}$$

In HOL we did not use the Curry-Howard isomorphism when defining cbw. Hence we have to introduce a new definition of arrow of the free monoidal category. Note however the parallel between this new definition and the definition of cbw in ALF (see Section 3.1). Moreover, we do not have primitive dependent types in HOL. Before defining the set of arrows between two object we must therefore first define a datatype of *raw arrows*:

```
arr = id of bw | oo of arr => arr | xO of arr => arr
    | alpha of bw => bw => bw | alphaI of bw => bw => bw
    | lambda of bw | lambdaI of bw
    | rho of bw | rhoI of bw
```

Arrow composition oo and arrow product xO are defined as infixes such that composition has a lower precedence: we can write the expression $(f \, \text{xO} \, g) \, \text{oo} \, (f' \, \text{xO} \, g')$ without parentheses as $f \, \text{xO} \, g \, \text{oo} \, f' \, \text{xO} \, g'$. Both operators associate to the right.

The domain (dom) and codomain (cod) functions of raw arrows are introduced by straightforward primitive recursive definitions; for instance, dom(id a) = a and dom(g oo f) = dom f, while cod(g oo f) = cod g.

Not all raw arrows are well-defined, due to restrictions on the way we can compose arrows. Hence, we introduce a predicate def, which yields true on all "primitive" arrows and behaves as follows on the composite arrows:

```
def(g oo f) = (cod f = dom g) /\ def f /\ def g
def(f xO g) = def f /\ def g
```

The arrows of our monoidal category are the defined raw arrows of type arr.

We also introduce the following notation for Hom sets in HOL:

```
Hom(a,b) = {f | def f /\ (dom f = a) /\ (cod f = b)} .
```

Note how this makes it possible to simulate counterparts of ALF's dependent types in HOL. With this definition we can write f IN Hom(a,b) instead of the more verbose def f /\ (dom f = a) /\ (cod f = b). In the HOL formalization we do not use Hom sets often, because proofs can usually be performed more easily by using the constants def, dom and cod directly.

4.2 Equality of Arrows

To make these collections of objects and arrows into a category (and a free monoidal category) we also have to define when two arrows are equal. In set theory one would then define the Hom set as a quotient. But since neither HOL nor ALF has a primitive way of forming quotient types, a Hom set is represented as a set together with an equivalence relation.

In HOL we introduce the relation of equality of raw arrows

```
==:arr->arr->bool
```

whereas in ALF we only compare arrows with the same source and target, that is, we introduce a binary family of relations:

$$== \in (f, g \in \mathrm{Hom}(A, B)) \; \mathbf{Set}$$

However, though not visible in the type of ==, two equal arrows will also have the same source and target in HOL. In fact, we can prove

```
|- !f g.
    (f == g) ==>
    def f /\ def g /\ (dom f = dom g) /\ (cod f = cod g) .
```

The arrow equality is introduced as an inductive definition with 25 rules, which fall into five classes

- Basic laws of category theory (three rules),
- Bifunctoriality of arrow product (two rules),
- Rules for the natural isomorphisms witnessing associativity and unit laws (twelve rules),
- Coherence equations (three rules) stated in Mac Lane [11] (pp. 158–159),
- Equivalence and congruence rules (five rules),

Examples of these rules can be found in [3] and [1].

4.3 The Coherence Theorem

The main theorem of the development is the coherence theorem, which states that any two arrows in a Hom set are equal. In HOL this theorem is written

```
|- ! a b f g. f IN Hom(a,b) /\ g IN Hom(a,b) ==> (f == g),
```

and in ALF

$$\mathrm{Coherence} \in (A, B \in \mathrm{Obj}; f, g \in \mathrm{Hom}(A, B)) == (f, g)$$

The proof of the coherence theorem is a corollary of a commuting diagram

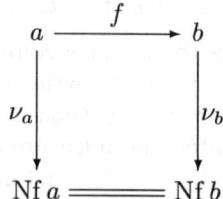

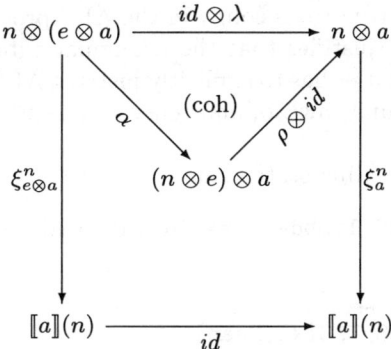

Fig. 1. The λ case of the naturality proof for ξ.

Since Nf can be extended to a functor by letting Nf $f = id$ this diagram expresses naturality of ν. The diagram entails that an arbitrary arrow f between a and b must be equal to $\nu_b^{-1} \circ \nu_a$.

The ALF definitions of ν and the auxiliary ξ were presented already in Section 3.2; ν witnessed the fact that an arbitrary binary word is congruent with its normal form.

Since we did not use the Curry-Howard analogy for the proof of normalization in HOL, we instead introduce a new analogous notion by the definition

```
nu a = xi a e oo lambdaI a
```

where

```
xi e n = rho n
xi(Var x)n = id (n Ox Var x)
xi(a Ox b)n = xi b(Mean a n) oo xi a n xO id b oo alpha n a b.
```

Naturality of ν (nu) is proved as a corollary of naturality of ξ (xi), which in turn is proved by induction on arrows of the free monoidal category. Each of the nine cases of this induction consists in proving that a certain diagram commutes. This is the main part of the proof of the coherence theorem. It makes heavy use of equality reasoning, involving the derived equality on arrows, which is a real challenge to theorem provers.

ALF does not have any support for equality reasoning and the proof of naturality of ξ therefore became quite a tedious exercise, where many trivial steps had to be performed manually. For example, consider the ALF proof

nat$\xi\lambda \in (n \in \text{Obj}; A \in \text{Obj}) ==$(o($\iota$(Mean($n, A$)), $\xi(n, \otimes(e, A)$)), o($\xi(n, A), \otimes(\iota(n), \lambda(A))$))
nat$\xi\lambda \equiv [n, A]$transE(λo, transE(αo, oE(refE, coh$_{\text{tri}}$)))

which implements the λ case in the proof of naturality of ξ, also shown as a diagram in Fig. 1. The only non-trivial step in this proof is the application of the coherence equation, referred to as coh$_{tri}$ in the ALF proof. This step is denoted "coh" in Fig. 1 where it justifies that the triangular subdiagram commutes. In addition to this step, the user has to explicitly instruct ALF to apply associativity of composition αo, the unit law λo, and reflexivity refE and transitivity transE of ==.

In HOL the corresponding statement is

```
xi b n oo id n x0 lambda b == id (nf b n) oo xi (e 0x b) n .
```

The HOL proof

```
PURE_ONCE_REWRITE_TAC[xi,Mean]
THEN APPLY COHERENCE THEN RAPPLY LEFT_ID THEN QED
```

is closer to the diagram in the sense that it doesn't mention things subsumed by the diagram notation such as associativity of composition. It first expands definitions, then applies coherence, then removes composition with identity, and finally uses reflexivity of arrow equality. Note that APPLY only applies an "arrow conversion" to the left-hand side of an arrow equality, i.e. to the upper path in a diagram, whereas RAPPLY is used to apply an "arrow conversion" on the Right-hand side.

The simplicity of the proof is due to the tool support for diagram proofs [1]. This greatly reduces the size and complexity of proofs, which become comparable to informal paper-and-pencil proofs. The tactics and conversions implemented to support this are each only a few lines long.

The tool support scales to more difficult proofs such as the induction step for arrow product, see Fig. 2. The ALF implementation of this diagram is too big to show in this paper. The HOL proof follows the diagram closely:

```
PURE_ONCE_REWRITE_TAC[xi]
THEN APPLY ASSOC_RIGHT
THEN APPLY ALPHA THEN PURE_ONCE_REWRITE_TAC[dom]
THEN APPLY PROD_COMP
THEN APPLY_ASM
THEN APPLY(RIGHT(LEFT(RIGHT(LEFT_ID THENA RIGHT_ID_INTRO))))
THEN APPLY PROD_COMP_RED
THEN APPLY ASSOC_RIGHT
THEN APPLY_ASM
THEN PURE_ONCE_REWRITE_TAC[Mean]
THEN COND_REWRITE1_TAC Mean_eq_dom_cod
THEN QED
```

This proof is just twelve lines long while the original HOL proof (developed without support for diagram chasing) was approximately 300 lines long. The reason for this big reduction is both that congruence, transitivity and associativity steps are performed behind the scenes and that side conditions on arrow equality rules, which are stated in terms of dom, cod and def, are proved automatically: this is built into APPLY and RAPPLY. The statement we prove in HOL is:

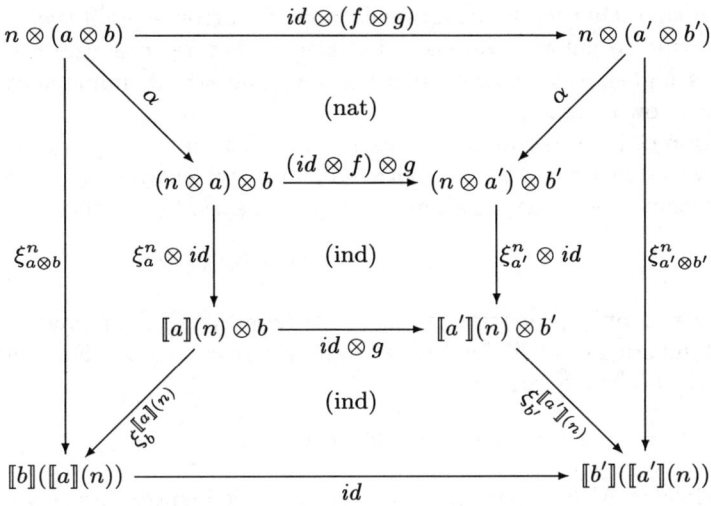

Fig. 2. The product case of the naturality proof for ξ.

```
xi (cod f Ox cod f') n oo id n xO (f xO f') ==
   id (Mean (cod f Ox cod f') n) oo xi (dom f Ox dom f') n
```

with two induction hypotheses:

```
xi (cod f') n oo id n xO f' ==
   id (Mean (cod f') n) oo xi (dom f') n
xi (cod f) n oo id n xO f ==
   id (Mean (cod f) n) oo xi (dom f) n .
```

Let us look at the HOL proof in some detail to see how close it is to the diagram proof. The main steps are expanding the definition of ξ, applying naturality of α (called ALPHA, and "nat" in the diagram) and applying the two induction hypotheses using APPLY_ASM (called "ind" in the diagram). APPLY_ASM applies an assumption somewhere in a goal. Note that we (usually) do not have to say where a rule must be applied. There are a number of strategies for finding the first suitable application.

The other tactics used are invisible in the diagram representation, which is highly "optimized". In fact, a few nontrivial steps are left unmentioned in the diagram proof, such as the steps exploiting the bifunctoriality of arrow product, which correspond to the use of PROD_COMP and PROD_COMP_RED in the HOL proof. Note that rules are applied modulo associativity but parentheses must be moved right first (using ASSOC_RIGHT). The QED tactic applies reflexivity. The other tactics just clean up terms by rewriting.

The support for diagram chasing in HOL is described in more detail in Agerholm [1]. The overall idea is to use a theorem stating that two arrows are equal as a rewrite rule. Due to the congruence rules for arrow equality we can apply the rewrite rule to subexpressions. This means that we can use ideas similar to Paulson's higher order conversions for rewriting [18] to implement the tool support for arrow equality.

In addition to tedious equality reasoning, another more subtle problem in the ALF proof was that we had to cope with a difficulty which arose as a consequence of the treatment of equality in intensional type theory. In the theorem

$$\text{nf} \in (A, B \in \text{BW}; f \in \text{cbw}(A, B)) = (\text{Nf}(A), \text{Nf}(B))$$

the equality = is only propositional and not definitional. As a consequence we cannot just substitute $\text{Nf}(B)$ for $\text{Nf}(A)$ in an arbitrary context. For example, we cannot build an identity arrow

$$\iota \in \text{cbw}(\text{Nf}(A), \text{Nf}(B)) \quad []$$

since this violates ALF's type checking. Instead this identity arrow will depend explicitly on the element witnessing that $\text{Nf}(A)$ and $\text{Nf}(B)$ are propositionally equal. This complicates the induction steps (composition and product) in the proof of naturality of ξ somewhat. The reader is referred to Beylin and Dybjer [3] for more details.

5 Conclusion

We can divide the comparison into a theory part (Church classical versus Martin-Löf intuitionistic type theory) and a system part (HOL versus ALF).

For the theory part we first notice that both systems implement not only the original type theories of Church and Martin-Löf, but practically justified extensions: HOL adds a polymorphic type system a la ML and ALF adds a general inductive definition mechanism and pattern-matching. We wish to emphasize that the example was initially developed to illustrate an interesting use of Martin-Löf type theory. In particular the coherence proof is constructive and classical reasoning plays no role. Moreover, the example makes essential use of proofs as programs, an aspect which is highlighted in the ALF development.

It was however possible to develop the proof in a very parallel way in HOL. The ALF proof uses the primitive notions of dependent type and inductive definition which are basic in Martin-Löf type theory. But there is support in HOL for deriving analogous notions. In particular essential use was made of the packages for inductive datatype and relation definitions.

The main theoretical advantage of HOL is probably its treatment of equality. When two things have been proved equal they can be substituted for each other everywhere. In ALF, which implements intensional type theory, there is a distinction between propositional and judgemental ("definitional") equality. It is only with respect to the latter that we have full substitutivity of equality. There is a rule of substitutivity of equality also for propositional equality, but this rule introduces witnesses. This somewhat complicates the checking of some diagrams in ALF. On the positive side, maintaining a distinction between propositional and judgemental equality means that the latter can be kept decidable and checked directly using the normalization procedure which is part of ALF's type checking mechanism. This gives a limited amount of automatic theorem proving capability to ALF, which does not have a counterpart in HOL.

We would like to end the discussion by emphasizing that we were mainly interested in translation from Martin-Löf type theory to higher order logic "in practice" and our paper is not intended as a theoretical contribution. The techniques we use for eliminating dependent types are for example well known.

For the system part one should first keep in mind that HOL is a mature system which has been developed over more than ten years, and has been used for many big developments. The ALF system is much more of a prototype and has mainly been used by people in Göteborg.

What more than anything else gave HOL an advantage was the possibility to write tactics for diagram chasing in Standard ML. After a few days of programming, tool support for diagram chasing was made available that gave considerable smaller proofs, e.g. a reduction from 300 to 10 lines! Furthermore, we think that it would be possible to write a graphical user-interface to this proof support and in this way allow a category theorist to have a sketchy paper-and-pencil proof mechanically checked by pointing and clicking, at the same time as the LaTeX source for the (correct!) diagram proof is produced automatically.

ALF, on the other hand, scores some positive points for its window interface and the possibility of doing "proof-by-clicking". This could however not compensate for the lack of support for equational reasoning. Moreover, the HOL development also used a graphical interface to HOL, developed by Syme [20], that allows different fonts and supports proof-by-clicking to a limited degree.

Acknowledgements

The work of the first author was financially supported by the Danish Natural Science Research Council. The work of the second author was partly supported by a grant from the Swedish Institute and partly by the Swedish Technical Science Research Council. The work of the third author was also partly supported by the Swedish Technical Science Research Council and partly by the Isaac Newton Institute for Mathematical Sciences.

The paper was typeset in LaTeX with LLNCS style using Paul Taylor's `diagrams` package and Tomas Rokicki's `epsf` macros.

References

1. Sten Agerholm. Formalizing a proof of coherence for monoidal categories. Draft manuscript, available from //ftp.ifad.dk/pub/users/sten, December 1995.
2. Michael Barr and Charles Wells. *Category Theory for Computing Science*. Prentice Hall, 1990.

3. Ilya Beylin and Peter Dybjer. Extracting a proof of coherence for monoidal categories from a formal proof of normalization for monoids. In Stefano Berardi and Mario Coppo, editors, *TYPES '95*, LNCS, 1996. To appear.
4. J. Camilleri and T. Melham. Reasoning with inductively defined relations in the HOL theorem prover. Technical Report No. 265, University of Cambridge Computer Laboratory, August 1992.
5. Thierry Coquand. Pattern matching with dependent types. In *Proceedings of The 1992 Workshop on Types for Proofs and Programs*, June 1992.
6. Peter Dybjer. Inductive families. *Formal Aspects of Computing*, 6:440–465, 1994.
7. M. J. C. Gordon. HOL: A proof generating system for higher order logic. In G. Birtwistle and P. A. Subrahmanyam, editors, *Current Trends in Hardware Verification and Automated Theorem Proving*. Springer-Verlag, 1989.
8. M. J. C. Gordon and T. F. Melham, editors. *Introduction to HOL: A Theorem-proving Environment for Higher-Order Logic*. Cambridge University Press, 1993.
9. E. Gunter. A broader class of trees for recursive type definitions for HOL. In J. J. Joyce and C. H. Seger, editors, *Proceedings of the 6th International Workshop on Higher Order Logic Theorem Proving and its Applications*, volume 780 of *Lecture Notes in Computer Science*. Springer-Verlag, 1994.
10. Gérard Huet and Amokrane Saibi. Constructive category theory. In *Proceedings of the Joint CLICS-TYPES Workshop on Categories and Type Theory, Göteborg*, January 1995.
11. Saunders Mac Lane. *Categories for the Working Mathematician*, volume 5 of *Graduate Texts in Mathematics*. Springer-Verlag, New York, 1971.
12. Lena Magnusson. *The Implementation of ALF — a Proof Editor for Martin-Löf's Monomorphic Type Theory with Explicit Substitution*. PhD thesis, Chalmers T. H., 1994.
13. Per Martin-Löf. An intuitionistic theory of types: Predicative part. In *Logic Colloquium '73*, pages 73–118. North-Holland, 1975.
14. Per Martin-Löf. Constructive mathematics and computer programming. In *Logic, Methodology and Philosophy of Science, 1979*, pages 153–175. North-Holland, 1982.
15. Per Martin-Löf. *Intuitionistic Type Theory*. Bibliopolis, 1984.
16. T. Melham. A package for inductive relation definition in HOL. In M. Archer, J. J. Joyce, K. N. Levitt, and P. J. Windly, editors, *Proceedings of the 1991 International Workshop on the HOL Theorem Proving System and its Applications, Davis, August 1991*. IEEE Computer Society Press, 1992.
17. Bengt Nordström, Kent Petersson, and Jan Smith. *Programming in Martin-Löf's Type Theory: an Introduction*. Oxford University Press, 1990.
18. L. C. Paulson. A higher order implementation of rewriting. *Science of Computer Programming*, 3, 1983.
19. B. Pierce. *Basic Category Theory for Computer Scientists*. MIT Press, 1991.
20. D. Syme. A new interface for HOL – ideas, issues and implementaion. In E. T. Schubert, P. J. Windley, and J. Alves-Foss, editors, *Proceedings of the 8th International Workshop on Higher Order Logic Theorem Proving and its Applications*, volume 971 of *Lecture Notes in Computer Science*. Springer-Verlag, 1995.

Modeling a Hardware Synthesis Methodology in Isabelle

David Basin and Stefan Friedrich

Max-Planck-Institut für Informatik
Im Stadtwald, D-66123, Saarbrücken, Germany

Abstract. *Formal Synthesis* is a methodology developed at Kent for combining circuit design and verification. We have reinterpreted this methodology in ISABELLE's theory of higher-order logic so that circuits are synthesized using higher-order resolution. Our interpretation simplifies and extends Formal Synthesis both conceptually and in implementation. It also supports integration of this development style with other synthesis methodologies and leads to techniques for developing new classes of circuits, e.g., recursive descriptions of parameterized circuits.

1 Introduction

Verification by formal proof is time intensive and this is a burden in bringing formal methods into software and hardware design. One approach to reducing the verification burden is to combine development and verification by using a calculus where development steps either guarantee correctness or, since proof steps are in parallel to design steps, allow early detection of design errors. We present such an approach here, using resolution to synthesize circuit designs during proofs of their correctness. Our approach is based on modeling a particular methodology for hierarchical design development within a synthesis by resolution framework.

Our starting point is a novel methodology for hardware synthesis, called *Formal Synthesis*, proposed by the VERITAS group at Kent [7]. In Formal Synthesis, one starts with a *design goal*, which specifies the behavioral properties of a circuit to be constructed, and interactively refines the design using a small but powerful set of *techniques*, which allows the developer to hierarchically decompose specifications and introduce subdesigns and library components in a natural way. Internally, each technique consists of a pair of functions: a *subgoaling function* and a *validation function*. The former decomposes a specification *Spec* into subspecifications $Spec_i$. The latter takes proofs that some $Circ_i$ achieve $Spec_i$, and constructs an implementation *Circ* with a proof that *Circ* achieves *Spec*. When the refinement is finished, the system composes the validation functions that were applied and this constructs a theorem that a (synthesized) circuit satisfies the original design goal.

To make the above clearer, consider one of the simpler techniques, called *Split*. The subgoaling function reduces a design goal $Spec_1 \land Spec_2$ to two new design goals, one for each conjunct. The validation function is based on the rule

$$\frac{\vdash Imp_1 \rightarrow Spec_1 \quad \vdash Imp_2 \rightarrow Spec_2}{\vdash Imp_1 \land Imp_2 \rightarrow Spec_1 \land Spec_2}, \tag{1}$$

which explains how to combine implementations that achieve the subgoals into an implementation which achieves the original goal. This kind of top-down problem decomposition is in the spirit of (LCF style) tactics and designing a good set of techniques is analogous to designing an appropriate set of tactics for a problem domain. However, tactics decompose a goal into subgoals from which a validation proves the original goal, whereas with techniques, the validation proves a different theorem altogether. Formal Synthesis separates design goals and techniques from ordinary theorem-proving goals and tactics. This is a conceptual separation: they are different sorts of entities with different semantics. Moreover they are treated differently in the implementation too and Formal Synthesis required extending the VERITAS theorem prover itself.

We show how to reinterpret Formal Synthesis as deductive synthesis based on higher-order resolution. To do this, we begin with the final theorem that the Formal Synthesis validations should deliver: a circuit achieves a stated specification. However the circuit is not given up front; instead it is named by a metavariable. Proof proceeds by applying rules in ISABELLE which correspond to VERITAS validation rules like (1). Because rule application uses resolution, the metavariable is incrementally instantiated to the synthesized circuit. Let us illustrate this with the Split technique. Our goal (simplifying slightly) is to prove a theorem like $?Circ \rightarrow Spec$, where the question mark means that $Circ$ is a metavariable. If $Spec$ is a conjunction, $Spec_1 \wedge Spec_2$, then we can resolve $?Circ \rightarrow Spec_1 \wedge Spec_2$ with the rule in (1). The result is a substitution, $?Circ = ?Imp_1 \wedge ?Imp_2$, and the subgoals $?Imp_1 \rightarrow Spec_1$ and $?Imp_2 \rightarrow Spec_2$. Further refinement of these subgoals will generate instances of the $?Imp_i$ and hence $?Circ$.

There are a number of advantages to this reinterpretation. It is conceptually simple. We do away with the subgoal and validation functions and let resolution construct circuits. It is simple to implement. No changes are required to the Isabelle system; instead, we derive rules and program tactics in an appropriate extension of Isabelle's theory of higher-order logic (HOL). Moreover, the reinterpretation makes it easy to implement new techniques that are compatible with the Formal Synthesis style of development. In the past, we have worked on rules and tactics for the synthesis of logic programs [1] and these programs are similar to circuits: both can be described as conjunctive combinations of primitive relations where existential quantification is used to 'pass values'. By moving Formal Synthesis into a similar theorem proving setting, we could adapt and apply many of our tactics and rules for logic programming synthesis to circuit synthesis. For example, rules for developing recursive programs yield new techniques for developing parameterized classes of circuits. We also exhibit (cf. Section 6) that the kind of rules used for resolution-based development in the LAMBDA system are compatible with our reinterpretation.

2 Background

We assume familiarity with ISABELLE [10] and higher-order logic. Space limitations restrict us to reviewing only notation and a few essentials.

ISABELLE is an interactive tactic-based theorem prover. Logics are encoded in ISABELLE's metalogic by declaring a theory, which is a signature and set of axioms. For example, propositional logic might have a type *Form* of formulae, with constructors like $\wedge$ and a typical proof rule would be !! $A, B : Form. A \Longrightarrow (B \Longrightarrow A \wedge B)$. Here $\Longrightarrow$ and !! are implication and universal quantification in ISABELLE's metalogic. Outermost (meta-)quantifiers are often omitted and iterated implication is written in a more readable list notation, e.g., $[\![A; B]\!] \Longrightarrow A \wedge B$. These implicitly quantified variables are treated as metavariables, which can be instantiated when applying the rule using higher-order unification. In our work we use ISABELLE's theory of higher-order logic, extended with theories of sets, well founded recursion, natural numbers, and the like.

Isabelle supports proof construction by higher-order resolution. Given a proof state with subgoal ψ and a proof rule $[\![\phi_1; \ldots; \phi_n]\!] \Longrightarrow \phi$, we unify ϕ with ψ. If this succeeds, then unification yields a substitution σ and the proof state is updated by applying σ to it and replacing ψ with the subgoals $\sigma(\phi_1), \ldots, \sigma(\phi_n)$; since unification is used to apply rules, the proof state itself may contain metavariables. We use this to synthesize circuits during proofs. Note that a proof rule can be read as an intuitionistic sequent where the ϕ_i are the hypotheses. Isabelle's resolution tactics apply rules in a way that maintains this illusion of working with sequents and we will often refer to the ϕ_i as assumptions.

VERITAS [8] is a tactic based theorem prover similar to ISABELLE and HOL, but its higher-order logic is augmented with constructions from type theory, e.g., standard type constructors such as (dependent) function space, product, and subtype. When used to reason about hardware, one proves theorems that relate circuits, *Circ*, to specifications, *Spec*, e.g.,

$$\forall p_1 \ldots p_n. \, Circ(p_1, \ldots, p_n) \to Spec(p_1, \ldots, p_n),$$

where $p_1 \ldots p_n$ are the external port names. As is common, hardware is represented relationally, where primitive constructors (e.g., transistors, gates, etc.) are relations combined with conjunction and 'wired together' using existential quantification [3]. The variables in $p_1 \ldots p_n$ are typed, and the primary difference between VERITAS and similar systems is that one can use richer types in such specifications. For example, if we are defining circuits operating over 8-bit words, we would formalize this requirement on word-length in the types.

3 Formal Synthesis

Formal Synthesis is based on *techniques*. As previously indicated, each technique combines a subgoaling function with a validation. The validations are executed when a derivation is completed to build a proof that a circuit achieves an implementation. Before giving the techniques we introduce some notation used by Formal Synthesis. A design specification *spec* which is to be implemented is written $\Box spec$ ($\Box$ is just a symbol; there is no relationship to modal logics). A formula *Thm* to be proved is written as $\vdash Thm$. A term *Tm* to be demonstrated well typed is written as $\Diamond Tm$. A proof in VERITAS is associated with a support

Name	Subgoaling rule	Validation rule
Claim	$\dfrac{\Box Spec}{\Box Spec' \quad \vdash Spec' \to Spec}$	$\dfrac{\vdash Imp \to Spec' \quad \vdash Spec' \to Spec}{\vdash Imp \to Spec}$
Split	$\dfrac{\Box(Spec_1 \wedge Spec_2)}{\Box Spec_1 \quad \Box Spec_2}$	$\dfrac{\vdash Imp_1 \to Spec_1 \quad \vdash Imp_2 \to Spec_2}{\vdash Imp_1 \wedge Imp_2 \to Spec_1 \wedge Spec_2}$
Reveal	$\dfrac{\Box \exists dec.\ Spec}{[\![dec]\!] \Box Spec}$	$\dfrac{[\![dec]\!] \vdash Imp \to Spec}{\vdash (\exists dec.\ Imp) \to (\exists dec.\ Spec)}$
Inside	$\dfrac{\Box Circ \mathrel{\widehat{=}} \lambda dec.\ Spec}{[\![dec]\!]\Box Spec}$	$\dfrac{[\![dec]\!] \vdash Imp \to Spec}{[\![Circ \mathrel{\widehat{=}} \lambda dec.\ Spec]\!] \vdash \forall dec.\ Imp \to Circ\ ports}$
Library	$\dfrac{\Box library\_part\ args}{}$	$\dfrac{}{\vdash library\_part\ args \to library\_part\ args}$
Subdesign	$\dfrac{\Box Spec}{[\![c \mathrel{\widehat{=}} \lambda dec.\ s']\!] \Box Spec \quad \Box c \mathrel{\widehat{=}} \lambda dec.\ s'}$	$\dfrac{[\![c \mathrel{\widehat{=}} \lambda dec.\ s']\!] \vdash Imp \to Spec}{\vdash (\text{let } c \mathrel{\widehat{=}} \lambda dec.\ s' \text{ in } Imp) \to Spec}$
Design	$\dfrac{\Box subdesign\ args}{}$	$\dfrac{}{\vdash subdesign\ args \to subdesign\ args}$
Proof	$\vdash Thm \quad \diamond Tm$	$\vdash Thm \quad \vdash Tm$

Table 1. VERITAS design techniques

signature, which contains the theory used (e.g., datatype definitions), definitions of predicates, and the like. In VERITAS, the signature can be extended dynamically during proof. If a technique extends the signature of a goal *Spec*, then this is written as $[\![extension]\!] Spec$. Finally, the initial design goal must be of the form $Circ \mathrel{\widehat{=}} \lambda dec.\ Spec$ where *Circ* is the name assigned to the circuit.

There are eight techniques and these are adequate to develop any combinational circuit in a structured hierarchical way. The subgoaling and validation functions are both based on rules. The subgoaling rules should be read top-down: to implement the goal above the line, solve all subgoals below the line. The validation rules operate in the opposite direction: the goal below the line is established from proofs of the goals above the line. These rules behave as follows:

Claim: The subgoaling function yields a new design goal *Spec'* and a subgoal that this suffices for *Spec*. The validation is based on the transitivity of $\to$.

Split: The subgoaling function decomposes a conjunctive specification into the problem of implementing each conjunct. The validation constructs a design from designs for the two subparts.

Reveal: Shifts internally quantified ports into the signature, allowing further refinement.

Inside: Since the initial design goal is given as a lambda abstraction, a technique is needed to remove this binding. To implement the circuit *Circ* means to implement the specification *Spec*. The validation theorem states that the implementation is correct for all port-values that can appear at the ports *port*, which are declared in the declaration *dec*.

Library: The subgoaling function imports components from a database of predefined circuits. The validation corresponds to the use of a lemma.

Subdesign: A subdesign, submitted as a lambda abstraction, may be introduced. It must be implemented (second subgoal) and may be used in proving the original design goal (first subgoal).

Design: Like the Library technique, this enables the user to apply a design that has already been implemented. In this case, the implementation is a subdesign introduced during the design process.

Proof: VERITAS is used to prove a theorem.

4 Implementation and Extension

We now describe our reinterpretation of Formal Synthesis in ISABELLE. We divide our presentation in several parts, which roughly correspond to ISABELLE theories that we implemented: circuit abstractions and 'concrete technology', technique reinterpretation, and new techniques.

4.1 Circuit Abstractions and Concrete Technology

We represent circuits as relations over port-values. We model kinds of signals as sets and we use ISABELLE's set theory (in HOL) to imitate the dependent types of the VERITAS logic; hence quantifiers used in our encoding are the bound quantifiers of ISABELLE's set theory. We name the connectives that serve as constructors for circuits with the following definitions.

$$C \text{ Sat } S \equiv C \rightarrow S \qquad \text{Port } p:P.\ \phi(p) \equiv \forall p:P.\ \phi(p)$$
$$C_1 \text{ Join } C_2 \equiv C_1 \wedge C_2 \qquad \text{Wire } w:W.\ Circ(w) \equiv \exists w:W.\ Circ(w)$$

We will say more about the types P and W shortly. After definition, we derive rules which characterize these operators and use these rules in subsequent proofs (rather than expanding definitions). For example, using properties of conjunction and implication we derive (hence we are guaranteed of their correctness) in ISABELLE the following rules which relate Join to Sat.

$$Circ_1 \text{ Sat } Spec \Longrightarrow (Circ_1 \text{ Join } Circ_2) \text{ Sat } Spec$$
$$Circ_2 \text{ Sat } Spec \Longrightarrow (Circ_1 \text{ Join } Circ_2) \text{ Sat } Spec$$

Associating definitions with such characterization theorems increases comprehensibility and provides some abstraction: if we change definitions we can reuse our theories provided the characterization theorems are still derivable. Definitions have a second function: they distinguish circuit constructors from propositional connectives and this restricts resolution (e.g., Join does not unify with $\wedge$) and makes it easier to write tactics that automate design.

Using the above abstractions, we can express the correctness of a (to be synthesized) circuit with respect to a specification as

$$\text{Port } p_1:P_1 \ldots p_n:P_n.\ ?Circ(p_1,\ldots,p_n) \text{ Sat } Spec(p_1,\ldots,p_n).$$

This is not quite adequate though to simulate Formal Synthesis style proofs. A small problem is that the application of the techniques Reveal, Inside, and Subdesign extend the VERITAS signature. As there is no direct analog of dynamic signature extension in ISABELLE, we model Reveal and Inside using bounded quantification and achieve an effect similar to signature extension by adding declaration information to the assumptions of the proof state according to the rules of set theory. Subdesign, which in VERITAS extends the signature with a new definition is slightly trickier to model; we regard the new definition as a condition necessary for the validity of the correctness theorem and we express this condition using ordinary implication. Of course, when stating the initial goal we do not yet know which definitions will be made, so again we use a metavariable to leave this as an unknown. Thus a theorem that we prove takes the initial form

Def($?Definitions$) → Port $p_1 : P_1 \ldots p_n : P_n.\ ?Circ(p_1,\ldots,p_n)$ Sat $Spec(p_1,\ldots,p_n)$.

Analogous to the Formal Synthesis design goals we call such a theorem a *design theorem*. The constant Def is simply the identity function and Def($?Definitions$) serves as a kind of 'definition context'. Each time a definition is made, $?Definitions$ is instantiated to a conjunction of the definition and a fresh metavariable, which can be instantiated further with definitions made later.

In the design theorem, wires are bound to types P_i; these types are defined in theories about 'concrete technologies', e.g., representations of voltage, signals, and the like. A simple instance is where port values range over a datatype bin where voltage takes either the value Lo or Hi. We use ISABELLE's HOL set theory to define a *set* bin containing these two elements. Afterwards, we derive standard rules for bin, e.g., case-analysis over binary values.

We extend our theory with tactics that automate most common kinds of reasoning about binary values. For example, we have tactics that perform exhaustive analysis on Port values quantified over bin and tactics that combine such case analysis with ISABELLE's simplifier and heuristic based proof procedure (`fast_tac`) for HOL. These tactics automate almost all standard kinds of reasoning about combinational circuits.

We extend our voltage theory with a theory of gates that declares the types of gates and axiomatizes their behavior. For example, the and gate is a predicate of type [bin, bin, bin] ⇒ bool whose behaviour is axiomatized as

Port a:bin b:bin c:bin. and(a,b,c) Sat $(c = \text{Hi}) \leftrightarrow (b = \text{Hi}) \wedge (c = \text{Hi})$.

Such axioms form part of a library of circuit specifications and are used to synthesize parts of circuits.

4.2 Reinterpreting Techniques

We have implemented tactics that simulate the first seven techniques (for the Proof technique we simply call the prover). Table 2 lists the tactics and the derived rules corresponding to the VERITAS techniques. The tactics are based on

Claim: claim_tac *str i*
            [[*Circ* Sat *Spec'*; *Spec'→Spec*]] $\Longrightarrow$ *Circ* Sat *Spec*

Split: split_tac *i*
            [[*Circ*$_1$ Sat *Spec*$_1$; *Circ*$_2$ Sat *Spec*$_2$]] $\Longrightarrow$ (*Circ*$_1$ Join *Circ*$_2$) Sat (*Spec*$_1$ $\wedge$ *Spec*$_2$)

Reveal: reveal_tac *i*
            [[!! *w*. *w:T* $\Longrightarrow$ *Circ*(*w*) Sat *Spec*(*w*)]] $\Longrightarrow$ (Wire *w:T*. *Circ*(*w*)) Sat ($\exists w{:}T.Spec(w)$)

Inside: inside_tac *i*
            [[!! *p*. *p:T* $\Longrightarrow$ *Circ*(*p*) Sat *Spec*(*p*)]] $\Longrightarrow$ Port *p:T*. *Circ*(*p*) Sat *Spec*(*p*)

Library: library_tac *dels elims thm i*
            [[Def(*H*)→*P*; Def(*H*); *P* $\Longrightarrow$ *R*]] $\Longrightarrow$ *R*

Subdesign: subdesign_tac *str i*
            [[Def(Port $p_1:P_1 \ldots p_n:P_n$. $Name(p_1,\ldots,p_n) = Circ'(p_1,\ldots,p_n)$));
            Port $p_1:P_1 \ldots p_n:P_n$. $Circ'(p_1,\ldots,p_n)$ Sat $Spec'(p_1,\ldots,p_n)$;
            Port $p_1:P_1 \ldots p_n:P_n$. $Name(p_1,\ldots,p_n)$ Sat $Spec'(p_1,\ldots,p_n)$ $\Longrightarrow$ *Circ* Sat *Spec*
            ]]$\Longrightarrow$ *Circ* Sat *Spec*

Design: design_tac *dels elims i*
            [[Port *x* : *T*. *P*(*x*); *P*(*x*) $\Longrightarrow$ *R*; *x:T*]] $\Longrightarrow$ *R*

Table 2. ISABELLE Techniques (Name, Tactic, and Rule)

the derived rules, which correspond to the validation rule associated with the VERITAS technique; their function is mostly self explanatory.

Claim and **Split**: The tactics apply rules that are direct translations of the corresponding VERITAS validation rules. The specification *Spec'* is supplied as a string *str* to claim_tac.

Reveal and **Inside**: These are identical to the VERITAS techniques except that internal wiring, or quantification over ports, is converted to quantification at the metalevel and type constraints (expressed in ISABELLE's HOL set-theory) become assumptions. The rules state that if an implementation satisfies its specification, then we can wire a signal to a port with Inside or we can hide it as an internal wiring with Reveal. reveal_tac and inside_tac apply their respective rule as many times as possible to the specified goal.

Library: library_tac solves the design goal *R* by using a previously implemented design, supplied by the user. The rule is a form of implication elimination. The first subgoal is instantiated with the component's design theorem. The second is solved by extending the definition context of the overall design. The third establishes the correctness of the design using the specification of the library component. This involves type checking for the ports, which depends on the concrete technology of the designs; hence we supply the tactic with additional 'elimination' and 'deletion' rules, which solve the type checking goals.

Subdesign: We have given an informal rule schema (the others are formally derived) that represents infinitely many rules (there are arbitrarily many quantifiers, indicated by the ellipses). subdesign_tac simulates the effect of applying such a rule. The user gives the design goal of the subdesign to be implemented in the same form as the initial goal. Three new subgoals are generated by the tactic. The first corresponds to the subdesign definition and the tactic discharges this by adding it to the definition context of the main design. The second subgoal commits us to implement the subdesign. The third allows us to use the subdesign when proving the original subgoal.

Design: This solves a design goal by using a previously introduced subdesign. The subdesign has been given as a design goal, which is part of the assumptions of the goal to be solved. The tactic removes all port quantifiers for the assumption by repeatedly applying the associated rule. For each port a new subgoal is generated, which concerns the type of that port, and is solved as in the Library technique; hence we provide lists of type checking rules to the tactic.

Proof: General proof goals arise from the application of the Claim rule. These are undecidable in general and must be proven by the user.

4.3 Extensions of the Calculus

The techniques defined by Formal Synthesis are effective for developing combinational circuits. However, nontrivial circuits are often best developed as instances of parametric designs. For example, rather than designing a 16-bit adder it is preferable to develop one parameterized by word-length and afterwards to compute particular instances. We have developed new techniques that are compatible with our reinterpretation of Formal Synthesis and construct such parameterized circuits. Structural induction is used to build parameterized linear circuits and more generally n-dimensional grids, and course-of-values induction is used to build general recursively defined designs. We will consider course-of-values induction below and later apply it to build a tree-structured addition circuit.

The idea for such extensions is motivated by previous work of ours on calculi in ISABELLE for synthesizing recursive logic programs [1]. There we developed rules and tactics based on induction which extend definition contexts with templates (a function or predicate name with a metavariable standing in for the body) for recursive definitions and leave the user with a goal to prove where use of the induction hypothesis builds a recursive program. This past work has much in common with our technique-based calculus for circuits. Syntactically, logic programs and circuits are similar: both can be described as conjunctive combinations of primitive relations where existential quantification 'passes values'. It turns out that we could (with very minor adaptation) directly use the rules and tactics we developed for synthesizing logic programs to build recursive circuits. We find this kind of reuse, not just of concepts but also of actual rules and tactics, an attractive advantage of interpreting different synthesis methodologies in a common framework. We will address this point again in Section 6 where we consider how techniques developed for the LAMBDA system can also be applied in the Formal Synthesis setting.

We construct a parameterized circuit by proving a parameterized design theorem, which is a design theorem where the outermost quantifier (or quantifiers) ranges over an inductively defined datatype like the natural numbers, e.g.,

$$\forall n{:}N.\; \mathsf{Port}\, p_1{:}P_1\; \ldots\; p_m{:}P_m.\; \mathit{Circ}(n, p_1, \ldots, p_m)\; \mathsf{Sat}\; \mathit{Spec}(n, p_1, \ldots, p_m)$$

specifies an implementation whose size depends on the number n. We use induction to instantiate Circ to a recursively specified design. ISABELLE's HOL comes with a theory of natural numbers (given as an inductive definition) from which it is easy to derive the following course-of-value induction rule.

$$[\![\, !!\, n.\; [\![\, n{:}N;\; \forall k{:}N.\; k < n \to P(k)\,]\!] \Longrightarrow P(n)\,]\!] \Longrightarrow \forall n{:}N.\; P(n)$$

We use this rule as the basis for a tactic, cov_induct_tac, which functions as if the following rule schema were applied.

$[\![\, \mathsf{Def}(\,\forall n{:}N.\; \mathsf{Port}\, p_1{:}P_1\; \ldots\; p_m{:}P_m.\; \mathit{Circ}(n, p_1, \ldots, p_m) = \mathit{Definition}(n, p_1, \ldots, p_m)\,);$

$!!\, n.[\![n{:}N;$

$\qquad \forall k{:}N.\, k < n \to \mathsf{Port}\, p_1{:}P_1\; \ldots\; p_m{:}P_m.\; \mathit{Circ}(k, p_1, \ldots, p_n)\; \mathsf{Sat}\; \mathit{Spec}(k, p_1, \ldots, p_m)$

$\;]\!] \Longrightarrow\; \mathsf{Port}\, p_1{:}P_1\; \ldots\; p_m{:}P_m.\; \mathit{Definition}(n, p_1, \ldots, p_m)\; \mathsf{Sat}\; \mathit{Spec}(n, p_1, \ldots, p_m)$

$]\!] \Longrightarrow \forall n{:}N.\; \mathsf{Port}\, p_1{:}P_1\; \ldots\; p_m{:}P_m.\; \mathit{Circ}(n, p_1, \ldots, p_m)\; \mathsf{Sat}\; \mathit{Spec}(n, p_1, \ldots, p_m)$

When this rule is applied by higher-order resolution, Spec will unify with the specification of the design theorem, and Circ with the metavariable standing for the circuit. The first subgoal sets up a parameterized circuit definition: an equality for Circ is defined, where it is equal to $\mathit{Definition}$, which will be a metavariable. Our tactic discharges this subgoal by adding it to the definitions of the main design. This leaves us with the second goal, which is the design goal. However, now we build a circuit named $\mathit{Definition}$ and instantiating this in subsequent proof steps will instantiate our definition for Circ. Moreover, we now have a new assumption (the induction hypothesis), which states that Circ achieves Spec for smaller values of k. If we can reduce the problem to implementing a smaller design, then we can resolve with the induction hypothesis and this will build a recursive design by instantiating the definition body of Circ with an instance of Circ. How this works in practice will become clearer in Section 5.2.

Parameterized specifications require parameterized input types, e.g., rather than having input ports, we have parameterized input busses. To support this we develop a theory of busses encoded as lists of specified lengths (there are other possibilities, but this allows us to directly use ISABELLE's list theory). Some of our definitions are as follows.

$$B\; \mathsf{bus}\,(n) \equiv \{b.\; \mathsf{length}(b) = n\; \wedge\; (\mathsf{Alls}\, x{:}b.\; x{:}B)\}$$

$$\mathsf{upper}(n, b) \equiv \epsilon u.\, (\mathsf{length}(u) = n) \wedge (\exists l.\, (l@u) = b)$$

$$\mathsf{lower}(n, b) \equiv \epsilon l.\, (\mathsf{length}(l) = n) \wedge (\exists u.\, (l@u) = b)$$

$$l\,`n \equiv (\mathsf{nat\_rec}(n,\; \mathsf{hd},\; \lambda u\, v\, r.\; v(\mathsf{tl}(r)))) l$$

A bus (B bus (n)) is a list of length n and whose members are from the set B. The functions upper and lower return the upper and lower n bits (when they exist) from a bus b and $l`n$ returns the nth element of a bus. We have proven many standard facts about these definitions, e.g., we can decompose busses.

$$[\![k = n + m;\ b{:}(B\text{ bus }(k))]\!] \Longrightarrow b = \text{lower}(n,b)@\text{upper}(m,b)$$

5 Examples

We present two examples. Our first example is a comparator developed using Formal Synthesis in [7]. It illustrates how we can directly mimic Formal Synthesis to create hierarchical designs. Our second example uses induction to construct a parameterized tree-shaped adder.

5.1 Comparator Cell

A comparator takes as input two words A and B, representing numerals, and determines their relative order, i.e., which of the three cases $A < B$, $A = B$ or $A > B$ holds. Such a circuit can be built in a ripple-carry fashion from comparator cells. These cells (left-hand figure in Figure 1) compare a bit from A and B and also have as input three bits (one for each case) which are the result of previous comparisons (*grin*, *eqin*, and *lsin*), as output three bits (*grout*, *eqout*, *lsout*). The behavioral specification is:

```
CompCellS(a, b, grin, eqin, lsin, grout, eqout, lsout) ==
    (vl(a) < vl(b) --> grout = Lo    & eqout = Lo    & lsout = Hi) &
    (vl(a) = vl(b) --> grout = grin & eqout = eqin & lsout = lsin) &
    (vl(b) < vl(a) --> grout = Hi    & eqout = Lo    & lsout = Lo)
```

The function vl is defined in our theory of binary values: $\text{vl}(\text{Lo}) = 0$ and $\text{vl}(\text{Hi}) = 1$. We can now submit the following design theorem to ISABELLE.

```
Def(?H) -->
 (Port a:bin b:bin grin:bin eqin:bin lsin:bin grout:bin eqout:bin lsout:bin.
    ?Circ(a, b, grin, eqin, lsin, grout, eqout, lsout) Sat
    CompCellS(a, b, grin, eqin, lsin, grout, eqout, lsout))
```

We apply an initialization tactic which sets up the design goal by implication introduction (which moves the definition context into the assumption list) and applies the Inside tactic; this yields the following proof state (we elide some information about the typing of the ports).

```
Def(?H) --> (Port a:bin b:bin grin:bin eqin:bin lsin:bin grout:bin eqout:bin lsout:bin.
    ?Circ(a, b, grin, eqin, lsin, grout, eqout, lsout) Sat
    CompCellS(a, b, grin, eqin, lsin, grout, eqout, lsout))

 1. !! a b grin eqin lsin grout eqout lsout.
       [| a:bin; ...; lsout:bin; Def(?H) |] ==>
         ?Circ(a, b, grin, eqin, lsin, grout, eqout, lsout) Sat
         CompCellS(a, b, grin, eqin, lsin, grout, eqout, lsout)
```

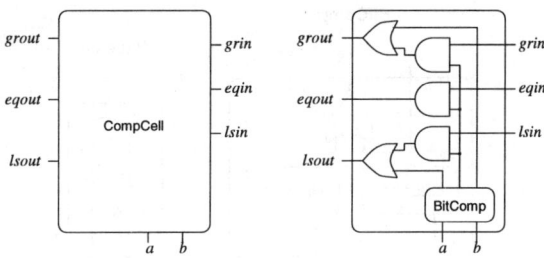

Fig. 1. Comp-cell and its claimed implementation

The original goal is given on the first-three lines; it contains two metavariables *?Circ* and *?H* standing for the implementation of CompCellS and the definitions that will be made, respectively. These are also present in the following lines (which contain the subgoal that must be established to prove the original goal) and will be instantiated when proving this subgoal. Next we introduce a new subdesign *BitComp* that we will use as a component. The idea is that we first compare the two bits *a* and *b* representing the current digit. Then we combine the result of this comparison with information coming from comparisons of less significant bits to give the result. The specification of the *BitComp* subdesign is the following.

```
BitCompS(a, b, gr, eq, ls) ==
   (vl(a) < vl(b) --> gr = Lo & eq = Lo & ls = Hi) &
   (vl(a) = vl(b) --> gr = Lo & eq = Hi & ls = Lo) &
   (vl(b) < vl(a) --> gr = Hi & eq = Lo & ls = Lo)
```

We apply subdesign_tac and this yields two subgoals. At the top, we see our original goal where the definition context *?H* was extended with a definition for *BitComp* and there is a new metavariable *?G* for further definitions. The first subgoal is a design theorem for the subdesign. The second is the original design theorem but now with an additional assumption that there is an implementation of the subdesign which satisfies the specification BitCompS.

```
Def(?G &
     (Port a:bin b:bin gr:bin eq:bin ls:bin. ?BitComp(a,b,gr,eq,ls) = ?C(a,b,gr,eq,ls)))  -->
  Port a:bin b:bin grin:bin eqin:bin lsin:bin grout:bin eqout:bin lsout:bin.
     ?Circ(a, b, grin, eqin, lsin, grout, eqout, lsout) Sat
     CompCellS(a, b, grin, eqin, lsin, grout, eqout, lsout)

  1. Def(?G) ==>
     Port a:bin b:bin gr:bin eq:bin ls:bin. ?C(a,b,gr,eq,ls) Sat BitCompS(a,b,gr,eq,ls)

  2. !! a  b grin eqin lsin grout eqout lsout.
     [| a:bin; ...; lsout:bin; Def(?G);
        Port a:bin b:bin gr:bin eq:bin ls:bin.
           ?BitComp(a, b, gr, eq, ls) Sat BitCompS(a, b, gr, eq, ls) |] ==>
     ?Circ(a, b, grin, eqin, lsin, grout, eqout, lsout) Sat
     CompCellS(a, b, grin, eqin, lsin, grout, eqout, lsout)
```

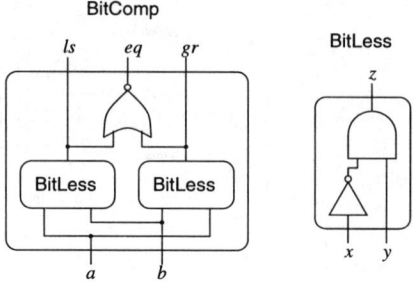

Fig. 2. Claimed implementations of bit-comp and bit-less

Given this subdesign we use claim_tac to state that the following specification entails the original goal (Figure 1).

```
EX gr:bin eq:bin ls:bin x:bin y:bin.
    BitCompS(a, b, gr, eq, ls) & andS(eq, grin, x) & orS(gr, x, grout) &
    andS(eq, eqin, eqout) & andS(eq, lsin, ls) & orS(ls, y, lsout)
```

Due to space limitations we will just sketch the remaining proof. First we show that the claimed specification entails the original one; we prove this automatically using a tactic that performs case-analysis and simplification. After, we implement it by using reveal_tac to strip existential quantifications (and introduce internal wires) and then use split_tac to break up the conjunctions (and Join together subcircuits). The components are each implemented either by introducing simpler subdesigns and implementing those (see below), or using library_tac, which accesses appropriate library parts, and using design_tac, to apply developed subdesigns.

Let us sketch one of these remaining tasks: implementing the subdesign BitCompS. We proceed in the same manner as earlier and we introduce a new subdesign *BitLess*; then we build the above *BitComp* using *BitLess* twice as shown in Figure 2. Finally, BitLessS is so simple that we can claim a direct implementation consisting of components from the library. After these steps, the design theorem that we have proved is the following.

```
Def((Port a:bin b:bin gr:bin eq:bin ls:bin.
        ?BitComp(a,b,gr,eq,ls) = ?BitLess(a,b,ls) Join ?BitLess(b,a,gr) Join nor(ls,gr,eq)) &
    (Port x:bin y:bin z:bin. ?BitLess(x, y, z) = Wire 1. inv(x, 1) Join and(1, y, z))) -->
Port a:bin b:bin grin:bin eqin:bin lsin:bin grout:bin eqout:bin lsout:bin.
    ( Wire gr:bin eq:bin ls:bin x:bin y:bin.
        ?BitComp(a, b, gr, eq, ls) Join and(eq, grin, x) Join or(gr, x, grout) Join
        and(eq, eqin, eqout) Join and(eq, lsin, ls) Join or(ls, y, lsout))
    Sat CompCellS(a, b, grin, eqin, lsin, grout, eqout, lsout)
```

The metavariable ?*H* has become instantiated with the conjunction of the definitions of the subdesigns used in the implementation of the main design goal, i.e., *BitComp* and *BitLess*. In the main goal, the unknown ?*Circ* has become a predicate that represents the structure of the desired circuit and is built from these subdesigns and additional and and or gates. Overall, our proof builds the same circuit and uses the identical sequence of technique applications as that presented in [7]. The difference is not so much in the techniques applied, but rather the underlying conceptualization and implementation: in the VERITAS system, the implementation is constructed at the end of the proof by the validation functions, whereas in our setting, the design takes shape incrementally during the proof. We find this advantageous since we can directly see the effects of each design decision taken.

5.2 Carry Lookahead Adder

Our second example illustrates how our well-founded induction technique synthesizes parameterized designs: We synthesize a carry lookahead adder (henceforth, cla-adder) that is parametric in its bit-width n. For n-bit numbers, such an adder has a height that is proportional to $\log(n)$ and thus computes the sum s and a carry c_o from two numbers a, b and an incoming carry c_i in $O(\log(n))$ time. Instead of propagating the carry from digit to digit as it is done in a ripple-carry adder, we compute more detailed information (c.f. [9]). A *generate bit* g indicates when a carry is generated by adding the digits of a and b and a *propagate bit* p indicates if an incoming carry is handed through. From this information we obtain the *carry bit* c_o for the adder in the following way: it is Hi if $g = $ Hi or if both $p = $ Hi and $c_i = $ Hi. A carry lookahead adder is implemented, roughly speaking, by recursively decomposing it in two adders, each half the size of the original. The propagate and generate bits for the overall adder are obtained by combining the corresponding bits of the subparts with the incoming carry c_i. In the case of adding single digits of a and b (the base case of the recursion) the propagate bit corresponds to the logical or and the generate bit corresponds to the logical and of the digits.

The adder we synthesize is built from two components (Figure 3). The first, *cla*, computes the sum s, the propagate bit p and the generate bit g from the numbers a and b and the incoming carry c_i. The second is an auxiliary component *aux*, which is used to combine the propagate bit, the generate bit and the incoming carry to the outgoing carry c_o. This component consists of two gates and can be derived in a few steps. We focus here on the development of the more interesting component *cla*. We can specify its behavior using data abstraction by an arithmetic expression.

```
claS(n,a,b,s,p,g,ci) ==
  case g of
    Lo => val(s) + 2^n*vl(p)*vl(ci) = val(a) + val(b) + vl(ci)
  | Hi => val(s) +                    2^n = val(a) + val(b) + vl(ci)
```

Note that numbers are represented by busses (i.e. bit vectors) and the value of a bus as a natural number is given by val. We assume busses to have a nonzero length. Hence in the following design theorem, we restrict induction to the set nnat(1) of natural numbers greater than zero.

```
Def(?H)-->
  ( !n:nnat(1). Port a:(bin bus n) b:(bin bus n) s:(bin bus n) p:bin g:bin ci:bin.
     ?cla(n,a,b,s,p,g,ci) Sat claS(n,a,b,s,p,g,ci))
```

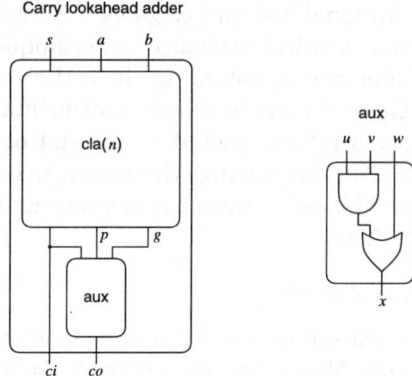

Fig. 3. Implementation of a cla-adder from two components

As before, we begin by shifting the definition environment Def(?H) to the assumptions. After, we apply the course-of-values induction tactic which yields the following proof state

```
Def((ALL n:nnat(1). Port a:(bin bus n) b:(bin bus n) s:(bin bus n) p:bin g:bin ci:bin.
        ?cla(n, a, b, s, p, g, ci) = ?D16(n, a, b, s, p, g, ci)) &
     ?Q2) -->
(ALL n:nnat(1). Port a:(bin bus n) b:(bin bus n) s:(bin bus n) p:bin g:bin ci:bin .
        ?cla(n, a, b, s, p, g, ci) Sat claS(n, a, b, s, p, g, ci))

 1. !!n. [| Def(?Q2); n:nnat(1);
            ALL k:nnat(1). k < n -->
                (Port a:(bin bus k) b:(bin bus k) s:(bin bus k) p:bin g:bin ci:bin .
                    ?cla(k, a, b, s, p, g, ci) Sat claS(k, a, b, s, p, g, ci)) |] ==>
        Port a:(bin bus n) b:(bin bus n) s:(bin bus n) p:bin g:bin ci:bin .
            ?D16(n, a, b, s, p, g, ci) Sat claS(n, a, b, s, p, g, ci)
```

As previously described, ?H is extended with a definition template and further definitions are collected in $?Q_2$. The metavariable at the left-hand side of the definition serves as the name for the design being defined. The metavariable on the right-hand side will be instantiated with the implementation when we prove subgoal 1. The induction hypothesis is added to the assumptions of that subgoal: We may assume that we have an implementation for all k less than n and that we can use this to build a circuit of size n.

We proceed by performing a case analysis on n. We type by(if_tac "n=1" 1), which resolves subgoal 1 with the rule

$$[\![P \Longrightarrow \textit{Circ} \text{ Sat } \textit{Spec}; \neg P \Longrightarrow \textit{Circ}' \text{ Sat } \textit{Spec}]\!] \Longrightarrow \text{if}(P, \textit{Circ}, \textit{Circ}') \text{ Sat } \textit{Spec},$$

where P is instantiated with $n = 1$. ISABELLE responds:

```
Def((ALL n:nnat(1). Port a:(bin bus n) b:(bin bus n) s:(bin bus n) p:bin g:bin ci:bin .
        ?cla(n, a, b, s, p, g, ci) =
            if(n = 1, ?C23(n, a, b, s, p, g, ci), ?C'23(n, a, b, s, p, g, ci))) & ?Q2) -->
```

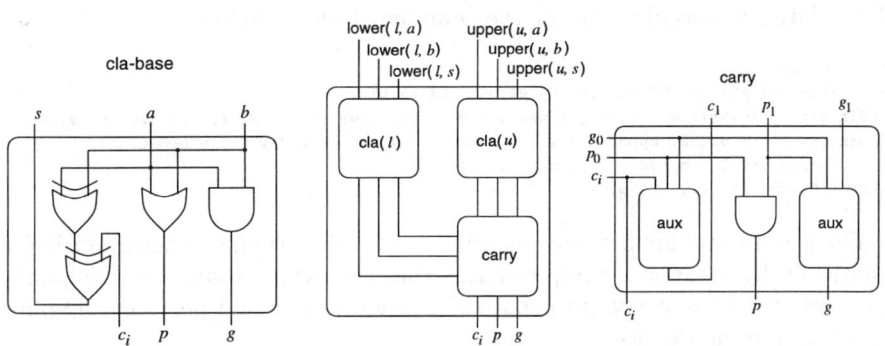

Fig. 4. Base case (left) and recursive decomposition (middle/right) where $l = k \,\mathrm{div}\, 2$ and $u = k \,\mathrm{div}\, 2 + k \,\mathrm{mod}\, 2$

```
(ALL n:nnat(1). Port a:(bin bus n) b:(bin bus n) s:(bin bus n) p:bin g:bin ci:bin .
     ?cla(n, a, b, s, p, g, ci) Sat claS(n, a, b, s, p, g, ci))

 1. !!n a b s p g ci.
      [| Def(?Q2); n:nnat(1);
         ALL k:nnat(1). k < n -->
            (Port a:(bin bus k) b:(bin bus k) s:(bin bus k) p:bin g:bin ci:bin .
                cla(k, a, b, s, p, g, ci) Sat claS(k, a, b, s, p, g, ci));
         a:bin bus n; b:bin bus n; s:bin bus n; p:bin; g:bin; ci:bin; n = 1 |] ==>
      ?C23(n, a, b, s, p, g, ci) Sat claS(n, a, b, s, p, g, ci)

 2. !!n a b s p g ci.
      [| Def(?Q2); n:nnat(1);
         ALL k:nnat(1). k < n -->
            (Port a:(bin bus k) b:(bin bus k) s:(bin bus k) p:bin g:bin ci:bin .
                cla(k, a, b, s, p, g, c) Sat claS(k, a, b, s, p, g, ci));
         a:bin bus n; b:bin bus n; s:bin bus n; p:bin; g:bin; ci:bin; n ~= 1 |] ==>
      ?C'23(n, a, b, s, p, g, ci) Sat claS(n, a, b, s, p, g, ci)
```

In the overall design goal, the right-hand side of the definition has been instantiated with a conditional whose alternatives $?C_{23}$ and $?C'_{23}$ are the respective implementations for the base and the step case. The former is to be implemented by proving subgoal 1 under the assumption $n = 1$ the latter by proving subgoal 2 under the assumption $n \neq 1$. The base case is solved by the subdesign *cla_base* (Figure 4) in a few simple steps.

In the step case we build the adder from two smaller adders of half the size: We decompose the busses a, b and s each in a segment lower($n \,\mathrm{div}\, 2, \_$) containing the inferior $n \,\mathrm{div}\, 2$ bits of the bus and a segment upper($n \,\mathrm{div}\, 2 + n \,\mathrm{mod}\, 2, \_$), containing the remaining bits. The lower (upper) segments of a and b can be added by an adder of bit-width $n \,\mathrm{div}\, 2$ ($n \,\mathrm{div}\, 2 + n \,\mathrm{mod}\, 2$) yielding the lower (upper) segment of the sum s. The propagate, generate and carry bit for the overall adder and the carry flowing from the lower to the upper part of the adder are computed by some additional circuitry collected in a new subdesign. Accordingly we can reformulate the specification for the component *cla* as follows.

```
by( claim_tac
" EX p0:bin. EX g0:bin. EX p1:bin. EX g1:bin. EX c1:bin.
  claS(n div 2, lower(n div 2, a), lower(n div 2, b), lower(n div 2, s), p0, g0, c) &
  claS(n div 2 + n mod 2, upper(n div 2 + n mod 2, a), upper(n div 2 + n mod 2, b),
       upper(n div 2 + n mod 2, s), p1, g1, c1 ) &
  carryS(p0, g0, p1, g1, c1, p, g, ci)" 1 );
```

The proof that this new specification entails the original is accomplished automatically by a tactic we implemented that performs exhaustive case analysis on the values of the carry, propagate and generate bits and performs simplification of arithmetic expressions.

We decompose the new specification into its subparts and implement the recursive occurrences of the specification claS by using the induction hypothesis. This is done by applying design_tac twice. The remaining specification carryS is solved by a new subdesign, which is implemented as shown in Figure 4. Note that we reuse the formerly developed *aux* here. Thus, all design goals are solved and after 39 steps we are finished.

```
Def((ALL n:nnat(1). Port a:(bin bus n) b:(bin bus n) s:(bin bus n) p:bin g:bin ci:bin.
      ?cla(n, a, b, s, p, g, ci) =
      if(n = 1, ?cla_base(a'0, b'0, s'0, p, g, ci),
         Wire p0:bin g0:bin p1:bin g1:bin c1:bin.
         ?cla(n div 2, lower(n div 2,a), lower(n div 2,b), lower(n div 2,s), p0, g0, ci) Join
         ?cla(n div 2 + n mod 2,upper(n div 2 + n mod 2, a),upper(n div 2 + n mod 2, b),
              upper(n div 2 + n mod 2,s),p1,g1,c1) Join ?carry(p0,g0,p1,g1,c1,p,g,ci))) &
   (Port a:bin b:bin s:bin p:bin g:bin ci:bin.
      ?cla_base(a, b, s, p, g, ci) =
      (Wire w:bin. xor(a, b, w) Join xor(w, ci, s) Join or(a, b, p) Join and(a, b, g))) &
   (Port p0:bin g0:bin p1:bin g1:bin c1:bin p:bin g:bin  ci:bin .
      ?carry(p0, g0, p1, g1, c1, p, g, ci) =
      (?aux(p0, ci, g0, c1) Join ?aux(g0, p1, g1, g) Join and(p0, p1, p))) &
   (Port u:bin v:bin w:bin x:bin .
      ?aux(u, v, w, x) = (Wire wa:bin. and(u, v, wa) Join or(wa, w, x)))) -->
 (ALL n:nnat(1). Port a:(bin bus n) b:(bin bus n) s:(bin bus n) p:bin g:bin ci:bin .
      ?cla(n, a, b, s, p, g, ci) Sat claS(n, a, b, s, p, g, ci)))
```

In our definition context, $?H$ has become instantiated by four definitions. There is a one to one correspondence between the implementations shown in Figures 3 and 4 and the predicates defining them.

6 Comparison and Conclusion

We have combined two development methodologies: Formal Synthesis as implemented in VERITAS and resolution based synthesis in ISABELLE. The result is a simple realization of Formal Synthesis that is compatible with other approaches to resolution based synthesis. Moreover, our implementation supports structural, behavioral, and data-abstraction as well as independence from the concrete circuit technology. Our implementation is based on a series of extensions to higher-order logic and we were able to directly utilize standard Isabelle theories in our work as well as Isabelle's simplification tactics. Most of our derived rules were proven in one step proofs by ISABELLE's classical prover.

The idea of using first-order resolution to build programs goes back to Green in the 1960s [6]. More recently, within systems like ISABELLE, interactive proof by higher-order resolution has been used to construct verified programs and hardware designs [1, 4, 10]. The work most closely related to ours is that of Mike Fourman's group based on the LAMBDA system, which is a proof development system that supports synthesis based on second-order resolution [5]. Motivated by ISABELLE, they too use rules in order to represent the design state. The difference lies in the particular approach they use in proof construction;[1] instead of using general purpose techniques as in Formal Synthesis, they derive (introduction) rules for each component from its definition. These rules are applied to the proof state in order to simplify the specification and thereby refine the implementation. The specialized form of their rules supports a higher degree of automation than general purpose techniques. Conversely, the generality of the Formal Synthesis techniques provides a more abstract view of the design process and better supports hierarchical development.

Just as we have extended Formal Synthesis with techniques for induction, it is possible to adapt their methodology within our setting. We have carried out some initial experiments which indicate both that we can use our Formal Synthesis techniques to synthesize LAMBDA style design rules and that such rules can be combined with techniques in circuit development. As a simple illustration, suppose we have an axiom for an adder circuit given by

$$\text{Port } a:\text{nat } b:\text{nat } s:\text{nat. addc}(a, b, s) \text{ Sat } s = a + b.$$

We can apply our techniques to synthesize a partial implementation for a schematic specification *Spec*, which contains a subexpression of the form $a + b$.

```
Port a:nat b:nat. ?Circ(a,b) Sat Spec(a+b)
```

After applying the techniques Inside, Claim, Reveal and Split, and solving the proof obligation from Claim, we arrive at the following intermediate proof state.

```
Port a:nat b:nat.
   (Wire s:nat. addc(a, b, s) Join ?C2(a, b, s)) Sat Spec(a + b)
1. !!a b s. [| a:nat; b:nat; s:nat |] ==> ?C2(a, b, s) Sat Spec(s)
```

After discharging the first subgoal by assuming it and removing the Port quantifiers, we arrive at a LAMBDA-style design rule.

$$[\![\;!!\; a\, b\, s.[\![\; a:\text{nat}; \; b:\text{nat}; \; s:\text{nat} \;]\!] \Longrightarrow ?C(a, b, s) \text{ Sat } ?Spec(s);$$
$$?a:\text{nat}; \; ?b:\text{nat}$$
$$]\!] \Longrightarrow (\text{Wire } s:\text{nat. addc}(?a, ?b, s) \text{ Join } ?C(?a, ?b, s)) \text{ Sat } ?Spec(?a+?b)$$

Explaining this as a technique, it says that we can reduce a specification involving the addition of $a + b$ to one instead involving s. The validation tells us that the

---

[1] In the end, there are only so many (700?) proof development systems, but there are many more strategies for constructing proofs.

circuit built for *Spec(s)*, when hooked up appropriately to an adder, builds a circuit for the original specification. It would not be difficult to build tactics that enable us to integrate such LAMBDA techniques with the others we developed, taking advantage of the different strengths of these two approaches.

We conclude with a brief mention of deficiencies and future work. Currently, the amount of information present during proof can be overwhelming. A short-term solution is to instruct Isabelle to elide information; however, a graphical interface, like that in LAMBDA would be of tremendous value both in displaying designs and giving specifications. Another weakness is automation. We have automated many simple kinds of reasoning by combining Isabelle's simplifiers with case-analysis over binary values. The resulting tactics are effective, but their execution is slow. There are decision procedures based on BDDs that can more effectively solve many of these problems. We have started integrating one of these with our synthesis environment, namely, a decision procedure for a decidable monadic logic that is well-suited for modeling hardware [2]. We hope that this is a step towards a synthesis framework in which different verification methodologies may be integrated.

References

1. David A. Basin. Logic frameworks for logic programs. In *4th International Workshop on Logic Program Synthesis and Transformation, (LOPSTR'94)*, volume 883 of *LNCS*, pages 1–16, Pisa Italy, June 1994. Springer-Verlag.
2. David A. Basin and Nils Klarlund. Hardware verification using monadic second-order logic. In *Computer-Aided Verification (CAV '95)*, volume 939 of *LNCS*, pages 31–41. Springer-Verlag, 1995.
3. A. J. Camilleri, M. J. C. Gordon, and T. F. Melham. Hardware verification using higher-order logic. In D. Borrione, editor, *From HDL Descriptions to Guaranteed Correct Circuit Designs*. North Holland, September 1986.
4. Martin David Coen. Interactive program derivation. Technical Report 272, Cambridge University Computer Laboratory, Cambridge, November 1992.
5. S. Finn, M. P. Fourman, M. Francis, and R. Harris. Formal system design — interactive synthesis based on computer-assisted formal reasoning. In Dr. Luc Claesen, editor, *IMEC-IFIP International Workshop on Applied Formal Methods for Correct VLSI Design*, volume 1, pages 97–110, Amsterdam, 1989. Elsevier Science publishers B.V. (North-Holland).
6. Cordell Green. Application of theorem proving to problem solving. In *Proceedings of the IJCAI-69*, pages 219–239, 1969.
7. F. Keith Hanna, Neil Daeche, and Mark Longley. Formal synthesis of digital systems. In *IMEC-IFIP International Workshop on: Applied Formal Methods For Correct VLSI Design*, volume 2, pages 532–548, Leuven, Belgium, 1989.
8. F.K. Hanna, N. Daeche, and M. Longley. VERITAS+: A specification language based on type theory. In *Hardware Specification, Verification and Synthesis: Mathematical Aspects*, Ithaca, New York, 1989. Springer-Verlag.
9. John L. Hennessy and David A. Patterson. *Computer Architecture, a Quanitative Approach*. Morgan Kaufmann, 1990.
10. Lawrence C. Paulson. *Isabelle : a generic theorem prover; with contributions by Tobias Nipkow*, volume 828 of *LNCS*. Springer, Berlin, 1994.

Inference Rules for Programming Languages with Side Effects in Expressions

Paul E. Black* and Phillip J. Windley

Computer Science Department, Brigham Young University, Provo UT 84602-6576, USA

Abstract. Much of the work on verifying software has been done on simple, often artificial, languages or subsets of existing languages to avoid difficult details. In trying to verify a secure application written in C, we have encountered and overcome some semantically complicated uses of the language.

We present inference rules for assignment statements with pre- and post-evaluation side effects and while loops with arbitrary pre-evaluation side effects in the test expression. We also discuss the need to abstract the semantics of program functions and present an inference rule for abstraction.

1 Introduction

In 1969 Hoare published axiomatic semantics for proving programming languages [14] which Gordon implemented in HOL [9]. More recently there has been work such as Hale's Reasoning About Software [11], Agerholm's Mechanizing Program Verification [1], Curzon's Verified Compiler [7], work on distributed systems [13, 20], embedding TLA [4, 18], etc.

Most work has been on relatively simple languages. Verifying programs written in commonly used production languages, such as C, has lagged. Production languages are generally very rich, with many overlapping features instead of a minimal set, to express different kinds of algorithms and data structures succinctly. Models which completely describe the semantics of such production languages are very complicated.

The object of this work is thttpd, a secure http daemon written in C. It is engineered to provide information to the World Wide Web and to be free of security flaws, even in the presence of a few operating system bugs or administrative errors. The code has a five page informal proof of correctness, and has been reviewed and critiqued by dozens of experts. The code for thttpd seems like an ideal candidate for a formal proof, and the proof would add real value.

In order to prove thttpd, we began with code from Harrison [13], but soon extended it with rules from Gordon [9] and wrote goal-directed tactics instead of using Harrison's forward inference functions. We also express code in the rules

---

* This work was sponsored by the National Science Foundation under NSF grant MIP–9412581

as an abstract syntax tree, rather than HOL strings. A parser [19] converts C code into equivalent abstract syntax trees.

Sect. 2 summarizes the design goals of thttpd. Sect. 3 presents our inference rules for pre-evaluation side effects, post-evaluation side effects, `while` loops with side effect tests, and function calls. We include an example from thttpd in Sect. 4 to demonstrate some of the complexities we encountered. Finally Sect. 5 lists some future work, and Sect. 6 is our conclusions.

2 A Secure HTTP Daemon

In June 1995, Management Analytics wrote a secure http[2] daemon, thttpd. The code is about 100 lines of C. They explain [5, 6] the goals of the daemon and why they believe it to be secure:

> The main risk to providers of [web] services is that someone might be able to fool their server software into doing something it is not supposed to do, thus allowing an attacker to break into their server ...

They continue that their

> ... solution to the security problem with servers is to design a secure server with security properties that can be explicitly demonstrated.

They then list the properties.

> The general properties of interest to us are (in order of highest to lowest priority):
> - Information Integrity - We want to assure that the information residing on the server is not corrupted by the actions of outside users as a result of their use or abuse of the secure service.
> - Availability of Service - We want to assure that outside users of the service cannot make the server unusable for other users as a result of their use or abuse of the secure service.
> - Confidentiality - We want to assure that the service only provides information to outside users that is explicitly authorized for outside access.
>
> We would also like to assure these properties to an even higher degree for information not explicitly designated for outside use than for information that is explicitly designated for outside use.

They verified the properties by argument, but we want to formally verify them. Trying to verify a program which was written in a semantically complex language, C, led us to face problems which are typically avoided such as

- side effects,
- function calls in tests, and
- function semantics abstraction.

---

[2] HTTP, the HyperText Transfer Protocol, is the most commonly used protocol for the World Wide Web on the Internet. HTTP is operational similar to FTP, but has been enhanced and optimized for Web interaction.

3 Inference Rules for Side Effects

We use Hoare axiomatic semantics to express the correctness of program statements. The representation for partial correctness is

$$\vdash \{P\} \text{ code } \{Q\}$$

This means, if predicate P is true of the current state, executing code results in a state in which predicate Q is true.[3]

3.1 A Rule for Pre-evaluation Side Effects

The assignment axiom for v = expr; is

$$\vdash \{Q^v_{expr}\} \text{ v = expr; } \{Q\}$$

as long as expr doesn't have any side effects ([9], pp. 15-17). That is, if Q^v_{expr}[4] is true and the assignment is executed, Q will be true. Since C statements may have side effects, this does not apply. As a simple example, the semantics of a = 2 * ++b; is well defined [12] (it is equivalent to the compound statement ++b; a = 2 * b;), but the statement modifies the values of b as well as a.

To reason about it, we introduce a rule to separate pre-evaluation side effects, that is, side effects which take place *before* the expression is evaluated.

$$\frac{\vdash \text{SEM\_EQ (PreEval expr s1) s2} \quad \vdash \{\text{pre}\} \text{ expr; } \{\text{interim}\} \quad \vdash \{\text{interim}\} \text{ s1 } \{\text{post}\}}{\vdash \{\text{pre}\} \text{ s2 } \{\text{post}\}} \quad (1)$$

Informally, if

- s1 is semantically equivalent[5] to s2 with expr removed and pre-evaluated,
- one can prove {pre} expr; {interim}, and
- one can prove {interim} s1 {post}

one can conclude {pre} s2 {post}.

Why add another inference rule just to separate side effects? Homeier's language, Sunrise [15], has an operator with a side effect, increment, which can occur in test expressions. He handles this by embedding the semantics of the operator in the inference rules. However functions, which have arbitrary semantics including side effects, can occur in loop or test expressions in C. Even statements without function calls can have multiple side effects using, say, increment and assignment operators. We take this more general approach to be able to separate a side effect from the expression in which it occurs.

---

[3] That is, if code terminates. A total correctness theorem ⊢ [P] code [Q] (note square brackets) means if P is true and code executes, Q is true and code *always* terminates.

[4] The notation Q^v_{expr} denotes Q with all free occurrences of v replaced by expr. For instance, $(g * (h+1))^g_{(i-1)}$ is $((i-1)*(h+1))$.

[5] We have shallowly embedded our logic for now. That is, SEM_EQ is defined as ∀s1 s2 . SEM_EQ s1 s2 = T and an ML function checks equivalence and specializes the definition. Eventually we will prove this from a definitional semantics.

3.2 A Rule for Post-evaluation Side Effects

C allows post-evaluation side effects in expressions in addition to pre-evaluation side effects. The statement a = 2 * b++; is well defined, just as the pre-evaluation case. The statement can be broken down into the equivalent compound statement a = 2 * b; b++;.

The following rule is similar to the pre-evaluation side effect rule (1).

$$\frac{\vdash \text{SEM\_EQ (PostEval expr s1) s2} \\ \vdash \{\text{pre}\}\ \text{s1}\ \{\text{interim}\} \\ \vdash \{\text{interim}\}\ \text{expr;}\ \{\text{post}\}}{\vdash \{\text{pre}\}\ \text{s2}\ \{\text{post}\}} \quad (2)$$

Briefly if

- s1 is semantically equivalent to s2 with expr removed and post-evaluated,
- one can prove {pre} s1 {interim}, and
- one can prove {interim} expr; {post}

one can conclude {pre} s2 {post}.

3.3 A Rule for While Loops with Side Effects

In simple languages the inference rule for a while loop, or backward jump, is straight forward:

$$\frac{\vdash \text{IS\_VALUE expr test} \\ \vdash \{\text{invariant} \land \text{test}\}\ \text{body}\ \{\text{invariant}\}}{\vdash \{\text{invariant}\}\ \text{while expr body}\ \{\text{invariant} \land \sim\text{test}\}} \quad (3)$$

IS_VALUE means that test is the HOL equivalent of expr.

When test expressions can have side effects, the rule is more complex. Note that the pre-evaluation side effect rule (1) does *not* apply. Otherwise one could prove

```
while (pre-eval side-effects in expr)
    body
```

by proving

```
pre-eval side-effect;
while (expr)
    body
```

but the side effect is not executed every loop! (One purpose of the SEM_EQ (PreEval ...) condition in rule 1 is to prevent the rule from being applied incorrectly to while loops, for loops, etc.) Conceptually a while loop with pre-evaluation side effects must be verified as

```
while-begin-tag:
  {interim}
    pre-eval side-effect;
  {invariant}
    if (expr)
      {invariant /\ test}
        body
        goto while-begin-tag
  {invariant /\ ~test}
```

The inference rule for while statements is then

$$\frac{\begin{array}{l} \vdash \text{SEM\_EQ (PreEval sexpr testExpr;) expr;} \\ \quad \lor (\text{testExpr} = \text{expr}) \land (\text{sexpr} = \text{expr}) \\ \vdash \{\text{interim}\}\ \text{sexpr};\ \{\text{invariant}\} \\ \vdash \text{IS\_VALUE testExpr test} \\ \vdash \{\text{invariant} \land \text{test}\}\ \text{body}\ \{\text{interim}\} \end{array}}{\vdash \{\text{interim}\}\ \text{while expr body}\ \{\text{invariant} \land \sim\!\text{test}\}}$$

In other words if

– *Either* testExpr is semantically equivalent to expr with sexpr removed and pre-evaluated *or* expr is used for sexpr and testExpr,
– one can prove {interim} sexpr; {invariant},
– test is the HOL equivalent of testExpr, and
– one can prove {invariant ∧ test} body {interim}

one can conclude {interim} while expr body {invariant ∧ ~test}.

We allow (testExpr = expr) ∧ (sexpr = expr) in the first condition in case expr has no side effects. Note that, when expr has no side effects, testExpr = sexpr = expr and interim = invariant, reducing this rule to the standard rule (3).

3.4 Function Call Abstraction

The program thttpd uses 15 library and operating system functions. All of them are important to the correct operation of the program. When the correctness of the program depends on these functions, how can the program be verified?

Often the approach is to essentially include the body of the called function in the code to be verified ([8], page 151). However this won't work for several reasons. First, we don't have access to the operating system code. Even if we did, thttpd should be independent of any one operating system. Second, we do not want to repeat all the work of verifying operating system or library functions every time they are used; verification would never scale up to large projects.

Finally, programmers rarely use all the functionality of an operating system call in one piece of code. So we only need partial semantics to prove the correctness of code. For instance, thttpd calls time to log when actions are taken.

Since the properties of interest mentioned in Sect. 2 don't include the log, returning the wrong time does not violate the top-level security policy! That is, the reliable operation of thttpd is not dependent on *which* time is returned, only that some time is. Hence we need some way to abstract the semantics of a function call. Jones [17] treats this problem by using extended type checking, but non-rigorously.

Similarly to Homeier [15] we declare axioms to express the operation of library and system functions. For example, we declare time as

$$\vdash \{T\} \text{ time(int *tloc)}$$
$$\{(C\_Result = 0) \land (\exists TIME\_errno.errno = TIME\_errno) \lor$$
$$\exists some\_time.(C\_Result = some\_time) \land C\_Result > 0 \land$$
$$\sim(tloc = NULL) \longrightarrow (deref(tloc) = some\_time)\}$$

That is, either C_Result[6] is 0 and errno is set, or C_Result is set to some non-zero time and that time is also put in tloc if tloc is not null.

The inference rule to abstract function call semantics from a function declaration is (after [10])

$$\frac{\vdash \text{DECLARE type funcName formals body} \\ \vdash \{\text{pre}\} \text{ body } \{\text{post}\} \\ \vdash \text{formalsToActuals formals actuals} \\ \vdash \text{SUBST pre formals actuals preSub} \\ \vdash \text{SUBST post formals actuals postSub}}{\vdash \{\text{preSub}\} \text{ funcName(actuals) } \{\text{postSub}\}} \quad (4)$$

That is, if

– funcName is declared with formals and body,
– one can prove {pre} body {post},
– one can prove that the actuals are equivalent to the formals,
– preSub is pre with the formals replaced by the actuals, and
– postSub is post with the formals replaced by the actuals

one can conclude that calling funcName with the actuals in a state satisfying preSub results in a state satisfying postSub.

4 An Example From thttpd

This section presents one function from thttpd and outlines a verification showing how system calls and side effects are handled.

---

[6] The special variable C_Result is the result or return value of the function call.

4.1 Code and Operation

We present the function `logfile`. This function records an access by a remote user from a remote machine by writing the user's and machine's name along with the time to a log file. Here is the code and applicable declarations:

```
char timestamp[64], remotehost[BUFSIZE], remoteuser[BUFSIZE];
```

```
void logfile(F)
FILE *F;
{time_t t;
t=time(NULL);strftime(timestamp, 20, "%Y/%m/%d %T",localtime(&t));
fprintf(F,"%s %s %s ",remotehost,remoteuser,timestamp);}
```

In detail, the function `logfile` gets the current "time in seconds since the Epoch" [16] with the call of `time`. The call of `localtime` returns the time converted to the local time zone, and `strftime` formats the time as "yyyy/mm/dd hh:mm:ss" and saves it in `timestamp`. Finally `logfile` writes the `timestamp` to the file F along with the requesting user's name and the name of the user's computer by calling `fprintf`.

Two of the system calls used in thttpd have complicated semantics: `time`, whose semantics we gave in Sect. 3.4, and `strftime`, which is

$\vdash \{T\}$ strftime(char *s, int maxsize, char *format, tm *timeptr)
$\{((strlen(strftimeSpec(format, timeptr)) < maxsize \Rightarrow$
$\quad (C\_Result = strlen(s) \land strcmp(s, strftimeSpec(format, timeptr)) = 0)$
$\quad | (C\_Result = 0))$
$\land (\forall index.accessed(s, index) \longrightarrow index \geq 0 \land index < maxsize)\}$

The function `strftime` returns the length of the string which it placed in s. If the string exceeds `maxsize`, zero is returned and the contents of s are indeterminate. Note that this is not the full semantics of `strftime`; no mention is made, for instance, of mapping from the month number and locale into a full month name for the %B format or the hundreds of other details of `strftime`. But this level of detail is sufficient for thttpd.

4.2 Verifying System Calls and Side Effects

We begin verifying `logfile` by using the function rule (4) with the assumption

`!s. strlen(strftimeSpec('%Y/%m/%d %T',s)) < 20`

(informally: any time formatted with %Y/%m/%d %T yields a string less than 20 characters long) and take care of the call of `time`. Then we use an inference rule for sequential statements to separate the `strftime` statement from the `fprintf`. The condition after `strftime` must be that `timestamp` has a formatted time string:

`"strcmp(timestamp, strftimeSpec('%Y/%m/%d %T',tsptr))=0"`

Notice that the call of localtime is embedded in the call of strftime. We separate it with the pre-evaluation side effect rule (1) and prove it with an axiom for its semantics.

Now we prove the call of strftime. In the following, CALL_TAC sets a goal to weaken the postcondition from the that given in the axiom. We rewrite with the definition of conditional ($\Rightarrow$) from COND_CLAUSES to simplify it. Next a selector function, chooseStrcmp (generated by find_filter [3]), selects the initial assumption (that the output is always less than the maximum size) and a rewrite solves the subgoal.

```
e(CALL_TAC SYS_strftime THEN
    ASM_REWRITE_TAC [COND_CLAUSES] THEN
    REPEAT STRIP_TAC THEN
    let chooseStrcmp (t:term) =
      (fst o dest_var o rator o rand o rator)t='strcmp'?false in
    ASSUM_LIST (\thl .
      UNDISCH_TAC (find chooseStrcmp (map (snd o dest_thm) thl)))
    THEN ASM_REWRITE_TAC []);;
```

The proof of the final statement follows quickly and this verification is done.

5 Future Work

We plan to prove our inference rules and predicates, such as SEM_EQ, from a denotational semantic of C. We also plan to rework from the current HOL88 to HOL90.

For post-hoc verification of C programs to be practical, we must improve the various tactics so they will prove more subgoals automatically. Even better would be to change to a verification condition generator style.

We need to add inference rules to handle straight-forward array accesses. Notice that thttpd uses arrays very conservatively. We believe an approach such as [9] (page 31) will suffice.

Currently Hoare style axiomatic semantics only allow for a single postcondition. Since we are mostly concerned with partial correctness, both return and exit are modeled with a postcondition of F. That is, since control never flows from a return or exit to an immediately subsequent statements, any condition is allowed. However verifying total correctness with these jumps may be difficult. Therefore we plan to introduce multiple exit conditions as suggested in [2]. This will allow us to reason about continue and break statements in loops as well as return statements.

Much work is needed on function calls. How can we supply different levels of abstraction of operating system and library function calls for different needs?

Finally, we have the skeleton of the proof for thttpd almost done. With arrays we can finish the proof, but the properties proved are quite weak. We will go back through the proof with formalizations of the properties of interest and flesh out the proof. Then we believe that we can fairly claim to have formally verified the program. Management Analytics has several other servers which should be simple to prove once the infrastructure is done.

6 Conclusions

Our work on a well-engineered production program written in a complex language exposed a number of technically interesting problems, such as side effects and abstracting function semantics. We propose a number of new inference rules to handle

- pre-evaluation side effects more generally than before,
- post-evaluation side effects, and
- pre-evaluation side effects in loop tests.

We also point out the importance of abstracting the semantics of functions in doing large proofs or proofs involving operating system or library calls. Work on simple or primarily academic languages is fruitful, but verifying "industrial" languages is possible and useful. With the computer performance and theorem provers available today, the complexities found in industrial programs need not be a barrier to real verifications.

Acknowledgements

We are grateful to Dr. Frederick B. Cohen of Management Analytics for allowing and encouraging us to use thttpd as a test case for verification. We thank William L. Harrison whose code we used to begin our implementation. We also thank the reviewers that pointed out errors in our understanding of C semantics and a fatal flaw in an earlier version of the post-evaluation rule.

References

1. Sten Agerholm, "Mechanizing Program Verification in HOL," in *Proceedings of the 1991 International Workshop on the HOL Theorem Proving System and Its Applications (HOL '91)*, edited by Myla Archer, Jeffrey J. Joyce, Karl N. Levitt, and Phillip J. Windley, IEEE Computer Society Press, Los Alamitos, California, 1991, pp. 208–222.
2. Michael A. Arbib and Suad Alagić, "Proof Rules for Gotos," *Acta Informatica*, Vol. 11, No. 2, 1979, pp. 139–148.
3. Paul E. Black and Phillip J. Windley, "Automatically Synthesized Term Denotation Predicates: A Proof Aid," in *Higher Order Logic Theorem Proving and Its Applications (HOL '95)*, edited by E. Thomas Schubert, Phillip J. Windley, and James Alves-Foss, Springer-Verlag, Berlin, Germany, 1995, pp. 46–57.
4. Holger Busch, "A Practical Method for Reasoning about Distributed Systems in a Theorem Prover," in *Higher Order Logic Theorem Proving and Its Applications (HOL '95)*, edited by E. Thomas Schubert, Phillip J. Windley, and James Alves-Foss, Springer-Verlag, Berlin, Germany, 1995, pp. 106–121.

5. Frederick B. Cohen, "Why is thttpd Secure?" http://all.net/ManAl/white/whitepaper.html or http://all.net/ → Products → Secure http and gopher daemons (14 March 1996).
6. Frederick B. Cohen, "A Secure World Wide Web Daemon," *Computers and Security*, submitted, 1995.
7. Paul Curzon, "Deriving Correctness Properties of Compiled Code," in *Higher Order Logic Theorem Proving and Its Applications (HOL '92)*, edited by Luc Claesen and Michael Gordon, Elsevier Science Publishers, 1992, pp. 97–116.
8. Nissim Francez, *Program Verification.* Addison-Wesley, 1992.
9. Michael J. C. Gordon, *Programming Language Theory and its Implementation.* Prentice-Hall, Inc., New Jersey, 1988.
10. David Gries & Gary Levin, "Assignment and Procedure Call Proof Rules," *ACM Transactions on Programming Languages and Systems*, vol. 2, no. 4, Oct 1980, pp. 564–579.
11. Roger Hale, "Reasoning About Software," in *Proceedings of the 1991 International Workshop on the HOL Theorem Proving System and Its Applications (HOL '91)*, edited by Myla Archer, Jeffrey J. Joyce, Karl N. Levitt, and Phillip J. Windley, IEEE Computer Society Press, Los Alamitos, California, 1991, pp. 52–58.
12. Samuel P. Harbison and Guy L. Steele Jr., *C, A Reference Manual.* Prentice-Hall, Inc., 1991.
13. William L. Harrison, Karl N. Levitt, Myla Archer, "A HOL Mechanization of The Axiomatic Semantics of a Simple Distributed Programming Language," in *Higher Order Logic Theorem Proving and Its Applications (HOL '92)*, edited by Luc Claesen and Michael Gordon, Elsevier Science Publishers B.V., 1992, pp. 117–126.
14. C. A. R. Hoare, "An Axiomatic Basis for Computer Programming," *Communications of the ACM*, vol. 12, October 1969, pp. 576–583.
15. Peter Vincent Homeier, *Trustworthy Tools for Trustworthy Programs: A Mechanically Verified Verification Condition Generator for the Total Correctness of Procedures*, Ph.D. Dissertation, University of California, Los Angeles, 1995.
16. HP-UX on-line manual, HP-UX Release 9.0: August 1992, Hewlett-Packard Company, Palo Alto, California.
17. Derek Jones, "Checking an Application's use of API's," http://www.knosof.co.uk/apichk.html (14 March 1996).
18. Sara Kalvala, "A Formulation of TLA in Isabelle," in *Higher Order Logic Theorem Proving and Its Applications (HOL '95)*, edited by E. Thomas Schubert, Phillip J. Windley, and James Alves-Foss, Springer-Verlag, Berlin, Germany, 1995, pp. 214–228.
19. John P. Van Tassel, "The HOL parser Library," http://lal.cs.byu.edu/lal/holdoc/library.html (13 June 1996).
20. Cui Zhang, Brian R. Becker, Mark R. Heckman, Karl Levitt, and Ron A. Olsson, "A Hierarchical Method for Reasoning about Distributed Programming Languages," in *Higher Order Logic Theorem Proving and Its Applications (HOL '95)*, edited by E. Thomas Schubert, Phillip J. Windley, and James Alves-Foss, Springer-Verlag, Berlin, Germany, 1995, pp. 385–400.

Deciding Cryptographic Protocol Adequacy with HOL: The Implementation

Stephen H. Brackin*

Arca Systems, Inc.
ESC/AXS
Hanscom AFB, MA 01731-2116

Abstract. *Cryptographic protocols* are sequences of message exchanges, usually involving encryption, intended to establish secure communication over insecure networks. Whether they actually do so is a notoriously subtle question. This paper describes a proof procedure that automatically proves desired properties of cryptographic protocols, using a HOL formalization of a "belief logic" extending that of Gong, Needham, and Yahalom [9], or precisely identifies where these proof attempts fail. This proof procedure is not a full decision procedure for the belief logic, but it proves all theorems that have been of interest. This proof procedure has quickly shown potential deficiencies in published protocols, and is a significant application for HOL90 and SML.

1 Introduction

People often need to communicate securely over insecure networks, e.g., when they exchange financial information. In the worst case, every message on a network can originate from, or be read, recorded, modified, or deleted by, an enemy.

Communication might be impossible under these circumstances. If the following conditions are all true, though, it is possible to communicate securely even over insecure networks: messages do get through; these messages are encrypted; the encryption is strong enough to prevent those without the needed keys from decrypting the messages quickly; the messages' recipients can identify them as being of expected forms and not being replays of earlier messages; and the distribution of needed keys is known.

Cryptographic protocols are sequences of message exchanges intended to establish secure communication over insecure networks. They typically use encryption in one or both of two forms: *symmetric-key* encryption, in which data encrypted with a key is decrypted with the same key; or *public-key* encryption, in which data encrypted with a key is decrypted with a different, mathematically related key. In public-key encryption, one of the two keys in a pair, the

---

* The author wishes to thank Grace Hammonds, Randy Lichota, Shiu-Kai Chin, and Jack Wool for their assistance. This work was supported by Air Force Materiel Command's Electronic Systems Center/Software Center (ESC/AXS), Hanscom AFB, through the Portable, Reusable, Integrated Software Modules (PRISM) contract.

public key, is made widely available, while the corresponding private key is kept secret. It is computationally infeasible to compute a private key from the corresponding public key. See [18] for more information.

Cryptographic protocols also often use time stamps, nonces (newly created random values unlikely to have been created before), hash functions (functions computing seemingly random bit strings as functions of their inputs, with the property that it is computationally infeasible to create an input with a given hash or find two inputs with the same hash), message authentication codes (values produced by key-dependent hash functions), and key-exchange functions (functions allowing two users to compute a key only they share by combining their own private keys and each others' public keys). See [18] for more information.

There are many instances, though, of published protocols, recommended by experts, that are vulnerable to attack [7]. Most such attacks involve replacing parts of messages with recorded parts of old messages, particularly old messages using encryptions that the attacker has been able to break. Current approaches to avoiding such failures either attempt to construct possible attacks, using algebraic properties of the algorithms in the protocols [15, 17, 16, 12, 13, 14], or attempt to construct proofs, using specialized logics based on a notion of "belief", that protocol participants can confidently reach desired conclusions, regardless of what an adversary might be able to do [9, 21].

The basic issues for protocols are *authentication* (i.e., whether participants can determine who sent the messages they receive), and *security* (i.e., whether those not meant to receive information can obtain it). The attack-construction approaches address both authentication and security, but suffer from a combinatorial explosion in the number of possibilities they must consider. The proof-construction approaches do not address security, but avoid a combinatorial explosion. The work described here uses the proof-construction approach, but might be extensible to cover both authentication and security; see [3].

In an earlier paper [1], the author gave an axiom-free HOL theory protocol formalizing the full Gong, Needham, Yahalom (GNY) belief logic [9], including the "rationality rule" for making deductions about arbitrarily highly nested levels of belief, and extensions by Gong [8] to eliminate impossible protocols. That theory also added explicit pairing and conjunction operators and the ability to make deductions about individual protocol stages. The paper promised a future paper on using this theory to prove desired properties of the ticket-granting-service Kerberos [20] protocol, and sketched an algorithm by which all theorems provable from this theory could be proved automatically.

Since then, though, the author has developed the *Automatic Authentication Protocol Analyzer* (AAPA), consisting of a simple *Interface Specification Language* (ISL) for specifying protocols and their desired properties, a Yacc/Lex translator aapa, and a HOL90.7 executable phol with HOL theory bgny and SML function ProveGoals already loaded.

Theory bgny implements everything that theory protocol does, and also extends the GNY logic to cover multiple encryption and hash operations, message authentication codes, hash codes used as keys, and key-exchange algorithms.

Translator `aapa` translates ISL specifications into SML code defining HOL theories, having `bgny` as their parent, of individual protocols and their properties. This code also calls `ProveGoals` with arguments appropriate to these theories.

`ProveGoals` automatically proves all provable protocol properties that are likely to be of interest, and produces output files giving ISL versions of proved theorems and failed proof attempts.

The AAPA makes HOL-based formal analyses of cryptographic protocols available to users who know *nothing* about HOL. It's also fast. The author used the AAPA to specify the ticket-granting-service Kerberos protocol and its desired properties in about an hour, then prove those properties in less than 5 minutes. He also used the AAPA, with a few hours of specification time and about 20 minutes of proof time, to analyze three published SPX protocols [22] and point out potential failures in two of them.

A precursor to `bgny` — named `protocol` like the theory in [1], but differing greatly from that theory and differing in relatively minor, technical ways from `bgny` — is described in [3]. ISL and `aapa` are described in [4] and in Sections 3 and 4. `ProveGoals` and its outputs are described in Section 6. The results on the SPX protocols are given in [2]; those on the ticket-granting-service Kerberos protocol are given in Section 4.

Theory `bgny` uses the "deep embedding" technique of defining a polymorphic concrete recursive type that serves as a language for describing cryptographic protocols and their properties, then inductively defining a function that assigns truth values to the elements of this type. For a detailed description, see [1, 5].

This paper is organized as follows: Section 2 describes the network model assumed by `bgny`'s extension of the GNY logic, and Section 3 describes the ISL constructs that define protocols and their desired properties. Section 4 gives and discusses an ISL specification of the ticket-granting-service Kerberos protocol, and describes how the AAPA automatically proves that it has all its desired properties. Section 5 describes the HOL theorems that `ProveGoals` proves. Section 6 describes the algorithm that `ProveGoals` follows, and relates this algorithm to other techniques for proving results in the GNY logic. Finally, Section 7 describes `ProveGoals`' limitations, and how they might be reduced in the future.

This paper assumes basic understanding of inductively defined relations [6] and mutually recursive, polymorphic concrete recursive types [10].

2 Computational Model

This section gives assumptions and terminology for AAPA models of protocols.

A *network*, or distributed environment, consists of *principals* connected by communication links. Messages on these links constitute the only communications between principals. Any principal can place a message on any link and can see or modify any message on any link.

A *protocol* is a distributed algorithm, carried out by principals acting as state machines; the protocol determines which messages the principals send as functions of their internal states. A protocol is divided into *stages* by message transmissions; the number of stages is always finite. A run, or *session*, is a particular execution of a protocol. This paper will only consider sessions that seem to end successfully, not considering possible alarm, adaptation, or retry provisions for other sessions.

At each stage of a protocol session, each principal has a finite set of *received* data items, which the principal either had before the session began or extracted from messages sent to it during the session. Each principal also has a potentially infinite set of *possessions*, which are pieces of information that the principal is able to compute from the data items it has received. (This treatment of received and possessed items is slightly different from their treatment in [1].)

At each stage of a protocol session, each principal also has a set of *beliefs*. Principals believe a proposition if they can be confident, even though this confidence is not absolute, that this proposition is true. Belief can be based, for instance, on the near impossibility of quickly decrypting encrypted information without having the needed key.

Every principal starts a session with initial sets of received items and beliefs, and expands these sets by receiving messages. Every principal's received-items and belief sets increase monotonically during a session, but a principal need not still have a data item or belief that this principal had in an earlier session.

3   ISL Statements

ISL statements describe assumed or desired properties of data items and protocol principals. The main ISL statement constructors, with informal descriptions of their intended meanings, follow:

- Fresh: Applied to a data item, if asserts that the item was created for the current run of the protocol.
- NeverMalFromSelf: Applied to two principals and a data item, it asserts that either the two principals are equal, in which case the data item is supposed to be a replay of something the first principal sent out earlier, or the two principals are not equal and the first principal has adequate reason to believe that the data item is not all or part of any message that this principal sent out at any past time. (The name means, "never maliciously from self".)
- Trustworthy: Applied to a principal, it asserts that if the principal was the source of a data item with an associated statement, and the protocol assumes this data item would not have been released unless the principal releasing it believed this associated statement, then this is adequate reason for believing that this associated statement is true.
- PrivateKey: Applied to a principal, an encryption algorithm, and a key, it asserts that the key, for the algorithm, is one of the principal's private keys.
- PublicKey: Applied to a principal, an encryption algorithm, and a key, it asserts that the key, for the algorithm, is one of the principal's public keys.
- Conveyed: Applied to a principal and a data item, it asserts that the principal was the creator and source of this item.

- **Possesses**: Applied to a principal and a data item, it asserts that the principal received the item or is able to compute it from items that it has received.
- **Received**: Applied to a principal and a data item, it asserts that the principal received the item before the current session, received the item as, or as part of, some message sent earlier in the current session, or can compute the item without using any other item identified in the protocol.
- **Recognizes**: Applied to a principal and a data item, it asserts that the principle can identify the item as being of an expected form.
- **SharedSecret**: Applied to two principals and a data item, it asserts that if two principals possess the item, or come to possess it through secure means, then they are or will be the only ones other than principals trusted by both of them who possess it.
- **Believes**: Applied to a principal and a statement, it asserts that the principal has adequate reason to believe the statement.

4 Example

An ISL specification of the ticket-granting-service Kerberos protocol [20], with interspersed explanations, follows.

The first section of the specification identifies the protocol's principals and the algorithms and data items they use in the protocol. For this protocol, there are four principals, A, B, C, and S. S acts as a global server, and B acts as a local server. The principals will use DES [18] encryption, a standard form of symmetric-key encryption. They will use five keys that are, or will be, shared between A and B, A and C, A and S, B and C, and B and S. They will also use four time stamps, two initially sent by A, one initially sent by B, and one initially sent by S. The protocol will also involve decrementing data items to change them slightly, so it uses the functions of adding and subtracting 1. ISL requires that keys be identified as PUBLIC, PRIVATE, or SYMMETRIC. As explained in Section 6 below, ProveGoals uses this information to guide its proving process.

```
DEFINITIONS:
PRINCIPALS: A,B,C,S;
SYMMETRIC KEYS: Kab,Kac,Kas,Kbc,Kbs;
OTHER: Tab,Tac,Tb,Ts;
ENCRYPT FUNCTIONS: Des;
OTHER FUNCTIONS: Minus1,Plus1;
```

The specification next tells which functions have inverses and what those inverses are. **HASINVERSE** tells which algorithm computes the inverse of which other algorithm, or which algorithm-with-key combination computes the inverse of which other algorithm-with-key combination. The **ANYKEY** construct is used with symmetric-key encryption functions, and denotes that any value can be used as a key for encryption and decryption as long as the same value is used for both. The specification assumes that the same algorithm is used for DES encryption and decryption, a slight and harmless simplification of the truth.

Des WITH ANYKEY HASINVERSE Des WITH ANYKEY;
Minus1 HASINVERSE Plus1;

The next section of the specification gives the protocol's assumed initial conditions. A trusts S, B, and C, B trusts S and A, and C trusts B and A. A and S, and no others, share the key Kas; the same is true for B and S with Kbs, and for B and C with Kbc. Further, S has a (presumably newly generated) key Kab that A and B can use to communicate securely between themselves, if the protocol succeeds in communicating it to both of them and no others that they do not both trust, and B has a similar key Kac for A and C. The principals also know how to perform various computations, know their own and other principals' names (i.e., network addresses), have the time stamps they will need, are able to recognize time stamps as created for the current protocol session, and can identify their own or other principals' names as meaningful information. The ISL verbs expressing these conditions were discussed briefly in Section 3 above.

```
INITIALCONDITIONS:
A Received Des,Minus1,Plus1,A,B,C,Tab,Tac,Kas;
A Believes
 (Fresh Tac; Fresh Tb; Fresh Ts;
  A Recognizes A; A Recognizes B; A Recognizes C;
  SharedSecret A S Kas;
  Trustworthy B; Trustworthy C; Trustworthy S);
B Received Des,Minus1,Plus1,Tb,Kac,Kbc,Kbs;
B Believes
 (Fresh Tab; Fresh Ts;
  B Recognizes A;
  SharedSecret A C Kac; SharedSecret B S Kbs;
  Trustworthy A; Trustworthy S);
C Received Des,Minus1,Plus1,Kbc;
C Believes
 (Fresh Tac; Fresh Tb;
  C Recognizes A;
  SharedSecret C B Kbc;
  Trustworthy A; Trustworthy B);
S Received Des,Minus1,Plus1,Ts,Kab,Kas,Kbs;
S Believes SharedSecret A B Kab;
```

The protocol works as follows:

1. A sends S a plaintext request to communicate with B.
2. S sends A Kab, along with a time stamp and information A can identify as meaningful, encrypted with Kas. S also sends A a similar collection of items, encrypted with Kbs, that A can forward to B; B will use this information to obtain Kab and know that it came from S.
3. A sends B the information that it was to forward from S. A also sends B a time stamp and information that B can identify as meaningful, encrypted

with Kab; B will use this information to confirm that A has also obtained Kab and believes that it can be treated as a secret shared between A and B. A also sends B a plaintext request to communicate with C.
4. B sends A Kac, along with a time stamp and information A can identify as meaningful, encrypted with Kab. A will use this information to both obtain Kac and confirm that B has also obtained Kab and believes it can be treated as a secret shared between A and B. B also sends A a similar collection of items, encrypted with Kbc, that A can forward to C; C will use this information to obtain Kac and know that it came from B.
5. A sends C the information that it was to forward from B. A also sends C a time stamp and information that C can identify as meaningful, encrypted with Kac; C will use this information to confirm that A has also obtained Kac and believes that it can be treated as a secret shared between A and C.
6. C decrements the time stamp in the information it just received originating with A, encrypts the result with Kac, and sends it back to A; A will use this information to confirm that C has also obtained Kac and believes that it can be treated as a secret shared between A and C.

The next section of the specification gives the protocol itself. In ISL, -> means "sends", {X}Des(K) means X encrypted with DES encryption using key K, and || binds a message to a list of statements that the protocol assumes the message's sender must have had adequate reason to believe or else it would not have sent the message.

```
PROTOCOL:
1. A->S:  A,B;
2. S->A:  {A,B,Kab,Ts}Des(Kbs)||(SharedSecret A B Kab),
          {Kab,B,Ts}Des(Kas)   ||(SharedSecret A B Kab);
3. A->B:  {A,B,Kab,Ts}Des(Kbs)||(SharedSecret A B Kab),
          {A,Tab}Des(Kab)      ||(SharedSecret A B Kab),
          C;
4. B->A:  {A,C,Kac,Tb}Des(Kbc)||(SharedSecret A C Kac),
          {Kac,C,Tb}Des(Kab)   ||(SharedSecret A B Kab;
                                  SharedSecret A C Kac);
5. A->C:  {A,C,Kac,Tb}Des(Kbc)||(SharedSecret A C Kac),
          {A,Tac}Des(Kac)      ||(SharedSecret A C Kac);
6. C->A:  {A,Minus1(Tac)}Des(Kac)||(SharedSecret A C Kac);
```

The final section of the specification gives the *user-set goals* — the protocol's expected properties. These assert that the principals obtain keys, have adequate reason to believe these keys can be treated as shared secrets, and have adequate reason to believe that appropriate other principals have also obtained these keys and also have adequate reason to believe that they can be treated as shared secrets. Each goal has an associated stage, the protocol stage at which the goal is expected to become true. A stage can have zero or more associated goals. Again, the ISL verbs expressing these expected properties were discussed briefly in Section 3 above.

GOALS:
2. A Possesses Kab;
   A Believes SharedSecret A B Kab;
3. B Possesses Kab;
   B Believes SharedSecret A B Kab;
   B Believes A Possesses Kab;
   B Believes A Believes SharedSecret A B Kab;
4. A Possesses Kac;
   A Believes SharedSecret A C Kac;
   A Believes B Possesses Kab;
   A Believes B Believes SharedSecret A B Kab;
5. C Possesses Kac;
   C Believes SharedSecret A C Kac;
   C Believes A Possesses Kac;
   C Believes A Believes SharedSecret A C Kac;
6. A Believes C Possesses Kac;
   A Believes C Believes SharedSecret A C Kac;

As an example of what the AAPA can do, if the ISL specification just given is put into a file named kerberos2.isl, executing aapa kerberos2 produces 21 lines of terminal output describing the process of creating the theory kerberos2 and proving that it satisfies all its user-set goals.

ProveGoals normally generates ISL output files giving proved theorems and failed goals only if it fails to prove a user-set goal. For those of little faith, setting the environment variable ISLOUTPUTLEVEL to isl forces ProveGoals to produce such output files even without a user-goal failure. For the example, this causes ProveGoals to produce output files kerberos2.prvd and kerberos2.fail. File kerberos2.prvd contains ISL versions of 133 theorems.

For those of no faith, setting ISLOUTPUTLEVEL to hol causes ProveGoals to also export the HOL theory kerberos2. Theory kerberos2 then contains 4 type constants, 32 term constants, 0 axioms, 32 definitions, and 137 theorems, the last 2 of which follow. Section 5 explains how to interpret these theorems.

```
S6BP5__'
|- BGNY__ kerberos2FEnv__' kerberos2Protocol__' 6
     (Believes__ A__' (Possesses__ C__' (Td__ Tac__')))
S6BBS1__'
|- BGNY__ kerberos2FEnv__' kerberos2Protocol__' 6
     (Believes__ A__'
       (Believes__ C__'
         (SharedSecret__ A__' C__' (Tk__ Kac__'))))
```

If ISLOUTPUTLEVEL is undefined, AAPA execution for the kerberos2.isl example takes 293.7 seconds on a T-SPARC 1+ with 32 megs of RAM. If ISLOUTPUTLEVEL is hol, the corresponding execution takes 331.7 seconds. AAPA execution can be up to 4 times faster on a SPARC 10 with 64 megs of RAM.

5 What ProveGoals proves

This section describes the theorems that ProveGoals actually proves, giving their HOL formalization. To simplify subsequent discussion, the section begins with a description of naming conventions used inside the AAPA.

To avoid name conflicts, every identifier in bgny ends with __ and every identifier in the HOL theory of an individual protocol ends with __'. These naming conventions, which aapa enforces by forbidding ISL identifies from ending with __' then appending __' to ISL identifiers when defining HOL identifiers, make it safe to use, say, S and K as ISL identifiers, even though they are also the names of HOL combinators. ProveGoals removes the __' characters in producing its ISL output files. To avoid clutter, this paper omits the final __ or __' in describing identifiers defined in bgny or the HOL theories of individual protocols.

As described in Section 1, theory bgny uses the "deep embedding" technique of defining a polymorphic concrete recursive type that serves as a language for describing cryptographic protocols, then inductively defining a function that assigns truth values to the elements of this type. More specifically, it defines the polymorphic, mutually recursive, concrete recursive types :Function, :Term, and :Statement, then inductively defines the function BGNY assigning truth values to :Statement elements.

Types :Function, :Term, and :Statement are all polymorphic in the type variables 'algorithm, 'data, 'key, and 'principal, which are instantiated by the types of algorithms, non-key data, keys, and principals for individual protocols. For the example described in Section 4, for instance, theory kerberos2 instantiates 'algorithm with the concrete recursive enumerated type whose elements are Des, Minus1, and Plus1. This theory instantiates 'data with the type whose elements are Tab, Tac, Tb and Ts, instantiates 'key with the type whose elements are Kab, Kac, Kas, Kbc, and Kbs, and instantiates 'principal with the type whose elements are A, B, C, and Svr.

In the following descriptions of the types :Function, :Term, and :Statement, a, d, k, p, f, t and s denote elements of type :'algorithm, :'data, 'key, 'principal, :Function, :Term, and :Statement, respectively.

Type :Function has 2 constructors. NoF is the "null" function; the inverse of non-invertible functions. (Fk a t) is the function computed by algorithm a using term t as a key.

Type :Term has 8 constructors. NoT is the "null" term, the key used by functions that do not use a key. (Ta a), (Tk d), (Tf f), (Ta k), and (Ta p) denote the values a, d, f, k and p considered as terms. (App p f t s) is the result, computed by principal p, of applying function f to term t, with the associated statement s. The protocol assumes that p will not send this result in a message unless p has adequate reason to believe s. (t1 ;; t2) is the ordered pair of terms t1 and t2; ;; corresponds to the comma in ISL.

Type :Statement has 14 constructors. Most directly correspond to the ISL statement constructors with the same names described in Section 3, though some of the ISL statement constructors are infix operators while all the :Statement constructors are prefix operators. NoS is the "null" statement associated with an App term that has no associated statement. (Fresh t), (Trustworthy p), (PrivateKey p a k), (PublicKey p a k), and (Conveyed p t), as well as (Possesses p t), (Received p t), (Recognizes p t), (NeverMalFromSelf p1 p2 t), (SharedSecret p1 p2 t), and (Believes p s), are as described in Section 3. (Sends p1 p2 t) asserts that principal p1 sends principal p2 term t; Sends corresponds to the -> operator in ISL. (s1 && s2) is the conjunction of statements s1 and s2; && corresponds to the semicolon used as a statement separator in ISL. Type :Statement does *not* include a concept of negation.

In addition to a :Statement element, BGNY takes three other arguments that give the dependence of the statement's meaning on properties of the algorithms and keys used in the protocol, on the protocol itself, and on the protocol stage. Descriptions of these arguments follow:

- A *function environment* is a tuple of functions that express the assumed properties of the algorithms and keys used in the protocol. These functions tell which algorithms give secure encryption, hash, and key-exchange functions, which functions are other functions' inverses, and which terms have the same binary values as which other terms. For readability, bgny defines the functions FeCrypt, FeHash, FeKeyEx, FeInvOf, and FeEquals to extract the different elements of a function environment.
- A *protocol* is a list of :Statement elements. The 0th element in this list gives the protocol's assumed initial conditions. The other elements in the list are Sends statements giving the protocol's message transfers.
- A *protocol stage* is a :num. When used as an index into a protocol, it gives either the protocol's initial conditions or the Sends statement identifying the sender, receiver, and message transferred at that stage.

The theory of an individual protocol defines the protocol's function environment and the protocol itself. For the example given in Section 4, theory kerberos2 defines the function environment kerberos2FEnv and the protocol kerberos2Protocol. The definitions of these constants are too long to include in this paper, but the following fragment of the definition of the FeInvOf part of kerberos2FEnv gives the flavor of these definitions:

(kerberos2InvOf NoF = NoF) /\
(kerberos2InvOf (Fk a t) =
((a = Des) => (Fk Des t) | (a = Minus1) => (Fk Plus1 NoT) | NoF))

BGNY is defined in terms of the subrelation NotMalFromSelf ("not maliciously from self"), which takes a protocol, a current stage, two principals, and a term as arguments. It checks that either the two principals are equal, in which case it assumes that the term is supposed to be a replay of something sent earlier by the first principal, or else that the term is not contained in anything sent by the first principal earlier in the protocol.

BGNY is also defined in terms of EquivInf, a relation that holds for two lists of statements if they describe equivalent inferences, assuming principals are able to draw rational conclusions from their beliefs and know that other principals are able to do likewise. Here, a list of statements corresponds to the inference that the final statement in the list follows from the earlier statements in the list. Two lists are equivalent if the lists are equal, or if, for some principal, every statement in one list is the statement that this principal believes the corresponding statement in a list equivalent to the other list.

Each "inference rule" for BGNY is really an infinite schema of rules in which each member of the schema is equivalent to the schema's base rule. The relation EquivInf describes the members of such a schema. This fully implements the GNY "rationality rule". See [3] for examples.

The inductive definition of BGNY uses 43 rules, many of them equivalent to, or generalizing, one of the 44 axioms in the GNY logic. Even so, 20 of the 43 BGNY rules implement functionality not found in the GNY logic. Except for minor differences, these rules are described in [3]. Because of length constraints, this paper will only describe three of these rules, and only give one of them.

"Initial conditions" rule GI1 asserts that the statements in the definition of a protocol become true at the stage indexing these statements. The rule has no hypotheses or side conditions, but the conclusion

BGNY fenv protocol stage (EL stage protocol)

"Monotonicity" rule GM1 asserts that statements remain true at all stages after they have become true.

The base rule of the schema defined by "believes conveyed" rule GBC2 says that if the following conditions hold — p has received data encrypted by q using the function given by algorithm a2 with tk as key, the function given by a2 with tk as key has as inverse the function given by a1 with tk as key, p possesses this inverse function, p and q have tk as a shared secret, p can identify the decrypted result as meaningful, and p can identify the encrypted data as being created for the current session but not being a replay of something sent earlier in the current session by p — then p has adequate reason to believe that this encrypted data was conveyed by q.

```
{hypotheses = [--'BGNY fenv protocol stage S1'--],
 side_conditions =
  [--'EquivInf
        [((Received p (App q (Fk a2 tk) t s)) &&
          (Possesses p (Tf (Fk a1 tk))) &&
          (Believes p (SharedSecret p q tk)) &&
          (Believes p (Recognizes p t)) &&
          (Believes p (Fresh (App q (Fk a2 tk) t s))));
         (Believes p (Conveyed q (App q (Fk a2 tk) t s)))]
        [S1; S2]'--,
   --'FeCrypt fenv (Fk a2 tk)'--,
   --'FeInvOf fenv (Fk a2 tk) = (Fk a1 tk)'--,
   --'NotMalFromSelf protocol stage p q
        (App q (Fk a2 tk) t s)'--],
 conclusion = --'BGNY fenv protocol stage S2'--}
```

Rule GBC2, is central to the analysis of the example in Section 4. Note that the issue is not whether the encrypted data was actually encrypted and released by q, but whether p has adequate reason to believe this.

These rules indicate how bgny implements a belief logic extending the GNY logic, and the sorts of theorems that ProveGoals proves. Since type :Statement has no concept of negation, the logic is "sound" in the sense that BGNY never has value "true" on both a statement and its negation. Since the mutrec library proves that type :Statement exists [10], and the ind_def library proves that function BGNY exists [6], there is no question as to what the theorems proved by BGNY *are*. The bgny formalization does not resolve the deeper question, though, which arises for all belief logics [19], of exactly what these theorems *mean*.

6 Proof Procedure Algorithm

This section describes the ProveGoals algorithm. While less general than the algorithm sketched in [1], this algorithm is much more efficient, and it is general enough to suffice for all practical purposes to which it has been applied so far.

The rest of this paper will treat the "function environment" and "protocol" arguments to BGNY as implicit, and describe BGNY as if it were only a function of a stage and a :Statement. *Goal* will denote a :Statement element; *proving* a goal will denote proving that BGNY has value "true" on this goal for some stage. The goal *holds* for a stage if BGNY has value "true" on this goal for this stage.

The two basic concepts of the ProveGoals algorithm are *default goals* and *failure records*. The default goals that hold at a stage typically have all the user-set goals that hold for that stage as trivial consequences. (Note that the default goals are "default" only in the sense that they have as consequences all the properties that the user will probably want; user-set goals do not override default goals.) Failure records allow ProveGoals to retry default goals that do not hold at a stage at the first later stage where they might hold.

The default goals are defined as follows. If a protocol's stage n has principal p sending message M to principal q, let the *submessages* of M be M itself, the terms t1 and t2 for each pair t1 ; ; t2 that is a submessage of M, and the term t for each function application App r f t s that is a submessage of M. The default goals for stage n which often never hold, are the following:

- p possesses M. Although this is a stage n goal, ProveGoals checks that it holds for stage $n - 1$, guaranteeing that p is capable of sending M at stage n. Goals of this form are called *feasibility* goals.
- For each submessage m of M:
    - q received m.
    - If m is App r f t s and f is an encryption or hash function, then q can believe that r conveyed m.
    - If m is App r f t s and q can believe that r conveyed m, then q can believe s and believe that r believes s.

The default goals include conveyance goals for only encrypted values and hash codes because only these are protected against being surreptitiously modified in transit; the rules defining BGNY do not allow other forms of conveyance conclusions. If m is App r f t s and q can believe that r conveyed m, though, and s asserts only that r is capable of creating m, then the rules defining BGNY do allow q to believe s, even if q does not trust r.

If a default goal fails to hold at a stage, it often begins to hold at a later stage, say when a principal who was earlier unable to perform a decryption receives the necessary key. The *retryable* default goals are all the default goals except the feasibility goals. For the retryable default goals, ProveGoals prepares for retrying these goals even before it tries to prove them. It prepares a "failure" record for each retryable default goal of type:

```
{FromStage:int, GoalType:int,
 MainGoal:term list, ProofFun:thm list -> tactic,
 SubGoals:(term * tactic) list, SubThms:thm list}
```

whose entries are interpreted as follows:

- FromStage is the first stage having the goal as one of its default goals.
- GoalType identifies the goal as a reception, conveyance, or "believes associated statement" goal.
- MainGoal is an implicitly conjuncted list of :Statement elements expressing the "failed" goal.
- ProofFun is an SML function that proves the main goal for a stage, if possible, by matching it against one of the rules defining BGNY, then rewriting with the "failure" record's SubThms list of proved subgoals.
- SubGoals is a list of :(Statement,tactic) pairs that give subgoals (e.g., key possession, recognition, freshness, trust) sufficient for proving the main goal, and ProveGoals tactics that prove these subgoals, if they hold, for the current stage.
- SubThms is a list of proved subgoals and/or needed theorems proved before the goal was set (e.g., a reception theorem for a conveyance goal).

Given these preliminaries, here is the ProveGoals algorithm, which proceeds by induction on stage starting with stage 1. For stage n, ProveGoals does the following:

1. It computes the feasibility goal for stage n and tries to prove that this goal holds for stage $n-1$. If this proof attempt fails, it aborts with a "protocol feasibility failure at stage n" error message.
2. It creates "failure" records for each of the retryable stage-n default goals, and adds them to the stack of "failure" records it associates with the principal receiving the message sent at stage n.
3. It tries to prove that the main goal for each "failure" record associated with the principal receiving the message sent at stage n holds at stage n. For each record, it does this by attempting to prove that each of the record's SubGoals statements holds at stage n. It considers these subgoals one at a

time, starting with the first on the SubGoals list. If it proves that a subgoal holds for stage n, it removes the corresponding SubGoals entry and adds the theorem it has proved to the record's SubThms list. If it fails to prove a subgoal, it moves to the next record. If it proves all of a record's subgoals, it uses this record's ProofFun entry to prove the record's main goal.
4. As long as it is able to prove the main goal for one or more such "failure" records, it repeats trying to prove the main goal for all remaining such records.
5. After proving all the "default" goals that hold at stage n, it tries to prove all the user-set goals for stage n. If one of these proof attempts fails, it aborts with a "user-goal failure at stage n" message.

If it aborts, or if the environment variable ISLOUTPUTLEVEL is set to isl or hol, ProveGoals produces .fail and .prvd files containing ISL versions of all the "failure" records and all the theorems it has proved. If ISLOUTPUTLEVEL is set to hol, it also calls export_theory to save the HOL theory it has constructed.

The default goals have the very convenient property that if they hold there is only one top-level BGNY rule that makes them hold. Because of length constraints, this paper cannot demonstrate this, but it can be deduced by examining the BGNY rules in [3], particularly the conveyance rules GBC1 through GBC7. ProveGoals uses lists of public, private, and symmetric keys derived from the ISL specifications to decide which conveyance rule to apply. Rule GBC2, given in Section 5 above, for example, is the only conveyance rule applying to symmetric-key encryption when no NeverMalFromSelf statement holds.

In contrast to the ProveGoals algorithm's direct attempt to prove a collection of theorems that are very likely to be useful, the algorithm sketched in [1] is based on trying to prove what are essentially all possible theorems, since for the GNY logic a finite set of statements has only a "practically finite" set of consequences. (See [1] for an explanation of why this is true.) Mathuria, Safavi-Naini, and Nickolas use essentially the same idea in [11].

The portion of the ProveGoals algorithm that proves user-set goals from proved default goals can be replaced by a general decision procedure if that ever seems desirable. If this is done, having the highly relevant proved default goals available can be expected to make the decision procedure find proofs of provable results much more quickly.

Using a general decision procedure to derive user-set goals from proved default goals currently seems likely to have the major effect, though, of slowing the proof of useful goals. As an example of a provable protocol property that ProveGoals cannot prove, it is true that if

P Believes Q Possesses K

is a consequence of the default goals that hold at a particular stage, then

P Believes Q Believes Q Believes Q Possesses K

is also such a consequence, but ProveGoals cannot prove it. (The rules defining BGNY assume that principals can always believe that they can compute what they can compute, though what they compute might be meaningless to them. This contrasts with the treatment of belief in the belief logic by Syverson and van Oorschot [21].) The complicated statement that ProveGoals cannot prove contains no real information, though, in addition to that in the simple statement that ProveGoals can prove.

7 Limitations and Future Extensions

This section describes limitations in the AAPA and how these limitations might be reduced or eliminated in the future.

As just described in Section 6, ProveGoals does not prove all potentially provable user-set goals. A decision procedure that constructs proofs by induction on statement complexity might remove this limitation without making the ProveGoals proof process significantly less efficient. The author does not believe that this AAPA limitation is serious enough to bother addressing unless a specific need to reduce or eliminate it arises.

A much more serious AAPA limitation, common to all approaches using belief logics, is that an AAPA analysis address only authentication, not security. The ability of theory bgny to model separate protocol stages might be the basis for extending it to model security as well as authentication, perhaps by introducing a "known only to" statement constructor; see [3].

The AAPA also does not address attacks, even for authentication, based on exploiting algebraic properties of the encryption and decryption algorithms used. (See the TMN example in [19] for an example.) This limitation might be reduced by expanding the AAPA's use of bgny's FeEquals function-environment component, which is currently only used to describe keys computed by key-exchange functions.

Finally, the author will investigate whether the AAPA's use of a "belief logic" can be replaced by the use of an apparently more complete and rigorous formal theory of protocols, such as that by Snekkenes [19]. The essential issue is whether covering security and getting increased completeness and rigor intrinsically force combinatorial explosions like those faced by the attack-construction protocol analysis tools.

References

1. S. Brackin. Deciding cryptographic protocol adequacy with HOL. In *Higher Order Logic Theorem Proving and Its Applications*, number 971 in Lecture Notes in Computer Science, pages 90–105, Aspen Grove, UT, September 1995. Springer-Verlag.
2. S. Brackin. Automatic formal analyses of cryptographic protocols. To Appear in the 19th National Conference on Information Systems Security, Baltimore, MD, October 1996.
3. S. Brackin. A HOL extension of GNY for automatically analyzing cryptographic protocols. In *Proceedings of Computer Security Foundations Workshop IX*, County Kerry, Ireland, June 1996. IEEE.

4. S. Brackin. An interface specification language for cryptographic protocols and its translation into HOL. Submitted to the New Security Paradigms Workshop, Arrowhead, CA, September 1996.
5. S. Brackin and S-K Chin. Server-process restrictiveness in HOL. In *Higher Order Logic Theorem Proving and Its Applications*, number 780 in Lecture Notes in Computer Science, pages 454–467, Vancouver, BC, August 1993. Springer-Verlag.
6. J. Camilleri and T. Melham. Reasoning with inductively defined relations in the HOL theorem prover. Technical Report 265, University of Cambridge Computer Laboratory, Cambridge, UK, August 1992.
7. D. Denning and G. Sacco. Timestamps in key distribution protocols. *CACM*, 24(8):533–536, August 1981.
8. L. Gong. Handling infeasible specifications of cryptographic protocols. In *Proceedings of Computer Security Foundations Workshop IV*, pages 99–102, Franconia NH, June 1991. IEEE.
9. L. Gong, R. Needham, and R. Yahalom. Reasoning about belief in cryptographic protocols. In *Proceedings of the Symposium on Security and Privacy*, pages 234–248, Oakland, CA, May 1990. IEEE.
10. E. Gunter. Library `mutrec`. HOL90.7, `contrib` directory, 1994.
11. A. Mathuria, R. Safavi-Naini, and P. Nickolas. On the automation of GNY logic. *Australian Computer Science Communications*, 17(1):370–379, 1995.
12. C. Meadows. Using narrowing in the analysis of key management protocols. In *Proceedings of the Symposium on Security and Privacy*, pages 138–147, Oakland, CA, May 1989. IEEE.
13. C. Meadows. A system for the specification and analysis of key management protocols. In *Proceedings of the Symposium on Security and Privacy*, pages 182–195, Oakland, CA, May 1991. IEEE.
14. C. Meadows. Applying formal methods to the analysis of a key management protocol. *J. Computer Security*, 1(1):5–36, 1992.
15. J. Millen. The interrogator: A tool for cryptographic protocol analysis. In *Proceedings of the Symposium on Security and Privacy*, pages 134–141, Oakland, CA, May 1984. IEEE.
16. J. Millen. The Interrogator model. In *Proceedings of the Symposium on Security and Privacy*, pages 251–260, Oakland, CA, May 1995. IEEE.
17. J. Millen, S. Clark, and S. Freedman. The Interrogator: Protocol security analysis. *IEEE Trans. on Software Engineering*, SE-13(2):274–288, February 1987.
18. B. Schneier. *Applied Cryptography: Protocols, Algorithms, and Source Code in C*. John Wiley & Sons, New York, NY, 1995.
19. E. Snekkenes. *Formal Specification and Analysis of Cryptographic Protocols*. PhD thesis, University of Oslo, Oslo, Norway, January 1995.
20. J. Steiner, C. Neuman, and J. Schiller. An authentication service for open network systems. In *Proceedings of the USENIX Winter Conference*, pages 191–202, February 1988.
21. P. Syverson and P. van Oorschot. On unifying some cryptographic protocol logics. In *Proceedings of the Symposium on Security and Privacy*, pages 14–28, Oakland, CA, 1994. IEEE.
22. J. Tardo and K. Alagappan. SPX: Global authentication using public key certificates. In *Proceedings of the Symposium on Security and Privacy*, pages 232–244, Oakland, CA, 1991. IEEE.

Proving Liveness of Fair Transition Systems *

Holger Busch
SIEMENS AG, CORP. R&D, MUNICH

Abstract. A graph-based method for automatically synthesizing liveness proofs of non-finite concurrent systems is described. The implemented procedure is a significant extension of a previous mechanization of TLA[Lam94] in a higher-order logic theorem prover [Bus95].

1 Introduction

The complex behaviour of concurrent systems has caused a demand for verification methodologies which support early design stages and allow abstractions. Formalisms like Unity[ChM88], TLA[Lam94], TLR[MaP94], and TLT[CuW96] allow high-level specifications with parameterization, recursive datatypes and functions. Corresponding calculi are available for deductive proofs of temporal properties of abstract specifications. Theorem provers are appropriate tools for mechanizing these calculi and complementing BDD-based verification tools. However, the expertise typically required and proof complexity even for moderate examples restrict the applicability of deductive reasoning tools in industry. Our goal is to create a user-friendly proof environment for a methodology for the compositional specification and verification of distributed reactive systems[CWB94]: *Temporal Language of Transitions*. An embedding of TLA in the higher-order logic theorem prover LAMBDA(Sect. 2.1), called TLA-RISPE[Bus95], supports interactive proof development for TLT specifications, which are translatable into TLA formulas. In this paper, a new method for the automatic synthesis of TLA proofs of liveness properties is presented. The implemented procedure, which significantly reduces the required user interaction, reuses the infrastructure for interactive TLA reasoning.

The synthesis method works in two phases. First a graph is constructed which is labelled with the state predicates and transitions obtained by symbolic execution of a TLA program. In the second phase, instances of TLA calculus rules are extracted from the graph, which in total form a TLA proof. Exit conditions for cycles are derived from fairness assumptions or by induction.

Related work includes the embedding of Unity[APP94] and TLA[WrL93, EGL92] in theorem provers, as well as procedures for proving progress[Bra92, MaP94]. Diagrams for representing temporal proofs are proposed in [APP94, Lam94a, MaP94]. Our distinguishing result is the automatic synthesis of explicit deductive liveness proofs; an incomplete proof can be continued interactively at a high level while a significant portion of the proof has already been generated.

---
* Supported in part by a grant from the German Federal Ministry of Education, Science, Research and Technology under contract number 01IS519A (project KORSYS).

In Section 2 the setting for interactive development of TLA proofs is summarized. Then the automatic procedure for constructing a fair transition graph of a TLA program is presented. In Section 4 the analysis of the graph and the extraction of TLA proofs is detailed. A basic knowledge of TLA[Lam94] is supposed.

2 Interactive Development of TLA Proofs

This section is a recapitulation of TLA and its embedding in LAMBDA [Bus95]. It is well known that mechanical checks of TLA proofs helps to avoid errors easily made on paper. Moreover, the use of a theorem prover is an essential prerequisite for automating the construction of TLA proofs.

2.1 The Theorem Prover LAMBDA-RISPE

LAMBDA is a theorem prover for classical higher-order logic with polymorphic types like in ML. A rule-based calculus with higher-order unification is available for interactive proof development in the style of natural deduction, where goals are reduced to simpler subgoals. Rule schemes (e.g. Fig. 1) are instances of abstract ML datatypes, which are defined as axioms or derived by deduction. Tactics and rewrite functions are facilities for automating and combining proof steps. An extended library includes building bricks for safely augmenting the inference machinery with sophisticated rewrite strategies and tactics. Those automate proof decisions on the basis of term analysis and success or failure of preceding automatic proof steps. Transformation of logical terms is equally supported. It is possible to implement decision procedures relying on safe deductions; they are often reasonably efficient. Algorithmic verification procedures and external tools can be integrated at any level with deductive proof functions.

$$
\begin{array}{l}
3: \exists_{me} \forall x \in S.\ Q[x] \Rightarrow \mathsf{me}(d(x)) <_{lex} \mathsf{me}(x) \\
2: \forall x \in S.\ Q[x] \wedge P[d(x)] \Rightarrow P[x] \\
1: \forall x \in S.\ \neg Q[x] \Rightarrow P[x] \\
\hline
\forall x \in S.\ P[x]
\end{array}
\qquad
x <_{lex} y \stackrel{\Delta}{=} \begin{array}{l} x_{hd} < y_{hd} \vee \\ x_{hd} = y_{hd} \wedge x_{tl} <_{lex} y_{tl} \end{array}
$$

(false if x or $y = nil$)

When the induction rule is applied to a proof goal, the predicate P is instantiated to the current context of the induction argument. The instantiatable predicate Q denotes the step case, d is a destructor function such that a well-founded descending order is achieved. An appropriate measure function me has to be determined which maps the recursion arguments to a list of natural numbers. RISPE includes heuristics for automatic measure function computation and proving the well-foundness premise containing lexicographic comparison.

Fig. 1. Well-Founded Induction Rule Scheme.

We extended the reasoning capabilities of LAMBDA by creating the reduced instruction set environment RISPE for a higher level of user interaction. The accessibility, safe user-extensibility and flexibility of a tactic-based theorem prover is crucial for providing application-specific proof infrastructure. Although comparable theorem provers could have been used, we chose LAMBDA because of advantageous features such as its efficient conversion mechanism, its library of utilities, its ML syntax and the automatic proof tactics of RISPE.

2.2 TLA

TLA is a simple yet expressive formalism for modeling and reasoning about concurrent programs above the BDD-level. Verification of temporal properties, but also refinement and composition are supported. TLA is well-suited for mechanization in LAMBDA: First, algorithms are specified as logical formulas, as illustrated in Fig. 3. State predicates are Boolean expressions with unprimed program variables, constants and quantification. Actions are transition predicates over state dependent program variables, the primes symbolizing their values in the state after the transition. Thus state and transition predicates can be manipulated in a theorem prover for predicate logic. Second, TLA comprises a rule-based calculus for deductive proofs according to the linear time semantics of TLA. As the deductive proof calculus of LAMBDA is based on rules also, it is possible to mechanize deductive TLA reasoning and apply subgoaling strategies. Third, the calculus rules of TLA reduce conjectures about temporal formulas to either temporal subgoals or predicate logic proof obligations. Hence predicate logic reasoning functions can be reused for a large amount of subgoals.

$$\begin{array}{ll}
\text{WF1.} & \text{SF1.} \\
3: P \land [\mathcal{N}]_z \Rightarrow (P' \lor Q') & 3: P \land [\mathcal{N}]_z \Rightarrow (P' \lor Q') \\
2: P \land \langle \mathcal{N} \land \mathcal{A} \rangle_z \Rightarrow Q' & 2: P \land \langle \mathcal{N} \land \mathcal{A} \rangle_z \Rightarrow Q' \\
1: P \Rightarrow \text{Enabled } \langle \mathcal{A} \rangle_z & 1: \Box P \land \Box[\mathcal{N}]_z \land \Box F \Rightarrow \Diamond \text{Enabled } \langle \mathcal{A} \rangle_z \\
\hline
\Box[\mathcal{N}]_z \land \text{WF}_z(\mathcal{A}) \Rightarrow (P \rightsquigarrow Q) & \Box[\mathcal{N}]_z \land \text{SF}_z(\mathcal{A}) \land \Box F \Rightarrow (P \rightsquigarrow Q)
\end{array}$$

$$\begin{array}{l}
\text{LATTICE. } 2: \succ \text{ a well-founded partial order on a set S} \\
\phantom{\text{LATTICE. }} 1: F \land (c \in S) \Rightarrow (P_c \rightsquigarrow (Q \lor \exists d \in S : (c \succ d) \land P_d)) \\
\hline
\phantom{\text{LATTICE. }} F \Rightarrow ((\exists c \in S : P_c) \rightsquigarrow Q)
\end{array}$$

$$\begin{array}{lll}
\text{LTIMP.} & \text{LTDISJ.} & \text{LTTRANS.} \\
1: \ldots \Rightarrow (P \Rightarrow Q) & 2: \ldots \Rightarrow (P \rightsquigarrow Q) & 2: \ldots \Rightarrow (R \rightsquigarrow Q) \\
\hline
\ldots \Rightarrow (P \rightsquigarrow Q) & 1: \ldots \Rightarrow (R \rightsquigarrow Q) & 1: \ldots \Rightarrow (P \rightsquigarrow R) \\
\hline
& \ldots \Rightarrow ((P \lor R) \rightsquigarrow Q) & \ldots \Rightarrow (P \rightsquigarrow Q)
\end{array}$$

where F is a TLA formula P, Q, R are predicates $[\mathcal{A}]_z$: action or z may be unchanged
$\mathcal{A}, \mathcal{N}$ are actions z is a state function $\langle \mathcal{A} \rangle_z$: action and z must be changed

Fig. 2. The TLA Rules for Liveness Proofs.

Fairness ensures that an enabled transition eventually happens. TLA has the notions of *weak* and *strong fairness*. Premises 1 of Rules WF1 and SF1 in Fig. 2 show the difference: a strongly fair transition allows other transitions to disable it, provided it is sometime enabled.

The algorithm DC displayed in Fig. 3 is an example of a specification not expressible in terms of finite states. It is used to illustrate the methods developed in the next sections. We will investigate the generation of liveness proofs including induction with the following conjecture for the program DC: $\text{true} \rightsquigarrow y \geq l$.

2.3 Embedding TLA in LAMBDA

We chose shallow embedding, i.e. TLA terms are directly mapped to logical terms in LAMBDA and the semantic interpretation is kept outside of the logic. While TLA is untyped, we exploit the static type checking in LAMBDA.

$Init_{DC} \triangleq x = 0 \wedge y = 0 \wedge pc = \alpha$

$\mathcal{I}_1 \triangleq pc = \alpha \wedge pc' = \beta \wedge$ Unchanged $\langle x, y \rangle$

$\mathcal{I}_2 \triangleq pc = \alpha \wedge n < x \wedge pc' = \beta \wedge x' = 0 \wedge y' = y + 1$

$\mathcal{I}_3 \triangleq pc = \beta \wedge pc' = \gamma \wedge x' = x + 1 \wedge$ Unchanged $\langle y \rangle$

$\mathcal{I}_4 \triangleq pc = \gamma \wedge pc' = \alpha \wedge$ Unchanged $\langle x, y \rangle$

$\mathcal{N} \triangleq \mathcal{I}_1 \vee \mathcal{I}_2 \vee \mathcal{I}_3 \vee \mathcal{I}_4,$

$z \triangleq \langle pc, x, y \rangle$

$DC \triangleq Init_{DC} \wedge \Box[\mathcal{N}]_z \wedge \text{WF}_z(\mathcal{I}_1) \wedge \text{SF}_z(\mathcal{I}_2) \wedge \text{WF}_z(\mathcal{I}_3) \wedge \text{WF}_z(\mathcal{I}_4)$

Two variables x, y are incremented. Variable y may only be altered when x is greater than a limit n. The algorithm could be a model of a conference session, where a speaker (y) is permitted to continue a presentation for a nondeterministic period of time after the regular speaking time (n) is over. Strong fairness is required: otherwise a particularly eloquent speaker would always succeed in continuing for some more time.[a] (In the simplified model the time x used by speaker y does not affect the time of subsequent presentations.)

---

[a] The specification has not been minimized in order to illustrate of the proof construction procedure.

Fig. 3. The TLA Specification of an Incrementer.

Various proof functions of RISPE are reused and customized for the mechanization of TLA. In particular the automation of first-order predicate-logic reasoning, arithmetical reasoning, and proving termination by well-founded induction (Fig.1) are crucial. Additional TLA tactics exploit the structure of subgoals frequently arising in TLA proofs; e.g., tactics for simplifying *Enabled* - predicates ($Enabled\ \mathcal{B} \triangleq \exists_{a,b}\ \mathcal{B}[a/u', b/v']$) and for using *Unchanged* -predicates for program variables (including index-dependent replacements: $Unchanged\ \langle u, v_{j:F} \rangle \triangleq u' = u \wedge \forall_j\ F(j) \Rightarrow v'_j = v_j$) are called by tactics for discharging TLA proof obligations.

Previous case studies demonstrated that the embedding is suitable for stepwise checking and development of TLA proofs. The experience that there is typically a large amount of routine suggested further automation.

3 Generating Fair Transition Graphs

A directed fair transition graph is generated from a TLA specification. The method follows a principle of *provability by construction*: the way the graph is constructed ensures that each step through the graph is provable with a TLA calculus rule instantiated with graph node predicates. Consequently, a proof is complete if all paths starting from the root lead to nodes which satisfy the liveness predicate under investigation. The generated graph, for which we chose a two-level representation, closely matches the structure of a deductive TLA proof.

3.1 Two-Level Graphs

Interactive development of TLA proofs typically follows an operational understanding of the specified system's behaviour, which is controlled by the enabledness of program transitions in the individual states. An abstraction of program variables, possibly with an infinite range, to finite-state variables encoding the enabledness of individual program transitions results under certain constraints in the same control behaviour as the original program. This can be exploited for model checking classes of infinite systems. If the control behaviour depends on values of infinite-state variables, however, this abstraction fails.

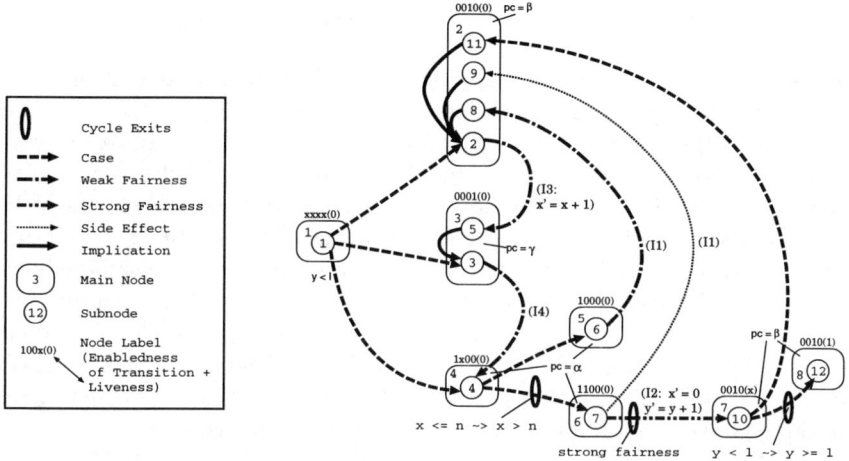

Fig. 4. Transition Graph of the Incrementer Problem: $true \rightsquigarrow y \geq l$

The importance of the enabledness of transitions in TLA liveness proofs led us to a two-level model of a fair transition system. The detailed level mirrors a symbolic execution of the program transitions starting from the predecessor state (P) of the $\rightsquigarrow$ relation to be proven $(P \rightsquigarrow L)$. The generated state predicates are assigned to subnodes, which are classified according to the enabledness of each specified transition. At the coarse level just these enabledness patterns are visible and subnodes with same enabledness pattern are collected in nodes. The enabledness patterns simplify the selection of transitions for generating successors of subnodes by symbolic execution and guarantee the enabledness premises (1) of the TLA fairness rules to be provable.

Accordingly, the fair transition graph is defined as $FTG = (M, S, E)$, where M is the set of main nodes, S is the set of subnodes, and E is the set of edges between subnodes. Each subnode uniquely belongs to exactly one main node. A function $m(s)$ returns the (main) node of a subnode s. The relation between the fair transition system corresponding to a TLA specification, (z, I, N, A), and FTG is defined through predicates labelling the elements of FTG:

z	Tuple of state variables.		I	Set of initial states.
N	Set of all program transitions.[1]		A	Set of fair transitions.
q_m	Predicate of a main node $m \in M$.		p_s	Predicate of a subnode $s \in S$.
p_{s_1}	Predicate of initial subnode s_1.			
t_e	Labeling of Edge e:			
	weakly, strongly fair or side effect transition predicate, case analysis or implication.			

---

[1] Relations are used: TLA *actions*.

Nodes and subnodes are both labelled with state predicates; the predicate of a node is implied by the predicates of its subnodes. The subnode predicates p_s characterize the (sets of) states evolving during symbolic program execution. Subnodes are assigned to nodes by testing the enabledness of each program transition with their subnode predicates:

$$En(p_s, t) = \begin{cases} 1 & \text{if } p_s \Rightarrow \text{Enabled } t \\ 0 & \text{if } p_s \Rightarrow \neg(\text{Enabled } t) \\ x & \text{if none of the above is provable} \end{cases}.$$

The tests are performed automatically by TLA-RISPE tactics. In addition the test is performed with the liveness predicate $L(= \text{Enabled } L)$ instead of $\text{Enabled } t$. The value x is not only assigned if neither of enabledness or disabledness follows from p_s, but also in the case that the implemented tactics are too weak to prove that one of those follow logically. The pattern of the results of these tests performed for all transitions and the liveness predicate in a fixed order can be used as unique labellings of the nodes. Fig.4 displays these labellings on top of the round boxes. For instance, the enabledness pattern of Subnode 4 in Fig.4 $(En(p_4, \mathcal{I}_1), En(p_4, \mathcal{I}_2), En(p_4, \mathcal{I}_3), En(p_4, \mathcal{I}_4), (En(p_4, y \geq l)) = 1, x, 0, 0, (0))$ denotes that Transition $\mathcal{I}_1$ is enabled, Transitions $\mathcal{I}_3, \mathcal{I}_4$ are disabled, and Transition $\mathcal{I}_2$ cannot be decided; the liveness predicate $L := l \leq y$ is not satisfied. Table 1 shows the enabledness predicates for each transition of Program DC.

Transition	$\mathcal{I}_1$	$\mathcal{I}_2$	$\mathcal{I}_3$	$\mathcal{I}_4$	(L)
Enabled $\mathcal{I}$	$pc = \alpha$	$pc = \alpha \land n < x$	$pc = \beta$	$pc = \gamma$	$l \leq y$

Table 1. Enabledness Predicates for Program DC.

The predicate of a node $m(s)$ characterizes the set of states in which the transitions are enabled and the liveness predicate is satisfied according to the enabledness pattern:

$$qm(s) = \bigwedge_{t \in A \cup N \cup \{L\}} \begin{cases} \text{Enabled } t & \text{for } (En(p_s, t)) = 1 \\ \neg(\text{Enabled } t) & \text{for } (En(p_s, t)) = 0 \\ \text{true} & \text{for } (En(p_s, t)) = x \end{cases}.$$

X-enabled transitions do not contribute to the node predicate. Reversely, the enabledness pattern of a node is obtained if the enabledness and liveness test is performed with the node predicate. Hence enabledness pattern and node predicate are interchangeable.

The edge labels (cf. legend Fig.4) record the way subnodes are generated from their predecessors during the construction phase. Successors are generated by case analysis, implicative approximation, symbolic execution of a strongly or weakly fair transition, or a side effect (Sect.3.2). By construction, a subnode has at most one outgoing (weak or strong) fairness edge to a successor. All edges to successors generated through other transitions are labelled as side effect transitions. In the analysis phase (Sect.4), the edge labels influence the selection of the appropriate TLA rules to prove the relation between each subnode and its predecessors and the analysis of cycles.

3.2 Recursive Graph Construction

Corresponding to local model checking, the construction of FTG starts with $p_{s_1} := P$ and does not proceed beyond subnodes satisfying L, rather than generating all reachable states, if $P \leadsto L$ is to be proven.

A fair transition graph is constructed recursively for the conjecture $P \leadsto L$. In each recursion step the successors of a selected subnode through an enabled fair transition (or case analysis) are computed and included in FTG. The predicates computed for our example in this first phase are listed in Table2.

m	En	q_m		s	p_s	
1	xxxx(0)		$y < l$	1		$y < l$
2	0010(0)	$pc = \beta \wedge$	$y < l$	2	$pc = \beta \wedge$	$y < l$
				8	$pc = \beta \wedge x \leq n \wedge$	$y < l$
				9	$pc = \beta \wedge n < x \wedge$	$y < l$
				11	$pc = \beta \wedge x = 0 \wedge 0 < y < l$	
3	0001(0)	$pc = \gamma \wedge$	$y < l$	3	$pc = \gamma \wedge$	$y < l$
				5	$pc = \gamma \wedge 0 < x \wedge$	$y < l$
4	1x00(0)	$pc = \alpha \wedge$	$y < l$	4	$pc = \alpha \wedge$	$y < l$
5	1000(0)	$pc = \alpha \wedge x \leq n \wedge$	$y < l$	6	$pc = \alpha \wedge x \leq n \wedge$	$y < l$
6	1100(0)	$pc = \alpha \wedge n < x \wedge$	$y < l$	7	$pc = \alpha \wedge n < x \wedge$	$y < l$
7	0010(x)	$pc = \beta$		10	$pc = \beta \wedge x = 0 \wedge 0 < y < l+1$	
8	0010(1)	$pc = \beta \wedge$	$l \leq y$	12	$pc = \beta \wedge x = 0 \wedge y = l$	

Table 2. Node and Subnode Predicates for Program DC.

Initialization Before the procedure is started, the trivial possibility $P \Rightarrow L$ is checked. Otherwise, the initial subnode predicate is set to $p_{s_1} \triangleq P \wedge \neg L$. The initial node predicate $q_{m_{s_1}}$ is computed as given above.

The initial set of edges is empty. s_1 is the subnode for starting the recursive extension of FTG. Each transition $\langle \mathcal{A} \rangle_z \in \mathcal{A}$ is checked for its ability to render L true:[2] $L' \Rightarrow \exists_z \neg L \wedge \mathcal{A}$. If the test is positive, we call the transition *liveness transition*. The formulas for the enabledness of the individual program transitions are minimized once for all.

Generating Next States From a given state predicate p_{s_i}, the next state predicate $p_{s_{i'}}$ is computed by symbolically executing a transition. As liveness is to be proven, only an enabled fair transition $\mathcal{A} \in \mathcal{A}$ with $En(p_{s_i}, \mathcal{A}) = 1$ is selected. At this stage, the treatment of strong and weak fairness is completely identical. If several fair transitions are enabled, a liveness transition is preferred, or strong to weak fairness. Using the selected transition $\mathcal{A} \in \mathcal{A}$, the successor state $p_{s_{i'}}$ is computed as the *strongest post-condition*:[3]

$$p'_{s_{i'}} := (SP(p_{s_i}, \mathcal{A}))' := \exists_z \, p_{s_i}[z/z] \wedge \mathcal{A}[z/z]$$

---

[2] Explicitly written as $L[z'/z] \Rightarrow \exists_z \neg(L[z/z]) \wedge \langle \mathcal{A} \rangle_z[z/z, z']$. Throughout this paper, state predicates are assumed to have occurrences of the symbolic state variable z, actions additionally of z', primed state predicates only of z'.

[3] For example, $p_{10}[pc'/pc, x'/x, y'/y] = \exists_{pc,x,y} \, x' = 0 \wedge y' = y + 1 \wedge pc' = \beta \wedge \mathsf{pc} = \alpha \wedge n < \mathsf{x}$, which is evaluated to $x' = 0 \wedge 0 < y' \wedge pc' = \beta$ through specific tactics.

Side Effects. If apart from the selected fair transition $\mathcal{A}$, other transitions are 1- or X-enabled, they may blur the effect of $\mathcal{A}$. In order to ensure the provability of Premise 3 of Rule SF1 or WF1, the side effects of all these extra transitions $\mathcal{A}_{sd}$ are taken into account by adding further subnodes s_{sd} for each extra successor with $\quad p'_{s_{sd}} = \neg p'_{s_{i'}} \wedge \neg p'_{s_i} \wedge \exists_{\mathbf{z}}\, p_{s_i}[z/z] \wedge \mathcal{A}_{sd}[z/z]$.

Case Analysis. If all fair transitions are disabled $\forall_{\mathcal{A} \in A}.En(p_{s_i}, \mathcal{A}) = 0$ for a subnode s_i, then $P \rightsquigarrow L$ is not true if p_{s_i} cannot be excluded by way of reachability considerations (Sect.4.4). If no fair transition is 1-enabled, but there exists at least one X-enabled transition,[4] then a case split is performed such that $p_{s_i} \Rightarrow \bigvee_k p_{s_{ik}}$. The approach $p_{s_i} \Rightarrow p_{s_i} \wedge c \vee p_{s_i} \wedge \neg c$ ensures exhaustiveness. Instantiating $c \triangleq Enabled\; \langle \mathcal{A} \rangle_z$ yields two successors of s_i, of which at least the positive case ensures an enabled fair transition. The negative case will often imply the enabledness of other transitions; otherwise another case analysis has to be performed. As each case analysis enforces the decision for one more of the finite set of x-enabled transitions, the case splitting is guaranteed to terminate. Typically some of the resulting cases are outside of the reachable state space, which are excluded if appropriate invariants are available. Note that the three-valued enabledness coding helps to avoid unnecessary case analyses.

Cases are sometimes split even though a fair transition is 1-enabled: before adding a side effect edge, a case split $c \triangleq Enabled\; \langle \mathcal{N}_i \rangle_z$ is inserted for X-enabled transitions $\mathcal{N}_i$ causing the side effect. This strategy simplifies reasoning about cycle exits (Section 4.3) by decoupling well-foundedness from fairness, e.g. Subnodes 4 and 7 in Fig.4. Another possibility is a liveness transition which does not enforce liveness: $p_{s_{i'}} \Rightarrow L$ is not provable for the successor $s_{i'}$ (cf. Fig.4, Subnode 10). Here the case split $c \triangleq L$ is appended. The resulting subnode with predicate $p_{s_{i'}} \wedge L$ is the potential exit of a cycle.

If p_{s_i} is already given as a disjunction, an alternative to to case analysis is just splitting the disjunction, which may lead to subpredicates in which fair transitions are enabled. Similarly to the variable ordering in BDD-based algorithms, the case analysis influences the complexity of the further graph construction, which may even concern the number of inductions to be performed.

Updating the Fair Transition Graph After a successor $s_{i'}$ of Subnode s_i has been generated, FTG is augmented. If it already contains a subnode $\tilde{s}_{i'}$ with $p_{\tilde{s}_{i'}} \Leftrightarrow p_{s_{i+1}}$, just an edge is inserted: $E := E \cup \{s_i \to \tilde{s}_{i'}\}$, $M := M$, $S := S$. The edge is labelled as discussed in Section 3.1. If $s_{i'}$ is new, it is included, $S := S \cup \{s_{i'}\}$, and the enabledness pattern is computed. If FTG does not yet contain a node with equivalent predicate, a new node is added $M := M \cup \{m(s_{i'})\}$. If $s_{i'}$ is a new subnode of an existing node it is checked whether implicative relations to previous subnode predicates pertaining to the same node exist. If so, additional implicative edges are added.

---

[4] Hence the correctness of the construction is not affected if enabledness is not recognized, but cases are added which have to be proven irrelevant (cf.Sect.3.1,4.4).

If $s_{i'}$ is a new subnode of an existing node, the possibility of a new cycle is checked:
$$cyc(s_{i'}) \triangleq \exists_{\tilde{s}_i \in S}. m(s_{i'}) = m(\tilde{s}_i) \wedge m(\tilde{s}_i) \in \{m(s) | s \stackrel{*}{\to} s_{i'} \in E^*\} .$$
The symbol $\stackrel{*}{\to}$ denotes a non-zero path between two (possibly identical) subnodes in FTG, as far as that one has been constructed. The symbol E^* is used for the set of all paths between subnodes in FTG. All members of the new cycle are tagged with the same cycle identifier.

Continuation A subnode $s_i \in S$ from which the iterative construction of FTG is continued is selected from a subset of S determined as:
$$S_i := \{s_i \in S | \neg(\exists_{s_{i'}}, s_i \to s_{i'} \in E) \wedge \neg(p_{s_i} \Rightarrow L) \wedge \neg cyc(s_i)\}.$$
As all paths must eventually lead to a state satisfying the liveness condition, any member of S_i may be selected for the next iteration. The recursion stops when $S_i = \emptyset$. The condition $\neg(\exists_{s_{i'}}, s_i \to s_{i'} \in E)$ means that no successors have been computed yet. The second condition excludes subnodes which imply the liveness predicate without temporal reasoning. As in an infinite transition system the generation of new subnodes does not necessarily stop even though the number of nodes is by definition finite,[5] the membership in a cycle disqualifies a continuation candidate. In the case missing invariants or errors, subnodes for which all fair transitions are disabled can occur; they are also excluded from S_i.

If FTG contains a subnode $\tilde{s}_i$ which is transitive predecessor (at the level of nodes) and pertains to the same node, s_i is excluded from reiteration. Hence the iteration is stopped after one cycle; a well-foundedness or fairness argument is derived in the next phase (Sect. 4.3). As the cycle is determined through node relationships the number of which is limited under the assumption of finite representability of the algorithm, the construction phase always terminates.

Approximation. The FTG construction terminates immediately after one round of a cycle, which, however, is not sufficient sometimes to render cycle exit conditions true. If, for instance, the parameter n in Program DC is instantiated to a value, say 5, and the goal is to prove $x = 0 \leadsto y \geq l$, after one round the condition for leaving the inner cycle $n < x := 5 < 1$ is still false.

The following check is therefore performed for all cycle nodes. If the main node predicate q_m instead of a subnode predicate would lead to a successor with a different enabledness pattern then any other successor of all subnodes in the same main node, the cycle is incomplete. An approximative subnode s_g is added with
$$m(s_g) = m \wedge (p_{s_g} \Rightarrow q_m) \wedge \forall_{s_i \in \{s | m(s) = m\}} p_{s_i} \Rightarrow p_{s_g} .$$
All previous subnodes of m are made predecessors through implicative edges to s_g. The coarsest approximation is $p_{s_g} \Leftrightarrow q_m$. In the example above we obtain $p_{s_g} \Leftrightarrow pc = \alpha \wedge x \leq 5 \wedge y < l$.

---

[5] Finite representability of a TLA program is assumed; even infinite index spaces of parametric transitions can usually be partitioned into finitely many subspaces, depending on the property to be proven.

A more precise characterization of the locally reachable state set can be constructed for classes of problems by symbolically computing the limit value in dependence on the number of rounds until the exit condition becomes true.
Completeness. All reachable main nodes are generated by the procedure extended with approximation. All locally reachable states are included in FTG, except for those occurring in cycles. Thus the construction of FTG is complete under the assumption that in the cycle analysis phase arguments for leaving cycles are found. Final open nodes which do not satisfy the liveness predicate and do not belong to cycles characterize counterexamples, unless they are excluded by way of invariants (cf. Sect.4.4).

All further computations referring to the evolution of program variable values are performed during cycle analysis (Sect.4.3).

4 Synthesizing TLA Proofs from Fair Transition Graphs

After the construction phase, FTG contains the information needed to generate a proof in terms of TLA rules. Walking through FTG from the beginning, the original liveness goal is gradually split into smaller goals, or in terms of graphics, the distance between the starting point and the final node with the liveness predicate fullfilled is split into chunks by inserting intermediate nodes (Fig.5,6). The edges of FTG between a subnode and its successors directly correlate with the TLA rules to use for proving the corresponding leadsto sub-relation. The subnode predicates are the intermediate assertions to instantiate the predicate variables of the TLA rules. With these instantiations, the premises of the TLA rules are ensured to be provable.

Counter examples follow directly from subpaths not leading to liveness, if a reachability analysis is not able to exclude the subpaths. Cycles in FTG are reduced through fairness assumptions or well-foundedness.

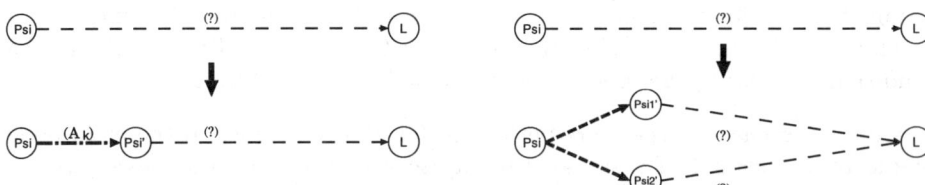

Fig. 5. Insertion of WF Transition   **Fig. 6.** Insertion of Case Analysis

4.1 Linear Paths

If a subnode s_i has just one successor $s_{i'}$, the corresponding TLA step is trivially LTTRANS($P := p_{s_i}$, $Q := L$, $R := p_{s_{i'}}$), since the overall goal in s_i is $p_{s_i} \leadsto L$ (Fig.5). The proof of $p_{s_i} \leadsto p_{s_{i'}}$ follows directly for $t_{s_i \to s_{i'}} = \mathcal{A}_k$ with the instantiations $P := p_{s_i}$, $Q := p_{s_{i'}}$, $\mathcal{A} := \mathcal{A}_k$ in the TLA rules WF1 or SF1, depending on whether $\mathcal{A}_k$ is weakly or strongly fair. The premises of WF1 or SF1 are true by construction. The goal $p_{s_{i'}} \leadsto L$ is left for the next recursion step.

4.2 Forks

A subnode s_i may have several successors, e.g., $s_{i'_1}$ and $s_{i'_2}$. The **TLA** steps:
LTTRANS($P := p_{s_i}$, $Q := L$, $R := p_{s_{i'_1}} \lor p_{s_{i'_2}}$), LTIMP($P := p_{s_i}$, $Q := p_{s_{i'_1}} \lor p_{s_{i'_2}}$),
LTDISJ($P := p_{s_{i'_1}}$, $R := p_{s_{i'_2}}$, $Q := L$)
leave the goals $(p_{s_{i'_1}} \rightsquigarrow L) \land (p_{s_{i'_2}} \rightsquigarrow L)$ if the successors are due to case splitting, as shown in Figure 6.

If a subnode has one successor $s_{i'_1}$ by a fair transition $\mathcal{A}_k$, and in addition one (or more) successor(s) $s_{i'_2}$ by side effect(s), the instantiations of WF1 or SF1 are $P := p_{s_i}$, $Q := p_{s_{i'_1}} \lor p_{s_{i'_2}}$, $\mathcal{A} := \mathcal{A}_k$. Then Rule LTDISJ($P := p_{s_{i'_1}}$, $R := p_{s_{i'_2}}$, $Q := L$) is applied. For instance, in Figure 4, $\mathcal{A}_k := \mathcal{I}_2, s_i := 7, s_{i'_1} := 10, s_{i'_2} := 9$.

4.3 Cycles

Let a cycle contain the subnodes $C(s_i) = \{s | s_i \xrightarrow{*} s \in E^* \land s \xrightarrow{*} s_i \in E^* \lor s = s_i\}$ and the exit subnodes be $X(s_i) = \{s | s \notin C(s_i) \land \exists_{s_j} \ s_j \in C(s_i) \land s_j \rightarrow s \in E\}$; thus a cycle may have several exit subnodes. A cycle is reduced by way of the fairness or the induction approach when a fork subnode s_i in FTG is its own transitive successor $s_i \xrightarrow{*} s_i \in E^*$ and has a direct successor $s_i \rightarrow s_x$ outside of the cycle. Both approaches have in common that in order to prove the liveness goal $p_{s_i} \rightsquigarrow L$, an argument is constructed for leaving the cycle such that $p_{s_i} \rightsquigarrow p_{s_x} \lor p_{s_i}$ is reduced to $p_{s_i} \rightsquigarrow p_{s_x}$.

Fairness According to the first premise of rule WF1 and SF1, the predicate P is instantiated in such a way that it is preserved within all cycle subnodes ($C(s_i)$) until the fair exit transition $\mathcal{A}_x$ is executed; predicate Q is composed of all exit predicates ($X(s_i)$).

$$P := \bigvee_{s \in C(s_i)} p_s, \ Q := \bigvee_{s_x \in X(s_i)} p_{s_x}, \ \mathcal{A} := \mathcal{A}_x \ (= t_{s_i \rightarrow s_x}, s_x \in X(s_i))$$

The only difference between strong and weak fairness concerns the enabledness of the fair exit transition $\mathcal{A}_x$: the additional constraint for weak fairness $\forall_{s \in C(s_i)} \cdot p_s \Rightarrow Enabled \ \langle \mathcal{A}_x \rangle_z$. By construction the enabledness constraint for strong fairness follows trivially from p_{s_i}, which ensures that $\mathcal{A}_x$ becomes enabled as long as the cycle is not left, even if it is temporarily disabled.

It turns out that the case splitting performed in the course of the construction of FTG for x-enabled transitions causing side effects (Sect. 3.2) guarantees that only 1-enabled transitions cause forks from which fair cycles are continued: $\forall s. \ s_i \rightarrow s \in E \land s \in C_F(s_i) \Rightarrow En(p_{s_i}, t_{s_i \rightarrow s}) = 1$;
wellfoundedness is required for case-split forks. This additional constraint on the selection of s_i thus decouples well-foundedness and fairness exits. The decision of which approach to use is hence computationally inexpensive.

Induction Case-split successors, and by construction only those, are potential exits of a cycle for which fairness does not apply. A well-founded measure must be derived which guarantees that eventually the condition for leaving the cycle is fulfilled. The problem of deriving such a measure is separated from the temporal relations between the subnode predicates by way of the transformation

$$\frac{2: \forall_{z_0,z_1}\ P[z_0/z] \wedge P[z_1/z] \wedge \neg Q[z_0/z] \wedge \neg Q[z_1/z] \wedge R(z_0,z_1) \Rightarrow \exists_{me}\ me(z_1) <_{lex} me(z_0)}{1: \forall_{z_0,z_1}\ (P[z] \wedge z = z_0) \rightsquigarrow (P[z] \wedge z = z_1 \wedge R(z_0,z_1) \vee Q[z])}{(\exists_{z_0}\ P[z] \wedge z = z_0) \rightsquigarrow Q[z]}$$

which is an adaptation of the TLA lattice rule. The variable R is an instantiatable predicate describing the relation between symbolic values of the state variable z before (z_0) and after (z_1) one round along the cycle to be reduced.

The problem posed in the first subgoal is related to the termination of recursive functions, so that a variety of previously created automatic proof functions for well-founded induction (Sect. 2.1) are reusable.

The proof of the second subgoal follows from predicates P and Q which are extracted from FTG and augmented with symbolic value predicates for z_0 and z_1. The augmentation of FTG proceeds as follows. A symbolic value z_0 is assigned to the state variable z in the fork subnode s_i. The initial augmented predicate $p^0_{s_i}[z, z_0] := p_{s_i} \wedge z = z_0$ is assigned to Subnode s_i. The transitions of one round of a cycle c to be proven well-founded are executed one after another in terms of strongest post-conditions. Implicative and case analysis edges are treated as stutter transitions which do not change the symbolic values. Finally, the augmented predicate $p^c_{s_i}[z, z_0]$ is obtained. In general, Subnode s_i may be member of several cycles c_k with different effects upon the symbolic value z_0, which are described by the disjunction $\bigvee_k p^{c_k}_{s_i}[z, z_0]$. The occurrence of several exits is treated analogously to fairness cycles. The induction rule is instantiated as follows:

$$P := p_{s_i},\ Q := \bigvee_{s_x \in X(s_i)} p_{s_x},\ R(z_0, z_1) := \bigvee_k p^{c_k}_{s_{ik}}[z_1/z, z_0],\ \text{where}\ t_{s_i \to s_x} = Case\ (s_x \in X(s_i))$$

If the relation R is functional for a cycle c_k, i.e., $\exists_{f_{c_k}}\ R_k(z_0, z_1) = (z_1 = f_{c_k}(z_0))$, methodologies for proving the termination of recursive functions apply. It is often possible to abstract away components of the compound state variable z which evolve non-functionally but do not affect the well-foundedness of the cycle. For instance, if program variables have constant or unchanged values in Subnodes s_i and s_{ik}, they do not contribute to a diminution of a well-founded measure. Our heuristics for synthesizing the measure function me try to abstract away program variables for which measures like the construction depth of recursive datatypes do not apply.

Symbolic Value Propagation along Nested Cycles In order to illustrate the computation of symbolic values, the 3 nested cycles of Program DC (Fig. 4) are analysed. According to the general scheme given above, we have to show: $p_4 \rightsquigarrow p_7$ for the inmost cycle (:= Cycle 1), $p_7 \rightsquigarrow p_{10}$ for the next cycle (:= Cycle 2), and $p_{10} \rightsquigarrow p_{12}$ for the outmost cycle (:= Cycle 3).

Well-foundedness is chosen to reduce Cycle 1, because Subnode 7 is the result of a case analysis, such that $p_4 \leadsto p_7$. A preliminary analysis is made which ones of the program variables are constants: $pc(=\alpha)$, and which ones are unchanged within Cycle 1: y (because the only transition able to change y, $\mathcal{I}_2$, is not invoked in Cycle 1). The symbolic initial value, x_0, of the remaining program variable x is propagated as given above; the new symbolic value is extracted after each cycle transition. For Cycle 1 the following sequence of values attached to cycle subnodes is obtained, with unchanged values omitted: $4(x_0)$, 6, 8, 2, $5(x_0+1)$, 3, $4(x_0+1)$; hence $x_1 := x_0 + 1$. Obviously, the termination of the function
fun $lim_{c_1}(x) =$ if $n < x$ then $lim_{c_1}(x+1)$ else x is provable with the required well-foundedness measure, which is the function: fun $me(x) = [n + 1 - x]$.
The well-foundedness goal is reduced to $x_1 \le n \Rightarrow n+1-x_1 < n+1-x_0$.

For Cycle 2, the theorem $p_4 \leadsto p_7$ allows Cycle 1 to be neglected; the reduction lemma for Cycle 2, $p_7 \leadsto p_{10}$ is obtained from the goal $(p_{10} \vee p_2 \vee p_3 \vee p_4 \vee p_7) \leadsto p_{12}$ through strong fairness of transition $\mathcal{I}_2$.

The reduction of Cycle 3 illustrates an additional difficulty for nested cycles: In general, the reduction lemmas of inner cycles, $p_{s_i} \leadsto p_{s_x}$, are not directly usable to reduce an outer cycle through a well-founded measure, if they do not take the symbolic value predicates into account. The augmented reduction lemma for an inner cycle requires an expression for a symbolic limit value taken after the inner cycle as been rounded a number of times until either its exit condition has become true or it is left through a fair transition. However, it is not always possible to specify the limit value in terms of a function. For instance, the value of the program variable x when Cycle 2 is left is not known. If the unknown number of rounds is functionally related to the symbolic value afterwards, a limit function with the number of rounds as additional paramater can be synthesized, which is then used in a limit predicate such as $\exists_r\ x_1 = lim_{c2}(r, x_0)$ with
fun $lim_{c_2}(r,x) =$ if $r = 0$ then x else $lim_{c_2}(r-1, lim_{c_1}(x))$. Implicit descriptions of symbolic values in terms of limit functions or predicates can be used to analyse the well-foundedness of outer cycles. Unfortunately, the heuristics for automatic measure function generation hardly succeed if applications of synthetic limit functions are not replaced with explicit symbolic limit values, such as
$\exists_r\ x_1 = lim_{c_2}(r,x_0)$ with $\exists_d\ x_1 = max(x_0, n) + d$.

In our example, the transformation of Variable x through Cycles 1 or 2 to prove the well-foundedness of Cycle 3 is unessential, because $p_{10} \Rightarrow x = 0$. In order to propagate the symbolic value y_0 through the inner cycles, the reduction lemmas of those are enhanced with the information that y is unchanged: $p_4 \wedge y = y_0 \leadsto p_7 \wedge y = y_0$ for Cycle 1 and $p_7 \wedge y = y_0 \leadsto p_{10} \wedge y = y_0 + 1$ for Cycle 2. Only then the result $y_1 = y_0 + 1$ is obtained, with which the well-foundedness of Cycle 3 is proved similarly to Cycle 1.

The example illustrates that nested cycles are better reduced top-down, because the relevance of symbolic values for outer cycles is known, which determines whether limits are needed when reduction lemmas of inner cycles are derived.

4.4 Reachability Analyses

Open nodes without continuation, i.e., where all fair transitions are disabled, and which do not satisfy the liveness property do not necessarily indicate that the liveness property is not provable: The $\leadsto$-precondition from which the construction starts, and consequently the generated successor predicates, need not be explicitly restricted to the reachable part of the state space; an approximation can also introduce unreachable states. If the state sets pertaining to all open nodes are excluded by way of appropriate invariants, the liveness property is proven. Although open nodes provide users with valuable information to find invariants, an automatic procedure for extracting invariants from predicates of open nodes under certain conditions frees the user from any interaction.

In the following we discuss the use of an iterative procedure called *invariant strengthening*, which has been given by Manna et alt. [BBM95]. Invariant strengthening is required, if a property of a program is always true within the reachable state space, but is not preserved for unreachable states within the full state space corresponding to the property. Instead of having to compute the reachable state space predicate it is sufficient to determine a weaker invariant from which the always-true property follows by implication, which is in general less complex.

The procedure, which requires computing weakest pre-conditions,

$$WP(\mathcal{A}, p_{s_i})[z] = \forall_{z'} \langle \mathcal{A} \rangle_z [z, z'/z'] \Rightarrow p_{s_i}[z'/z]$$

works as follows. In Iteration n, the invariant candidate φ_n, is tested for being satisfied in the initial set of states $I \vdash \varphi_n$ and being preserved by each program transition $\mathcal{A}_i$: $\mathcal{A}_i \wedge \varphi_n \vdash \varphi_n'$. For each program transition which violates the second condition the weakest precondition is computed in order to strengthen the invariant candidate:

$$\varphi_{n+1} := \varphi_n \wedge \bigwedge_{\mathcal{A}_i \in \{\mathcal{A} | \neg(\mathcal{A} \wedge \varphi_n \vdash \varphi_n')\}} WP(\mathcal{A}_i, \varphi_n)$$

The iteration terminates when all program transitions satisfy the strengthened invariant. If the global initial state predicate of the proram under investigation does not satisfy an invariant candidate, the whole sequence of invariant candidates does not belong to the set of always-true properties of the transition system.

Assuming the sets of states of open nodes to be characterized by a predicate B, we try to prove that B is in contradiction to reachability. Hence an invariant φ_n is needed with $\varphi_n \Rightarrow \neg B$. The procedure is started with the first invariant candidate $\varphi_0 := \neg B$. A straightforward implementation of the procedure as given above would result in an explosive growth of the predicates to be transformed. Therefore the predicate transformation is performed on separate subterms, which are only combined when necessary.

In addition to invariant strengthening, we have implemented another method [Gou95] for completing annotations which are pre- and post-conditions attached to individual program transitions specifiable in TLT. The completion procedure iteratively augments incomplete annotations with additional conditions, also using a *weakest pre-conditions* approach. The annotations, which can be regarded as local invariants, allow users to insert valuable design knowledge in order to simplify proofs. Correspondingly the generated *FTG* becomes smaller if transitions can be shown to be disabled by exploiting the annotations.

5 Conclusion

We discussed a new procedure for automatically generating liveness proofs for non-finite concurrent systems. The most important benefits are: (1) a user is not required to be expert in TLA and theorem proving, (2) a user is freed from tedious TLA reasoning, (3) user-supplied invariants accelerate the construction, but are not generally required, as the procedure includes invariant computation, (4) even if the underlying tactics are not able to decide a formula, case analyses allow meaningful continuation of the automatic construction, (5) generated proofs are expansible to explicit TLA proofs, (6) paths in the generated transition graph can be traced back to derive debugging information, (7) the generated state predicates are readable, since they use the same notions as the specification (cf. Table 2), (8) similarly to Figure 4, the graph is visualizable through appropriate softeware (e.g. [San95]), (9) a user can switch between interactive and automatic proof construction in any stage of the generation process.

For reasons of space, TLA-RISPE proof scripts are not shown in this paper. Nevertheless, in several non-trivial examples the fully-automatically generated scripts actually discharge the liveness conjectures. Without using the procedure, a student with basic knowledge in formal methods for distributed systems had investigated those examples before. The student needed several days for each of the examples; most of the elapsed time was spent on developing the TLA proofs on paper, before checking these interactively in TLA-RISPE. The synthesized proofs are not significantly longer than the interactively developed ones.

The language TLT allows a very wide range of systems to be specified. However, the decidability of predicate logic formulæ in proof obligations is a principle limitation, if fully automatic verification is asked for. In practice, interesting classes of problems beyond the *BDD*-level have to be identified which are handled by the automatic theorem proving functions of TLA-RISPE[Bus95]. What happens if these functions fail? Undecided subpredicates lead to superfluous nodes in the graph for which unnecessarily the liveness property or unreachability has to be shown. This overhead slows down the proof construction or even results in unmanageable expressions. In general, large internal formulæ are a bottleneck. Therefore and in favour of better readability, much effort went into tactics for normalization and minimization of state predicates. State predicates are the medium through which a user is supposed to communicate with the proof tools at a reasonable level which does not differ from the specification level: without performing TLA calculus reasoning, the user can enter invariants and annotations of program transitions in terms of state predicates, and reversely the state predicates assigned to the graph nodes provide valuable debugging information.

Future work will address improvements of the reachability analysis, which is still restricted to invariants which are found without inductive argument. The extraction of the TLA proof does not depend on how the transition graph has been generated. Therefore, the user interface could well be furnished with a feature for interactive manipulation of transition graphs.

References

[APP94] F. Andersen, K.D. Petersen, and J.S. Pettersson, 'A Graphical Tool for Proving UNITY Progress', in *Higher Order Logic Theorem Proving and its Applications, 7th International Workshop*, T. Melham, and J. Camilleri (Eds.), pp. 17–32, Springer, LNCS 859, 1994.

[BBM95] N. Bjørner, a. Browne and Z. Manna, 'Automatic Generation of Invariants and Intermediate Assertions', in *First International Conference on Principles and Practice of Constraint Programming*, U. Montanari (Ed.), LNCS, Cassis, France, September 1995.

[Bra92] J.C. Bradfield, 'Verifying Temporal Properties of Systems', Birkhäuser, *Progress in Theoretical Computer Science*, 1992.

[Bus92] H. Busch, 'Rule-Based Induction', in *FORMAL METHODS IN SYSTEM DESIGN - Special Issue on HOL'92*, Kluwer, Vol. 5, Issue 1 & 2, July/August 1994.

[Bus95] H. Busch, 'A Practical Method for Reasoning About Distributed Systems in a Theorem Prover', in *Higher Order Logic Theorem Proving and its Applications, 8th International Workshop*, E.T. Schubert, P.J. Windley, J. Alves-Foss (Eds.), pp. 106–121, Springer, LNCS 971, 1995.

[ChM88] K.M. Chandy and J. Misra, 'Parallel Program Design - A Foundation', Addison-Wesley, 1988.

[CWB94] J. R. Cuéllar, I. Wildgruber, and D. Barnard, 'Combining the Design of Industrial Systems with Effective Verification Techniques', in *Proc. of FME'94, Barcelona, Spain*, pp. 639–658, Springer LNCS 873, M. Naftalin, T. Denvir, and M. Betran (Eds.), October 1994.

[CuW96] J. R. Cuéllar and I. Wildgruber, 'The Dagstuhl Steam-Boiler Controller-Problem - The TLT Solution', Presented at Dagstuhl Seminar 9523, 1995, in *The Steam-Boiler Case Study*, Springer LNCS, J.-R. Abrial, E. Börger, and H. Langmaack (Eds.), 1996.

[EGL92] U. Engberg, P. Grønning, and L. Lamport, 'Mechanical Verification of Concurrent Systems with TLA', in *CAV'92, 4th International Workshop*, G.v. Bochmann and D.K. Probst (Eds.), pp. 44–55, Springer, LNCS 663, 1993.

[Gou95] G. Gouverneur, 'Korrekter Entwurf und Verifikation verteilter Systeme', Dissertation, University of Kaiserslautern, 1995.

[Lam94] L. Lamport, 'The Temporal Logic of Actions', *ACM Transactions on Programming Languages and Systems*, Vol.16, No.3, pp. 872–923, May 1994.

[Lam94a] L. Lamport, 'TLA in Pictures', in *DIMACS Workshop on Specification of Parallel Algorithms, Princeton, May 1994*, 1994.

[MaP94] Z. Manna and A. Pnueli, 'Verification of Parameterized Programs', in *Specification and Validation Methods*, E.Börger (Ed.), Oxford University Press, 1994.

[San95] G. Sander, 'VCG - Visualization of Compiler Graphs', User Documentation V.1.30, Saarbrücken University, 1995.

[WrL93] J. v. Wright and T. Långbacka, 'Using a Theorem Prover for Reasoning about Concurrent Algorithms', in *CAV'92, 4th International Workshop*, G.v. Bochmann and D.K. Probst (Eds.), Springer, LNCS 663, 1993.

Program Derivation Using the Refinement Calculator

Michael Butler[1] and Thomas Långbacka[2]

[1] Dept. of Electronics & Computer Science, University of Southampton
[2] Dept. of Computer Science, University of Helsinki

Abstract. The refinement calculus provides a theory for the stepwise refinement of programs and this theory has been formalised in HOL. TkWinHOL is a powerful graphical user interface (GUI) that can be used to drive the HOL window Library. In this paper, we describe a tool called the Refinement Calculator which combines TkWinHOL and the HOL Refinement Calculus theory, to provide support for formal program development. The tool improves the usability of the HOL Refinement Calculus theory considerably through its window-inference based GUI and by supporting a conventional programming syntax.

1 Introduction

The refinement calculus [2, 3, 14, 15] is a formalisation of the stepwise refinement method of program construction. The required behaviour of a program is specified as an abstract, possibly non-executable, program which is then refined by a series of correctness-preserving transformations into an efficient, executable program.

When using formalisms like the refinement calculus, the derivations are usually long and error-prone. Therefore the use of a proof assistant such as HOL to increase the level of trust in proofs is a natural step. Work on using HOL for constructing refinement calculus proofs has been carried out for number of years at Åbo Akademi University:

- Work by Back on the refinement calculus [2, 3].
- Work on formalising the refinement calculus theory in HOL [4, 21, 22]
- More recently work on building a software environment supporting program development using the refinement calculus formalisation in HOL.

Although much work has been done on formalising programming logics (e.g. [1, 6]), working with complex objects like programs using a purely textual interface to HOL is difficult, especially for non-HOL users. As well as having a sound proof engine, an important goal of this work on providing tool support for the refinement calculus was that it could be used by people who may be familiar with program refinement, but not so experienced in using HOL.

Usually a linear calculational style is used when deriving programs by hand in the refinement calculus. A specification is transformed in a series of steps to an executable program by applying transformation rules which preserve a refinement relation between the stages. Many of the steps involve sub-derivations on sub-components. It is precisely this style of reasoning that is supported by

Grundy's HOL window Library [10]. But the HOL window Library is difficult to work with directly especially since accessing sub-components of large programs is quite cumbersome.

TkWinHOL [13] is a GUI tool that can be used to drive the HOL window Library. Rather than having to provide complex parameters in order to access sub-terms when starting a sub-derivation, the user can focus on a sub-term using the mouse. It also supports the menu-driven application of transformation rules to terms.

In this paper, we describe a tool called the Refinement Calculator which combines TkWinHOL and the Refinement Calculus formalisation (refcalc for short), to provide support for formal program development. As well as providing all the features of TkWinHOL, the Refinement Calculator supports a conventional programming syntax by providing a parser and pretty-printer layer between the user and the syntax of the refcalc theory. This is especially important since it allows the user to work with program variables, as is usual in the refinement calculus, whereas program variables don't appear in the syntax of the refcalc theory. Another feature of the refcalc theory is that it is a *shallow embedding*, i.e., programming constructs are defined semantically rather than through the syntax. This means any existing theories about data-types can be readily used in the Refinement Calculator. For example, in Section 4, we show how an existing theory of arrays can be used to work with arrays in program refinement.

2 Background

In this section we will give a brief description of those parts of the refinement calculus and its formalisation in HOL that are relevant to this paper.

2.1 The Refinement Calculus

The refinement calculus is based on Dijkstra's weakest precondition semantics for programs [8]. Dijkstra's work is extended by introducing a refinement relation between programs. Another change to Dijkstra's original work is the (partial) relaxation of the required *healthiness conditions* of program statements, to allow the introduction of useful (yet unimplementable in practice) specification statements into the set of statements.

The programming notation used in this paper is basically Dijkstra's guarded commands language, extended with *assertions* and *nondeterministic assignment*. An assertion has the form $\{p\}$, where p is a predicate on the program state; occurrence of an assertion in a program allows us to assume that p holds at that point in the program. A nondeterministic assignment has the form $x := x' \bullet p$ and specifies that x is assigned some value x' satisfying predicate p.

The refinement relation is defined in terms of the weakest preconditions of the related programs. For program S and postcondition P, $wp(S,P)$ represents the weakest precondition under which S is guaranteed to terminate in a state satisfying P. Program S_0 is refined by S_1, denoted $S_0 \sqsubseteq S_1$, iff

$$\forall P \cdot wp(S_0, P) \Rightarrow wp(S_1, P)$$

which states that program S_1 must preserve the total correctness of program S_0. The refinement relation is a preorder. Thus programs can be developed in a *linear* fashion as in the following sequence

$$S_0 \sqsubseteq S_1 \sqsubseteq \ldots \sqsubseteq S_n$$

which establishes the refinement $S_0 \sqsubseteq S_n$, because of the transitivity of the relation.

An important property is that one can refine sub-components of programs without affecting the total correctness of the whole program. Formally this means that if we have proven that $T \sqsubseteq T'$ then we have in fact established the refinement $S[T] \sqsubseteq S[T']$ provided the program context $S[\ldots]$ is monotonic.

2.2 Formalising The Refinement Calculus in HOL

Program state When formalising a (state-based) programming logic in HOL an important decision one has to make is how to represent the state of a program. Since refcalc is a shallow embedding, one natural way to deal with states is to represent them as tuples. On the general level, states are defined using the polymorphic type. In individual programs, each program variable is represented as one component in the tuple. This means that program variables are anonymous.

This could be seen as a disadvantage of the approach, making the practical use of the theory unmanageable. The Refinement Calculator described in section 3, hides the state representation under a surface syntax allowing program variables to be used. In the translation, program variables are modelled using *projection functions*. Assume the state has type *num × bool*, then the variables indicate positions in the state tuple as follows:

let x = FST in let y = FST ∘ SND in...

The main advantage in using a shallow embedding is that one can reuse HOL data types in the state. So program variables ranging over, e.g., natural numbers can be treated by means of existing theorems and proof procedures without difficulty. In case new data types are defined, these can also be used without having to worry about incompatibility (see section 4 for an example of this). In a *deep embedding* data types have to be embedded into the theory every time new types are to be used.

Predicates, predicate transformers and refinement The semantics used is weakest precondition semantics where the meaning of programs are defined using predicates over the program state. Thus, predicates (*pred*) in refcalc are functions of type *state → bool*, where the state is represented as described above.

Program statements are *predicate transformers*, i.e., functions of type *pred → pred*. For predicate transformer S and predicate q, $S\,q$ corresponds to $wp(S,q)$ in Dijkstra's formalism. Assuming q is a predicate, s is a program state, f is a state function (*state → state*) and $S1$ and $S2$ are predicate transformers, we can define the following three predicate transformers:

$$skip\ q = q$$
$$assign\ f\ q = \lambda s \cdot q(f\ s)$$
$$seq\ S1\ S2\ q = S1(S2\ q)$$

representing *skip*, *variable assignment* and *sequential composition*. An assignment statement in the surface syntax of the Refinement Calculator having the form $x := e$, would be translated into the form $assign(\lambda x \cdot e)$.

The refinement relation is defined as follows (here *implies* is HOL implication lifted to the level of predicates):

$$S1 \sqsubseteq S2 \ = \ \forall q \cdot (S1\ q)\ implies\ (S2\ q).$$

2.3 The use of window inference as proof engine

The refinement calculus has characteristics that make window inference [16] and more specifically the implementation of window inference in HOL, the HOL window Library [10, 11] an attractive environment to work in when carrying out proofs. In this section we will very briefly describe the HOL window Library and discuss why it is a useful proof engine in the context of the refcalc theory and the Refinement Calculator. We will also very briefly describe the steps taken in [21] to adapt the refcalc theory for use with the Window Library.

Window inference In window inference, one transforms a term preserving a *preorder*. An expression p transformed to q preserving the relation R, gives a proof of $p\ R\ q$.

A transformation of a sub-term is a transformation of the whole term, if certain *context-dependent monotonicity* conditions hold. For example, in the expression $A \wedge B$, the sub-term A can be transformed under assumption B (assuming implication is the relation preserved) since

$$\frac{B \vdash A \Rightarrow A'}{\vdash (A \wedge B) \Rightarrow (A' \wedge B)}$$

The HOL implementation supports window reasoning under *implication*, *equality* and *reverse implication*. New relations can be added to the system by proving that the relation is *reflexive* and *transitive* (i.e., a preorder) and adding rules that govern how windows can be opened and closed.

Users work with so called *window stacks*. Windows consist of a *focus* (the term being transformed), the relation preserved and a *context* (assumptions, lemmas and conjectures). Windows also hold a *window theorem* which records the current proof state. Windows can be opened on sub-terms of the focus using the operators *rand*, *rator* and *body* to identify the sub-term one is interested in. The activity of proof is carried out using two main categories of operations. One can either change the scope of interest by opening or closing windows or one can transform the current focus using pre-proved theorems, conversions, rewrite rules etc[3].

---

[3] The HOL window Library also contains a hybrid command *AT* which can be used to perform a transformation of a sub-term as if a new window had been opened.

Why window inference? From the refinement point of view the HOL window Library is well suited as the basic inference mechanism since:

- The refinement relation is a preorder.
- Refinement steps are often refinements of sub-components of programs (i.e. refinement in context)
- In a program derivation one doesn't necessary know exactly in what direction the program development might go so being able to reason transformationally, as in window inference, is an advantage.

Thus, the window inference mechanism offers a suitable environment for structuring proofs within the refcalc theory.

The HOL window Library has other attractive features:

- Window inference encompasses both *forward* and *backward* reasoning.
- The user has full control over at which level of detail to reason.
- The transformations one performs to the current focus are usually simple (in HOL terms); typically they are simple rewrites.
- Contextual information that might prove useful in transformations is made available when windows are opened.

Furthermore, as Grundy phrased it in [9], "The window inference system suggests a graphical user interface"; in particular, it suggests the use of a mouse to select sub-terms and access contextual information, and the use of menus, buttons and dialogs to enter commands.

Refinement using window inference In order to use window inference in program development, one has to prove that the refinement relation is transitive and reflexive, and add rules for opening and closing of windows. The first part is trivial.

A program may be specified with a statement of the form $\{pre\}; v := v' \bullet post$. Derivation of a program satisfying this involves transforming the specification while preserving the refinement relation. Monotonicity of the program constructs allows us to focus on sub-programs as, for example, the following rules for sequential composition show:

$$\frac{\vdash S \sqsubseteq S'}{\vdash S;T \sqsubseteq S';T} R1 \qquad \frac{\vdash T \sqsubseteq T'}{\vdash S;T \sqsubseteq S;T'} R2$$

The refinement calculus provides a set of rules for transforming the current focus in order to introduce extra program structure to the focus, or to simplify the focus. The refinement rules supported by the refinement calculator are all derived from the HOL semantics of programs and refinement described above. Some of these rules are listed in Section 3.

Assertion statements are used to carry context information around. This context can be used directly when we focus on the right hand side of an assignment or the postcondition of a nondeterministic assignment:

$$\frac{p \vdash e1 = e2}{\vdash \{p\}; x := e1 \sqsubseteq \{p\}; x := e2} R3$$

$$\frac{p \vdash q1 \Leftarrow q2}{\vdash \{p\}; x := x' \bullet q1 \sqsubseteq \{p\}; x := x' \bullet q2} R4$$

Notice that, in the first case, the relation to be preserved on the expression is equality while, in the second case, the relation to be preserved on the postcondition is backwards implication.

Context information is propagated using theorems such as the following:

$\vdash$ **if** G **then** S **else** T **fi** $\sqsubseteq$ **if** G **then** $\{G\}; S$ **else** $\{\neg G\}; T$ **fi**

$\vdash x := e \sqsubseteq x := e; \{x = e\}$

3 The Refinement Calculator

The Refinement Calculator consists of:

- TkWinHOL[12, 13] – an extendable general purpose interface to the HOL window Library .
- Extensions to TkWinHOL dealing with refinement specific issues.

As mentioned above we have tried to make the base tool TkWinHOL general and extendable in nature. Unlike the work in e.g. [18, 19] we have not attempted to build a general purpose user interface to the HOL system rather we restricted ourselves to window inference. Also our approach has all the time been aiming at a specific extension, the Refinement Calculator.

In [18] Syme also discusses HCI aspects of user interface building as an influence on his work. Our approach is less structured though it is strongly influenced by the calculational style of proof and by window inference. We have tried to build a tool with a user interface that looks and feels like any "typical" modern graphic user interface. Our aim is to minimize the amount of typing a user has to do and allow the user to interact in a "select – operate" type of way.

3.1 TkWinHOL

Once TkWinHOL is started the user is presented with three windows. Two of these are of less importance, offering a simple editor and a session window for entering commands directly to HOL. The third window presents a structured view of the current state of the active window stack. Thus different types of information (i.e., focus, context, window theorem) about the current stack is displayed in different regions of the window.

Working with TkWinHOL Accessing sub-terms using the mouse instead of textual path information (when opening windows) is an important feature in a system such as this. In fact the selection of parts of the focus stretches beyond this. One can select a segment of the focus that wouldn't normally be accessible without reordering the associations in the focus, and have the interface perform the required reordering automatically.

The commands of the HOL window Library are bound to buttons and menus. For many commands the information needed is typically made available using the mouse to select relevant information already present on the screen. If that is not enough, information is entered through dialogs. The dialogs in TkWinHOL are implemented in such a way that they don't have to be destroyed (e.g. by clicking the OK button) before data in other windows can be selected. Thus, it is possible to use information already present on the screen to fill some of the fields in any given dialog.

Customisability Emphasis has been placed on making TkWinHOL easily customisable (for the purpose of building the Refinement Calculator). When using TkWinHOL together with a specific theory (as in the Refinement Calculator) one usually wants to use a higher level notation for the interaction. This can be achieved through two steps

- By constructing a translator from the higher level notation to the corresponding HOL representation.
- By constructing a pretty-printer doing the opposite. Actually the pretty-printer has to do more since all output from the HOL window Library commands have to be augmented with some instructions to the interface on which sub-terms of the focus are to be selectable etc. Because of this the base TkWinHOL system has a default pretty-printer although there is no translator.

Another way in which one wants to customise TkWinHOL is by adding theory specific menu choices bound to specialised commands for transformations etc. To support this there is a very general prompting procedure (prompting is done through dialogues) built into TkWinHOL. The idea is that through this prompting procedure one can easily build the glue by which to tie an add-on menu choice to a command built on the HOL level.

For parameterless commands such glueing is not necessary (the menu choice is bound more or less directly to the command) but for any command needing parameters, it is convenient to have some method to specify how this parameter passing should take place, and not force the user to patch the user interface source code directly.

The way this is accomplished is through binding add-on menu choices to a call to the above mentioned prompting procedure. As parameters to this procedure one can describe how many fields the prompting dialogue should have and possibly what individual fields have as default value. To construct the actual commands to be sent to HOL one provides so called command specifications as additional parameters to this procedure. These are simply text strings where one can refer to the contents of the different fields (as well as the current screen selection) through symbolic names. Once the dialogue is ok'd by the user, a substitution takes place. One can provide several command specifications to the procedure. Which one is used is determined by where on the screen the selection sits (i.e., if information from the context of the focus is selected do one thing, if a part of the focus is selected do something else etc).

3.2 The Refinement Calculator Extension

The Refinement Calculator supports a customised program notation with the following syntax:

$Prog ::=$ **program** $Name$ **var** $v : Type \cdot Com$
$Com ::= Com; Com \mid \{BTerm\} \mid v := Term \mid v := v' \bullet BTerm$
$\mid$ **if** $BTerm$ **then** Com **else** Com **fi** $\mid$ **do** $BTerm \to Com$ **od**
$\mid \mid [$ **var** $v : Type \cdot Com]\mid$

Here, variable name v may represent a list of variable names, *Type* represents any HOL type, *Term* represents any HOL term, and *BTerm* represents any boolean-valued HOL term. Thus the standard HOL notation is embedded in the program-specific notation. The last construct, $\mid [$ **var** $v : Type \cdot Com]\mid$, represents a block with local variable v.

Cond Introduction:
$\vdash S \sqsubseteq$ **if** G **then** S **else** S

Block Introduction:
$\vdash v := v' \bullet post \sqsubseteq \mid [$ **var** $x : T \cdot v, x := v', x' \bullet post]\mid$

Loop Introduction:
$\vdash pre \Rightarrow inv$
$\vdash (inv \wedge G \wedge E = e) \ll Body \gg (inv \wedge 0 \leq E < e)$
$\vdash inv \wedge G \Rightarrow post[v' := v]$
v not free in *post*
$\overline{\vdash \{pre\}; v := v' \bullet post \sqsubseteq \textbf{do } G \to Body \textbf{ od}}$

Assignment Introduction:
$\dfrac{\vdash pre \Rightarrow post[v' := E]}{\vdash \{pre\}; v := v' \bullet post \sqsubseteq v := E}$

Leading Assignment Introduction:
$\dfrac{x \text{ not free in } post}{\vdash v := v' \bullet post \sqsubseteq x := E; v := v' \bullet post}$

Fig. 1. Refinement Transformations

The Refinement Calculator provides a menu of transformations for refining programs; some of these are represented as rules in Figure 1. Each of the rules requires the user to provide some arguments before the transformation is applied; for example, the *Cond-Introduction* rule requires the guard G to be supplied. Some of the rules have side-conditions (about free variables) and assumptions. The assumptions are added to the current window as conjectures that can be established later. A transformation will fail if its side-conditions are not satisfied.

Note that a term of the form $pre \ll prog \gg post$, as used in the *Loop-Introduction* rule of Figure 1, describes a total-correctness assertion stating that *prog* is guaranteed to establish *post* when executed in initial state satisfying *pre*.

The Refinement Calculator provides transformations for converting assertions of this form into boolean terms. The calculator also provides a transformation for propagating context information as assertion statements and for applying rewrite theorems to the focus. All the transformations supported by the calculator are provided by glueing menus to the appropriate HOL command as outlined previously.

The role of the rules of Figure 1 will be described more clearly in the next section where they are applied to some examples. The full program syntax and list of refinement rules supported by the Refinement Calculator may be found in [5].

4 Some Example Derivations

We present three example derivations that may be carried out using the refinement calculator: a program that finds the maximum of two numbers, a program that finds the maximum value in an array of numbers, and a program that sorts an array of numbers. In the following, we write a window as

$R * f$

where R is the relation to be preserved and f is the term to be transformed. Also, we write context assumptions as ! p and conjectures as ? p.

4.1 Finding the Maximum of Two Numbers

This first example illustrates the use of a structure introduction rule, refinement of program sub-components, and the use of context information in refinement.

Assume the operator *max*, is defined by:

$m \ max \ n = if \ (m \geq n) \ then \ m \ else \ n$

The program we wish to derive is then specified as follows:

program *maximum* **var** $m, n, x : num. \quad x := m \ max \ n$

The derivation commences by opening a window on this specification, with $\sqsubseteq$ as the relation to be preserved:

$\sqsubseteq *$ **program** *maximum* **var** $m, n, x : num. \quad x := m \ max \ n$

Let us assume that the *max* operator is not available in our target programming language so that we have to implement this specification by comparing m and n. We proceed by introducing an if-statement: using the Refinement Calculator, we focus on the assignment statement and then apply the *Cond-Introduction* transformation with $m \geq n$ (i.e., the guard) as an argument. This results in the focus:

$\sqsubseteq *$ **if** $m \geq n$ **then** $x := m \ max \ n$ **else** $x := m \ max \ n$ **fi**

When refining the first branch of this statement, we can assume that $m \geq n$ holds. To make this assumption available in the branches, the guard and it's negation are propagated as assertions:

$\sqsubseteq *$ **if** $m \geq n$ **then** $\{m \geq n\}; \ x := m \ max \ n$ **else** $\{\neg(m \geq n)\}; \ x := m \ max \ n$ **fi**

Focusing on the first branch yields:

$\sqsubseteq * \{m \geq n\}; x := m \; max \; n$

Now, focusing on the right hand side of the assignment yields the window (see rule R3, page 6):

! $m \geq n$
$= * m \; max \; n$

Notice that the relation to be preserved on this new focus is equality. Using the definition of *max*, the focus may be rewritten to

$= * if \; (m \geq n) \; then \; m \; else \; n$

Selecting the assumption $m \geq n$, and rewriting the focus results in

$= * m$

Closing the window yields:

$\sqsubseteq * \{m \geq n\}; x := m$

Since the assertion is no longer required, it can be dropped resulting in:

$\sqsubseteq * x := m$

Closing this window yields:

$\sqsubseteq *$ **if** $m \geq n$ **then** $x := m$ **else** $\{\neg(m \geq n)\}; x := m \; max \; n$ **fi**

The second branch[4] may be refined in a similar manner, so that we end up with the program

$\sqsubseteq *$ **if** $m \geq n$ **then** $x := m$ **else** $x := n$ **fi**

Thus, we have constructed the theorem:

$\vdash x := m \; max \; n \sqsubseteq$ **if** $m \geq n$ **then** $x := m$ **else** $x := n$ **fi**

4.2 Finding the Maximum of an Array of Numbers

Let s represent a set of numbers, i.e., $s : num \rightarrow bool$. The maximum element of s is selected by *Max*:

$Max \; s = (\varepsilon \; m \cdot s \; m \wedge (\forall n \cdot s \; n \Rightarrow m \geq n))$

Arrays may be modelled by defining a new HOL type; assume $(\alpha)array$ is the type of arrays that are polymorphic on α, and that the following functions on arrays have been defined:

asize : $(\alpha)array \rightarrow num$
lookup : $(\alpha)array \rightarrow num \rightarrow \alpha$
swap : $(\alpha)array \rightarrow num \rightarrow num \rightarrow (\alpha)array$

An array a is indexed from 0 to $(asize \; a) - 1$. The ith element of a is given by $(lookup \; a \; i)$, while $(swap \; a \; i \; j)$ represents the array a with the values at positions i and j swapped around.

For array a, $(elems \; a \; n)$ represents the set of values of a in the range $0..n-1$:

---

[4] Note that the two branches could have been refined in either order.

$elems\ a\ n = (\lambda x \cdot \exists i \cdot i < n \wedge (lookup\ a\ i) = x)$

A program that finds the maximum value of a non-empty array is specified as follows:

program *Maximum* **var** $a : (num)array;\ m : num \cdot$
$\{(asize\ a) > 0\};\ m := Max\ (elems\ a\ (asize\ a))$

We will implement this using a loop that traverses the array starting at position zero. The array will be traversed using an indexing variable $k : num$. In order to apply the *Loop-Introduction* rule, we require a loop guard (G), a loop body (*Body*), a loop invariant (*inv*), and a loop variant (E), as follows:

Guard : $k < (asize\ a)$
Body : $m, k := m', k' \bullet m' = Max(elems\ a\ k') \wedge k' = k+1$
Invariant : $k \leq (asize\ a) \wedge m = Max\ (elems\ a\ k)$
Variant : $(asize\ a) - k$

The invariant says that m is the maximum value in the array slice $a[0..k-1]$. The body increments k and re-establishes the invariant.

Before introducing the loop, we introduce k as a local variable by applying *Block Introduction* to the specification:

$\sqsubseteq * |[\ \textbf{var}\ k : num \cdot m, k := m', k' \bullet m' = Max(elems\ a\ (asize\ a))\ ]|$

Both m and k will have to be initialised to establish the invariant, so we add an assignment before the body of this block (using *Leading Assignment Introduction*):

$\sqsubseteq * |[\ \textbf{var}\ k : num \cdot m, k := (lookup\ a\ 0), 1;$
$\qquad m, k := m', k' \bullet m' = Max(elems\ a\ (asize\ a))\ ]|$

Propagating assertions through appropriately and focusing on the second statement of the block yields:

$\sqsubseteq * \{(asize\ a) > 0 \wedge m = (lookup\ a\ 0) \wedge k = 1\};$
$\qquad m, k := m', k' \bullet m' = Max(elems\ a\ (asize\ a))$

Now applying the *Loop-Introduction* rule with the guard, body, invariant, and variant described above as arguments, yields the focus

$\sqsubseteq *\ \textbf{do}\ k < (asize\ a) \rightarrow$
$\qquad k < (asize\ a) \wedge m = Max(elems\ a\ k);$
$\qquad m, k := m', k' \bullet m' = Max(elems\ a\ k') \wedge k' = k+1\ \textbf{od}$

This step also generates a number of conjectures. These state respectively that the invariant should hold initially (1), the body should preserve the invariant and decrease the variant (2), and the invariant and the negated guard should establish the original postcondition (3):

? $\forall a, m, k \cdot (asize\ a) > 0 \wedge m = (lookup\ a\ 0) \wedge k = 1$
$\qquad \Rightarrow k \leq (asize\ a) \wedge m = Max(elems\ a\ k)$ (1)

? $\forall k, a, m, e \cdot$
$\qquad k < (asize\ a) \wedge m = Max(elems\ a\ k) \wedge ((asize\ a) - k) = e$
$\qquad \ll m, k := m', k' \bullet m' = Max(elems\ a\ k') \wedge k' = k+1 \gg$ (2)
$\qquad k \leq (asize\ a) \wedge m = Max(elems\ a\ k) \wedge 0 \leq ((asize\ a) - k) < e$

$$? \forall k, a, m \cdot (k < (asize\ a)) \land k \leq (asize\ a) \land m = Max(elems\ a\ k) \\ \Rightarrow m = Max(elems\ a\ (asize\ a)) \qquad (3)$$

At this stage in the derivation, we can either attempt to discharge the conjectures[5], or continue refining the body of the program. Conjectures such as above may be established by opening windows on the respective conjectures with backward implication as the relation to be preserved, and then transforming the focus to true. The second conjecture above can be simplified by applying a transformation which reduces a correctness assertion (of the form $pre \ll assignment \gg post$) to a boolean term. When proceeding with the loop body above, we focus on the sub-term $Max(elems\ a\ k')$ of the assignment statement yielding the window:

! $k < (asize\ a)$! $m = Max(elems\ a\ k)$! $k' = k+1$
$= * Max(elems\ a\ k')$

We can rewrite the focus using the assumption $k' = k+1$ to

$= * Max(elems\ a\ (k+1))$

Using a theorem about Max and the assumption $k < (asize\ a)$, we rewrite this to

$= * (Max\ (elems\ a\ k))\ max\ (lookup\ a\ k)$

and using the assumption $m = Max(elems\ a\ k)$, this is rewritten to

$= * m\ max\ (lookup\ a\ k)$

Closing the current window and focusing on the assignment statement, we now have:

$\sqsubseteq * m, k := m', k' \bullet m' = m\ max\ (lookup\ a\ k) \land k' = k+1$

This is easily transformed to

$\sqsubseteq * m := m\ max\ (lookup\ a\ k);$
$\quad k := k+1$

In the manner of the previous section, the first of these may be refined to

$\sqsubseteq *$ **if** $m < (lookup\ a\ k)$ **then** $m := (lookup\ a\ k)$ **else** *skip* **fi**

By closing several windows we have arrived at the program

$\sqsubseteq * \;|[\; \textbf{var}\ k : num \cdot m, k := (lookup\ a\ 0), 1;$
$\quad \textbf{do}\ k < (asize\ a) \rightarrow$
$\quad\quad \textbf{if}\ m < (lookup\ a\ k)\ \textbf{then}\ m := (lookup\ a\ k)\ \textbf{else}\ skip\ \textbf{fi};$
$\quad\quad k := k+1\ \textbf{od}\;]|$

---

[5] If we postpone establishing the conjectures, then what we have is a refinement theorem with the conjectures as assumptions.

4.3 Sorting an Array

Let *sorted* and *perm* be defined as follows:

$sorted\ (a:(num)array)\ (r:num \to bool) =$
$(\forall i \cdot (r\ i) \Rightarrow i < (asize\ a)) \land$
$(\forall ij \cdot (r\ i) \land (r\ j) \land i < j \land j < (asize\ a) \Rightarrow (lookup\ a\ i) \leq (lookup\ a\ j))$

$perm\ (a1:(num)array)\ (a2:(num)array) =$
$(asize\ a1) = (asize\ a2) \land$
$(\exists f \cdot (injective\ f) \land$
$(\forall i \cdot i < (asize\ a1) \Rightarrow (f\ i) < (asize\ a1) \land$
$(lookup\ a1\ i) = (lookup\ a2\ (f\ i))))$

The term $(sorted\ a\ (\lambda i \cdot p))$ states that that projection of array a whose indices satisfy p is sorted, e.g., $(sorted\ a\ (\lambda i \cdot 5 \leq i \leq 10))$ says that the array slice 5..10 of a is sorted. The term $(perm\ a1\ a2)$ specifies that array $a2$ is a permutation of array $a1$. The sorting program is then specified as:

program *Sort* **var** $a:(num)array \cdot$
$a := a' \bullet (sorted\ a'\ (\lambda i \cdot i < (asize\ a))) \land (perm\ a\ a')$

We shall implement this using an insertion sort algorithm which requires a pair of nested loops. The body of the outer loop will ascend the array one step at a time ensuring that the array slice $0..k-1$, where k is the current position, is sorted. This suggests the following guard, invariant, and variant for the outer loop:

Guard : $k < (asize\ a)$
Invariant : $k \leq (asize\ a) \land (sorted\ a\ (\lambda i \cdot i < k)) \land (perm\ a0\ a)$
Variant : $(asize\ a) - k$

Before transforming the specification using loop introduction, k is introduced as a local variable, and $a0$ (the initial value of a) is introduced using an assertion:

$\sqsubseteq\ *\ |[\ \textbf{var}\ k:num \cdot k := 0;$
$\{(k=0) \land (a=a0)\};$
$a, k := a', k' \bullet (sorted\ a'\ (\lambda i \cdot i < (asize\ a0))) \land (perm\ a0\ a')\ ]|$

Now, using
$\{k < (asize\ a) \land (sorted\ a\ (\lambda i \cdot i < k)) \land (perm\ a0\ a)\};$
$a, k := a', k' \bullet (sorted\ a'\ (\lambda i \cdot i < k')) \land (perm\ a0\ a') \land k' = k+1$

as the body of the outer loop, application of loop introduction yields:

$\sqsubseteq\ *\ \textbf{do}\ k < (asize\ a) \to$
$\{k < (asize\ a) \land (sorted\ a\ (\lambda i \cdot i < k)) \land (perm\ a0\ a)\};$
$a, k := a', k' \bullet (sorted\ a'\ (\lambda i \cdot i < k')) \land (perm\ a0\ a') \land k' = k+1\ \textbf{od}$

This step generates some conjectures similar to those shown in the previous example derivation.

The body of this loop is easily transformed to:

$\sqsubseteq\ *\ \{k < (asize\ a) \land (sorted\ a\ (\lambda i \cdot i < k)) \land (perm\ a0\ a)\};$
$a, k := a', k' \bullet (sorted\ a'\ (\lambda i \cdot i < k+1)) \land (perm\ a0\ a');$
$k := k+1$

Thus, under the assumption that the array is sorted between 0 and $k-1$, this body must establish that the array is sorted between 0 and k. It will do this by shuffling the element of a at position k down the array to the appropriate position. The following arguments will be used to perform the introduction of the inner loop:

Guard: $\quad l > 0 \wedge (lookup\ a\ l) < (lookup\ a\ (l-1))$
Body: $\quad a := (swap\ l\ (l-1)\ a);\ l := l-1$
Invariant: $l \leq k \wedge (sorted\ a\ (\lambda i \cdot i < l \vee (i > l \wedge i \leq k))) \wedge$
$\qquad l < k \Rightarrow (lookup\ a\ l) < (lookup\ a\ (l+1)) \wedge$
$\qquad (perm\ a0\ a)$
Variant: $\quad l$

This invariant states that the array slice $0..k$, with the lth position excluded, is sorted, and that the value at the lth position is less than the value at the $(l+1)^{th}$ position. Loop introduction yields

$\sqsubseteq * \textbf{do}\ l > 0 \wedge (lookup\ a\ l) < (lookup\ a\ (l-1)) \rightarrow$
$\qquad a := (swap\ l\ (l-1)\ a);\ l := l-1\ \textbf{od}$

Closing several windows gives us the implementation:

$\sqsubseteq * |[\ \textbf{var}\ k : num \cdot k := 0;$
$\qquad \textbf{do}\ k < (asize\ a) \rightarrow$
$\qquad\quad |[\ \textbf{var}\ l : num \cdot l := k;$
$\qquad\quad\ \textbf{do}\ l > 0 \wedge (lookup\ a\ l) < (lookup\ a\ (l-1)) \rightarrow$
$\qquad\qquad a := (swap\ l\ (l-1)\ a);\ l := l-1\ \textbf{od}\ ]|\ \textbf{od};$
$\qquad k := k+1\ ]|$

5 Conclusions

We have described the Refinement Calculator, a tool for the derivation of provably correct programs. The Refinement Calculator makes it feasible to do program refinement using HOL by providing a powerful GUI and a high level surface syntax on top of a Refinement Calculus HOL formalisation and the HOL window Library. The window inference mechanism supports the transformational style of reasoning used in program refinement. The ability to perform sub-derivations by simply focusing using the mouse makes the tool scalable to larger programs.

An important feature of the system is that it effectively hides the underlying representation of the program state, thus minimizing the possible drawbacks of using a shallow embedding of the formalism. That way one can also make use of the advantages of shallow embeddings by using existing theories of data types. This was demonstrated in the examples provided in section 4.

Independently of the tool described here, a refinement tool called PRT has been developed by a group at the University of Queensland [7]. PRT is built on top of the Ergo theorem prover [20] which also supports the window inference style of reasoning. This tool is quite similiar to the Refinement Calculator though an important difference is that PRT uses a purpose-built logic in which commands, predicates, program variables, and logic variables are treated as separate syntactic classes. In contrast, the Refinement Calculator uses only a conservative extension of the HOL logic giving us a higher degree of confidence in its soundness. However, having a closer match between the programming syntax and the underlying logic, as PRT has, may also have advantages, and the comparison between the approaches deserves a more thorough evaluation.

Much work can still be done on improving the Refinement Calculator. The underlying HOL theory still needs more proof support. There is also a need to extend the theory (and consequently the high level language) to support data refinement and action systems (to deal with reactive programs). Furthermore, the interface itself needs to be improved upon to deal, for example, with theorem retrieval, re-running proofs, etc.

Acknowledgments

Jim Grundy, Rymvidas Rukšėnas, and Jockum von Wright were also directly involved in the development of the Refinement Calculator.

References

1. F. Andersen. *A Theorem Prover for UNITY in Higher Order Logic*. PhD thesis, Technical University of Denmark, Lyngby, 1992.
2. R. Back. *Correctness Preserving Program Refinements: Proof Theory and Applications*, volume 131 of *Mathematical Center Tracts*. Mathematical Centre, Amsterdam, 1980.
3. R. Back. A calculus of refinements for program derivations. *Acta Informatica*, 25:593–624, 1988.
4. R. Back and J. von Wright. Refinement concepts formalized in higher order logic. *Formal Aspects of Computing*, 2:247–272, 1990.
5. M. Butler, T. Långbacka, R. Rukšėnas, and J. von Wright. Refinement Calculator tutorial and manual. Draft – available upon request.
6. A. Camillieri. Mechanizing CSP trace theory in Higher Order Logic. *IEEE Transactions on Software Engineering*, 16(9):993–1004, 1990.
7. D. Carrington, I. Hayes, R. Nickson, G. Watson, and J. Welsh. A tool for developing correct programs by refinement. For presentation at *7th BCS–FACS Refinement Workshop*, July 1996.
8. E. W. Dijkstra. *A Discipline of Programming*. Prentice–Hall International, 1976.
9. J. Grundy. Window inference in the HOL system. In Myla Archer, Jeffrey J. Joyce, Karl N. Levitt, and Phillip J. Windley, editors, *Proceedings of the International Tutorial and Workshop on the HOL Theorem Proving System and its Applications*, pages 177–189, University of California at Davis, August 1991. ACM-SIGDA, IEEE Computer Society Press.
10. J. Grundy. A window inference tool for refinement. In Jones et al, editor, *Proc. 5th Refinement Workshop*, London, Jan. 1992. Springer–Verlag.
11. J. Grundy. HOL90 window library manual. 1994.
12. T. Långbacka. TkWinHOL users guide. Draft – available upon request.
13. T. Långbacka, R. Rukšėnas, and J. von Wright. TkWinHOL: A tool for doing window inference in HOL. In Schubert et al. [17], pages 245–260.

14. C.C. Morgan. *Programming from Specifications (2nd Edition)*. Prentice–Hall, 1994.
15. J.M. Morris. A theoretical basis for stepwise refinement and the programming calculus. *Sci. Comp. Prog.*, 9(3):298–306, 1987.
16. P.J. Robinson and J. Staples. Formalising the hierarchical structure of practical mathematical reasoning. *Journal of Logic and Computation*, 3(1):47–61, February 1993.
17. E. Thomas Schubert, Phillip J. Windley, and James Alves-Foss, editors. *Higher Order Logic Theorem Proving and Its Applications: Proceedings of the 8th International Workshop*, volume 971 of *Lecture Notes in Computer Science*, Aspen Grove, Utah, September 1995. Springer-Verlag.
18. D. Syme. A new interface for HOL – ideas, issues and implementation. In Schubert et al. [17], pages 324–339.
19. L. Théry. A Proof Development System for the HOL Theorem Prover. In Jeffrey J. Joyce and Carl-Johan H. Seger, editors, *Higher Order Logic Theorem Proving and Its Applications – 6th International Workshop, HUG '93 Vancouver, B. C., Canada, August 1993*, volume 780 of *Lecture Notes in Computer Science*, pages 115–128. Springer Verlag, 1993.
20. M. Utting and K. Whitwell. Ergo user manual. Technical Report 93-19, Software Verification Research Centre, University of Queensland, 1994.
21. J. von Wright. Program refinement by theorem prover. In *BCS FACS Sixth Refinement Workshop – Theory and Practise of Formal Software Development. 5th – 7th January, City University, London, UK.*, 1994.
22. J. von Wright, J. Hekanaho, P. Luostarinen, and T. Långbacka. Mechanising some advanced refinement concepts. *Formal Methods in Systems Design*, 3:49–81, 1993.

A Proof Tool for Reasoning About Functional Programs

Graham Collins

Department of Computing Science, University of Glasgow, Glasgow,
Scotland, G12 8QQ. email: grmc@dcs.gla.ac.uk

Abstract. This paper describes a system to support reasoning about lazy functional programs. The system is based on a combination of a deep embedding of the language in HOL with a set of proof tools to raise the level of interaction with the theorem prover. This approach allows meta-theoretic reasoning about the semantics and reasoning about undefined programs while still supporting practical reasoning about programs in the language.

1 Introduction

It is often claimed that functional programming languages, and in particular pure functional languages, are suitable for formal reasoning. This claim is supported by the fact that many people in the functional programming community do reason about their programs in a formal or semi-formal way. Depending on the nature of the problem, different styles of reasoning, such as equational reasoning, induction and co-induction, are used.

This paper discusses some of the technical issues involved in constructing a proof tool, using HOL [8], for reasoning about a small, lazy functional language. To be usable the system must support a variety of proof styles and work at a high enough level to make the proof process practical. The aim of the tools discussed here is to raise the interaction of the user with the proof tools to a similar logical level as would be typical of a proof on paper.

A deep embedding [2] of the syntax is used and the semantics of the language are defined in an operational style. This combination allows meta-theoretic proof about language constructs. This is important in the development of proof tools. It is common for functions in a lazy functional programming language to consume and generate infinite data structures and the approach used to embed the language allows such functions and data structures to be introduced directly without the need to encode infinite data structures as functions. Equivalence of programs is defined as applicative bisimulation and so in addition to supporting proofs by equational reasoning and induction the system will support proofs by co-induction. Some aspects of this are discussed elsewhere [4].

The resulting hierarchy of theories provides the basis for the rest of the system. The theories alone are not useful for reasoning about actual programs; proof tools need to be developed to raise the level of interaction, and significant amounts of the proof process must be automated.

The next section gives a brief overview of applicative bisimulation and co-induction and gives a simple example to illustrate the type of problem we wish to solve. Sections 3 and 4 discusses the embedding of the syntax and semantics of the language and the definition of the equivalence relation. The next three sections discuss how to raise the level of interaction with the theorem prover to the desired level.

2 Applicative bisimulation and co-induction

This section gives a brief introduction to the ideas involved in defining the equivalence of programs by a bisimulation relation that is defined co-inductively [1, 6, 7]. The formalisation of co-induction and applicative bisimulation used is in terms of relations represented by functions and is given in section 4. An informal introduction using set notation is given below.

2.1 Co-induction

The definition depends on two concepts: monotonic functions and F-dense sets. A function F is monotonic if

$$\forall X\ Y.\ (X \subseteq Y) \supset (F(X) \subseteq F(Y))$$

and a set X is F-dense if for the function F

$$X \subseteq F(X)$$

The greatest fixpoint of F (gfp F) is defined to be the union of all F-dense sets. For any monotonic function F this can be proved to be the largest F-dense set and a fixpoint.

The principle of co-induction is then:

$$X \subseteq F(X) \supset X \subseteq \text{gfp } F$$

for any X. For the definition of equivalence of programs we find some monotonic function $F_{==}$ capturing the meaning of the equivalence and define $==$ to be gfp $F_{==}$. To prove that $x == y$ by co-induction we need to find some relation X such that $(x, y) \in X$ and show that X is $F_{==}$-dense.

A second principle, sometimes referred to as strong co-induction [6], can be derived from co-induction:

$$X \subseteq F(X \cup \text{gfp } F) \supset X \subseteq \text{gfp } F$$

This variation can simplify the choice of the relation X in a proof by co-induction.

2.2 A labelled transition system

In order to define applicative bisimulation co-inductively we must first formulate the function $F_{==}$ which captures the meaning of applicative bisimulation. The definition of this function is based on a labelled transition system for the language.

The labelled transition system captures the idea of observable properties of programs. For a function, you can observe what the function is applied to and then observe the behaviour of the resulting program, and for any other type you can observe some facts about the value of the program if it has one. For example the labelled transitions for a program of list type with a cons cell as the outermost constructor are

$$\text{cons } a \ b \xrightarrow{\text{Hd}} a$$
$$\text{cons } a \ b \xrightarrow{\text{Tl}} b$$

It is possible to carry out a transition to either the head or tail of the list. These transitions have labels Hd and Tl indicating that we can make observations of either the head or tail of the list.

2.3 Applicative bisimulation

Informally, two programs e_1 and e_2 are bisimilar if they can make the same observable transitions to terms that are also bisimilar. More formally, bisimulation, $==$, is defined as the fixed point of a function $F_{==}$ that is defined to have the property

$$\forall S \ a \ b. \ (a,b) \in (F_{==} \ S) = \\ (\forall a'. \forall \alpha. \ a \xrightarrow{\alpha} a' \supset (\exists b'. \ b \xrightarrow{\alpha} b' \wedge (a',b') \in S)) \wedge \\ (\forall b'. \forall \alpha. \ b \xrightarrow{\alpha} b' \supset (\exists a'. \ a \xrightarrow{\alpha} a' \wedge (a',b') \in S))$$

The relation $==$ is defined to be gfp $F_{==}$. It can be proved to be an equivalence relation and a congruence. These results are sufficient to develop an equational reasoning system for the language within HOL.

Because the relation is co-inductively defined it also allows the possibility of co-inductive proof.

2.4 A simple example

This example illustrates the proof of a property of two functions by co-induction. The details of the definition and proof in HOL are not given in this section but are described later. The proof is intended to illustrate the level of interaction that the proof tools described later allow within the theorem prover.

Two functions map and iterate can be defined to have the following equational properties.

$$\text{map } f \text{ nil } == \text{ nil}$$
$$\text{map } f \text{ (cons } x \text{ } xs) == \text{ cons } (f \text{ } x) \text{ (map } f \text{ } xs)$$
$$\text{iterate } f \text{ } x == \text{ cons } x \text{ (iterate } f \text{ } (f \text{ } x))$$

The theorem we want to prove is that, for all correctly typed programs, two infinite lists generated in different ways are equivalent.

$$\vdash \forall f \text{ } x. \text{ iterate } f \text{ } (f \text{ } x) == \text{ map } f \text{ (iterate } f \text{ } x)$$

This is true because both sides are equal to the list

$$[f \text{ } x, \text{ } f \text{ } (f \text{ } x), \text{ } f \text{ } (f \text{ } (f \text{ } x)), \text{ ...}]$$

The proof is by strong co-induction using the relation S with definition

$$S = \{(a, b) \mid \exists f \text{ } x.(a = \text{ iterate } f \text{ } (f \text{ } x)) \wedge (b = \text{ map } f \text{ (iterate } f \text{ } x))\}$$

It is easy to prove that

$$(\text{iterate } f \text{ } (f \text{ } x), \text{map } f \text{ (iterate } f \text{ } x)) \in S$$

It remains to prove that S is included in $==$. By strong co-induction we need only show that S is included in $F_{==}(S \cup ==)$. That is for any $(a, b) \in S$

1. $\forall a' \text{ } \alpha. \text{ } a \xrightarrow{\alpha} a' \supset (\exists b'. \text{ } b \xrightarrow{\alpha} b' \wedge ((a', b') \in S \vee a' == b'))$
2. $\forall b' \text{ } \alpha. \text{ } b \xrightarrow{\alpha} b' \supset (\exists a'. \text{ } a \xrightarrow{\alpha} a' \wedge ((a', b') \in S \vee a' == b'))$

The proofs of (1) and (2) are similar and we look only at (1). Since a and b are related by S we know something of their form. For some f and x,

$$a = \text{ iterate } f \text{ } (f \text{ } x)$$
$$b = \text{ map } f \text{ (iterate } f \text{ } x)$$

so taking arbitrary a' and α we need to prove that

$$\text{iterate } f \text{ } (f \text{ } x) \xrightarrow{\alpha} a' \supset$$
$$(\exists b'. \text{ map } f \text{ (iterate } f \text{ } x) \xrightarrow{\alpha} b' \wedge ((a', b') \in S \vee a' == b'))$$

A useful property of the labelled transition system is that we can evaluate a and b without affecting the transitions they can make. After evaluating we only need to show:

$$\text{cons } (f \text{ } x) \text{ (iterate } f \text{ } (f(f \text{ } x))) \xrightarrow{\alpha} a' \supset$$
$$(\exists b'. \text{ cons } (f \text{ } x) \text{ (map } f \text{ (iterate } f \text{ } (f \text{ } x))) \xrightarrow{\alpha} b' \wedge$$
$$((a', b') \in S \vee a' == b'))$$

There are only two possible transitions for a cons cell so we can assume that either $a' = f \text{ } x$ and $\alpha = \text{Hd}$, or $a' = \text{iterate } f \text{ } (f \text{ } (f \text{ } x))$ and $\alpha = \text{Tl}$.

The first case is solved by letting $b' = (f \text{ } x)$, since $(f \text{ } x) == (f \text{ } x)$ by reflexivity. The remaining case is solved by letting $b' = (\text{map } f \text{ } ((\text{iterate } f) \text{ } (f \text{ } x)))$. The result follows since the values for a' and b' are related by S.

This proof is typical of examples where the programs are generating lists. In this case we do not have to reason about undefined lists. There are added complications when this simplification cannot be made. These are discussed in a later example.

3 An embedding of call by name PCF plus streams

The work described here is a based on a deep embedding of both the expressions and types of a simple functional programming language. The language discussed here is PCF plus streams. The formulation of the syntax and semantics and the definition of program equality is based on work by Andrew Gordon that provides a rigorous development, on paper, of a theory for a number of languages [6].

This paper does not provide all the details of the system developed. It concentrates on how a system can be structured and proof tools developed to allow proofs to be completed at the same logical level as the example in the previous section.

The syntax of PCF plus streams is given in figure 1. The important features include function abstraction, recursive functions, and lists. The case analysis function for lists takes a list on which to perform the case analysis, the list to be returned if the first list is empty, and a function that can be applied to the head and tail of the list if it is not empty. The syntax given in figure 1 is not exactly the syntax that is embedded in HOL where a slightly more abstract syntax is used. In particular, function application cannot be represented by juxtaposition as this is used for function application in the HOL logic. For the sake of clarity the concrete syntax shown here is used throughout the rest of this paper.

ty ::= Num exp ::= n (Natural number)
     | Bool | true
     | List ty | false
     | $ty_1 \to ty_2$ | nil ty (Empty list of type t)
                             | cons exp_1 exp_2
                             | x   (Identifier)
                             | succ exp
                             | pred exp
                             | iszero exp
                             | if exp_1 then exp_2 else exp_3
                             | lambda x ty exp
                             | exp_1 exp_2 (function application)
                             | rec x ty exp
                             | scase exp_1 exp_2 exp_3
                                    (case analysis for lists)

Fig. 1. The syntax of PCF plus streams

The static semantics are formalised by an inductively defined relation [11]

$$\text{Type} : (string, ty) fmap \to exp \to ty \to bool$$

where *fmap* is the type of finite maps [5]. An example of a defining rule for this relation is the rule for function application:

$$\frac{\mathsf{Type}\ C\ e_1\ (t_1 \to t_2) \quad \mathsf{Type}\ C\ e_2\ t_1}{\mathsf{Type}\ C\ (e_1\ e_2)\ t_2}$$

Typically we consider only well typed programs so it is useful to introduce a second relation, Prog, which holds of a type t and an expression e only if e has type t in the empty context.

The dynamic semantics are formalised with a relation between syntactic objects. This has similarities to other work on deep embeddings of ML in HOL [3, 10, 12, 13]. Here, a different style of operational semantics relating expressions to expressions instead of values and using substitution instead of environments is used. A substitution function, Sub, is defined in such a way that it deals correctly with variable capture. If we consider substituting only well-typed terms then variable capture can be ignored and simpler properties of the substitution function can be used in proofs.

A relation $\longrightarrow\ :\ exp \to exp \to bool$ between well typed expressions is introduced where $e_1 \longrightarrow e_2$ states that under the rules defining the relation e_1 can reduce to e_2. This reduction relation is a small step reduction. It may be possible to perform a series of reductions of an expression. This series may be infinite.

Two examples of the defining rules for this relation are the rules for function application:

$$\frac{e_1 \longrightarrow e_3}{(e_1\ e_2) \longrightarrow (e_3\ e_2)} \quad (1)$$

$$(\mathsf{lambda}\ y\ t\ e)\ e_1 \longrightarrow (\mathsf{Sub}\ e\ (y,\ e_1)) \quad (2)$$

A large step evaluation relation, Eval, can be defined in terms of $\longrightarrow$. This can be thought of as a specification of an interpreter for the language and is important for efficient automation. The rule for function application is:

$$\frac{\mathsf{Eval}\ e_1\ (\mathsf{lambda}\ y\ t\ e) \quad \mathsf{Eval}\ (\mathsf{Sub}\ e\ (y, e_2))\ c}{\mathsf{Eval}\ (e_1\ e_2)\ c}$$

This rule is obtained by repeated applications of rule 1 above until the function being applied is reduced to a lambda abstraction, followed by a single application of rule 2.

The expected meta-theoretic results such as the fact that evaluation is deterministic can be proved.

4 Mechanising applicative bisimulation

General support for co-inductive definitions has not yet been implemented in HOL although there is support for the related concept of inductive definitions [9, 11]. Relations in HOL are normally represented by functions rather than sets. For the definition of applicative bisimulation only binary relations are required and so this is all that has been implemented. The theory required is related to the theory behind one of the inductive definitions package [9] and more general co-inductive definition package similar to this could be developed. With our approach a binary relation is represented by a function of type $\alpha \to \alpha \to bool$ rather than a set of pairs. For example, the definition of the predicate formalising F-dense relations is

$$\text{Dense } F\ X\ =\ (\forall a\ b.\ X\ a\ b \supset (F\ X)\ a\ b)$$

and the fixpoint of a function F is defined as

$$(\text{gfp } F)\ a\ b\ =\ \exists X.\ \text{Dense } F\ X \wedge X\ a\ b$$

instead of as the union of the appropriate sets. These are the equivalent definitions in this formalisation using functions of the definitions in section 2. Both co-induction and strong co-induction can be derived.

The labelled transition system is introduced as a relation

$$\text{LTS} : exp \to exp \to act \to bool$$

where act is the type of possible labels. LTS $e_1\ e_2\ a$ means that under the rules for LTS the expression e_1 can make a transition to e_2 with label a.

The equivalence relation, ==, is introduced as the greatest fixpoint of a function, F, with the property

$$\begin{aligned}&\forall S\ a\ b.\ (F\ S)\ a\ b\ = \\ &\quad (\exists t.\ \text{Prog } t\ a \wedge \text{Prog } t\ b) \wedge \\ &\quad (\forall a'\ act.\ \text{LTS } a\ a'\ act \supset (\exists b'.\ (\text{LTS } b\ b'\ act) \wedge S\ a'\ b')) \wedge \\ &\quad (\forall b'\ act.\ \text{LTS } b\ b'\ act \supset (\exists a'.\ (\text{LTS } a\ a'\ act) \wedge S\ a'\ b'))\end{aligned}$$

The type judgements restrict this relation to well-typed terms. A theory of equality for programs has been developed from this definition. As this relation is an equivalence relation and a congruence, rewriting tools can be developed to allow the rewriting of programs with == as well as =. The proof of congruence is long and involves the introduction of an additional inductively defined relation that is easily proved to be a congruence, and then proving that the two relations are equal. The mechanised proof mirrors very closely the proof on paper [6].

5 Automating evaluation and type inference

The work described in the previous section establishes the theoretical foundations for a system to support reasoning about PCF programs. The rest of this paper deals with the practical aspects of reasoning about actual programs.

Results such as a proof that an expression evaluates to a specific value can be obtained by working out which rules to apply by hand or by conducting a long goal-directed proof. However, the number of rules to be applied may be very large and so applying all the rules by hand may not be practical. Similar problems occur when proving many results about programs. There are often a large number of obvious or trivial proof steps to be carried out.

The solution is to take advantage of the fact that the meta-language for HOL is ML. This can be used to write higher level proof functions that can be applied to perform many proof steps at once.

The simplest such functions are tactics that perform some simple manipulation of goals before and after the application of theorems. One example is induction tactics that manipulate the goal into the form required to apply the induction theorem, use the theorem and then simplify the goal into the base and step cases.

In this section we concentrate on a more substantial piece of automation. Many of the small steps in a proof will arise from calculating the type or value of a program and so tools have been developed to automate these proofs. The Type and Eval relations can be thought of as specifications of how to type or evaluate expressions on an abstract machine. It is possible to write, in ML, a program that implements this specification.

For the evaluation and typing relations, these programs and the relations will both be deterministic. The way in which the programs calculate the types or values will correspond exactly to the way in which the rules need to be applied to prove the same result. Because of this the programs can be used to return information about which rule from the relations need to be applied where and this can be used to generate the proof. This method provides a structured proof, following the definition precisely, rather than trying to solve a search problem or attempting the exhaustive application of rewrite rules.

Although HOL and the interpreter are both implemented in ML, they are treated as separate systems with an interface between them. A translator converts from the HOL types for the syntax of expressions and types to the ML types used in the interpreter. This allows the interpreter to be developed and tested separately from rest of the system. The ML types contain additional type constructors. For example we represent HOL variables and HOL constants with separate type constructors as discussed in the next section.

One advantage of developing an external system to the find proofs and then using a theorem prover such as HOL to check the proof and then store and manipulate the results is that checking a proof is may be more efficient than searching for a proof. The resulting system is still guaranteed sound. If the interpreter is not correct then an incorrect proof will fail when checked. There are also applications for which discovering the type of the term may be enough without completing the proof. This is useful when solving existential goals where we only wish the find the witness and delay the proof to later.

6 Reasoning with HOL variables and constants

Before we look at the use of the evaluator and type checker discuss how the tools handle variables and constants. The ML evaluator and type checker discussed in the previous section will work provided they are only given closed program fragments. In practice we also need to evaluate and type programs that have have some subexpression denoted by a HOL variable or constant. Typically, the information the tools need to deal with constants and variables will be derivable from the definition of a constant and from some information about a constant or variable that appears on the assumption list. There are three useful facts that may be known: the expression to which the variable or constant is equal; a theorem stating the type of the variable or constant; and a theorem stating what the variable or constant evaluated to.

The proof tools provided must handle these in a uniform and efficient way. The solution to this problem relies on building up a context in which the proof tools can be run which contains all the information we know about the variables and constants. The context for constants is built up as a globally accessible value while the context for variables is built up from the assumption list and other information for each call of the evaluator or type checker.

6.1 Constants

In order to handle constants we build up a context in a reference value which the proof tools can access. To do this a new constant definition mechanism is defined. This function, define_constant, calls new_definition but also derives the evaluation and typing theorems for the constants and stores them all in a record in the global state.

This state is then accessed whenever a proof tool needs to type or evaluate a constant. This means that the evaluation takes place only once instead of every time the constant is used.

The function map used in the example can be introduced in this way. As the language being considered is not polymorphic we cannot introduce a polymorphic map function directly but we can parameterise the constant with the necessary types so that it can then be instantiated to any particular map function. The definition of map is given in figure 2. The introduced constant, map, has type $ty \to ty \to exp$. Rather than write one instantiation of the map function as map t_1 t_2, as it would be entered into HOL, we write the types as subscripts for clarity.

$$\begin{aligned}
\mathsf{map}_{t\ t_1} = \ &\mathsf{rec}\ map\ ((t \to t_1) \to \mathsf{List}\ t \to \mathsf{List}\ t1) \\
&\mathsf{lambda}\ f\ (t \to t_1) \\
&\mathsf{lambda}\ x\ (\mathsf{List}\ t) \\
&\quad \mathsf{scase}\ x \\
&\qquad \mathsf{nil}_{t_1} \\
&\qquad (\mathsf{lambda}\ hd\ t \\
&\qquad \ \ \mathsf{lambda}\ tl\ (\mathsf{List}\ t) \\
&\qquad \ \ (\mathsf{cons}\ (f\ hd)\ (map\ f\ tl)))
\end{aligned}$$

Fig. 2. The definition of map

The theorems produced by this are the theorem storing the definition of map and the evaluation and typing theorems shown in figure 3. The evaluation theorem is one unwinding of the recursive function. An equational characterisation can be proved from the definition. The theorems produced by the definition are stored in the system's state. Currently the state is a list of these records but a more efficient implementation is possible.

$$\vdash \forall t\ t_1.\ \textsf{Eval map}_{t\ t_1}\ \textsf{lambda}\ f\ (t \to t_1)$$
$$\textsf{lambda}\ x\ (\textsf{List}\ t)$$
$$\textsf{scase}\ x$$
$$\textsf{nil}_{t_1}$$
$$(\textsf{lambda}\ hd\ t$$
$$\textsf{lambda}\ tl\ (\textsf{List}\ t)$$
$$(\textsf{cons}\ (f\ hd)\ (\textsf{map}_{t\ t_1}\ f\ tl)))$$

$$\vdash \forall t\ t_1.\ \textsf{Prog}\ ((t \to t_1) \to \textsf{List}\ t \to \textsf{List}\ t_1)\ \textsf{map}_{t\ t_1}$$

Fig. 3. The evaluation and typing theorems for map

6.2 Variables

HOL variables raise a more complex problem. The necessary information may not be known or may need to be deduced by the users. The proof tools should attempt to deduce as much as possible automatically but should raise appropriate proof obligations or fail sensibly if they cannot.

A goal of the form Prog Num a where the theorem $a = \text{num } 2$ occurs on the assumption list could be solved by first rewriting with the assumption and then calling the type-checker. This is not satisfactory because goals such as this should be solved automatically and rewriting unnecessarily with the assumption list could complicated the goal.

The solution adopted is to write "conversions" that take a list of terms representing known facts about the variables of the current term as an argument. These conversions have the type `[term] -> term -> thm`. The theorem produced will have the form

$$A_1, A_2, ..., A_n \vdash e_1 = e_2$$

where $A_1, A_2, ..., A_n$ are the assumptions made to prove $e_1 = e_2$. These assumptions can then be discharged from the assumption list or turned into new proof obligations by tactics provided to handle these assumptions. In normal use the assumptions of the theorem made will be a subset of the assumption list and can be discharged automatically. If the evaluator or type checker is unable to prove some theorem this may also be added to the assumptions of the theorem and eventually be turned into a new proof goal by the tactic applying the conversion. This goal may not be able to be proved and hence indicate an error in the proof attempt.

As an example of the use of these conversions, suppose we wished to prove the goal

$$\text{Prog } t_1 \ (f \ x)$$

under the assumptions

$$\vdash \text{Prog } (t \to t_1) \ f$$
$$\vdash \text{Prog } t \ x$$

The assumptions are essential because they are the only source of the information that the type of x is t. This would be proved by calling the function TYPE_CONV with the arguments [Prog $(t \to t_1)$ f, Prog t x] and Prog t_1 $(f \ x)$ and will return the theorem

$$\text{Prog } (t \to t_1) \ f, \text{Prog } t \ x \vdash \text{Prog } t_1 \ (f \ x) = \mathsf{T}$$

The type checking will have occurred by looking up the type of f and x from the assumptions.

This tool cannot make use of arbitrary facts from the assumption list. In the current system only theorems about the definition of expressions, equality of expressions, evaluation of expressions and the type of closed expressions will be used. This allows a large class of problems to simplified automatically while not generating too large a proof search. Tactics are provided that search the assumption list for suitable assumptions, call the conversions, and discharge the assumptions.

7 Higher level proof tools

The proof tools described in the previous sections can automate the proof of many facts about the evaluation and typing relations. As the language has a very simple type system, the types for most expressions should be able to be proved automatically. The only exception would be when the type could not be deduced for some HOL variables in the term.

The tools described are not sufficient to raise the level of interaction with the system to the level of the example in section 2. The typing and evaluation conversions must still be applied by hand as must the theorems we need to use about labelled transition systems. This section gives some examples of how the tools can be combined with commonly used meta-theorems to produce the higher level proof tools that give the desired level of interaction.

7.1 Co-induction

The tactics for co-induction and strong co-induction fall into the common pattern of manipulating the goal, applying a theorem and then simplifying the resulting subgoals. The tactics take a relation S and manipulate the goal, such as stripping away some universal quantifiers and assumptions, so that it is in the form $a == b$ suitable for the application of one of the principles of co-induction. The tactics then tidy up the resulting subgoals and will attempt to solve any subgoals involving only type checking and for some simple cases will prove that $S\ a\ b$.

7.2 Labelled Transition System

There are a number of theorems about the labelled transition system that depend on the evaluation and typing relations. Rather than force the user to apply the evaluation and type tactics explicitly, higher level proof tools are provided to apply the lower level tools automatically.

An example of a tool to apply a result about the labelled transition system is the conversion that applies the following theorem stating that evaluating a term does not affect the possible transitions.

$$\forall e_1\ e_2.\ \mathsf{Eval}\ e_1\ e_2 \supset (\forall a.\ \mathsf{LTS}\ e_1\ e\ a\ =\ \mathsf{LTS}\ e_2\ e\ a)$$

The conversion `LTS_EVAL_CONV : term list -> conv` takes a list of terms, typically derived from the assumption list as above, and tries to prove the equation $\mathsf{LTS}\ e_1\ e\ a\ =\ \mathsf{LTS}\ e_2\ e\ a$ by evaluating e_1, then instantiating the theorem above to the appropriate terms and using the evaluation theorem to remove the antecedent of the implication.

A second conversion, `LTS_CASE_CONV`, which can often be used in conjunction with the previous one, performs a case analysis on the structure of an expression to determine the possible transitions. For the example given in section 2 one step of the mechanised version of the proof involves case analysis of the term

$$\mathsf{LTS}\ \mathsf{cons}\ (f\ x)\ (\mathsf{iterate}_t\ f\ (f(f\ x)))\ a'\ \alpha$$

The crucial theorem about any cons expression is that

$$\forall x\ y\ a\ \alpha\ t.\ \mathsf{Prog}\ (\mathsf{List}\ t)\ (\mathsf{cons}\ x\ y) \supset \\ \mathsf{LTS}\ (\mathsf{cons}\ x\ y)\ a\ \alpha\ =\ (((a = x) \land (\alpha = \mathsf{Hd})) \lor ((a = y) \land (\alpha = \mathsf{Tl})))$$

The conversion performs the necessary specialisation and type checking to discharge the typing condition.

7.3 Strictness of functions

There are some programs that cannot be evaluated and additional information must be derived before the tools described above can be applied. For example, an application of the map function to some unknown list.

$$\text{map}_{t\ t_1}\ f\ x$$

If x is nil then the program will evaluate to nil and if x is a cons then the program will evaluate to a cons. But, x may be undefined in which case we cannot evaluate the application of map. This is because map is strict in its second argument.

A goal of the form

$$\text{LTS}\ (\text{map}_{t\ t_1}\ f\ x)\ \alpha \supset P$$

does give sufficient information to evaluate x and perform the case analysis described above. We can prove that if any program of list type makes a transition then it must also evaluate to some value. So $\text{map}_{t\ t_1}\ f\ x$ can be evaluated and because map is strict in its second argument we know that x evaluates. Inspection of the type of x will give the possible values of x and then LTS_EVAL_CONV and LTS_CASE_CONV can be applied.

8 Application of the tools

8.1 map-iterate

The tools described in section 7.1 and 7.2 are sufficient to complete the proof of the example in section 2 in HOL. The level of detail of the HOL proof is similar to the level of detail given in the example with the co-induction tactic and the tactics corresponding to LTS_EVAL_CONV and LTS_CASE_CONV providing automation of the main proof steps. There is no need for the user to reason about types or evaluation.

8.2 map-compose

This example involves the interaction of two functions, map and compose. The theorem we aim to prove is that

$$\forall f\ g\ x\ t\ t_1\ t_2.$$
$$\text{Prog}\ (t_1 \to t_2)\ f\ \wedge\ \text{Prog}\ (t \to t_1)\ g\ \wedge\ \text{Prog}\ (\text{List}\ t)\ x \supset$$
$$\text{map}_{t\ t_2}\ (\text{compose}_{t\ t_1\ t_2}\ f\ g)\ x\ ==\ \text{map}_{t_1\ t_2}\ f\ (\text{map}_{t\ t_1}\ g\ x)$$

The definition of map is given in figure 2 and the definition of compose in figure 4

The more natural equational definition can be derived from this definition and this will be automated in future. An equational theorem for compose is

$$(\text{compose}_{t\ t_1\ t_2}\ f\ g)\ x\ ==\ f\ (g\ x)$$

for all appropriately typed f, g, and x.

$$\text{compose}_{t\ t_1\ t_2} = \text{lambda } f\ (t_1 \to t_2)$$
$$(\text{lambda } g\ (t \to t_1)$$
$$(\text{lambda } x\ t$$
$$(f\ (g\ x))))$$

Fig. 4. The definition of compose

As with the map-iterate example the proof is by strong co-induction, using a relation S with definition

$S\ a\ b = \exists f\ g\ x.$
$\quad (a = \text{map}_{t\ t_2}(\text{compose}_{t\ t_1\ t_2}\ f\ g)\ x) \land$
$\quad (b = \text{map}_{t_1\ t_2}\ f\ (\text{map}_{t\ t_1}\ g\ x)) \land$
$\quad \text{Prog}\ (t_1 \to t_2)\ f\ \land\ \text{Prog}\ (t \to t_1)\ g\ \land\ \text{Prog}\ (\text{List}\ t)\ x$

The proof begins by applying the tactic for strong co-induction. In addition to applying co-induction this performs some automatic proof about the types of the terms and proves the theorem

$\vdash S\ (\text{map}_{t\ t_2}(\text{compose}_{t\ t_1\ t_2}\ f\ g)\ x)\ (\text{map}_{t_1\ t_2}\ f\ (\text{map}_{t\ t_1}\ g\ x))$

which states that the left and right hand sides of the original goal are included in the relation S. It remains to be shown that S is included in $F_{==}(S\ \cup\ ==)$. Two goals are generated.

1. $\forall a'\ act.\ \text{LTS}\ a\ a'\ act \supset (\exists b'.\ (\text{LTS}\ b\ b'\ act) \land (S\ a'\ b' \lor a' == b'))$
2. $\forall b'\ act.\ \text{LTS}\ b\ b'\ act \supset (\exists a'.(\text{LTS}\ a\ a'\ act) \land (S\ a'\ b' \lor a' == b'))$

where we know that

$a = \text{map}_{t\ t_2}(\text{compose}_{t\ t_1\ t_2}\ f'\ g')\ x'$
$b = \text{map}_{t_1\ t_2}\ f'\ (\text{map}_{t\ t_1}\ g'\ x')$

for some f', g', and x'. We cannot proceed as in the first example because we cannot evaluate either a or b unless we can evaluate x'. But, from the strictness of map, the assumption that a makes some transition and the type of x', the system can deduce that x' must evaluate to nil or some cons cell. Two goals are generated, with the new assumptions Eval x' nil_t and Eval x' (cons $h'\ t'$) for some h' and t' with the correct types.

With the possible values for x' known, a and b can be evaluated and the goal simplified using LTS_EVAL_TAC and LTS_DISCH_TAC as in the previous example. If x evaluates to nil then a and b evaluate to nil and the transition must be the Nil transition. If x evaluates to cons $h'\ t'$ then two goals corresponding to the Hd and Tl transitions are generated. Each of the goals is solved by choosing a witness for b' and followed by some simple equational reasoning.

9 Conclusions

The work described in this paper illustrates how a system with a high level of interaction can be built on top of a theory that in itself would not allow practical theorem proving about programs. The use of a deep embedding gives the ability to embed the required semantics and to perform meta-theoretic reasoning. The meta-theory, together with the proof tools, allows practical reasoning.

Each proof tool has a well defined task that corresponds to one logical step in the proof. They have been designed so that if they fail then the user will either be returned to the original proof state or will have some additional goals to prove that may complete the proof or show an error in the proof.

The tool produced will not be a HOL library. The aim has been to develop a tool that uses the HOL system as a theorem proving engine. The tools described create and modify a global state and in future more significant modifications of the HOL parser, pretty printer and goal stack will be made. One field of further work will be to develop a theory mechanism for this system to allow groups of theorems to be stored. While this will be built on top of the HOL theory mechanism there will be the additional need to re-establish the state of the system.

The tools described here have concentrated on supporting proof using coinduction but similar tools can be written to support proof by induction.

Little mention has been made of interfaces in this paper. It is hoped to address this issue at some point. One related issue that has been addressed is that the system has been designed to make significant use of the assumption list. All the proof tools will look at the assumption list for the definitions of variables, and for evaluation and typing theorems. For most proofs the user should never have to use these assumptions explicitly.

Acknowledgements

Thanks are due to Tom Melham for advice on all aspects of this work, and to the Engineering and Physical Sciences Research Council for financial support.

References

1. Samson Abramsky. The Lazy Lambda Calculus. In David Turner, editor, *Research Topics in Functional Programming*, pages 65–116. Addison-Wesley, 1990.
2. Richard Boulton, Andrew Gordon, Mike Gordon, John Harrison, John Herbert, and John Van Tassel. Experience with embedding hardware description languages in HOL. In V. Stavridou, T. F. Melham, and R. T. Boute, editors, *Theorem Provers in Circuit Design: Theory, Practice and Experience: Proceedings of the IFIP WG10.2 International Conference, Nijmegen*, pages 129–156. North-Holland, June 1992.
3. A. Cant and M.A. Ozols. A verification environment for ML programs. In *Proceedings of the ACM SIGPLAN Workshop on ML and its Applicatins*, San Francisco, California, June 1992.

4. Graham Collins. Supporting Reasoning about Functional Programs : An Operational Approach. In *1995 Glasgow Workshop on Functional Programming*, Electroninc Workshops in Computer Science. Springer-Verlag, 1996.
5. Graham Collins and Donald Syme. A Theory of Finite Maps. In E. Thomas Schubert, Phillip J. Windley, and Hames Alves-Foss, editors, *Higher Order Logic Theorem Proving and its Applications*, volume 971 of *Lecture Notes in Computer Science*, pages 122–137. Springer-Verlag, 1995.
6. Andrew D. Gordon. Bisimilarity as a Theory of Functional Programming. Technical Report NS-95-3, Basic Research in Computer Science, University of Aarhus, July 1995.
7. Andrew D. Gordon. A Tutorial on Co-induction and Functional Programming. In *1994 Glasgow Workshop on Functional Programming*, Workshops in Computer Science, pages 78–95. Springer-Verlag, 1995.
8. M. J. C. Gordon and T. F. Melham, editors. *Introduction to HOL: A theorem proving environment for higher order logic*. Cambridge University Press, 1993.
9. John Harrison. Inductive definitions: automation and application. In E. Thomas Schubert, Phillip J. Windley, and Hames Alves-Foss, editors, *Higher Order Logic Theorem Proving and its Applications*, volume 971 of *Lecture Notes in Computer Science*, pages 200–213. Springer-Verlag, 1995.
10. Savi Maharaj and Elsa Gunter. Studying the ML Module System in HOL. In Tom Melham and Juanito Camilleri, editors, *Higher Order Logic Theorem Proving and its Applications*, volume 859 of *Lecture Notes in Computer Science*, pages 346–361. Springer-Verlag, September 1994.
11. Tom F. Melham. A Package for Inductive Relation Definitions in HOL. In M. Archer, J. J. Joyce, K. N. Levitt, and P. J. Windley, editors, *Proceedings of the 1991 International Workshop on the* HOL *Theorem Proving System and its Applications, Davis, August 1992*, pages 350–357. IEEE Computer Society Press, 1992.
12. Donald Syme. Reasoning with the Formal Definition of Standard ML in HOL. In *Higher Order Logic Theorem Proving and Its Applications*, volume 780 of *Lecture Notes in Computer Science*, pages 43–60. Springer-Verlag, 1993.
13. Myra VanInwegen and Elsa Gunter. HOL-ML. In J. J. Joyce and C. J. H. Seger, editors, *Higher Order Logic Theorem Proving and its Applications*, volume 780 of *Lecture Notes in Computer Science*, pages 61–74. Springer-Verlag, 1993.

Coq and Hardware Verification: A Case Study

Solange Coupet–Grimal and Line Jakubiec*

Laboratoire d'Informatique de Marseille – URA CNRS 1787
39, rue F. Joliot–Curie 13453 Marseille France
e-mail:{Solange.Coupet,Line.Jakubiec}@lim.univ-mrs.fr

Abstract. We present several approaches to verifying a class of circuits with the Coq proof-assistant, using the example of a left-to-right comparator. The large capacity of expression of the Calculus of Inductive Constructions allows us to give precise and general specifications. Using Coq's higher order logic, we state general results useful in establishing the correctness of the circuits. Finally, exploiting the constructive aspect of the logic, we can show how a certified circuit can be automatically synthesized from its specification.

1 Introduction

During the past decade, intensive research has developed in designing mechanized theorem provers, resulting in a great deal of new proof assistants. Hardware verification was one of the original motivations and main application of this area. Two of the earliest and most significant achievements were the work of Gordon using HOL [14, 6] and the work of Hunt [17] using Nqthm [5]. On the one hand, using general purpose theorem provers to state circuit correctness has several advantages over ad hoc tools. These include the precision of the specifications, enhancing the reliability of the verification process, and an increased generality leading to reusable methodologies and libraries. Now, on the other hand, meeting the requirements of the hardware verification community has been a stimulating challenge for logicians, mostly for those working in computer-aided proof-checking. Thus, despite the fact that existing theorem provers are high-level and general-purpose and cover fields of application much wider than hardware verification, verifying hardware remains a challenging domain of experiences. Among recent investigations, let us quote the verification with PVS [19] of a part of a pipelined microprocessor, the AAMP5 [25] and various uses of the prover LP to verify circuits [2, 24].
In this paper, a case study allows us presenting the capabilities of Coq in verifying and synthesizing hardware.
Coq is a proof tool developed at INRIA-Rocquencourt and ENS-Lyon [9]. It provides a very rich and expressive typed language and a higher order constructive logic. Moreover it offers the possibility of extracting automatically functional programs from the algorithmic content of the proofs.

---

* This work was supported by the GDR-Programmation; it was partially done during a six-month visit of Solange Coupet-Grimal at ENS-Lyon, in the Coq group.

A lot of significant developments have been performed with Coq. They can be found in the library of the users' contributions delivered with the Coq release. However, few investigation has been done to verify circuits. A multiplier first introduced by M.Gordon in [13] has been proven in [11], later extended by C. Paulin-Morhing in [23] to a more general proof of this circuit, using a codification of streams in type theory as infinite objects.[3] is a verification of a multiplier specified at the bit vector level.

The first part of the research presented here is most similar to the work done by K. Hanna, N. Daeche and M. Longley with Veritas$^+$ [16]. We follow their approach, as exemplified by a comparator studied in their paper, to specifying and proving a circuit, by making heavy use of dependent types and higher order logic. We have produced several reusable Coq modules, providing expressive and precise specifications as well as general theorems applicable to a whole class of circuits.

Several other researchers have been investigating the use of dependent types for reasoning about hardware. For example, interesting results using Nuprl have been produced [1, 18]. Like Coq, and unlike to Veritas$^+$, Nuprl relies on an intuitionistic logic. Until now, however, the intuitionistic aspect of the underlying logic has not been exploited (at least, we are not aware of any work in this direction). For us, filling this gap is worthwhile and is the aim of the second part of our study. Indeed, working with a constructive logic presents some difficulties, since it disallows the excluded-middle principle. To begin with, it may require an effort of the user who is used to classical reasoning. That is the reason invoked by Hanna for choosing classical logic for Veritas$^+$ [16]. But the computational aspect of the proofs is a valuable asset that can be used. In our opinion, this highly compensates the drawbacks, if any, of this kind of logic. In this paper, we present a methodology for synthesizing a circuit from its specification, using the Coq program extractor. As an alternative, we also give a methodology, using the tactic "program" [20], which can be seen as a mid-point between proving that a circuit is correct with respect to its specification (both being expressed in the Coq language) and "blindly" extracting the circuit as a ML function from a proof of a theorem which, roughly, states the existence of an object verifying the specification. This method consists in giving the prover both the functional description of the circuit and its specification. Thus, the proof process is guided by the knowledge of the term extracted from the proof.

The rest of this paper is organized as follows. Section 2 briefly introduces Coq. Section 3 deals with the description and the verification of the comparator. In Section 4, we present our two approaches to synthesizing the circuit. We conclude with an analysis of our results and of the performances of Coq.

2 An Overview of Coq

The Coq system is a tactic oriented proof-checker, in the style of LCF [15]. Developments can be split into various parameterized modules to be separately verified. Thus, several developments can share modules that, being compiled once and for all, are loaded fast.

Coq's language implements a higher order typed lambda-calculus, the Calculus of Constructions [7, 8], enriched with inductive definitions [22].

Coq's logic is a higher order constructive logic and relies on the *propositions-as-types correspondence*. In Coq, a proposition is a type and a proof is a term inhabiting this type. Such a system provides an elegant unifying framework, since there is no fundamental difference between proofs and data, nor between propositions and datatypes. Therefore, proving amounts to type-checking.

However, there are two sorts of types : propositions are of sort *Prop* and sets are of sort *Set*. From a logical point of view, this distinction is not necessary, but it makes the system less confusing for the user. On the contrary, this distinction is highly significant when extracting programs from proofs, as we will show in the following.

Notations.
- $(A\ B)$ denotes the application of a functional object A to B.
- $[x:A]B$ denotes the abstraction of B with respect to a variable x of type A, (usually written $\lambda x \in A.B$).
- $(x:A)B$ as a term of type *Set*, denotes the cartesian product $\prod_{x \in A} B$. As a proposition, it corresponds to $\forall x \in A.B$. Moreover, if x does not occur in B, $A \to B$ is a shorter notation for the type of functions from A to B, or for a logical implication, depending on the sorts of A and B.

Induction and Recursion.
Selected parts of Coq specifications are depicted in Fig.1. The section *dependent_lists* is parameterized with respect to a term A of sort *Set*.

In this section is given a typical inductive definition involving dependent types, namely the definition of *list*. For each term n of type *nat*, *(list n)* is a type of sort *Set*, depending on the term n. *(list n)* denotes the type of the lists of elements of A whose length is n. This type is defined by means of two constructors, *nil* and *cons*. The type of *cons* expresses that it is a function which, given a natural number n, an element of A, and a length-n list, returns a length-$(n+1)$ list. Moreover, Coq automatically generates the induction principle corresponding to the type *list*.

When the section is closed, the parameter A is *discharged* in the sense that all the terms depending on A are abstracted with respect to A. Outside the section, the type of polymorphic length-n lists will be $\lambda A : Set\ (list\ A\ n)$.

Numerals are defined in the section *numerals* which requires the module *dependent_lists*. The word *Local* introduces local definitions of the current section.

In the Coq syntax, given a set A and a predicate P on A, $\{x:A|(P\ x)\}$ denotes the subtype of A corresponding to the elements for which the property P holds. Terms of this type are pairs consisting of an element x of A and a proof of $(P\ x)$. The function *Inj*, taking such a pair as argument, erases its logical component and returns x. This function is parameterized with respect to A and P. For ex-

```
Section dependent_lists .
Variable A:Set.
Inductive  list :nat->Set:= nil:(list 0)|
                            cons:(n:nat)A->(list n)->(list (S n)).
...
End dependent_lists.

Section numerals.
Require dependent_lists.

Definition BT:={b:nat|(lt 0 b)}.
Variable   BASE:BT.
Definition base:=(Inj nat [b:nat](lt 0 b) BASE).
Definition digit:={x:nat|(lt x base)}.
Definition val:digit->nat:=(Inj nat [x:nat](lt x base)).
Definition num:=(list digit).
Local Cons:=(cons digit).
Local Nil:=(nil digit).

Fixpoint Val[n:nat;X:(num n)]:nat:=<[m:nat]nat>Case X of
(*X=Nil*)            0
(*X=(Cons p d D)*) [p:nat][d:digit][D:(num p)]
                     (plus (mult (val d) (exp base p)) (Val p D)) end.
...
End numerals.
```

Fig. 1. "Dependent_lists" and "Numerals" Sections

ample, the variable $BASE$ is a pair of the form $(base, p)$ where $base$ is a natural number and p is a proof of $base > 0$. The function Inj, taking as arguments the set nat, the predicate $\lambda b.b > 0$, and the pair $BASE$, returns $base$.

We use subtypes to give precise specifications for systems of numeration such as base or digit definitions. For example, the type *digit* describes the set of natural numbers less than the base. The value *(val d)* of a digit d is the natural number obtained by keeping only the first component of its specification. A numeral is a list of digits, the length of which is specified.

On each concrete type inductively specified by constructors, it is possible to define functions recursively, by case analysis. The function *Val* is defined in such a way. Taking as arguments a natural number n and a length-n numeral X, it returns a natural number representing its value.

The expressions (*X=Nil*) and (*X=(Cons p d D)*) are just comments. The last line of the definition means that if X is the list whose length is $p+1$, whose head is d and whose tail is D then the function returns $(val\ d) * base^p + (Val\ p\ D)$ (note the recursive call in this last expression).

After this short presentation, we can move to the description and the verification of the comparator.

3 Verification

The particular example we choose is given in [16] as an illustration of an elegant and general methodology for specifying and proving iterative structures. As a first stage, it appeared to be an excellent benchmark in order to study the feasibility in Coq of already tested methods. But Coq's particular features lead us towards more powerful original approaches.

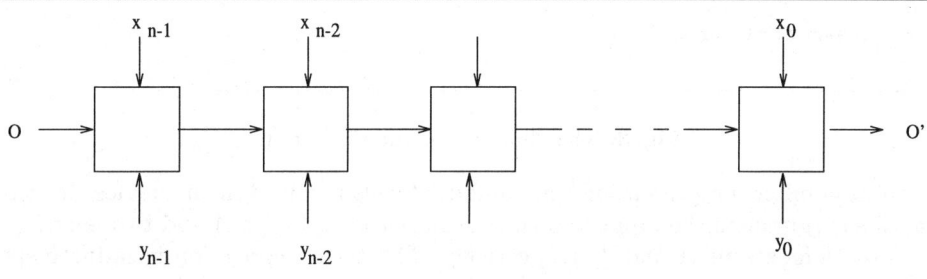

Fig. 2. A Comparator

The comparator (Fig.2) is a hardware device that accepts two numerals and determines their relative magnitude. It is composed of identical cells interconnected by a carry wire accepting comparison data in a 3-valued type. Each cell, from left to right, outputs a value that depends on the incoming carry and on the result of the comparison of two digit inputs.

3.1 Specifications

First of all, at the top level of genericity, there is the type *list* of dependent polymorphic lists presented in the previous section. It allows us to get high-level abstract specifications, more general than those in [16]. This type is particularly suitable in the framework of hardware specification where linear structures are prevalent. Numerals for example have been defined as particular lists. We have thus given in the *dependent_lists* section some additional definitions and properties that are not displayed on Fig.1 and that can be reused for any instance of *lists*. At this point, we do not go into more details about the contents of this module. A generic definition for connections of identical four ports cells is given in the *linear_structures* section (Fig. 3). It is parameterized with respect to the types A, B, C of the ports and to the relation *cell* implemented by the cells. Following the same idea as for the numerals, the type of a connection depends on a

```
Section linear_structures.

Require Dependent_lists.

Variables A,B,C:Set.
Variable cell:A->B->C->A->Prop.

Inductive connection :(n:nat)A->(list B n)->(list C n)->A->Prop:=
        C_O:(a:A)(connection O a (nil B) (nil C) a)|
        C_Sn:(n:nat)(a,a1,a':A)(b:B)(c:C)(lb:(list B n))(lc:(list C n))
             (cell a b c a1)->
             (connection n a1 lb lc a')->
             (connection (S n) a (cons B n b lb) (cons C n c lc) a').

End linear_structures.
```

Fig. 3. The "linear_structures" Section

natural number n representing the number of cells involved in the device. It also takes as arguments the input and the output carries of type A and two length-n lists of elements in A and B respectively. The term *connection* is inductively defined, in a typed Prolog style. With this analogy, the type of the constructors corresponds to the body of two Prolog rules (with reversed arrows) labeled C_O and C_Sn. The type of C_O states that, for all a in A, a connection with zero cells is just a wire carrying a. In this case, the two lists are the empty lists $(nil\ B)$ and $(nil\ C)$. The type of C_Sn states that any length-$(n+1)$ connection is obtained from a length-n connection whose port a_1 is connected to an additional cell. Figure 4 partially depicts a file in which various notions for comparing natural

```
Inductive order:Set:=L:order|E:order|G:order.

Definition comparison:=[v1,v2:nat]<order> Case (Lt_eq_Gt v1 v2) of
        [_:(lt v1 v2)]          L
        [_: v1=v2]              E
        [_:(gt v1 v2)]          G                               end.
```

Fig. 4. A Part of the "Compare_nat" Module

numbers are given. The set $order = \{L, E, G\}$ is denoted by the enumerated type *order*. The function *comparison* returns the value L, E or G depending on the relative magnitude of the natural numbers $v1$ and $v2$ it takes as arguments. This function is defined by case analysis on the term $(Lt\_eq\_Gt\ v1\ v2)$ which has been built before. This term is a proof of $(v1 < v2)$ *or* $(v1 = v2)$ *or* $(v1 > v2)$.

The second line of the definition of *comparison* must be interpreted by "given a proof of $(v1 < v2)$, return L". After that, several properties of *comparison* are established that are not shown on the figure.

All these tools having been defined, we are now able to describe the implementation and the expected behavior of the device. The section describing the com-

---

```
Section comparator.
(*system of numeration*)
Variable BASE: BT.
Local Digit:=(digit BASE).
Local ValB:=(Val BASE).
Local Num:=(num BASE).

(*semantics of the cells*)
Local f_cell:order->Digit->Digit->order:=
[o,x,y]<order>Case o of
(*o=L*)        L
(*o=E*)        (comparison (valB x) (valB y))
(*o=G*)        G                                        end.

Definition cell:order->Digit->Digit->order->Prop:=
[o, x, y, o'] o'=(f_cell o x y).

(*structure of the comparator*)

Local Connection:=(connection order Digit Digit cell).
Local Comparator:=[n:nat][o:order][X,Y:(Num n)](Connection n E X Y o).

(*behavior of the comparator*)

Local Specif:(n:nat)(inf n)->(inf n)->order:=[n,X,Y]
(comparison (val_inf n   X) (val_inf n Y)).
```

Fig. 5. Implementation and Behavior of the Comparator

parator (Fig. 5) requires the section *numerals*. The first argument of the terms *digit*, *num* and *Val* is instantiated with the current base, given as a parameter *BASE*. The functional specification of a cell is given by the function f_cell, taking three arguments o, x, and y and defined by case analysis on the value of o. The local notion of connection is specified by the general term *connection* in which the types of the ports are $A = order$ and $B = C = Digit$. Let us point out that $(inf\ n)$ denotes the interval $[0, n[$ and that val_inf is the natural injection of type $(n : nat)(inf\ n) \to nat$. It is worth noting that the circuit has only two inputs (the numerals to be compared) since the carry input value is constrained to be E.

```
Local f_circ:(n:nat)order->(Num n)->(Num n)->order:=
[n,o,X,Y]<order>Case o of
   (*o=L*)              L
   (*o=E*)              (comparison (ValB n X) (ValB n Y))
   (*o=G*)              G                                       end.

Lemma general_correct:(n:nat)(X,Y:(Num n))(o,o':order)
       (Connection n o X Y o')->o'=(f_circ n o X Y).

Induction 1.            (*Induction on (Connection n o X Y o')*)

Clear  H o' o Y X n.    (*Erasing the useless hypothesis*)

Intros o;Case o;Simpl;Auto.   (*base case, case analysis on o*)
Apply sym_equal;Auto.
Clear  H o' o Y X n.
Intros n o o1 o' x y X Y H_cell H_n H_rec.
Inversion_clear H_cell.
Rewrite -> H_rec;Rewrite -> H.
Cut (eq ? o o);Auto.
Pattern 2 3 o ;Case o;Intros e;Rewrite -> e;Unfold f_cell ;
Unfold f_circ ;Auto.
(Cut (eq ? (comparison (valB x) (valB y))
           (comparison (valB x) (valB y)));Auto).
Pattern 2 3 (comparison (valB x) (valB y)) ;
Case (comparison (valB x) (valB y));Intros C;Apply sym_equal;
Unfold ValB ;Unfold Digit ;Auto.
Save.

Lemma correctness:(n:nat)(X,Y:(Num n))(o:order)
(Comparator n o X Y)->
o=(Specif  (exp base n) (Val_bound n X) (Val_bound n Y)).

(Unfold Comparator ;Unfold Specif ).
Intros n X Y o H.Rewrite -> (general_correct n X Y E o H).
Auto.              (*automated resolution of the current goal*)
Save.
```

Fig. 6. Proofs of Correctness

3.2 Proving the correctness of the circuit

The theorem *correctness* in Fig.6 establishes that the implementation is correct with respect to the intended behavior and can be informally stated as follows :

For all n in nat, for all o in order, for all length-n numerals X *and* Y, *if* (comparator n o X Y) *then* $o = (Specif\ base^n\ \overline{X}\ \overline{Y})$ *where* $\overline{X}$ *is the value of the numeral* X *considered as a natural number of the interval* $[0,\ base^n[$.

However, because of the constant value of the input carry, the proof requires a generalization. Therefore a lemma *general_correct* is first established which sets forth the correct behavior of a connection, whatever value is given to the input carry. It is proven by induction on $(Connection\ n\ o\ X\ Y\ o')$. Fig.6 gives an idea of the length and the complexity of the proofs that have not to be read in detail.

4 Towards Synthesis

4.1 The Factorization Theorem

A more general approach oriented to verification as well to synthesis of 1-dimension arithmetic circuits is given in [16]. One can observe that, given a base b, each cell of the comparator implements a modulo-b version of the overall structure. This is also the case of right-to-left comparators, incrementors, circuits performing the multiplication of a numeral by a given natural number d, and so forth. Each cell of the latter, for example, performs the multiplication by d of a digit.

Let R be a relation of type: $(n : nat) A \to (inf\ n) \to (inf\ n) \to A \to Prop$.
We say that R is *proper* if

$$\forall n \in nat\ \forall a \in A\ (R\ 1\ a\ 0\ 0\ a).$$

We say that R is *factorizable* if the relation holds on two natural numbers x and x' as soon as it holds on the quotients and the remainders of the division of x and x' by any natural number n. More accurately, let n, m, x, x' be natural numbers such that:

$$x = nq + r\ ;\ x' = nq' + r'\ ;\ x, x' \in [0, mn[\ ;\ q, q' \in [0, m[\ ;\ r, r' \in [0, n[$$

R is factorizable if

$$\forall a, a1, a' \in A\ (R\ m\ a\ q\ q'\ a_1) \to (R\ n\ a_1\ r\ r'\ a') \to (R\ mn\ a\ x\ x'\ a').$$

The approaches presented in this section and in the following one apply to all linear structures whose cells implement such proper and factorizable arithmetic relation. The theorem of factorization states forth that for every relation R that is *proper* and *factorizable*, $(R\ b^n)$ is implemented by a connection of n cells implementing $(R\ b)$:

For all proper and factorizable relation R, *for all natural number* n, *for all length-n numerals* X *and* Y, *for all* a *and* a' *in* A, *if* $(Connection\ n\ a\ X\ Y\ a')$ *then* $(R\ b^n\ a\ \overline{X}\ \overline{Y}\ a')$.

The theorem is easily proven by induction on (*Connection n a X Y a'*). The proof of the comparator boils down to proving that the corresponding relation is proper and factorizable. This is done by case analysis on the variables a, $a1$ and a' occurring in the definition of *factorizable* and by using properties of the function *comparison*.

Indeed, this method is more general than that given in the previous section. However, although it is synthesis oriented, we have not really synthesized the circuit. We have established that a given linear structure satisfies a specification, but we have not obtained this structure from the specification. We show, in the following subsection, how to take advantage of the Coq proof extractor in an effective synthesis process.

4.2 Extracting the Circuit from its Specification

So far, we have used Coq as a powerful and expressive proof-checker. We intend now to take advantage of the constructive aspect of its logic.

Outline of the Coq Extraction Process Due to the *Curry-Howard isomorphism*, in Coq, proofs are λ-terms. They are thus objects of the underlying language, that can be displayed on the screen, stored, reused, exploited in various ways. Moreover, as λ-terms, proofs are nothing but functional programs.

In intuitionistic logic, a proof of a proposition of the form

$$\forall x \in A \; \exists y \in B \; (P \; x \; y)$$

necessarily contains an algorithm computing a function f of type $A \to B$ and a logical part certifying that for all x in A and y in B, if $y = (f\;x)$ then the proposition $(P\;x\;y)$ is verified. The Coq system involves a mechanism that is able to extract from such a proof a ML program computing the function f. By construction, this program is correct with respect to its specification P.

The distinction between the computational part and the logical part of a proof relies on the sorts, namely *Prop* and *Set*, in which the types are declared. A term is called *informative* if its type is of sort *Set* and *non-informative* if its type is of sort *Prop*. Analogously to program comments that are not taken into account by compilers, non-informative parts of the proof are erased during the extraction process. Moreover, all terms resulting from an extraction are terms in the system F_ω [12]. Informally, this means that the dependencies, if any, between types and terms, are lost during the extraction. For example, the term extracted from a "dependent" length-n list, as defined in this paper, is a usual list l, but with an additional parameter n of type *nat*. But l and n are not connected any more. Of course, the user gives the specifications and develops his proof in accordance to the term he wishes to obtain after extraction. As we mentioned in the section 2, the sorts *Prop* and *Set* are perfectly symmetrical and interchangeable. Let us outline, with an example, how the extraction works.

Let P be a predicate on the set of natural numbers. The proposition

$$\exists x \in nat \; (P \; x)$$

can be expressed in any of the three following terms T_1, T_2 or T_3:

- $T_1 = (ex\ nat\ P)$ of sort *Prop*, with $P : nat \to Prop$

- $T_2 = \{x : nat|\ (P\ x)\}$ of sort *Set*, with $P : nat \to Prop$

- $T_3 = \{x : nat\ \&\ (P\ x)\}$ of sort *Set*, with $P : nat \to Set$.

Now, what is the result of the extraction process on terms $t_1 : T_1$, $t_2 : T_2$ and $t_3 : T_3$?

- t_1 being non informative, it is erased by the extractor.
- As mentioned in section 2, t_2 is a pair consisting of a "witness" n of type *nat* and a non-informative proof of $(P\ n)$. The extracted term is n.
- For t_3, the result of the extraction is a pair (n, b) where b comes from the proof of $(P\ n)$ which, this time, is informative. For example, if $(P\ x)$ is a disjunction, b is the boolean *true* if the left part of the disjunction has been proven and *false* in the other case.

This is a very short and informal presentation. For more details, one can refer to [21].

Synthesizing the Comparator The extraction principles presented in the previous paragraph leads us to a new version of the factorization theorem, that can be stated as follows:

For all proper and factorizable relation R, *for all natural number* n, *for all length-n numerals* X *and* Y, *for all* a *in* A, *there exists* a' *in* A *such that* $(R\ b^n\ a\ \overline{X}\ \overline{Y}\ a')$.

It is organized so that a function f will be extracted from its proof. This function will take as arguments an element a of type A, a natural number n and two length-n numerals X and Y. It will return an element a' of type A. The function f is certified to be such that $(R\ b^n\ a\ \overline{X}\ \overline{Y}\ a')$. At this point, it is necessary to give a functional specification of the relation R, that is to say to state which are the input ports and the output ports of a circuit implementing R. This is done by defining the relation R as follows:

$(R\ n\ a\ x\ y\ a')$ if and only if $a' = (FR\ n\ a\ x\ y)$

where FR is of type:

$$FR : (n : nat)A \to (inf\ n) \to (inf\ n) \to A.$$

The function f will be defined by an algorithm which depends on the way the proof is developed. Here, we make an induction on n.

- If $n = 0$, we give the witness $a' = a$ and prove that $(R\ 1\ a\ 0\ 0\ a)$ using the fact that R is proper.

- Let us now consider a an element of type A, n a natural number and two length-$(n + 1)$ numerals $X = (Cons\ d\ D)$ and $Y = (Cons\ d'\ D')$. Let a_1 be $(FR\ b\ a\ d\ d')$. By induction hypothesis, there exists a' such that $(R\ b^n\ a_1\ \overline{D}\ \overline{D'}\ a')$. The relation being factorizable we can deduce that

$$(R\ b^{(n+1)}\ a\ \overline{(Cons\ d\ D)}\ \overline{(Cons\ d'\ D')}\ a').$$

The type $list$ extracted from the type $(n : nat)(list\ A\ n)$ is nearly the usual inductively defined type for polymorphic lists (the constructors take an additional argument of type nat). From FR a function fr of type $nat \to A \to nat \to nat \to A$ is obtained. Note that there are no dependencies between types and terms any longer and that the logical content of $(inf\ n)$ has disappeared. The extraction process, on the proof of the theorem, results in the function f of type $nat \to A \to list \to list \to A$ defined by

- $(f\ 0\ a\ D\ D') = a$
- $(f\ (n+1)\ a\ (Cons\ d\ D)\ (Cons\ d'\ D')) = (f\ n\ a_1\ D\ D')$
  where $a_1 = (fr\ b\ a\ d\ d')$

From the extracted term, a ML program can be automatically generated. It produces the expected result, when taking as inputs a natural number n and two length-n numerals X and Y. If one of the numerals is shorter than n, an exception is returned. Numerals longer than n are truncated.

Synthesizing the comparator is now extremely simple. It is sufficient to apply this theorem with the particular relation *cell* implemented by the cells of the comparator and defined in Fig.5.

4.3 A Mixed Approach using the Tactic "Program"

In the previous section, we showed how the user develops his proof according to the program he has in mind and he wants to synthesize. The *Program* tactic just implements the idea that the program to be extracted contains information about the structure of the proof and thus that it can be used as a guide during the proof process. This methodology can be viewed as dual to the extraction. Let us consider the function $Impl$ defined by :

$$(Impl\ 0\ a\ X\ Y) = a$$

and

$$(Impl\ (n+1)\ a\ X\ Y) = (Impl\ n\ (FR\ b\ a\ (Hd\ X)\ (Hd\ Y))\ (Tl\ X)(Tl\ Y))$$

In these equations Hd and Tl denote respectively the functions head and tail. This program is associated with the theorem

For all proper and factorizable relation R, *for all natural number* n, *for all length-n numerals* X *and* Y, *for all* a *in* A, *there exists* a' *in* A *such that* $(R\ b^n\ a\ \overline{X}\ \overline{Y}\ a')$

to be proven by the command *Realizer*. Then the tactic *Program_all* generates sequences of introduction, application and elimination tactics on every subgoal depending on the syntax of the program *Impl*. In particular, the induction scheme is found out by the system. Although the proof process is not fully automated, it is highly simplified.

5 Summary and Conclusions

Our aim in this paper was to demonstrate the capabilities of Coq in the field of hardware verification. We have given a general and illustrated presentation of the prover and we have investigated how to reap the greatest benefit of its particular features not only for proving circuit correctness but also for effectively synthesizing devices at the algorithmic level. Our results apply to arithmetic linear structures the cells of which implement proper and factorizable relations (incrementor, comparators, multipliers, $\cdots$). To sum up, precise and general specifications have been expressed in a natural way. Several reusable modules have been developed (for handling lists, numerals, repetitive arithmetic structures) in which generic properties and theorems have been proven. Several approaches have been investigated (verification of a particular circuit, verification of a class of circuits, synthesis of a class of circuits, intermediate approach). The synthesis methodology relies on the constructive aspect of the logic and, in practice, on the Coq extractor. A functional description of an implementation is automatically extracted from a proof of a statement of the form

$$\forall y \in A \, \exists x \in B \, (P \, x \, y).$$

P is a relation between the input x and the output y and represents the expected behavior (specification) of the circuit. In a third intermediate approach, specification and implementation are both given to the prover. As the proof process is guided by the syntactical structure of the implementation, it is more automated and thus easier to use.

Relying on the Curry-Howard isomorphism, Coq provides an elegant unifying framework for specifying and proving. Proof-checking and type-checking are the same process (in PVS for example a type checking step must precede the classical proof process). Let us also mention that, unlike in Nuprl, type-checking in Coq is decidable. Undoubtedly, Coq is a powerful tool, with advanced features, the most futuristic of them being the synthesis of certified programs. The drawback, in our point of view, is the lack of user friendliness and automation. Exploiting all Coq subtleties still requires skill and expertise.

However, various works are now in progress that will make Coq much easier to use in the future. A nice interface, CtCoq, is already available [4]. Moreover, a tool is being developed that, from the script of a Coq proof, automatically generates a text in natural mathematical language [10]. It will be of interest for analyzing, simplifying, and debugging proofs. New approaches are also being studied for improving the extraction process and the modularity. Finally, arithmetic decision procedures are about to be integrated in the system.

6 Acknowledgments

We would like to thank all the members of the Coq group at ENS- Lyon and INRIA-Rocquencourt for their stimulating seminar. We are particularly grateful to Cristina Cornes, Catherine Parent and Christine Paulin-Mohring for helpful discussions. Our thanks go also to the reviewers for their constructive comments.

References

1. M. Aagaard and M. Leeser. A Methodology for Reusable Hardware Proofs. In *International Workshop on Higher Order Logic Theorem Proving and its Applications*, 1992.
2. M. Allemand. *Modélisation Formelle et Preuve de Circuits avec LP*. PhD thesis, Université de Provence, July 1995.
3. L. Arditi. Formal Verification of Microprocessors : a First Experiment with the Coq Proof Assistant. Research Report I3S/Université de Nice - Sophia Antipolis. RR-96-31, 1996.
4. J. Bertot and Y. Bertot. Ctcoq : a System Presentation. In *CADE-13*, 1996.
5. R. S. Boyer and J. S. Moore. *A computational logic handbook*. Academic Press Inc., 1988.
6. A. Camilleri, M. Gordon, and T. Melham. Hardware Verification Using Higher Order Logic. In *From HDL Descriptions to Guaranteed Correct Circuit Designs*. Elsevier Scientific Publishers, 1987.
7. T. Coquand. *Une Théorie des Constructions*. PhD thesis, Université Paris 7, Janvier 1989.
8. T. Coquand and G. Huet. Constructions : A Higher Order Proof System For Mechanizing Mathematics. In *EUROCAL'85*, number 203 in LNCS. Springer-Verlag, 1985.
9. C. Cornes, J. Courant, J.-C. Filliâtre, G. Huet, P. Manoury, C. Muñoz, C. Murthy, C. Parent, C. Paulin-Mohring, A. Saïbi, and W. Benjamin. The Coq Proof Assistant Reference Manual. Technical report, INRIA-Rocquencourt, CNRS-ENS Lyon, Feb. 1996.
10. Y. Coscoy, G. Kahn, and L. Théry. Extracting Text from Proof. In *Typed Lambda-Calculi and Applications*, number 905 in LNCS. Springer-Verlag, April 1995.
11. S. Coupet-Grimal and L. Jakubiec. Vérification Formelle de Circuits avec COQ. In *Journées du GDR Programmation*, Sept. 1994.
12. J.-Y. Girard. The System F of Variable Types, Fifteen Years Later. *Theoretical Computer Science 45*, 1986.
13. M. Gordon. LCF-LSM. Technical Report 41, University of Cambridge, 1984.
14. M. Gordon. Why Higher-Order Logic is a Good Formalism for Specifying and Verifying Hardware. Technical Report 77, University of Cambridge Computer Laboratory, 1986. edited by G.Milne and P. A. Subrahmanyam, North Holland.
15. M. Gordon, R. Milner, and C. Wadsworth. *Edinburgh LCF : A Mechanized Logic of Computation*, volume 78 of *LNCS*. Sringer-Verlag, Department of Computer Science, University of Edinburgh, 1979.
16. F. Hanna, N. Daeche, and M. Longley. Specification and Verification Using Dependent Types. *IEEE Transactions on Software Engineering*, 16(9):949–964, Sept. 1990.

17. W. A. Hunt. Microprocessor Design Verification. *Journal of Automated Reasonning*, 5(4):429–460, 1989.
18. M. Leeser. Using Nuprl for the Verification and Synthesis of Hardware. In C. A. R. Hoare and M. J. C. Gordon, editors, *Mechanized Reasoning and Hardware Design*, International Series on Computer Science. Prentice Hall, 1992.
19. S. Owre, J. Rushby, N. Shankar, and M. Srivas. A Tutorial on Using PVS for Hardware Verification. In *2nd International Conference on Theorem Provers in Circuit Design*, number 901 in LNCS, pages 258–279. Springer Verlag, Sept. 1994.
20. C. Parent. *Synthèse de Preuves de Programmes dans le Calcul des Constructions Inductives*. PhD thesis, Ecole Normale Supérieure de Lyon, Janvier 1995.
21. C. Paulin. *Extraction de Programmes dans Coq*. PhD thesis, Université Paris 7, Janvier 1989.
22. C. Paulin-Mohring. Inductive Definitions in the System Coq: Rules and Properties. Research Report 92-49, Ecole Normale Supérieure de Lyon, 1992.
23. C. Paulin-Mohring. Circuits as Streams in Coq. Verification of a Sequential Multiplier. *Basic Research Action "Types"*, Juillet 1995.
24. J. B. Saxe, S. J. Garland, J. V. Guttag, and J. J. Horning. Using Transformations and Verification in Circuit Design. *Formal Methods in System Design*, (3):181–209, Dec. 1993.
25. M. K. Srivas and S. P. Miller. Applying Formal Verification to a Commercial Microprocessor. *IFIP International Conference on Computer Hardware Description Languages*, Aug. 1995.

Elements of Mathematical Analysis in PVS *

Bruno Dutertre

Department of Computer Science,
Royal Holloway, University of London,
Egham, Surrey TW20 0EX, UK

Abstract. This paper presents the formalization of some elements of mathematical analysis using the PVS verification system. Our main motivation was to extend the existing PVS libraries and provide means of modelling and reasoning about hybrid systems. The paper focuses on several important aspects of PVS including recent extensions of the type system and discusses their merits and effectiveness. We conclude by a brief comparison with similar developments using other theorem provers.

1 Introduction

PVS is a specification and verification system whose ambition is to make formal proofs practical and applicable to large and complex problems. The system is based on a variant of higher order logic which includes complex typing mechanisms such as predicate subtypes or dependent types. It offers an expressive specification language coupled with a theorem prover designed for efficient interactive proof construction.

In previous work we have applied PVS to the requirements analysis of a substantially complex control system [2]. This was part of the SafeFM project which aims to promote the practical use of formal methods for high integrity systems. We used PVS to formalise the functional requirements of the SafeFM case study and to verify several safety critical properties.

The main problem we had with PVS was the limited number of pre-defined notions and pre-proved theorems; a non-negligible part of the work was spent in writing general purpose "background knowledge" theories. In general, we found that PVS provides only the most elementary notions and that some effort must be directed towards constructing re-usable libraries extending the pre-defined bases. This has been recognised by others and the new version of the system (PVS2 [15, 1]) comes with a largely expanded prelude of primitive theories and with better support for libraries.

Our experiment with the SafeFM case study showed that elements of mathematical analysis could be extremely useful for modelling hybrid systems. The case study is a control application including both discrete and analogue elements and the modelling involves continuous functions of time which represent physical variables. Reasoning about such variables can be considerably simplified if

---
* Work partially funded by EPSRC Grant No GR/H11471 under DTI Project No IED/1/9013

standard notions and results of analysis are available. This paper presents the development of a PVS library introducing such notions. The library defines convergence of sequences, limits of functions, continuity, and differentiation, and contains various lemmas and theorems for manipulating these notions.

Applications to hybrid systems were our prime motivation for developing such a library but integrating mathematical analysis to theorem proving can have other interests. Harrison [8] cites applications in areas such as floating point verification [9] or the combination of theorem provers and computer algebra systems [10].

The work presented in this paper is an example of use of PVS in a slightly uncommon domain, different from the traditional computer related applications. It was not obvious from the start whether PVS was a practical tool for doing "ordinary mathematics". Writing the library showed us that PVS could cope without much difficulty with the form of specifications and reasoning encountered in traditional mathematical analysis. In particular, the rich PVS type system was convenient for defining limits, continuity, and derivatives in a fairly natural way, very close to conventional mathematical practice. The library also makes use of some of the most recent features of PVS such as judgements and conversions.

All the proofs were performed using only the pre-defined set of proof commands, without any attempt to define new rules or proof strategies, the equivalents of HOL tactics and tacticals [7]. The high level commands available were powerful enough to handle automatically a large proportion of the proofs.

The remainder of this paper gives a brief introduction to PVS focusing on the aspects most relevant to the library development and presents the main components of the library. Section 4 discusses the qualities and limits of PVS for the application considered and gives a comparison with similar work.

2 An Overview of PVS

PVS is an environment for the construction and verification of formal specifications. The system provides an expressive specification language, a powerful interactive proof checker, and various other tools for managing and analysing specifications. PVS has been applied to large and complex examples in domains such as hardware [14], fault-tolerant protocols [15], or real-time systems [11].

The PVS logic is largely similar to classic higher order logic but with several extensions. The PVS type system is richer than Church's theory of simple types and supports subtyping and dependent types. PVS also includes mechanisms for writing parametric specifications. These features are essential and are described in greater detail in the following sections. We also outline the main characteristics of the PVS proof checker which influenced the formulation of certain aspects of the specifications. A more complete descriptions of the language and prover can be found in [1, 17, 18] and a more formal presentation of the PVS logic is available in [16].

2.1 Type System

Simple Types. PVS includes primitive types such as the booleans or the reals, and classic constructors for forming functions and tuples types. For example,

- [real, real -> bool] is the type of functions from pairs of reals to the booleans,
- [nat, nat, nat] is the type of triples of natural numbers.

There are also other constructions for record types and built-in support for abstract data types [1, 17].

Subtypes. Given an arbitrary function p of type [t -> bool], one can define the subtype of t consisting of all the elements which satisfy p. This type is denoted by $\{x:t \mid p(x)\}$ or equivalently (p). More generally, subtypes can be constructed using arbitrary boolean expressions. For example,

```
nzreal : TYPE = {x : real | x /= 0}
```

declares the type nzreal whose elements are the non-null reals.

Subtypes can also be declared as follows

```
s : TYPE FROM t;
```

this defines s as an uninterpreted subtype of t. With this declaration PVS automatically associates a predicate s_pred:[t -> bool] characteristic of s: the two expressions s and (s_pred) denote the same subtype of t.

By default, PVS does not assume that types are non-empty but the user can assert that types are inhabited as follows:

```
s : NONEMPTY_TYPE FROM t.
```

This is sound as long as t itself is not empty.

Dependent Types. Function, tuple, or record types can be dependent: the type of some components may depend on the value of other components. For example, the function A below

```
A(x:real , (z : {y:real | y<x})) : real = 1 / (x - z)
```

has dependent type [x:real, {y:real | y<x} -> real].

Type Checking. Since arbitrary predicates can occur in type expressions, type checking is undecidable; the user may be asked to prove that specifications are well typed. In general, type correctness of an expression reduces to a finite number of proof obligations known as Type Correctness Conditions (TCCs) generated automatically by the system.

For example, the division operator has type [real, nzreal -> real] and type checking the definition of A above will produce the following TCC:

```
A_TCC1: OBLIGATION (FORALL (x, z: {y | y < x}): (x - z) /= 0).
```

Similarly, type checking an expression such as `A(2, 1)` requires to show that the arguments to `A` are of the right type. TCCs may be generated in various other situations, for example to ensure that recursive definitions are sound or to check that types are non empty when constants are declared [1, 17].

PVS treats the boolean operators and the `if then else` construction in a special way. Ordinary functions are strict: for an expression $f(t_1, \ldots, t_n)$ to type check, all the terms $t_1, \ldots, t_n$ must be type-correct. The boolean operators are not strict; definitions such as the following are type-correct:

```
a(x : real) : bool = x /= 0 AND 1/x > 2
```

The order of the arguments is important; the definition below

```
d(x : real) : bool = 1/x > 2 AND x /= 0
```

gives an unprovable TCC: `FORALL (x : real): x /= 0.`

2.2 Theories and Parameters

PVS specifications are organised in theories. A theory can contain type definitions, variable or constant declarations, axioms, and theorems. The primitive elements of PVS are introduced in the *prelude*, a collection of pre-defined theories. The following example is a fragment of a theory defining sets, extracted from the prelude.

```
sets [T:TYPE]: THEORY
  BEGIN

  set: TYPE = [T -> bool]

  x, y : VAR T

  a, b, c : VAR set

  member(x, a): bool = a(x)

  empty?(a): bool = (FORALL x : NOT member(x, a))

  ...

  END sets
```

The theory has one parameter T; it defines the type `set` (sets are represented by their boolean characteristic function) and the usual set-theoretic operations. Other theories can import `sets` and use the type `set`, the function `empty?` and any other type, constant, axiom, or theorem form `sets`[2]. The variables x, y, a, b, c are local to `sets` and are not exported.

---

[2] Prelude theories such as `sets` are implicitly imported; user-defined theories require an explicit `IMPORTING` clause.

One may import a specific instance of `sets` by providing actual parameters; this takes the following form

```
IMPORTING sets[real].
```

In this case, the identifier `set` refers unambiguously to the type `[real -> bool]`, `member` to a function of type `[real, [real->bool] -> bool]`, etc.

It is also possible to import theories without actual parameters and use names such as `set[real]`, `set[nat]`, `member[bool]` to refer to entities from different instances of `sets`. A more interesting possibility is to let PVS determine automatically the parameters. This provides a form of polymorphism as illustrated below:

```
F(A, B : set[real]) : set[[(A) -> (B)]] = {f : [(A) -> (B)] | true}

empty_function : PROPOSITION
    empty?(B) AND not empty?(A) IMPLIES empty?(F(A, B))
```

Since `set[real]` is `[real->bool]`, the two types (A) and (B) are subtypes of `real`. The function F has dependent type: F(A,B) is the set of all functions of type `[(A) -> (B)]`. In the proposition, the function `empty?` is polymorphic and PVS computes the parameter instantiation for the three occurrences according to the type of the arguments.

In the `sets` example, the parameter T is somewhat similar to a HOL type variable. Theories can also be parameterised by constants, and the user can impose conditions on the parameters. In the latter case, PVS may generate TCCs to check that actual parameters – either given in importing clauses or inferred by the type checker – satisfy the required conditions.

The constraints on parameters can be expressed using dependent types, for example, as follows:

```
intervals [a : real, b : {x : real| a <= x} ] : THEORY
    BEGIN
    J : NONEMPTY_TYPE = { x : real | a <= x AND x <= b}.
```

More complex conditions can be expressed as *assumptions*:

```
theo [T : TYPE FROM real] : THEORY
    BEGIN
    ASSUMING
    two_elements : ASSUMPTION  EXISTS (x, y : T) : x /= y.
```

2.3 Judgements and Conversions

Judgements have been introduced in PVS to solve a practical problem: the large number of TCCs that may be caused by subtyping. The following example, inspired by the PVS2 finite sets library [12], is typical of a very common situation:

- `finite_set` is a subtype of `set`,
- `union` has type `[set, set -> set]`,

– card has type [finite_set -> nat].

Assuming A and B are two constants of type finite_set, the following expression

    card(union(A, B))

generates a TCC: union(A, B) has type set; since card requires a finite_set argument, PVS asks the user to show that union(A, B) is in fact finite. Similar TCCs will appear every time union is applied to finite sets in a context where a result of type finite_set is expected.

A judgement allows one to suppress all these TCCs by indicating to the type checker that the union of finite sets is a finite set:

    JUDGEMENT union HAS_TYPE [finite_set, finite_set -> finite_set].

A proof obligation will be generated to verify that this judgement is valid, but it needs to be proved only once. Every time union is applied to finite sets, the type checker will recognise that the result is finite.

There is a different form of judgement to specify sub-type relations and PVS2 provides another extension to the type system: conversions. A conversion is a function of type [t1 -> t2] that the type checker may apply automatically to a term of type t1 in a context where a term of type t2 is expected.

For example, the prelude defines a conversion extend which transforms a term of type set[S] to a term of type set[T] when S is a subtype of T[3]. Such a conversion could be specified as follows:

    extend(E : set[S]) : set[T] = {x : T | S_pred(x) AND E(x)}
    CONVERSION extend

This allows, for example, to mix sets of reals and sets of natural numbers as in the following declarations:

    A : set[real]
    B : set[nat]
    C : set[real] = union(A, B).

The last expression is automatically transformed to union(A, extend(B)) by the type checker.

2.4 The PVS Prover

The PVS prover is based on sequent calculus and proofs are constructed interactively by developing a proof tree in a classic goal oriented fashion. A main characteristic of PVS is the high level of the proof commands available and the powerful decision procedures built in the prover. These procedures combine equational reasoning and linear arithmetic and include various rules (e.g. beta conversion) for simplifying expressions. It is possible to program proof strategies similar to HOL tactics and tacticals [7].

---

[3] The type set[S] defined as [S -> bool] is not a subtype of set[T].

The rewriting capabilities of PVS play an essential role in the analysis library. In their simplest form, rewrite rules are formulas of the form l = r where the free variables on the right-hand side of the equality occur free in the left-hand side. The prover can rewrite with such a formula by finding a term l' that matches l and replacing it by the corresponding substitution instance r' of r.

Other forms of formulas are accepted as rewrite rules (see [18]); examples taken from the prelude are given below:

```
div_cancel3: LEMMA x/n0z = y IFF x = y * n0z

union_subset2: LEMMA subset?(a, b) IMPLIES union(a, b) = b

surj_inv: LEMMA injective?(f) IMPLIES surjective?(inverse(f)).
```

In the first rule, boolean equivalence is used instead of equality. The second lemma is a conditional rewrite rule; when it is applied, a subgoal may be generated for proving that the premise holds. The last lemma is also a conditional rule, treated by the prover like the equivalent formula

```
injective?(f) IMPLIES surjective?(inverse(f)) = true.
```

Rewrite rules can be applied selectively by the user or can be installed as automatic rewrite rules. This gives a means of extending the set of built-in simplification rules. Once installed, the automatic rewritings can be activated explicitly but they are also used implicitly by many high level commands in combination with the decision procedures.

3 Main Elements of the Library

3.1 Low Level Theories

In PVS, the reals are built-in and constitute a primitive type. The other numerical types are defined as subtypes of real. The prelude contains an axiomatization of the reals which give the usual field and ordering axioms and a completeness axiom: *every non-empty set of reals which is bounded from above has a least upper bound*.

In addition to this axiomatization, a large set of rewrite rules are available in the prelude, useful for manipulating non-linear expressions that the decision procedures do not handle. The prelude also defines common functions such as absolute value, exponentiation, or the minimum or maximum of two numbers.

All these form a large basis of pre-defined theories for the manipulation of reals but it was necessary to extend these basic theories in several ways. The extensions include new lemmas about the absolute value and new properties of the reals, new functions such as the least upper bound or greatest lower bound of sets, and general operations and predicates on real-valued functions.

The definition of least upper bound (sup) illustrates a construction very common in the library. First, a subtype of set[real] defines the sets where sup makes sense, then the function is defined using Hilbert's epsilon operator:

```
U : VAR { S : (nonempty?[real]) | above_bounded(S) }

sup(U) : real = epsilon(lambda x : least_upper_bound?(x, U))
```

Thus **sup** is only defined for non-empty sets, bounded from above. As a consequence, the following equivalence holds:

```
sup_def : LEMMA sup(U) = x IFF least_upper_bound?(x, U).
```

PVS supports overloading; the low level theories define operations +, -, * on real-valued functions as follows:

```
real_fun_ops[T : TYPE] : THEORY
  BEGIN
  f1, f2 : VAR [T -> real]

  +(f1,f2): [T -> real] = lambda (x : T) : f1(x) + f2(x);
  ...
```

Due to the parametric definition, + is polymorphic and applies to sequences (functions of type [nat->real]), functions of type [real->real], etc.

3.2 Limits of Sequences

The theories of sequences are fundamental elements of the library. They define convergence and limits of sequences of reals and other standard notions such as Cauchy sequences or points of accumulations [13]. They also contain important results which are essential for developing the continuity theories. These include standard properties such as the uniqueness of the limit, the convergence of increasing or decreasing bounded sequences, the Bolzano-Weierstrass theorem: *every bounded sequence has a point of accumulation*, and the completeness of the reals: *every Cauchy sequence is convergent*. All the proofs are classic and translate without much difficulty to PVS. The completeness theorem follows from Bolzano-Weierstrass which is proved using a well known property: every sequence of reals contains a monotone sub-sequence.

PVS allows the function **limit** to be defined and used in a fairly standard way. The specification is similar to the definition of **sup**:

```
convergence(u, l) : bool =
   FORALL epsilon : EXISTS n : FORALL i :
      i >= n IMPLIES abs(u(i) - l) <= epsilon

convergent(u) : bool = EXISTS l : convergence(u, l)

limit(v : (convergent)) : real = epsilon(lambda l : convergence(v, l)).
```

The theories contain a collection of propositions – usable as conditional rewrite rules – for combining convergent sequences:

```
limit_sum : PROPOSITION
   convergence(s1, l1) AND convergence(s2, l2)
      IMPLIES convergence(s1 + s2, l1 + l2)

limit_diff : PROPOSITION
   convergence(s1, l1) AND convergence(s2, l2)
      IMPLIES convergence(s1 - s2, l1 - l2).
```

Installing these propositions as automatic rewrite rules makes trivial the proof of theorems such as the following:

```
test1 : LEMMA
   convergence(s1, l1) AND convergence(s2, l2) AND l2/=0
      IMPLIES convergence(s1 * (1/s2) - s2, l1 * (1/l2) - l2).
```

However, rules of the above form do not apply in the following situation:

```
test2 : LEMMA
   convergence(s1, 1) AND convergence(s2, 1)
      IMPLIES convergence(s1 - s2, 0).
```

This proposition is an immediate consequence of limit_diff but the latter cannot be used as a rewrite rule; it does not match convergence(s1 - s2, 0).

It is possible to do better using limit and convergent. First, we specify closure properties and judgements:

```
convergent_diff : PROPOSITION
   convergent(s1) AND convergent(s2) IMPLIES convergent(s1 - s2)

convergent_prod : PROPOSITION
   convergent(s1) AND convergent(s2) IMPLIES convergent(s1 * s2)
...
JUDGEMENT +, -, * HAS_TYPE [(convergent), (convergent) -> (convergent)]
...
```

then the following propositions provide more convenient rewrite rules:

```
v1, v2 : VAR (convergent)

lim_diff : PROPOSITION   limit(v1 - v2) = limit(v1) - limit(v2)
lim_prod : PROPOSITION   limit(v1 * v2) = limit(v1) * limit(v2)
...
```

Combined together all these rules are flexible enough to perform automatically a large class of simple limit computations. The two examples below are similar to test2 and can be proved by automatic rewriting:

```
test3 : LEMMA   limit(v1) = limit(v2) IMPLIES limit(v1 - v2) = 0

test4 : LEMMA   convergent(s1) AND convergent(s2)
   AND limit(s1) - 1 = limit(s2) * limit(s2)
      IMPLIES limit(s1 - s2 * s2) = 1
```

The first case is straightforward. The other requires slightly more work from the prover: the rules `lim_diff` and `lim_prod` apply but there is also a TCC to check that `s2 * s2` is of type `(convergent)`; this TCC is itself rewritten and reduced to `true` by `convergent_prod`.

Both lemmas are proved by a single command:

```
(GRIND :DEFS NIL :THEORIES ("convergence_ops")
       :EXCLUDE "abs_convergence").
```

This installs rewrite rules contained in theory `convergence_ops` then applies these rules and the decision procedures. The other parameters prevent the expansion of the definitions of `limit` and `convergent` and exclude a rewrite rule which would otherwise provoke infinite rewritings.

The original `test2` can be proved by exactly the same command with just an extra rule in `convergence_ops`:

```
limit_equiv : LEMMA
    convergence(s, 1) IFF convergent(s) AND limit(s) = 1.
```

3.3 Limits of Functions

The second main group of theories is concerned with pointwise limits of numeric functions. With the conventions used in [13], a limit is denoted:

$$\lim_{\substack{x \to a \\ x \in E}} f(x)$$

where E is a set in a metric space[4], f a real-valued function defined on E, and a a point adherent to E.

A similar PVS formulation is possible using dependent types but it presents certain inconveniences. If f is defined on a larger domain than E then

$$\lim_{\substack{x \to a \\ x \in E}} f(x)$$

still makes sense; we just have informally replaced f by its restriction to E. In PVS, function restrictions are not so easy; one can either introduce them explicitly (e.g. `lambda (x:(E)):f(x)`) or rely on automatic conversions. This tends to clutter specifications or make proofs less elegant.

For a simpler formulation, one could drop E and assume that x varies over the domain of f. This is less general and E is convenient for considering distinct limits of f at the same point a (for example on the left or on the right a).

After several attempts, we found the following definition sufficiently general and convenient.

```
convergence_functions [T : TYPE FROM real] : THEORY
    ...
    convergence(f, E, a, 1) : bool = adh(E)(a) AND
      FORALL epsilon : EXISTS delta :
        FORALL x : E(x) /\ abs(x - a) < delta => abs(f(x) - 1) < epsilon
```

---

[4] In our case, the metric space is $\mathbb{R}$.

with `f` of type `[T->real]`, E a set of reals, and a and l two reals. The variable x is of type T and `adh(E)(a)` holds iff a is adherent to $\{x:T \mid E(x)\}$.

This generic definition of convergence allows us to prove only once standard results: the limit is unique, the limit of a sum is the sum of the limits, etc. All these specialise easily to different types of functions by parameter instantiation. The argument E gives an extra level of flexibility; for example, the expression

```
convergence(f, {x|x<0}, 0, -1)
```

corresponds to the limit of `f` on the left of 0.

In the definition of convergence, the variable x may be equal to the adherence point a; this follows the convention of [13]. However, a is automatically excluded if it is not in the domain T of `f` or if it is not in the set E.

A separate theory develops the most common case of limits where E is the set of all reals, that is, where x can vary over the whole domain of `f`. This specialised theory defines a function `lim` as follows:

```
convergence(f, a, l) : bool = convergence(f, fullset[real], a, l)

convergent(f, a) : bool = EXISTS l : convergence(f, a, l)

lim(f, (x0 : {a | convergent(f, a)})) : real =
    epsilon(LAMBDA l : convergence(f, x0, l)).
```

Because of the dependent type, `lim(f, a)` is defined only if `f` is convergent at a. This function makes possible the specification of powerful rewrite rules, similar to those associated with the limit of sequences.

3.4 Continuity and Differentiation

Continuity of a function `f : [T -> real]` is defined easily:

```
continuous(f, x0) : bool = convergence(f, x0, f(x0))

continuous(f) : bool = FORALL x0 : continuous(f, x0).
```

Once again, the definition is parametric on a subtype T of the reals. Differentiation uses the Newton quotient `NQ` defined by:

```
A(x) : set[nzreal] = { u:nzreal | T_pred(x + u) }

NQ(f, x)(h : (A(x))) : real = (f(x + h) - f(x)) / h.
```

Dependent types and the predicate `T_pred` are essential here: `NQ(f, x)(h)` is only defined if h is non null and x+h is in the domain of `f`. Then `f` has a derivative at x iff `NQ(f, x)` has a limit at 0:

```
derivable(f, x) : bool = convergent(NQ(f, x), 0)

deriv(f, (x0 : {x | derivable(f, x)})) : real = lim(NQ(f, x0), 0)
```

This requires NQ(f, x) to be defined for h arbitrarily close to 0. In order to ensure that condition, we need assumptions on the parameter T:

```
connected_domain : ASSUMPTION
    FORALL (x, y : T), (z : real) : x <= z AND z <= y IMPLIES T_pred(z)

not_one_element : ASSUMPTION
    FORALL (x : T) : EXISTS (y : T) : x /= y
```

These two conditions ensure that T represents a possibly infinite real interval, not reduced to a single point.

The general properties of limits of functions are used to derive rewrite rules for proving continuity and computing derivatives. It is also convenient to introduce new types for continuous and derivable functions with adequate judgements. For our initial objective – reasoning about hybrid systems – the most important results are theorems which describe the behaviour of continuous or derivable functions on a closed interval:

- if f is continuous on $[a, b]$ then it is bounded and has a maximum and a minimum on $[a, b]$;
- for any y between $f(a)$ and $f(b)$ there is a point x in $[a, b]$ such that $y = f(x)$ (the intermediate value theorem).

These theorems and many similar properties such as the mean value theorem are included in the library.

3.5 An Example Proof

The proof of the mean value theorem is representative in its size and complexity of many proofs in the library. The theorem and a lemma are given below:

```
mean_value_aux : LEMMA
    derivable(f) AND a < b AND f(a) = f(b) IMPLIES
      EXISTS c : a < c AND c < b AND deriv(f, c) = 0

mean_value : THEOREM
    derivable(f) AND a < b IMPLIES
      EXISTS c : a < c AND c < b AND deriv(f, c) * (b-a) = f(b)-f(a).
```

The whole proof is the following:

```
(SKOSIMP)
(NAME-REPLACE "C" "b!1 - a!1" :HIDE? NIL)
(NAME-REPLACE "B" "f!1(b!1) - f!1(a!1)" :HIDE? NIL)
(ASSERT)
(AUTO-REWRITE-THEORY "derivatives[T]" :EXCLUDE ("derivable" "deriv"))
(USE "mean_value_aux" ("f" "f!1 - (B/C) * (I[T] - const_fun[T](a!1))"))
(GROUND)
(("1"
  (SKOSIMP)
```

```
(INST?)
(EXPAND "derivable")
(INST - "c!1")
(ASSERT)
(ASSERT)
(USE "div_cancel2")
(ASSERT))
("2" (DELETE -3 2) (GRIND) (USE "div_cancel2") (ASSERT))).
```

The proof applies lemma mean_value_aux to the function $f - (B/C) * (I[T] - a)$ where $B = f(b) - f(a)$, $C = b - a$, and $I[T]$ is the identity function. We have to show that the premises of the lemma hold and that, for the real c whose existence is asserted by mean_value_aux, we have $f'(c) * C = B$. All this is done using rewrite rules from theory derivatives[T], lemma div_cancel2 from the prelude, and the decision procedures.

4 Discussion and Related Work

The work presented in this paper represents a relatively large application of PVS. The library consists of around 3000 lines of specifications (including comments and blank lines) organised in 30 theories, and contains 519 theorems (including 156 TCCs). The amount of effort involved can be estimated at around 6 man-months. Most of the proofs are of a similar complexity as the proof of mean_value; there are a few larger proofs (up to 78 proof steps) but many propositions are proved in just one or two commands. Type checking the whole library and running all the proofs takes about 45 min (real time) on a Sparc 5 workstation with 64Mb of central memory.

The development gave us the opportunity to explore some of the most advanced features of PVS. The library relies extensively on the facilities offered by the rich type system: overloading of operators, subtypes, and dependent types. These are very comfortable for writing concise specifications, in a form very close to standard mathematical notations. The possibility to parameterise theories is at least as important; several of the notions developed could be specified without subtypes or dependent types but parameters are essential for re-usability and generality.

Type judgements are very effective in reducing the amount of effort spent on proof obligations. There are still some limitations: for example an expression such as lim(f1+f2,a) produces a TCC to check that f1+f2 is convergent at a. It would be convenient to be able to indicate to the type checker that $f_1 + f_2$ is convergent at a when both f_1 and f_2 are. In their present form, judgements do not give this possibility.

Unlike judgements, conversions did not appear extremely useful; very few are used in the library. The following one extends a real to a constant functions:

```
const_fun(a) : [T -> real] = LAMBDA (x : T) : a
CONVERSION const_fun.
```

We expected this conversion to make possible expressions such as

```
limit(s + 1) = limit(s) + 1,
```

with the first occurrence of 1 converted to a constant sequence. Unfortunately, this does not work; PVS applies a conversion but not the one we expected: `s+1` is transformed to `LAMBDA (x:nat):s(x)+1`. The rewrite rules do not match this lambda expression.

In general, automatic conversions can have unexpected effects. For example if A and B are of type `set[real]` and `set[nat]` respectively, then the "identity" `union(A, B) = union(B, A)` does not hold. The conversions inserted by PVS are not the same on both sides of the equality:

```
union(A, extend(B)) = extend(union(B, restrict(A))).
```

Because the user has no control on where conversions are introduced, other than making them explicit, they can only be used safely in very restricted situations.

Despite this last criticism, we think that PVS is a very powerful and practical tool. Its main qualities are the expressiveness of its specification language and type system, and the power and simplicity of use of its interactive prover. The library showed that relatively complex notions could be formalised easily and that proofs which sometimes rely on elaborate arguments could be performed without difficulty.

As far as we are aware, analysis is not a very common domain of application for mechanical theorem provers. The work the most closely related is due to Harrison who developed a large fragment of analysis in HOL [8]. There is also an extensive formalization of analysis and calculus in IMPS [4, 6]. Our own construction is modest in comparison: the HOL library for reals covers notions such as power series and transcendental functions and IMPS provides rich theories for metric and normed spaces.

There are important differences between the three systems in the way the reals are defined. In HOL, the positive rationals are first constructed from the natural numbers then the reals are constructed from the rationals using Dedekind cuts [8]. This corresponds to the HOL philosophy of having a small implementation of a basic logical kernel that users can extend in a safe way. IMPS adopts an axiomatic approach to mathematics [3] and the reals are specified as a complete ordered field. The emphasis of PVS is more on practicality issues and usability: the reals are axiomatized but a lot of knowledge is also embedded in the decision procedures.

Different approaches are used in the three systems for developing analysis. Both HOL and IMPS[5] define several notions with a general and abstract perspective [8, 5]. For example, convergence is defined in HOL using *convergence nets* instead of having two separate notions, one for sequences and one for functions. Economy is the main motivation; convergence nets avoids having to prove several theorems twice. IMPS is a system for doing mathematics and as such it

---

[5] Many thanks to the reviewers who signalled to us the IMPS work.

includes theories for abstract metric spaces or normed vector spaces which are of interest to mathematicians.

Our goal was more pragmatic and although abstract notions can be introduced in PVS we preferred a more direct approach. Furthermore, being too general may be counter-productive. On an abstract type such as convergence nets, the PVS decision procedures do not apply and proofs may get rapidly tedious and intricate. It is better to keep separate notions of convergence even if some of the theorems seem duplicated. With decision procedures, the proofs are not that difficult anyway, and the "hardest" parts which often involve manipulations of non-linear real expressions can be isolated in re-usable lemmas.

The three systems are different in the way specifications and theorems are introduced. They all allow interactive backwards proof construction (HOL also supports forward proofs) but the form of interactions are different. In PVS, the user first states theorems and then tries to prove them. High level proof commands are available and are sufficient for doing large proofs. In HOL and in IMPS, theorems are constructed with a functional language and the user is encouraged to define new functions for doing proofs. This gives HOL and IMPS some meta-theoretical possibilities not available in PVS, at least not to the ordinary user. For example, the HOL real library contains ML functions to build theorems from an equivalence relation R on a type σ and a list of theorems about representatives of the equivalence classes of R. The use of ML also makes HOL easier to interface with external tools as described in [10].

PVS and IMPS seem to provide much better support than HOL for modularity. IMPS uses a sophisticated technique based on theory interpretation. Although not as general, the PVS mechanism of parametric theories and parameter assumptions is extremely useful. The PVS definitions of limits, continuity, and derivability are parametric. This gives a superior level of generality and flexibility than the same notions from the HOL library which only applies to functions from $\mathbb{R}$ to $\mathbb{R}$.

5 Conclusion

This paper has presented an example of applying mechanical theorem proving to ordinary mathematics. The main result from this work is that PVS is a powerful and practical tool for this purpose. Due to the rich type system, specifications can be written in a very natural way. The PVS theorem prover is efficient for doing proofs which require more elaborate forms of reasoning than encountered in traditional computer-related areas. The proofs rely extensively on the decision procedures supplemented with user-defined rewrite rules but do not require any extension to the pre-existing proof commands.

The library covers the most fundamental elements of analysis and there are a lot of possibilities of extensions. The priority might be to include power series and define the common trigonometric functions, the exponential and logarithmic functions as done in HOL [8]. However, in its present state we hope the library is rich enough to provide adequate support for reasoning about hybrid systems.

References

1. J. Crow, S. Owre, J. Rushby, N. Shankar, and M. Srivas. A tutorial introduction to PVS. In *WIFT'95 Workshop on Industrial-Strength Formal Specification Techniques*, April 1995.
2. B. Dutertre. Coherent Requirements of the SafeFM Case Study. Technical Report SafeFM-050-RH-2, SafeFM project, September 1995.
3. W. M. Farmer, J. D. Guttman, and F. J. Thayer. Little theories. In D. Kapur, editor, *Automated Deduction—CADE-11*, volume 607 of *Lecture Notes in Computer Science*, pages 567–581. Springer-Verlag, 1992.
4. W. M. Farmer, J. D. Guttman, and F. J. Thayer. IMPS: An Interactive Mathematical Proof System. *Journal of Automated Reasoning*, 11:213–248, 1993.
5. W. M. Farmer, J. D. Guttman, and F. J. Thayer. The IMPS user's manual. Technical Report M-93B138, The MITRE Corporation, 1993.
6. W. M. Farmer and F. J. Thayer. Two computer-supported proofs in metric space topology. *Notices of the American Mathematical Society*, 38:1133–1138, 1991.
7. M.J.C. Gordon and T.F. Melham. *Introduction to HOL. A theorem proving environment for higher order logic.* Cambridge University Press, 1993.
8. J. Harrison. Constructing the real numbers in HOL. *Formal Methods in System Design*, 4(1/2):35–59, July 1994.
9. J. Harrison. Floating point verification in HOL. In E. T. Schubert, P. J. Windley, and J. Alves-Foss, editors, *Proceedings of the 8th International Workshop on Higher Order Logic Theorem Proving and Its Applications*, volume 971 of *Lecture Notes in Computer Science*, pages 186–199. Springer-Verlag, 1995.
10. J. Harrison and L. Théry. Extending the HOL theorem prover with a computer algebra system to reason about the reals. In J. J. Joyce and C.-J. H. Seger, editors, *Proceedings of the 6th International Workshop on Higher Order Logic Theorem Proving and its Applications (HUG'93)*, volume 780 of *Lecture Notes in Computer Science*, pages 174–184. Springer-Verlag, 1993.
11. J. Hooman. Correctness of Real Time Systems by Construction. In *Formal Techniques in Real-Time and Fault-Tolerant Systems*, pages 19–40. Springer-Verlag, LNCS 863, September 1994.
12. D. Jamsek, R. W. Butler, S. Owre, and C. M. Holloway. PVS finite sets library, 1995. Part of the standard PVS distribution.
13. S. Lang. *Analysis I*. Addison-Wesley, 1968.
14. S. P. Miller and M. Srivas. Formal Verification of the AAMP5 Microprocessor: A Case Study in the Industrial Use of Formal Methods. In *WIFT'95 Workshop on Industrial-Strength Formal Specification Techniques*, April 1995.
15. S. Owre, J. Rushby, N. Shankar, and F. von Henke. Formal verification for fault-tolerant architectures: Prolegomena to the design of PVS. *IEEE Transactions on Software Engineering*, 21(2):107–125, February 1995.
16. S. Owre and N. Shankar. The Formal Semantics of PVS. Technical report, Computer Science Lab., SRI International, June 1995.
17. S. Owre, N. Shankar, and J. M. Rushby. *The PVS Specification Language*. Computer Science Lab., SRI International, April 1993.
18. N. Shankar, S. Owre, and J. M. Rushby. *The PVS Proof Checker: A reference Manual*. Computer Science Lab., SRI International, March 1993.

Implementation Issues About the Embedding of Existing High Level Synthesis Algorithms in HOL*

Dirk Eisenbiegler[1], Christian Blumenröhr[1] and Ramayya Kumar[2]

[1] Institute for Circuit Design and Fault Tolerance (Prof. Dr.-Ing. D. Schmid),
University of Karlsruhe, Germany
[2] Forschungszentrum Informatik (Prof. Dr.-Ing. D. Schmid), Karlsruhe, Germany
e–mail: eisen@ira.uka.de, blumen@ira.uka.de, kumar@fzi.de

Abstract. This article describes the embedding of high level synthesis algorithms in HOL. For given standard synthesis steps, we describe, how its data can be mapped to terms in HOL and the synthesis process be expressed by means of a logical derivation. In contrast to post-synthesis verification techniques our approach is constructive in a sense that the proof is derived during synthesis rather than "guessed" afterwards. Therefore one does not get into the hardship of NP-completeness or undecidability. Our approach ensures correctness based on the HOL system and is also performed fully automatically.

1 Introduction

During the hardware design process of digital circuits, more and more complex tools are involved. Due to their complexity, guaranteeing the correctness of synthesis software is crucial. Bugs in the software may lead to incorrect hardware implementations.

One approach towards proving the correctness of implementations is by post-synthesis verification. An excellent overview of verification techniques is given in [Gupt92, Melh93]. However, full automation is only achievable for comparatively small sized circuits at lower levels of abstraction. For large sized circuits, verification algorithms either run into space/time hurdles or the user has to interact and perform some proofs by hand.

Formal synthesis is another approach towards hardware correctness. We consider formal synthesis as a derivation of the implementation from the specification by logical refinements.

We are developing a formal synthesis toolbox called HASH (Higher order logic Applied to Synthesis of Hardware) which exploits standard synthesis algorithms and is applicable to different abstraction levels. It is based on the HOL system, i.e. hardware is represented by means of HOL terms, and only rule applications are used to transform hardware descriptions. As opposed to conventional

---
* This work has been partly financed by the Deutsche Forschungsgemeinschaft, Project SCHM 623/6-1.

synthesis tools, where there is no restriction on how to compute the implementation, our approach can only produce correct hardware implementations. The reliability of our synthesis conversions only depends on the correctness of the implementation of the HOL core and is independent from the complexity of the conversions. In this article, we will present the high level synthesis component of HASH.

Other approaches in the area of formal synthesis are [Lars94, AHL92, HaLD89, John84, JoSh90]. All these above-mentioned techniques have one common drawback, namely they do not exploit the knowledge of the algorithms which abound in synthesis. Additionally, the interactions to be performed during synthesis are at the schematic level or from a logician's point of view. The novelty of our current approach is that no new synthesis algorithms (either formal or informal) are proposed, but a general scheme for logically embedding various existing synthesis algorithms within a formal set-up is presented.

The outline of this paper is as follows: We will first describe the high level synthesis procedure in an informal manner (section 2). Then the logical representations and the logical transformations corresponding to the synthesis process are introduced in sections 3, 4 and 5. Afterwards, we will present some experimental results (section 6) and finally discuss the embedding of existing high level synthesis techniques (section 7).

2 Our High Level Synthesis Process

The starting point of our approach is a so called basic block. Basic blocks are data flow graphs describing the input/output relation by a composition of atomic operations. The timing of the atomic operations is static in a sense that they can be executed in fixed time (see figure 1). The functional relation represents a pure algorithmic description without any timing information.

The result of high level synthesis is a structure at the RT-level. Our synthesis process consists of the following steps: scheduling, register allocation and binding, allocation and binding of functional units. Interface synthesis will not be considered in this paper.

Our implementation[1] does not yet allow pipelining, instead all hardware resources (functional units as well as registers) will be reused during different clock ticks of one evaluation period.

Also the synthesis approach currently does not support any control flow. For more details on high level synthesis see [GDWL94, CaWo91].

Scheduling

Scheduling determines the number of control steps k needed for the evaluation of the algorithm and assigns each operation to one particular control step $0, 1, \ldots, k$

---

[1] Currently, no chaining of functional units and no multi-cycle operation units are used.

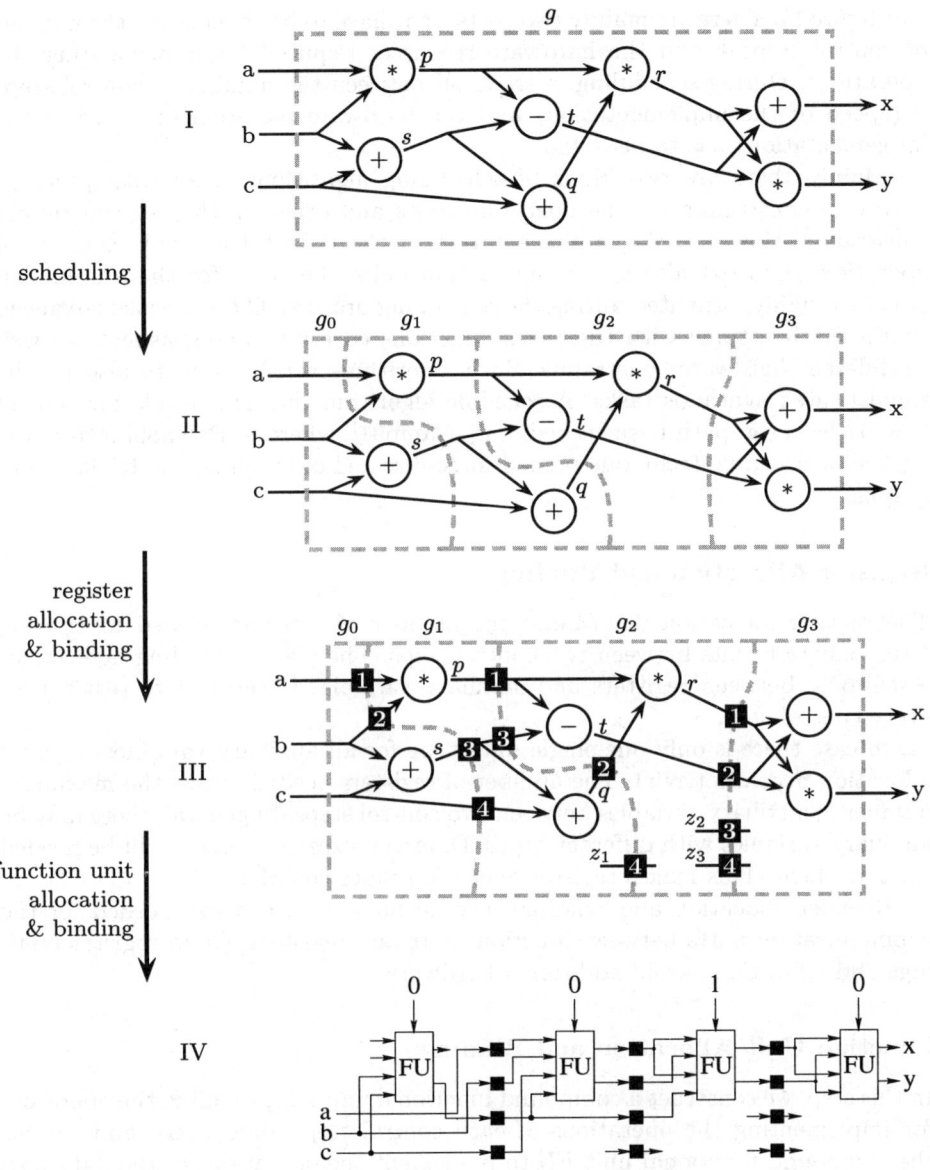

Fig. 1. High Level Synthesis Process

(see figure 1). There are mainly two costs, that have to be considered: the number of control steps k and the hardware resources required for implementing the operations. During scheduling, a trade off between the number of control steps k (speed of the implementation) and the hardware requirements (size of the implementation) has to be found.

Mainly there are two kinds of scheduling algorithms: ones with pregiven hardware constraints for the operation units and others with pregiven timing constraints. However, the implementation at the RT-level not only consists of operation units but also of communication units. The cost for these units can only be roughly estimated during the scheduling process. There are also advanced synthesis algorithms with their cost functions covering timing aspects as well as different hardware constraints. Such algorithms can be used to also handle sophisticated synthesis tasks. A schedule algorithm that is suitable for control flow paths is e.g. path-based scheduling [Camp91], whereas [PaKn89] introduces a possible schedule technique named force-directed only applicable to data flow graphs.

Register Allocation and Binding

The register allocation determines the number of registers needed for storing intermediate results between two control steps. The register binding determines a mapping between registers and auxiliary variables (intermediate results) for every control step.

In case there is only one single data type for all auxiliary variables, register allocation becomes trivial. The number of registers needed equals the maximum number of auxiliary variables between two control steps. In general, there may be auxiliary variables with different types. Different sizes of registers will be needed to store them. This makes register allocation more complex.

Register allocation and binding have an impact on the size needed for the communication parts between function units and registers. Good register bindings and allocations avoid additional hardware.

Function Unit Allocation and Binding

In this step, we construct a compound functional unit FU providing the operators for implementing the operations of each control step (allocation), and we use the compound functional unit FU to implement the operations of the data flow graph (binding). The function units are assumed to be given in a library. The library describes the mapping between its components and the operation(s) they can perform. There may be function units that are implementations of single operations as well as multi-purpose units with control input signals for selecting different operations. In our example, the function units consists of a multiplier implementing the ∗-operation and a multi-purpose unit implementing the + and − operation, where the operation is selected by a control signal having one of the values 0 and 1, respectively. Besides the functional aspects, the library also contains cost information such as area and power consumption.

3 Formal Representation of Data Flow Graphs

The efficiency of software strongly depends on the underlying data structures. In synthesis tools, suitable hardware representations have to be found. This also holds for our formal synthesis approach, where hardware is represented by means of HOL terms. In our approach, data flow graphs are represented as follows:

$\lambda(x_1, \ldots, x_m).$
  let $\langle outvars_1 \rangle = op_1 \langle invars_1 \rangle$ in
  let $\langle outvars_2 \rangle = op_2 \langle invars_2 \rangle$ in
    $\vdots$
  let $\langle outvars_l \rangle = op_l \langle invars_l \rangle$ in
  $(y_1, \ldots, y_n)$

The above structure describes its input/output function in terms of its basic operations. $x_1, x_2, \ldots, x_m$ are the inputs, $y_1, y_2, \ldots, y_n$ the outputs and op_1, $op_2, \ldots, op_l$ the operations of the data flow graph. let-terms are only used for a better readability of β-redices. Each let-term describes the connectivity of one operation. For all i, $\langle invars_i \rangle$ and $\langle outvars_i \rangle$ denote the inputs and outputs of operation op_i, respectively. The inputs and outputs of operations are tuples, with each operation having a specific arity of its input and output tuple.

Since these terms represent pure data flow graphs, i.e. no cycles are present, a partial ordering on the set of nodes is induced. This partial order corresponds to the fact, that some operation A must be executed before B if the output of A happens to be an input to B. This partially ordered data flow graph is represented as an arbitrarily ordered list, whereby the data dependency between the nodes is respected.

The following term gives an example for a data flow graph representation in HOL. The synthesis state in figure 1/I is formally represented as follows:

$\lambda(a, b, c).$
  let $p = a * b$ in
  let $s = b + c$ in
  let $q = s + c$ in
  let $r = p * q$ in
  let $t = p - s$ in
  let $x = r + t$ in
  let $y = r * t$ in
  (x, y)

A constructor function named `mk_dfg` and a destructor function `dest_dfg` have been implemented. In ML, dfg's are represented with the following type:

```
type dfg =
  {
    inputs:term list,
    outputs:term list,
```

```
   operations :
     {operator:term, invars:term list, outvars:term list} list
};
```

mk_dfg maps ML terms of type dfg to the corresponding HOL term. dest_dfg is the inverse function.

During scheduling, the function g is split into a concatenation of functions $g_1, g_2, \ldots, g_k$ with $g = g_k \circ \ldots \circ g_2 \circ g_1$ and each function again represents a data flow graph. The synthesis states described in figures 1/II and 1/III are formally represented as follows:

$$\langle dfg_k \rangle \circ \ldots \circ \langle dfg_2 \rangle \circ \langle dfg_1 \rangle$$

During the allocation and binding of the function units, a compound function unit FU is introduced as an abbreviation. This abbreviation is described by means of a β-redex. The synthesis state described in figure 1/IV is represented as follows:

let $FU = \langle dfg \rangle$ in
$\quad \langle dfg'_k \rangle \circ \ldots \circ \langle dfg'_2 \rangle \circ \langle dfg'_1 \rangle$
end

In this representation, each data flow graph $\langle dfg'_i \rangle$ consists of a single FU operator.

4 Transforming the Data Flow Graphs within HOL

This section describes, how the synthesis process described in figure 1 is implemented as a conversion in HOL. Our high level synthesis conversion is steered by external control information (the schedule, the register-allocation table, etc.). In this section we will only describe the logical aspects of formally deriving the synthesis result from the input data flow graph. The computation of the control information and invocation of the external heuristics will be discussed in section 7.

The approach is based on a conversion for normalizing functions. We will first describe this conversion and then describe, how the synthesis steps are realized using this conversion.

Function Normalization

All HOL representations corresponding to figure 1 are nothing but simple compositions of the same basic functions. In principle, normalizing such representations is pretty simple. The general algorithm looks as follows:

1. the original term g is converted to $\lambda(x_1, x_2, \ldots x_m).g(x_1, x_2, \ldots x_m)$ by applying a paired η-reduction in the inverse direction

2. the ∘ operations are expanded by rewriting with the definition of ∘ (if there are any) and the function unit abbreviation is expanded (provided there is one)
3. β-reductions and paired β-reductions are performed wherever possible

In all cases, the result looks as follows:

$$\lambda(x_1, x_2, \ldots x_m).v[x_1, x_2, \ldots, x_m]$$

In $v[x_1, x_2, \ldots x_m]$ there are no β-redices left and there is nothing but pure function applications.

A Universal Conversion

We will now introduce a simple conversion which is applicable to all synthesis steps (figure 2).

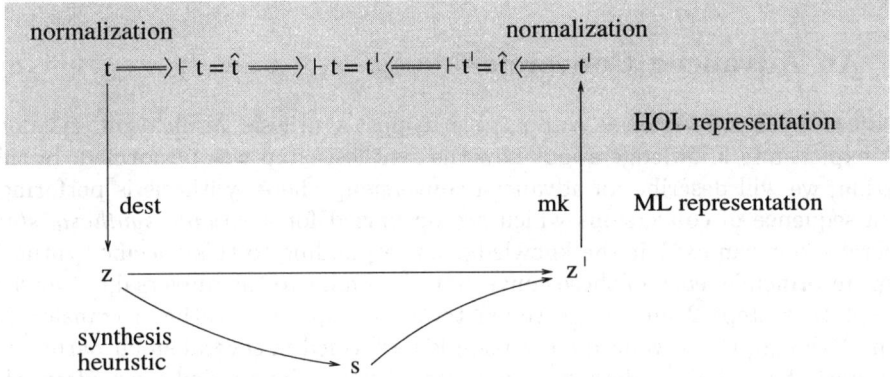

Fig. 2. Universal Conversion for all Synthesis Steps

1. The HOL term representation t is switched to its ML representation z. This is performed by applying some dest-function, which is based on `dest_dfg` (see section 3).
2. For the next step some external control information s (schedule, register allocation table, etc.) is required, which is produced by some arbitrary heuristic. According to s, z is then mapped to some new ML data structure z' corresponding to the result of the synthesis step under consideration. Step 2 is performed completely outside the logic.
3. The data structure z' is translated back to its HOL representation t'. This is performed by applying some mk-function, which is based on `mk_dfg` (see section 3).

4. Both t and t' are normalized by means of applying a normalization conversion. The results should be the same: $\vdash t = \hat{t}$ and $\vdash t' = \hat{t}$.
5. The equations $\vdash t = \hat{t}$ and $\vdash t' = \hat{t}$ are combined to $\vdash t = t'$ (symmetry and transitivity of equivalence).

The major drawback of this universal conversion is the complexity of step 4 when dealing with dfgs with a big depth, i.e. maximum number of operations on a path from some input to some output. Data flow graphs whose intermediate nodes have larger fanouts, i.e. the output of a node is used by many successor nodes as inputs, lead to a number of duplications during β-reduction. Since such β-redices can be nested, the term size and time consumption in step 4 may grow exponentially with the depth.

The universal conversion not only works for single synthesis step, but it is also possible to combine several of our synthesis steps within step 2 of the conversion. Applying the universal conversion mechanism to the entire synthesis process reduces the time consumption since step 4 has to be performed only once rather than thrice (scheduling, register allocation & binding and FU allocation & binding).

5 An Advanced Conversion

The universal conversion is comparable to post-synthesis verification and does not exploit any knowledge about how the synthesis step was performed. In this section, we will describe an advanced conversion, where synthesis is performed by a sequence of conversions which are optimized for a *specific synthesis step*. Thereby, one can exploit the knowledge corresponding to this specific synthesis step. In principle, each of these conversions is similar to the universal conversion except that steps 2 and 4 are tuned towards a specific synthesis transformation. Although the advanced conversion is performed in several small parts, and therefore the technique described in section 4 has to be applied more often, the overall cost is reduced due to the remarkably lower cost for step 4 within each part.

The Scheduling Conversion

The idea of our scheduling conversion is to split the data flow graph step by step rather than doing it all at once, as in the universal synthesis conversion. β-reduction is only applied to those variables whose corresponding nodes have been assigned to the current control step. Although some β-redices will remain, the terms achieved after normalization will be equal.

Other than in the universal synthesis conversion, $k - 1$ conversions (k – number of control steps) have to be applied successively rather than applying one single conversion. Hence, the exponential complexity associated with step 4 is avoided.

Figure 3 shows a HOL session performing the scheduling step applied to the example of figure 1. The HOL conversion SCHEDULING_CONV accomplishes the scheduling transformation according to the schedule which is determined by the scheduling heuristic. SCHEDULING_CONV gets the scheduling heuristic as a parameter. In this example, we applied the force-directed scheduling heuristic. Any other scheduling heuristic can be embedded as well (see section 7). For sake of readability, we used let-expressions rather than β-redices. EXPAND_LETS_CONV and ABBREVIATE_LETS_CONV have been applied to convert let-expressions to β-redices and vice versa.

```
- (
EXPAND_LETS_CONV THENC
(SCHEDULING_CONV force_directed) THENC
UNEXPAND_LETS_CONV
)
    (--'\(a,b,c).
        let p = a*b in
        let s = b+c in
        let q = s+c in
        let r = p*q in
        let t = p-s in
        let x = r+t in
        let y = r*t in
        (x,y)
    '--');
= = = = = = = = = = = = = = val it =
|- (\(a,b,c).
    let p = a * b
    in
    let s = b + c
    in
    let q = s + c
    in
    let r = p * q
    in
    let t = p - s in let x = r + t in let y = r * t in x,y) =
    (((\(r,t). let x = r + t in let y = r * t in x,y) o
      (\(p,q,s). let r = p * q in let t = p - s in r,t)) o
      (\(a,b,c,s). let p = a * b in let q = s + c in p,q,s)) o
      (\(a,b,c). let s = b + c in a,b,c,s) : thm
-
```

Fig. 3. HOL session performing a scheduling step

The Register Allocation and Binding Conversion

Register allocation and binding have one thing in common: they only have an effect on the interfaces between the slices. In our register allocation and binding conversion, the interfaces are changed step by step rather than all at once. The interfaces between $\langle dfg_i \rangle$ and $\langle dfg_{i+1} \rangle$ are changed by applying the universal synthesis conversion to $\langle dfg_i \rangle \circ \langle dfg_{i+1} \rangle$. Therefore in each step our universal synthesis conversion only has to be applied to a small subterm — the rest of the term remains unchanged. Again $k-1$ applications are needed to do the job, but it pays out since the data flow graph considered is significantly smaller.

To be able to apply the interface changing conversion to all subterms $\langle dfg_i\rangle \circ \langle dfg_{i+1}\rangle$, the associative law of the ∘-operation has to applied. The number of the associative law rule applications needed in our implementation is $2(k-2)$.

The Function Allocation and Binding Conversion

Function allocation and binding only convert slices to equivalent ones and the *FU* abbreviation is performed. Therefore, besides expanding the *FU* abbreviation, one can apply our general synthesis conversion scheme to each slice separately. k steps are needed rather than one, but again the data flow graphs considered have a smaller depth.

6 Experimental Results

We used a scalable data flow graph as a benchmark. It realizes the division of two polynomials with the given coefficients α_i and β_i:

$$\frac{\sum_{i=0}^{p+q} \alpha_i x^i}{\sum_{i=0}^{p} \beta_i x^i} = \sum_{i=0}^{q} \gamma_i x^i + \frac{\sum_{i=0}^{p-1} \delta_i x^i}{\sum_{i=0}^{p} \beta_i x^i}$$

The coefficients γ_i and δ_i should be computed. To facilitate the calculation, we assume that the divisor is normalized with respect to β_p. After a few algebraic transformations we get the following two formulas for the demanded coefficients:

$$\gamma_i = \alpha_{i+p} - \sum_{k=i+1}^{min\{i+p,q\}} \beta_{i+p-k} \cdot \gamma_k \qquad i = 0\ldots q$$

$$\delta_j = \alpha_j - \sum_{k=0}^{min\{j,q\}} \beta_{j-k} \cdot \gamma_k \qquad j = 0\ldots p-1$$

Using these formulas, the data flow graph can be realized very quickly. To illustrate the underlying structure, a data flow graph with $p = 3$ and $q = 4$ is shown in figure 4.

The data flow graph consists of $p+q$ subtractors, $p(q+1)$ multipliers and $q(p-1)$ adders, so there is a total of $2pq + 2p$ nodes. The critical path has a length of $3q+2$ nodes. In simplified terms, q controls the depth of the data flow graph whereas p determines the width.

We applied both the simple conversion presented in section 4 as well as the advanced conversion described in section 5. The runtimes[2] for the conversions

---
[2] All experiments have been run on a SUN ULTRA SPARC with 128MB.

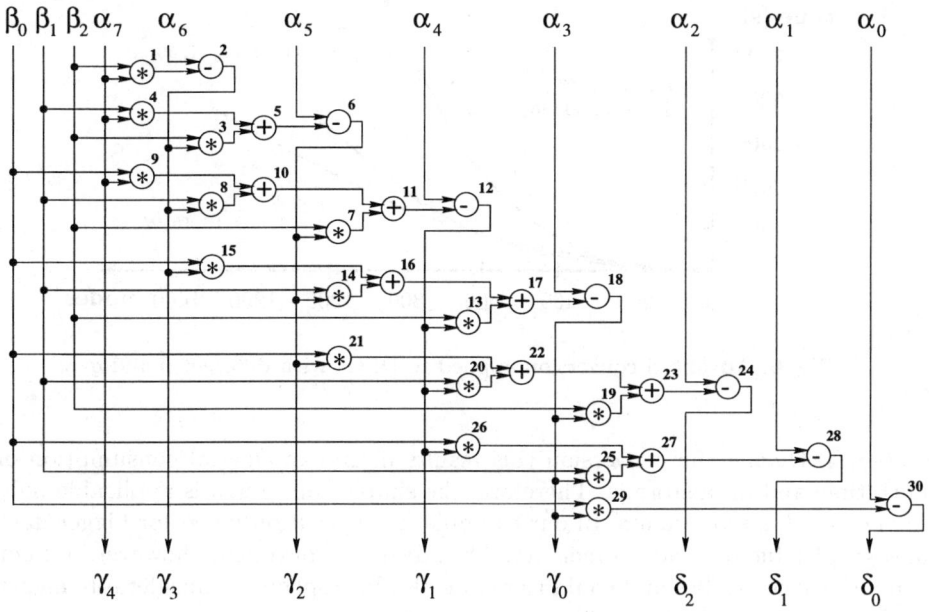

Fig. 4. A data flow graph with p=3 and q=4

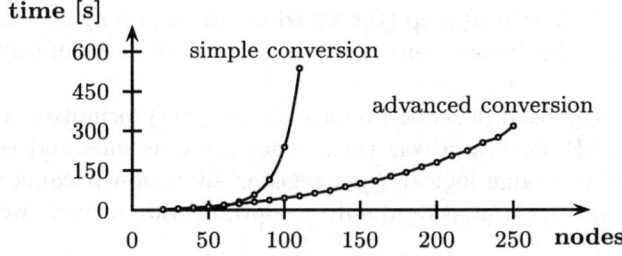

Fig. 5. Comparison simple/advanced conversion, $p = 5$, $q = 1, 2, \ldots$

are displayed in figure 5. It shows, that it pays out to interleave synthesis and logical derivation thereby exploiting the knowledge on how the implementation was derived, i.e. which synthesis steps have been applied and how they have been performed. The idea behind the technique of the simple conversion is pretty close to what one could do when performing post-synthesis verification. As can be seen in figure 4, some intermediate results ($\gamma_0 \ldots \gamma_q$) are used more often, which leads to an exponentially growth of β-redices in the universal conversion as shown in

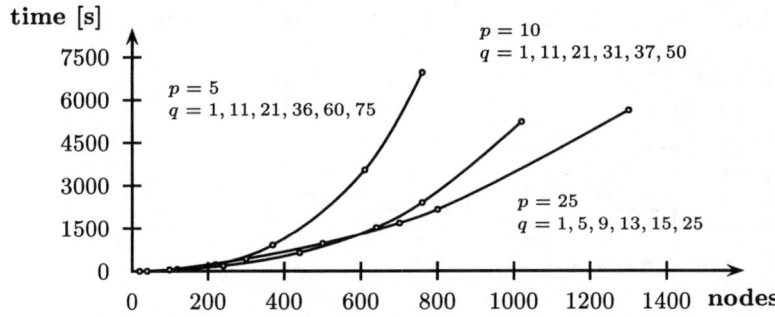

Fig. 6. Advanced conversion applied to DFGs with different p and q

section 4. During the conversion this results in an exponential consumption of both time and data storage. Therefore, the simple conversion is applicable only to very small sized circuits. In our example, the execution failed for bigger data flow graphs due to a lack of memory. The advanced conversion, however, did not run into space hurdles and could therefore also be applied to considerably bigger data flow graphs (see figure 6).

7 Embedding Existing High Level Synthesis Algorithms

The conversions described in the sections 4 and 5 are our basis for implementing synthesis tools in HOL. They are controlled by parameters telling them, how to perform the synthesis step (the schedule, the mapping between registers and variables etc.). Arbitrary heuristics can be invoked to compute this control information.

The heuristics invoked in section 6 have all been very primitive. For scheduling, a simple ASAP algorithm was used. Since the operands and results in all operations are of the same logical type, register allocation became trivial. The register binding was generated randomly — optimization aspects were not considered.

However, we also invoked more sophisticated synthesis heuristics. Table 7 shows different schedules achieved by different scheduling techniques. The schedules describe how the nodes (as numbered in the DFG in figure 4) are mapped to control steps. There are mainly two optimization goals for these algorithms: the number of control steps required and the number of operation units needed for the implementation.

In general, implementations with a big number of control steps can be realized with a small number of operation units whereas being restricted to a small number of control steps leads to a big number of operation units. There are mainly two kinds of scheduling algorithms: ones with hardware constraints and others with timing constraints. For a given restriction on the number of operation units, scheduling algorithms with hardware constraints try to find a schedule with a minimal number of control steps. Scheduling algorithms with timing constraints are the other way around: for a given limitation on the number of control steps, the algorithm tries to find a schedule with a minimal number of hardware requirements.

The ASAP/ALAP algorithm (as soon/late as possible) assigns the nodes to the earliest/latest control step according to the restrictions given by the data dependencies. The force directed heuristic [PaKn89] tries to minimize the hardware by distributing it uniformly over the control steps. The heuristic is modeled after the calculation of the equilibrium for a set of springs and weights which obey the Hooke's law. The ASAP, the ALAP and the force directed scheduling algorithm do not place any restriction on the hardware and produce a schedule with a minimal number of control-steps. The (static) list scheduling heuristic [JMSW91] has a given restriction on the hardware consumption and tries to minimize the number of control steps needed according to a precalculated priority list.

In our example, the ASAP produced a schedule with a total of 7 operation units (3 multipliers, 2 adders and 2 subtractors), the result of the ALAP required 8 operation units (3 multipliers, 2 adders and 3 subtractors), and the force-directed algorithm required 7 operation units (2 multipliers, 2 adders and 3 subtractors). For the list scheduling algorithm, the number of multipliers was limited to 1, the number of subtractors was limited to 2 and the number of adders was also limited to 2. However, it required two extra control steps compared to the other techniques. According to our experiments, the time for the logical transformation is independent from the synthesis algorithm invoked: 5.97s for the ASAP, 5.72s for the ALAP, 5.78s for the force-directed and 6.15s for the list scheduling algorithm.

In our approach, a synthesis step can be divided into two parts: computation of the control information and execution of the transformation within the logic (figure 8). Two important points are met independently with this strategy: quality and correctness of the implementation. The quality only depends on the algorithm that calculates the control information, whereas the correctness aspect is guaranteed due to the transformation being based on the HOL system.

Since the entire synthesis process is nothing but a HOL conversion, correctness is guaranteed implicitly. Faulty implementations *cannot* be achieved even if the control information produced by the external program is flawed, such as a schedule where the data dependencies are disregarded. In such cases, the transformation cannot be performed within the logic and an exception is raised. In conventional synthesis programs, such bugs could lead to faulty implementations. Our formal synthesis program either leads to correct implementations or to no implementation but an exception. In case of an exception, an information is produced telling the user in which synthesis step the error occurred.

The optimization tasks corresponding to high level synthesis steps are very complex and mutually depend on one another. Thus heuristics have to be involved. The major advantage of our approach is, that we can exploit the existing

C-Step	Heuristics			
	ASAP	ALAP	Force-Directed	List Scheduling
1	1,4,9	1	1	1
2	2	2	2,4	2,4
3	3,18,15	3,4	3	9
4	5,10	5	5,15	3
5	6	6,8,9	6,8,9	5,8
6	7,14,21	7,10	7,10,14	6,10,15
7	11,16	11	11,16	7
8	12	12,14,15	12	11,14
9	13,20,26	13,16	13	12,16,21
10	17,22	17	17,21	13
11	18	18,20,21	18,20,26	17,20
12	19,25,29	19,22,25,26	19,22,25	18,22,26
13	23,27,30	23,27,29	23,27,29	19
14	24,28	24,28,30	24,28,30	23,25
15	–	–	–	24,27,29
16	–	–	–	28,30

Fig. 7. Control information derived by different scheduling algorithms, $p=3$, $q=4$

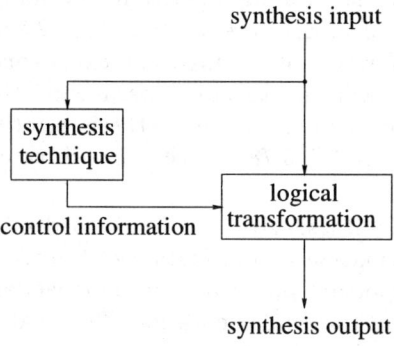

Fig. 8. The concept of our high level synthesis process

techniques. Our synthesis conversions offer the interface for embedding arbitrary conventional high level synthesis algorithms dedicated to the corresponding synthesis task. This has the effect, that – in contrast to most formal synthesis approaches — we do not have to invent new synthesis algorithms.

Although the conversions described in section 5 have to be performed in the given order, there is no restriction on how to compute the corresponding control information. It is possible to determine it step by step as sketched in the left side of figure 9 and one can as well determine it all at once as in the right side of figure 9. What really matters is that the control information is delivered to the conversions in the given order — the order in which they are computed is ambiguous. Therefore, it is possible to embed arbitrary external synthesis algorithms. This aspect is of big importance since there is no limit as to the achievable quality of synthesis tools based on our approach.

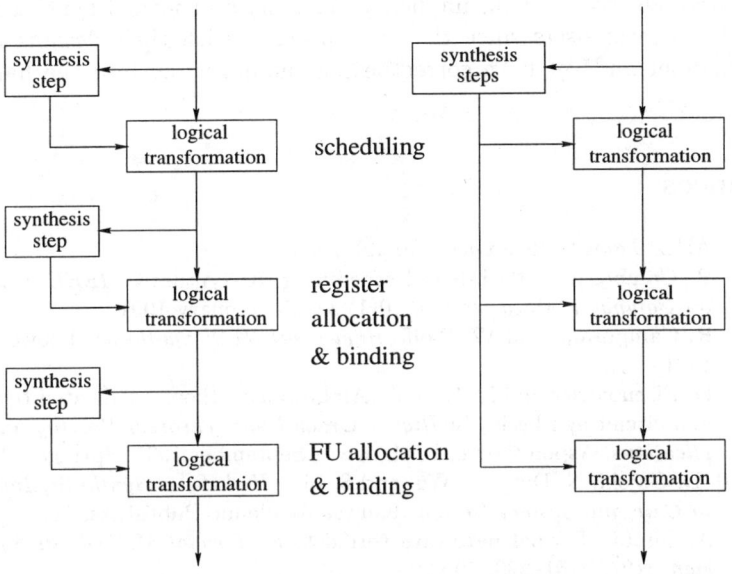

Fig. 9. Possibles schemes for the using of our synthesis conversions

8 Conclusion

We have described how high level synthesis can be performed by a sequence of logical transformations. The novelty of our approach lies in the exploitation of the existing knowledge in synthesis in a logically correct manner. As in conventional synthesis programs, finding suitable hardware representations and corresponding algorithms is essential for the efficiency. We have shown that it is possible to map algorithms and data of standard synthesis tools to logical conversions and representations in HOL.

Due to the expressiveness of HOL, general verification is an exacting goal. In our approach, however, the proof is constructed rather than "guessed" as in post-synthesis verification. Since our approach does not lead to NP-complete or undecidable problems, we believe, that formal synthesis is a well suited application for the HOL system.

In our recent work it turns out, that also in other abstraction levels of hardware design, formal synthesis can be a good alternative to the classical synthesis/post-synthesis verification approach [EiKu95]. It is our intention to provide a formal synthesis toolbox called HASH containing formally based synthesis steps that cover the entire synthesis from the algorithmic level down to the logical level.

For the hardware designer, there is no difference between using synthesis tools based on HASH and conventional synthesis tools. However, formal synthesis guarantees correctness, implicitly. This style of formal synthesis will be acceptable to most users since they can proceed with their designs in a customary manner and yet have correctness without getting into the hardship of logic.

References

[AHL92]   AHL. *Lambda Reference Manual*, 1989.
[Camp91] R. Camposano. Path-based scheduling for synthesis. *IEEE Transactions on Computer Aided Design*, 10(1):85–93, January 1991.
[CaWo91] R. Camposano and W. Wolf. *High-Level VLSI Synthesis*. Kluwer, Boston, 1991.
[EiKu95] D. Eisenbiegler and R. Kumar. An automata theory dedicated towards formal circuit synthesis. In *Higher Order Logic Theorem Proving and Its Applications*, Aspen Grove, Utah, USA, September 1995. Springer.
[GDWL94] D. Gajski, N. Dutt, A. Wu, and S. Lin. *High-Level Synthesis, Introduction to Chip and System Design*. Kluwer Academic Publishers, 1994.
[Gupt92] A. Gupta. Formal hardware verification. *Formal Methods in System Design*, 1(2/3):151–238, 1992.
[HaLD89] F.K. Hanna, M. Longley, and N. Daeche. Formal synthesis of digital systems. In *IMEC-IFIP Workshop on Applied Formal Methods for Correct VLSI Design*, pages 532–548, Leuven,Belgium, 1989. Elsevier Science Publishers B.V.
[JMSW91] R. Jain, A. Mujumdar, A. Sharma, and H. Wang. Empirical evaluation of some high-level synthesis scheduling heuristics. In *DAC '91*, pages 210–215, 1991.
[John84] S. Johnson. *Synthesis of Digital Designs from Recursion Equations*. MIT Press, 1984.
[JoSh90] G. Jones and M. Sheeran. Circuit design in Ruby. In J. Staunstrup, editor, *Formal Methods for VLSI Design*, pages 13–70. North-Holland, 1990.
[Lars94] M. Larsson. An engineering approach to formal system design. In Thomas F. Melham and Juanito Camilleri, editors, *Higher Order Logic Theorem Proving and Its Applications*, pages 300–315, Valetta, Malta, September 1994. Springer.
[Melh93] T. Melham. *Higher Order Logic and Hardware Verification*. Cambridge University Press, 1993.
[PaKn89] Pierre G. Paulin and John P. Knight. Force-directed scheduling for the behavioral synthesis of asic's. *IEEE Transactions on Computer Aided Design*, 8(6):661–679, June 1989.

Five Axioms of Alpha-Conversion

Andrew D. Gordon[1] and Tom Melham[2]

[1] University of Cambridge Computer Laboratory,
New Museums Site, Cambridge CB2 3QG, UK. adg@cl.cam.ac.uk
[2] Department of Computing Science, University of Glasgow,
Glasgow G12 8QQ, Scotland. tfm@dcs.gla.ac.uk

Abstract. We present five axioms of name-carrying lambda-terms identified up to alpha-conversion—that is, up to renaming of bound variables. We assume constructors for constants, variables, application and lambda-abstraction. Other constants represent a function Fv that returns the set of free variables in a term and a function that substitutes a term for a variable free in another term. Our axioms are (1) equations relating Fv and each constructor, (2) equations relating substitution and each constructor, (3) alpha-conversion itself, (4) unique existence of functions on lambda-terms defined by structural iteration, and (5) construction of lambda-abstractions given certain functions from variables to terms. By building a model from de Bruijn's nameless lambda-terms, we show that our five axioms are a conservative extension of HOL. Theorems provable from the axioms include distinctness, injectivity and an exhaustion principle for the constructors, principles of structural induction and primitive recursion on lambda-terms, Hindley and Seldin's substitution lemmas and the existence of their length function. These theorems and the model have been mechanically checked in the Cambridge HOL system.

The axioms presented in this paper are intended to give a simple, abstract characterisation of untyped lambda-terms, with constants, identified up to alpha-conversion, that is, renaming of bound variables. We were led to develop these axioms because we are interested in representing the syntax of programming languages with binding operators within a theorem prover. The difficulty of correctly defining substitution on lambda-terms is notorious. Previous experience with the pi-calculus (Milner, Parrow, and Walker 1992) in HOL (Melham 1994) suggests that developing substitution and binding operators directly is a tedious and error-prone business. Instead, to avoid error and repetition, we advocate first developing a metatheory of untyped lambda-terms, and secondly deriving syntax for a particular programming language as abbreviations for untyped lambda-terms. We will show in section 4 how to do this for a finitary pi-calculus.

Given higher-order logic, as implemented in the Cambridge HOL system (Gordon and Melham 1993), what we are after is a logical type $(\alpha)term$ that stands for the set of lambda-terms, where α is the type of constants. Terms are generated by the four constructors:

Con : $\alpha \to (\alpha)\mathit{term}$ (constants)
Var : $\mathit{string} \to (\alpha)\mathit{term}$ (variables)
App : $(\alpha)\mathit{term} \to (\alpha)\mathit{term} \to (\alpha)\mathit{term}$ (applications)
Lam : $\mathit{string} \to (\alpha)\mathit{term} \to (\alpha)\mathit{term}$ (lambda-abstractions)

Consider the concrete recursive type—the free algebra—generated by these constructors. Concrete recursive types are implemented in HOL using Melham's type definition package (Gordon and Melham 1993, Chapter 20). Given these constructors, the package proves the existence of a type characterized by the single axiom:

$\vdash \forall con : \alpha \to \beta.$
  $\forall var : string \to \beta.$
  $\forall app : \beta \to \beta \to (\alpha)term \to (\alpha)term \to \beta.$
  $\forall lam : \beta \to string \to (\alpha)term \to \beta.$
  $\exists! hom : (\alpha)term \to \beta.$
    $\forall k.\, hom(\mathsf{Con}\ k) = con\ k\ \wedge$
    $\forall x.\, hom(\mathsf{Var}\ x) = var\ x\ \wedge$
    $\forall t\, u.\, hom(\mathsf{App}\ t\ u) = app\ (hom\ t)\ (hom\ u)\ t\ u\ \wedge$
    $\forall x\, u.\, hom(\mathsf{Lam}\ x\ u) = lam\ (hom\ u)\ x\ u$

The axiom allows for the definition of functions by *primitive recursion*, where functions *con*, *var*, *app* and *lam* determine the outcome of the function when applied to each constructor, given access to the outcome of recursive calls and to the arguments of the constructor. In fact Melham's tool derives this axiom from a simpler *iteration* axiom. Iteration also allows for the definition of functions by recursion, but the functions *con*, *var*, *app* and *lam* have no direct access to the constructor arguments, only to the outcomes of recursive calls.

Here the type $(\alpha)\mathit{term}$ is a free algebra; all the constructors are injective. Two lambda-abstractions are equal just if their bound variables and their bodies are equal. Instead we are after a type in which terms are identified up to alpha-conversion, that is, in which two lambda-abstractions are equal just if their bodies are equal when the bound variables are renamed to a fresh variable.

1 The Axioms

The main contribution of this paper is to present five axioms for untyped lambda-terms identified up to alpha-conversion, to show how various reasoning principles derive from these axioms, and to show that the axioms are sound. The first three axioms are well-known (Curry and Feys 1958; Barendregt 1984); the fourth and fifth are new. The model we present in section 2 is based on earlier work (Gordon 1994) which showed how untyped lambda-terms could be modelled by de Bruijn terms.

Gordon was not concerned to specify axioms characterising the type of untyped lambda-terms, and did not consider how to define functions by recursion on these terms. Instead our new work shows the importance of an axiom for iteration. In section 3 it allows us to derive primitive recursion (analogous to the axiom displayed above), structural induction, finiteness of free variables, and the function returning the length of a lambda-term, which were all taken straight from the model in Gordon's earlier work. It also allows functions to be defined by recursion on lambda-terms, which was not previously considered.

In addition to the constructors introduced above, our axioms employ three further functions:

$$\begin{align}
\mathsf{Fv} &: (\alpha)term \to (string)set \\
\_[\_/\_] &: (\alpha)term \to ((\alpha)term \times string) \to (\alpha)term \\
\mathsf{Abs} &: (string \to (\alpha)term) \to (\alpha)term
\end{align}$$

$\mathsf{Fv}(u)$ returns the set of free variables in the term u and $u[t/x]$ produces the result of substituting the term t for the free occurrences of the variable x in the term u. (The partially curried type is needed to make substitution an infix operator in Cambridge HOL.) The function Abs, discussed in detail later, maps certain logical (meta) functions, namely those which carry out a variable-for-variable substitution on terms, to corresponding lambda-abstractions in $(\alpha)term$.

1.1 Free Variables

Axiom one defines the free variables of each constructor:

$$\vdash \forall k.\ \mathsf{Fv}(\mathsf{Con}\ k) = \{\} \land \qquad \text{(Axiom 1)}$$
$$\forall x.\ \mathsf{Fv}(\mathsf{Var}\ x) = \{x\} \land$$
$$\forall t\ u.\ \mathsf{Fv}(\mathsf{App}\ t\ u) = \mathsf{Fv}(t) \cup \mathsf{Fv}(u) \land$$
$$\forall x\ u.\ \mathsf{Fv}(\mathsf{Lam}\ x\ u) = \mathsf{Fv}(u) - \{x\}$$

We have expressed this axiom using set-theoretic notation and the type $(\alpha)set$ of sets whose elements are all of type α. This type is an abbreviation for the type of characteristic functions, $\alpha \to bool$. All the set-theoretic operators we use can easily be defined as operations on this type.

Our axioms need not assert that $\mathsf{Fv}(u)$ is always a finite set, for this follows as a theorem from the five axioms.

1.2 Substitution

Axiom two defines the interaction of substitution with each constructor:

$$\vdash \forall k\ u\ x.\ (\mathsf{Con}\ k)[u/x] = \mathsf{Con}\ k \land \qquad \text{(Axiom 2)}$$
$$\forall u\ x.\ (\mathsf{Var}\ x)[u/x] = u \land$$
$$\forall u\ x\ y.\ (x \neq y) \supset (\mathsf{Var}\ y)[u/x] = \mathsf{Var}\ y \land$$
$$\forall t\ u\ v\ x.\ (\mathsf{App}\ t\ u)[v/x] = \mathsf{App}\ (t[v/x])\ (u[v/x]) \land$$
$$\forall x\ t\ u.\ (\mathsf{Lam}\ x\ t)[u/x] = \mathsf{Lam}\ x\ t \land$$
$$\forall x\ y\ u.\ (x \neq y) \land y \notin (\mathsf{Fv}\ u) \supset \forall t.\ (\mathsf{Lam}\ y\ t)[u/x] = \mathsf{Lam}\ y\ (t[u/x])$$

Consider the situation of trying to push a substitution $\_[u/x]$ into Lam y t when the bound variable y occurs free in u. It is necessary to avoid the capture of any free occurrences of y in u. Our axiom two does not immediately apply in this situation. But since we identify lambda-terms up to alpha-conversion, we can use axiom three—alpha-conversion—to rename the bound variable y so that the last part of axiom two does apply.

In contrast, Curry and Feys (1958) incorporate this renaming into their definition of substitution—at the cost of an arbitrary choice of renamed variable—because they define substitution directly on the free algebra of lambda-terms and derive alpha-conversion later. On the other hand, Barendregt (1984) avoids this situation via his variable convention, which here permits us to assume that the bound variable y is different from any variable occurring free in u. Stoughton (1988) presents a definition of substitution by structural recursion—the definition by Curry and Feys is by recursion on the length of the term—which always renames bound variables.

1.3 Alpha-Conversion

Axiom three asserts the arbitrariness of bound variables:

$$\vdash \forall y\, u\, x.\, y \notin \mathsf{Fv}(\mathsf{Lam}\ x\ u) \supset (\mathsf{Lam}\ x\ u = \mathsf{Lam}\ y\ (u[\mathsf{Var}\ y/x])) \qquad \text{(Axiom 3)}$$

This is alpha-conversion; two lambda-abstractions may be equal but have distinct bound variables. A consequence is that no logical function is definable that distinguishes such terms.

There is a weaker version of the alpha-conversion axiom,

$$\vdash \forall y\, u.\, y \notin \mathsf{Fv}(u) \supset \forall x.\, \mathsf{Lam}\ x\ u = \mathsf{Lam}\ y\ (u[\mathsf{Var}\ y/x])$$

which in fact follows from the stronger axiom above. We adopt the stronger form because it immediately tells us that

$$\vdash \forall x\, u.\, \mathsf{Lam}\ x\ u = \mathsf{Lam}\ x\ (u[\mathsf{Var}\ x/x])$$

which turns out to be important in later proofs.

1.4 Unique Iteration

Axiom four asserts the unique existence of functions defined by iteration over the structure of terms:

$$⊢ ∀con : α → β. \qquad\qquad\qquad\text{(Axiom 4)}$$
$$∀var : string → β.$$
$$∀app : β → β → β.$$
$$∀abs : (string → β) → β.$$
$$∃!hom : (α)term → β.$$
$$∀k.\ hom(\mathsf{Con}\ k) = con\ k\ ∧$$
$$∀x.\ hom(\mathsf{Var}\ x) = var\ x\ ∧$$
$$∀t\ u.\ hom(\mathsf{App}\ t\ u) = app\ (hom\ t)\ (hom\ u)\ ∧$$
$$∀x\ u.\ hom(\mathsf{Lam}\ x\ u) = abs\ (λy.\ hom(u[\mathsf{Var}\ y/x]))$$

Suppose we want to define a function *hom* of type $(α)term → β$ on lambda-terms by recursion. Given functions *con*, *var*, *app* and *abs* that specify how *hom* treats each of the four constructors, the axiom asserts that such a *hom* exists and moreover is unique. It is analogous to the iteration axiom characterizing concrete recursive types, mentioned in the introduction. As we discuss in section 3 many properties, such as the fact that constructors yield distinct terms, follow from this axiom, in much the same way as analogous properties follow from the single axiom of concrete recursive types.

The difference between this axiom and the ones for concrete recursive types is in the recursion equation for lambda-abstractions. The value of $hom(\mathsf{Lam}\ x\ u)$ is determined (by the parameter *abs*) to be $abs(λy.\ hom(u[\mathsf{Var}\ y/x]))$ but not $abs(hom\ u)$. It cannot be the latter because that would fix the arbitrary bound variable to be x and allow us to distinguish alpha-equivalent terms. Instead the function *abs* is supplied with a function that will yield $hom(u[\mathsf{Var}\ y/x])$ for any $\mathsf{Var}\ y$ to be substituted for x in u. This function, rather than just u, should be regarded as the 'body' of the original lambda-term $\mathsf{Lam}\ x\ u$. So *abs* can work on $hom(u[\mathsf{Var}\ y/x])$ provided it chooses a name y for the arbitrary bound variable x. To employ this principle of iteration in practice, we appear to need one final axiom.

1.5 Abstraction

Axiom five asserts that from any function of type $string → (α)term$ that represents the body of a lambda-abstraction one can reconstruct 'the' lambda-abstraction itself:

$$⊢ ∀x\ u.\ \mathsf{Abs}(λy.\ u[\mathsf{Var}\ y/x]) = \mathsf{Lam}\ x\ u \qquad\qquad\text{(Axiom 5)}$$

where the constant Abs has the type $(string → (α)term) → (α)term$. (Remember that $λ$ signifies lambda-abstraction in the HOL logic itself, and that Lam signifies the lambda-abstraction of the type $(α)term$ of untyped lambda-terms embedded in HOL.)

This axiom could, of course, be taken as a *definition* of the Lam constructor. Our axiom set is therefore redundant (Lam is eliminable). But we retain

Lam because it clarifies the presentation and serves to highlight the correspondence between lambda-abstractions in $(\alpha)term$ and certain meta-level functions in $string \to (\alpha)term$.

The existence of Abs is of importance primarily because it lets us build lambda-abstractions from lambda-bodies. Consider, for example, the problem of defining a function that uses structural iteration to build a copy of any given lambda-term. We take

$$\begin{aligned} \beta &:= (\alpha)term \\ con &:= \mathsf{Con} \\ var &:= \mathsf{Var} \\ app &:= \mathsf{App} \\ abs &:= \mathsf{Abs} \end{aligned}$$

in the unique iteration axiom. This gives

$$\vdash \exists! hom : (\alpha)term \to (\alpha)term.$$
$$\forall k.\ hom(\mathsf{Con}\ k) = \mathsf{Con}\ k\ \wedge$$
$$\forall x.\ hom(\mathsf{Var}\ x) = \mathsf{Var}\ x\ \wedge$$
$$\forall t\, u.\ hom(\mathsf{App}\ t\ u) = \mathsf{App}\ (hom\ t)\ (hom\ u)\ \wedge$$
$$\forall x\, u.\ hom(\mathsf{Lam}\ x\ u) = \mathsf{Abs}(\lambda y.\ hom(u[\mathsf{Var}\ y/x]))$$

In the Lam equation we use Abs to reconstruct the abstraction. In the others, we can simply employ the appropriate constructor.

It is easy to see that this gives us a function that copies terms. The theorem states the unique existence of any function hom satisfying these equations. But the identity on terms, $\lambda u.\ u$, is just such a function (the actual proof makes use of the Abs axiom). Hence the function whose existence is asserted is itself the identity.

As will be seen in later sections, the Abs function may also be used more generally for getting fresh variables ('genvars') to supply to bodies of lambda-abstractions.

2 A Model of the Axioms

In this section we briefly recall the construction used by Gordon (1994), and discuss in some detail how to model axioms four and five. We begin with the free algebra of de Bruijn's nameless lambda-terms (de Bruijn 1972).

dCon	$: \alpha \to (\alpha)db$	(constants)
dVar	$: string \to (\alpha)db$	(free variables)
dBound	$: num \to (\alpha)db$	(bound variables)
dApp	$: (\alpha)db \to (\alpha)db \to (\alpha)db$	(applications)
dAbs	$: (\alpha)db \to (\alpha)db$	(lambda-abstractions)

Consider an occurrence of dBound i enclosed by j dAbs's in a term. If $i < j$ then it refers to the $(i+1)$'th enclosing dAbs. If $i \geq j$ then we say it is *dangling*, and that it is a *reference to parameter* $i - j$ of the term. We model $(\alpha)term$ by the *proper* de Bruijn terms, that is, those with no dangling indexes. Sometimes dangling indexes are used to represent free variables, but here we use the dVar constructor instead.

We can define dFv(d), the free variables of term d, by primitive recursion. Name-carrying lambda-abstraction and substitution can be defined as follows,

$$\vdash \text{dLam } x\ d = \text{dAbs}(\text{Abst } 0\ x\ d)$$

$$\vdash d[d'/x] = \text{Inst } 0\ (\text{Abst } 0\ x\ d)\ d'$$

where the term Abst $i\ x\ d$ is obtained by turning each occurrence of dVar x in d into a reference to parameter i, and the term Inst $i\ d\ d'$ is obtained by instantiating each reference to parameter i in d to the term d'. An important property is that the set inductively defined from the constructors dCon, dVar, dApp and dLam is exactly the set of proper de Bruijn terms. Given these definitions it is straightforward to model axioms one, two and three. See Gordon (1994) for a fuller discussion.

2.1 Soundness of the Iteration Axiom

First fix functions of the following types.

$$con : \alpha \to \beta$$
$$var : string \to \beta$$
$$app : \beta \to \beta \to \beta$$
$$abs : (string \to \beta) \to \beta$$

To model axiom four, iteration, it suffices to construct a function *hom* such that

$$\forall k.\ hom(\text{dCon } k) = con\ k\ \wedge$$
$$\forall x.\ hom(\text{dVar } x) = var\ x\ \wedge$$
$$\forall d\ d'.\ hom(\text{dApp } d\ d') = app\ (hom\ d)\ (hom\ d')\ \wedge$$
$$\forall x\ d.\ hom(\text{dLam } x\ d) = abs\ (\lambda y.\ hom\ (d[\text{dVar } y/x]))$$

and moreover to show that *hom* is the unique function on proper de Bruijn terms to satisfy these equations. We shall refer to these equations as (Hom Spec).

The substitution in the last part of (Hom Spec) prevents us from defining *hom* by primitive recursion. Instead we define *hom* indirectly in terms of another function, *chom*, which uses Landin's idea (1964) of a closure to represent the substitution in the last part of (Hom Spec). Let the *degree* of a term be 0 if it contains no dangling index, and otherwise one more than the greatest parameter referred to by a dangling index. Let a *closure* be a pair (ys, d) with d a possibly improper de Bruijn term, and ys:$(string)list$ a list of variable names of length no less than the degree of d. We can think of a closure $([y_0, \ldots, y_{n-1}], d)$ as standing for d with each reference to parameter i instantiated to dVar y_i. We now define *chom ys d*, where (ys, d) is intended to be a closure, by primitive recursion on de Bruijn terms.

$\forall k.\ chom\ ys\ (\mathsf{dCon}\ k) = con\ k\ \wedge$
$\forall x.\ chom\ ys\ (\mathsf{dVar}\ x) = var\ x\ \wedge$
$\forall i.\ chom\ ys\ (\mathsf{dBound}\ i) = var\ (i < \mathsf{Length}\ ys \Rightarrow \mathsf{El}\ i\ ys\ |\ \mathsf{Arb})\ \wedge$
$\forall d\ d'.\ chom\ ys\ (\mathsf{dApp}\ d\ d') = app\ (chom\ ys\ d)\ (chom\ ys\ d')\ \wedge$
$\forall x\ d.\ chom\ ys\ (\mathsf{dAbs}\ d) = abs\ (\lambda y.\ chom\ (\mathsf{Cons}\ y\ ys)\ d)$

The constant Arb has an arbitrary value, but provided that (ys, d) is a closure, *chom ys d* will not depend on Arb. We take *hom* to be $chom[\,]$. It is easy to see that this definition satisfies the first three equations in (Hom Spec). For the fourth, concerning dLam, we need a lemma that if any two closures (ys, d) and (ys', d') stand for the same lambda-term, in the sense given above, then *chom ys d* = *chom ys' d'*. This is proved by structural induction on d, and allows us to calculate the following, for any d.

$$hom(\mathsf{dAbs}\ d) = chom\ [\,]\ (\mathsf{dAbs}\ d)$$
$$= abs\ (\lambda y.\ chom\ [y]\ d)$$
$$= abs\ (\lambda y.\ chom\ [\,]\ (\mathsf{Inst}\ 0\ d\ (\mathsf{dVar}\ y)))$$
$$= abs\ (\lambda y.\ hom\ (\mathsf{Inst}\ 0\ d\ (\mathsf{dVar}\ y)))$$

By this, and the definitions of substitution and dLam the final part of (Hom Spec) follows.

$$hom(\mathsf{dLam}\ x\ d) = hom(\mathsf{dAbs}(\mathsf{Abst}\ 0\ x\ d))$$
$$= abs\ (\lambda y.\ hom\ (\mathsf{Inst}\ 0\ (\mathsf{Abst}\ 0\ x\ d)\ (\mathsf{dVar}\ y)))$$
$$= abs\ (\lambda y.\ hom\ (d[\mathsf{dVar}\ y/x]))$$

There does exist, then, a function *hom* satisfying (Hom Spec). Uniqueness follows by an induction on the length of the term, where the length of a de Bruijn term is the number of constructors it contains. (Length is definable by primitive recursion on de Bruijn terms.)

2.2 Soundness of the Abs Axiom

Here is a sketch of how to define a model for axiom five, concerning the Abs function. The essence of the proof is that Abs can be modelled by the function $abs : (string \to (\alpha)db) \to (\alpha)db$ defined by

$$abs(f) = \text{let } Y = \cap y.\ \mathsf{dFv}(f\ y) \text{ in}$$
$$\text{let } z = \text{New } Y \text{ in}$$
$$\mathsf{dLam}\ z\ (f\ z)$$

where New chooses a fresh string not in a given finite set of strings. The idea is that *abs* will be supplied with a function of the form $\lambda y.\ u[\mathsf{dVar}\ y/x]$ (that is, the body of a lambda-term). It then finds a fresh variable z and reconstructs the original lambda-term by building an alpha-equivalent one by substituting z into the body and abstracting over it using dLam.

The details are as follows. One can show that the free variables of dLam x u are a subset of the free variables of $u[\text{dVar } y/x]$ for any choice of y:

$$\vdash \forall x\, y\, u.\ \text{Proper}(u) \supset (\text{dFv}(\text{dLam } x\ u) \subseteq \text{dFv}(u[\text{dVar } y/x]))$$

The proof is by induction on length of the term u. Hence the free variables of dLam x u are contained in the intersection of the whole y-indexed family of free-variable sets:

$$\vdash \forall u.\ \text{Proper}(u) \supset (\text{dFv}(\text{dLam } x\ u) \subseteq \cap y.\ \text{dFv}(u[\text{dVar } y/x]))$$

Moreover, one can show that the containing set is finite, since it is the intersection of a family of finite sets. (The function dFv always produces finite sets.) Hence one can find a fresh variable, avoiding any variable free in dLam x u, by choosing a variable outside this finite set:

$$\vdash \forall u.\ \text{Proper}(u) \supset \text{New}(\cap y.\ \text{dFv}(u[\text{dVar } y/x])) \notin \text{dFv}(\text{dLam } x\ u)$$

Call this fresh variable z. But then by alpha conversion, we know that

$$\vdash \text{dLam } z\ (u[\text{dVar } z/x]) = \text{dLam } x\ u$$

giving us the required lambda term.

3 Theorems Provable from the Axioms

This section discusses some of the theorems derivable from our five axioms. We begin by deriving a recursion scheme for defining functions over lambda-terms and then use it to prove a new principle of structural induction for terms. We then illustrate the utility of these results by defining a length function on terms.

Also provable from our axioms are the theorems of Gordon (1994) stating distinctness, injectivity and an exhaustion principle for the constructors. The substitution lemmas 1.14 and 1.15 of Hindley and Seldin (1986) are also provable. Discussion of these theorems is omitted here.

3.1 Recursion Scheme

The unique iteration axiom allows us to define functions only by structural iteration over terms. A more general definition pattern is supplied by the *recursion scheme* theorem:

$\vdash \forall con : \alpha \to \beta.$
$\quad \forall var : string \to \beta.$
$\quad \forall app : \beta \to \beta \to (\alpha)term \to (\alpha)term \to \beta.$
$\quad \forall abs : (string \to \beta) \to (string \to (\alpha)term) \to \beta.$
$\quad \exists! hom : (\alpha)term \to \beta.$
$\qquad \forall k.\ hom(\text{Con } k) = con\ k\ \wedge$
$\qquad \forall x.\ hom(\text{Var } x) = var\ x\ \wedge$
$\qquad \forall t\, u.\ hom(\text{App } t\ u) = app\ (hom\ t)\ (hom\ u)\ t\ u\ \wedge$
$\qquad \forall x\, u.\ hom(\text{Lam } x\ u) = abs\ (\lambda y.\ hom(u[\text{Var } y/x]))\ (\lambda y.\ u[\text{Var } y/x])$

Here, hom is defined not only in terms of its values on the subterms of each kind of constructor, but also in terms of the subterms themselves. In the defining equation for $hom(\text{App } t\, u)$, the parameter app has access not just to $hom\, t$ and $hom\, u$ but also to t and u. Likewise, in the defining equation for $hom(\text{Lam } x\, u)$ the parameter abs may use the body.

The recursion scheme follows from axioms four and five. Suppose we have arbitrary parameter functions similar to those in the recursion except that they are (in part) paired:

$$\begin{aligned}
&\underline{con} : \alpha \to \beta \\
&\underline{var} : string \to \beta \\
&\underline{app} : (\beta \times (\alpha)term) \to (\beta \times (\alpha)term) \to \beta \\
&\underline{abs} : (string \to (\beta \times (\alpha)term)) \to \beta
\end{aligned}$$

Then instantiate the unique iteration axiom with

$$\begin{aligned}
\beta &:= (\alpha)term \times \beta \\
con &:= \lambda k.\,(\text{Con } k, \underline{con}\, k) \\
var &:= \lambda x.\,(\text{Var } x, \underline{var}\, k) \\
app &:= \lambda p\, q.\,(\text{App (Fst } p)\, (\text{Fst } q), \underline{app}\, p\, q) \\
abs &:= \lambda f{:}string{\to}((\alpha)term \times \beta).\,(\text{Abs}(\text{Fst} \circ f), \underline{abs}\, f)
\end{aligned}$$

to get a unique function hom of type

$$(\alpha)term \to ((\alpha)term \times \beta)$$

that produces a pair consisting of a rebuilt copy of its input, together with a recursively calculated result of type β. At each stage in the recursion, a copy of the 'lower' structures is available, having been delivered by the recursive call.

Now hom produces a pair, and so can be split into unique pair of functions:

$$\begin{aligned}
\vdash\ &\exists!(f, g) : ((\alpha)term{\to}(\alpha)term) \times ((\alpha)term{\to}\beta). \\
&\forall k.\, f(\text{Con } k) = \text{Con } k \wedge \\
&\forall x.\, f(\text{Var } x) = \text{Var } x \wedge \\
&\forall t\, u.\, f(\text{App } t\, u) = \text{App } (f\, t)\, (f\, u) \wedge \\
&\forall x\, u.\, f(\text{Lam } x\, u) = \text{Abs}(\lambda y.\, f(u[\text{Var } y/x])) \wedge \\
&\forall k.\, g(\text{Con } k) = \underline{con}\, k \wedge \\
&\forall x.\, g(\text{Var } x) = \underline{var}\, x \wedge \\
&\forall t\, u.\, g(\text{App } t\, u) = \underline{app}\, (f\, t, g\, t)\, (f\, u, g\, u) \wedge \\
&\forall x\, u.\, g(\text{Lam } x\, u) = \underline{abs}\, (\lambda y.\, f(u[\text{Var } y/x]), g(u[\text{Var } y/x]))
\end{aligned}$$

From these equations, one can easily see that f is copying the term by rebuilding it and g is computing the result using the copies produced by f along the way.

The next step is to observe that f must be the identity. As section 1.5 showed, any function satisfying the first four equations in this theorem equals $\lambda u.\, u$. We can therefore replace f by the identity in the defining equations for g, giving:

$\vdash \exists!g : (\alpha)term \to \beta.$
$\quad \forall k. \, g(\text{Con } k) = \underline{con} \; k \; \wedge$
$\quad \forall x. \, g(\text{Var } x) = \underline{var} \; x \; \wedge$
$\quad \forall t \, u. \, g(\text{App } t \, u) = \underline{app} \, (t, g \, t) \, (u, g \, u) \; \wedge$
$\quad \forall x \, u. \, g(\text{Lam } x \, u) = \underline{abs} \, (\lambda y. \, (u[\text{Var } y/x]), g(u[\text{Var } y/x]))$

But this is just the recursion scheme theorem, up to a little currying of the functions *app* and *abs*.

This construction resembles Church's definition of the predecessor on natural numbers in the lambda-calculus (Church 1941). The actual construction used here was inspired, in part, by the derivation in Lambek and Scott (1986) of a recursion scheme for a natural numbers object in a cartesian closed category.

3.2 Deriving Induction

Gordon's previous work produced two principles of induction for lambda-terms. The first involves the standard notion of the *length* of a term, and the second involves specification of finite sets of variables from which one may assume certain bound variables are distinct. Both principles are derivable in the present setting from a more primitive notion of induction, which itself follows from the recursion scheme as usual.

The derivation of this induction principle proceeds as follows. We suppose here that P is a fixed but arbitrary predicate on lambda-terms. Then take

$\beta \quad := bool$
$con := \lambda k. \, \mathsf{T}$
$var := \lambda x. \, \mathsf{T}$
$app := \lambda p \, q. \, \lambda t \, u. \, (p \wedge q) \vee P(\text{App } t \, u)$
$abs := \lambda f{:}string \to bool. \, \lambda g{:}string \to (\alpha)term. \, (\forall y. \, f \, y) \vee P(\text{Abs } g)$

in the recursion theorem to get

$\vdash \exists!hom : (\alpha)term \to bool.$
$\quad \forall k. \, hom(\text{Con } k) \; \wedge$
$\quad \forall x. \, hom(\text{Var } x) \; \wedge$
$\quad \forall t \, u. \, hom(\text{App } t \, u) = ((hom \, t) \wedge (hom \, u)) \vee P(\text{App } t \, u) \; \wedge$
$\quad \forall x \, u. \, hom(\text{Lam } x \, u) = (\forall y. \, hom(u[\text{Var } y/x])) \vee P(\text{Abs } \lambda y. \, u[\text{Var } y/x])$

Now, observe that $\lambda u. \, \mathsf{T}$ is just such a *hom* as is asserted to exist uniquely here. Hence any other function satisfying the above equations is constant true on the set of all lambda terms. In particular, the predicate P has this property, and so we have:

$$\vdash \forall k.\ P(\mathsf{Con}\ k) \land$$
$$\forall x.\ P(\mathsf{Var}\ x) \land$$
$$\forall t\ u.\ P(\mathsf{App}\ t\ u) = ((P\ t) \land (P\ u)) \lor P(\mathsf{App}\ t\ u) \land$$
$$\forall x\ u.\ P(\mathsf{Lam}\ x\ u) = (\forall y.\ P(u[\mathsf{Var}\ y/x])) \lor P(\mathsf{Abs}(\lambda y.\ u[\mathsf{Var}\ y/x]))$$
$$\supset$$
$$\forall u.\ P(u)$$

The **Abs** axiom lets us simplify this to

$$\vdash \forall k.\ P(\mathsf{Con}\ k) \land$$
$$\forall x.\ P(\mathsf{Var}\ x) \land$$
$$\forall t\ u.\ P(\mathsf{App}\ t\ u) = ((P\ t) \land (P\ u)) \lor P(\mathsf{App}\ t\ u) \land$$
$$\forall x\ u.\ P(\mathsf{Lam}\ x\ u) = (\forall y.\ P(u[\mathsf{Var}\ y/x])) \lor P(\mathsf{Lam}\ x\ u)$$
$$\supset$$
$$\forall u.\ P(u)$$

Finally, using the fact that $(A = B \lor A) = (B \supset A)$, we get our induction principle:

$$\vdash \forall P: (\alpha)term \rightarrow bool.$$
$$\forall k.\ P(\mathsf{Con}\ k) \land$$
$$\forall x.\ P(\mathsf{Var}\ x) \land$$
$$\forall tu.\ P(t) \land P(u) \supset P(\mathsf{App}\ t\ u) \land$$
$$\forall x\ u.(\forall y.\ P(u[\mathsf{Var}\ y/x])) \supset P(\mathsf{Lam}\ x\ u)$$
$$\supset$$
$$\forall u.\ P(u)$$

For the **Con**, **Var** and **App** constructors, the proof obligations are just the same as ordinary structural induction. But in the **Lam** case, we may assume the induction hypothesis that P holds under all substitutions of a variable for the specific bound variable involved.

Examples We can illustrate the induction principle just derived by using it to prove that the identity substitution has no effect:

$$\vdash \forall u\ z.\ u[\mathsf{Var}\ z/z] = u$$

This was an early lemma in Gordon's development and is part of one of Hindley and Seldin's substitution theorems.

The proof proceeds by induction on u. Only the **Lam** case is of any interest. The induction hypothesis is

$$\vdash \forall y\ z.\ (u[\mathsf{Var}\ y/x])[\mathsf{Var}\ z/z] = u[\mathsf{Var}\ y/x]$$

and we need to show

$$(\text{Lam } x\, u)[\text{Var } z/z] = \text{Lam } x\, u$$

The case where $z = x$ is trivial, so let us suppose $z \neq x$. Now, specialise the induction hypothesis to get

$$\vdash (u[\text{Var } x/x])[\text{Var } z/z] = u[\text{Var } x/x]$$

and apply $\text{Lam } x$ to both sides:

$$\vdash \text{Lam } x\, (u[\text{Var } x/x][\text{Var } z/z]) = \text{Lam } x\, (u[\text{Var } x/x])$$

Since $z \neq x$ we can draw the substitution for z outwards:

$$\vdash (\text{Lam } x\, (u[\text{Var } x/x]))[\text{Var } z/z] = \text{Lam } x\, (u[\text{Var } x/x])$$

But, as observed in section 1.3, our alpha-conversion axiom tells us immediately that $\text{Lam } x\, (u[\text{Var } x/x])$ is just $\text{Lam } x\, u$, and so we are finished.

The identity substitution theorem allows us to proceed to an inductive proof that free variables of a term are finite:

$$\vdash \forall u.\, \text{Finite}(\text{Fv } u)$$

This key theorem, whose actual proof we omit, lets us choose fresh variables not free in a given term u, since we know that there is always a string outside any finite set of strings (for example take a primed variant).

3.3 Definition of a Length Function

We now turn our attention to the problem of defining the standard notion of the *length* of a term in our theory. The length of a term is a count of the number of syntactic constructors in it. The concept is usually formalised by a function

$$\text{Lgh} : (\alpha)term \to num$$

with the property

$$\vdash \forall k.\, \text{Lgh}(\text{Con } k) = 1 \land$$
$$\forall x.\, \text{Lgh}(\text{Var } x) = 1 \land$$
$$\forall t\, u.\, \text{Lgh}(\text{App } t\, u) = (\text{Lgh } t) + (\text{Lgh } u) \land$$
$$\forall x\, u.\, \text{Lgh}(\text{Lam } x\, u) = (\text{Lgh } u) + 1$$

But the equation for Lam in this recursive 'definition' does not conform to the pattern of our recursion scheme. We must therefore make a somewhat indirect definition, from which the above theorem is derivable. First, we need some machinery for handling sequences of variable renamings. This turns out to have rather general utility, and so is presented in some detail.

General Renamings We are interested in arbitrary finite sequences of substitutions, which we may represent formally by an infix function

$$\mathsf{ISub} : (\alpha)term \to ((\alpha)term \times string)list \to (\alpha)term$$

defined by primitive recursion on lists as follows:

$$\vdash \forall u.\ u\ \mathsf{ISub}\ [\,] = u\ \land\ \forall u\,t\,x\,\sigma.\ u\ \mathsf{ISub}\ (\mathsf{Cons}\ (t,x)\ \sigma) = (u[t/x])\ \mathsf{ISub}\ \sigma$$

The function takes a term and a list of term-variable substitution pairs and applies all the substitutions in sequence. We call this an *iterated* substitution, a name which incidentally serves to distinguish it from the simultaneous parallel substitution commonly seen in other contexts.

An *iterated renaming* is an iterated substitution of variables for variables. We define the predicate Renaming inductively as follows:

$\vdash$ Renaming $[\,]$ always
$\vdash$ Renaming(Cons (Var x, y) σ)   if Renaming σ

As usual, the definition gives us rules (and HOL tactics) for the Renaming predicate, together with the corresponding rule induction principle.

Derivation of Length We may now proceed to derive the desired length function. Begin by taking

$$\begin{aligned}
\beta &:= num \\
con &:= \lambda k.\ 1 \\
var &:= \lambda x.\ 1 \\
app &:= \lambda n\,m.\ \lambda t\,u.\ n + m \\
abs &:= \lambda f.\ \lambda g.\ \text{let } v = \mathsf{New}(\mathsf{Fv}(\mathsf{Abs}\ g))\ \text{in}\ f(v) + 1
\end{aligned}$$

in the recursion scheme theorem. Applying the Abs axiom gives us:

$\vdash \exists!hom : (\alpha)term \to num.$
    $\forall k.\ hom(\mathsf{Con}\ k) = 1\ \land$
    $\forall x.\ hom(\mathsf{Var}\ x) = 1\ \land$
    $\forall t\,u.\ hom(\mathsf{App}\ t\ u) = (hom\ t) + (hom\ u)\ \land$
    $\forall x\,u.\ hom(\mathsf{Lam}\ x\ u) = \text{let } v = \mathsf{New}(\mathsf{Fv}(\mathsf{Lam}\ x\ u))\ \text{in}\ hom(u[\mathsf{Var}\ v/x]) + 1$

The key idea is that in the last equation, we have used Abs to reconstruct the term Lam $x\ u$. We can then generate a fresh variable Var v not free in this term, substitute this variable into the body, and then take the length of the result.

The next step is to show that the choice of variable to substitute into the body can, in fact, be made arbitrarily. We prove that if *hom* is as defined above, then

$$\vdash \forall u\,\sigma.\ \mathsf{Renaming}\ \sigma \supset (hom\ (u\ \mathsf{ISub}\ \sigma) = hom\ u)$$

So *hom* is invariant under iterated variable renaming. The proof is a straightforward induction on u, using alpha conversion to avoid variable name clashes in the Lam case.

Since a single substitution is also a renaming, we can replace $hom(u[\mathsf{Var}\ v/x])$ in our indirect definition with $hom\ u$. The fresh variable v then no longer plays a role and can be eliminated, giving equations for the length function in exactly the desired form.

4 An Application of the Axioms

This section shows how we can derive syntax for a particular programming language as a set of abbreviations of untyped lambda-terms. Our particular example is pi-calculus. The syntax of pi-calculus is built up from a denumerable set of *names*, x, y or z, and the set of *processes*, p or q, given by the following syntax.

$p ::=$		processes
	$\overline{x}y.p$	(send)
	$x(y).p$	(receive, y bound)
	$p \mid q$	(parallel composition)
	$(\nu x)p$	(restriction, x bound)
	$\mathbf{0}$	(zero process)

By convention binding occurrences are parenthesised. This is a particularly simple, finitary pi-calculus, but its syntax suffices to make our point, which is to demonstrate how this syntax can be encoded using lambda-terms. We shall not discuss the operational semantics of pi-calculus, though it too can be represented within HOL.

We shall encode pi-calculus by introducing a new syntactic constructor for each kind of process. We introduce a syntactic constructor k of arity n by the following definition scheme, where $\underline{k}$ is a *string* constant representing k:

$$\vdash k\ t_1\ t_2 \cdots t_n = \mathsf{App}\ (\cdots \mathsf{App}\ (\mathsf{App}\ \underline{k}\ t_1)\ t_2 \cdots)\ t_n$$

String constants and binary applications suffice to encode syntactic constructors of arbitrary arity. We wrote a simple tool to automate this scheme. Axioms one and two generalise to such constructors as follows.

$$\vdash \mathsf{Fv}(k\ t_1 \cdots t_n) = \mathsf{Fv}\ t_1 \cup \cdots \cup \mathsf{Fv}\ t_n$$

$$\vdash (k\ t_1 \cdots t_n)[u/x] = k\ (t_1[u/x]) \cdots (t_n[u/x])$$

To encode pi-calculus we introduce syntactic constructors Send, Recv, Par, Res and Zero, with arities 3, 2, 2, 1 and 0 respectively. We represent a free occurrence of a name x by the lambda-term Var x, and binding occurrences by Lam x. Given this preparation we can represent the syntax above by the following inductive definition of a predicate, Proc, on lambda-terms of type $(string)term$.

⊢ Proc(Send (Var x) (Var y) p)   if Proc p ($\overline{x}y.p$)
⊢ Proc(Recv (Var x) (Lam $y\,p$))   if Proc p ($x(y).p$)
⊢ Proc(Par $p\,q$)   if Proc p and Proc q ($p \mid q$)
⊢ Proc(Res (Lam $x\,p$))   if Proc p ((νx)p)
⊢ Proc(Zero)   always   (**0**)

Our axiomatised type of lambda-terms allows us to formalise the syntax given at the beginning of the section by this inductive definition within HOL. Rule induction on the Proc predicate formalises structural induction on pi-calculus processes.

For instance, to prove that the set of processes is closed under substitution of a name for a name,

$$\vdash \forall p.\ \text{Proc}\ p \supset \forall x\,y.\ \text{Proc}\ (p[\text{Var}\ x/y])$$

we prove the more general hypothesis that the set of processes is closed under iterated variable renaming,

$$\vdash \forall p.\ \text{Proc}\ p \supset \forall \sigma.\ \text{Renaming}\ \sigma \supset \text{Proc}\ (p\ \text{ISub}\ \sigma)$$

by rule induction on Proc, that is, structural induction.

5 Related Work

The idea of a metatheory of syntax has a long history, going back at least to Church's encoding of higher-order logic within simply-typed lambda-calculus. Martin-Löf's theory of arities is essentially the same idea (Nordström, Petersson, and Smith 1990). The idea is now widely used to represent syntax in theorem-provers such as Paulson's Isabelle (1994). Church, Martin-Löf and Paulson all encode syntax using simply-typed lambda-terms, identified up to alpha-beta-conversion. Types are needed to avoid meaningless divergent terms. We encode syntax using untyped lambda-terms, identified only up to alpha-conversion. We need to represent substitution as a separate function; in a system with beta-conversion it is represented simply as application of an abstraction to a term. On the other hand, our metatheory based on alpha-conversion supports structural induction more directly than one based on alpha-beta-conversion, where one would need to perform induction on the size of the normal-form of a term.

We have advocated representing syntax as a type within a mechanised logic. This is sometimes known as 'deep embedding' (Boulton, Gordon, Gordon, Harrison, Herbert, and Van Tassel 1992). Embeddings based either on de Bruijn terms or a free algebra of name-carrying terms are now quite common (see Gordon (1994) for a survey). We are aware of several recent strands of work on deep embedding that focus on the interaction between substitution and bound variables. Talcott (1993) proposed a generic theory of binding structures, now implemented in Isabelle by Matthews (1995). McKinna and Pollack (1993) proposed a scheme of binding based on two kinds of variables, that allows a straightforward definition of substitution and yet avoids the possibility of variable capture. They implemented it in Lego (Pollack 1994), and it has recently been re-implemented in Isabelle (Owens 1995). Our axiom five, which relates logical and embedded abstractions, is reminiscent of higher-order abstract syntax (Pfenning and Elliott 1988; Despeyroux and Hirschowitz 1994), in which variable binding in the embedded syntax is implemented via the lambda-abstraction in the logic itself.

6 Conclusions

We advocated untyped lambda-terms, identified up to alpha-conversion, as a metatheory suitable for representing the syntax of a formalism within a logic. An application would be proofs about the operational semantics of pi-calculus. Towards this end we proposed five axioms of such lambda-terms, showed them sound for de Bruijn terms, and derived a collection of useful theorems. All the proofs have been checked in HOL.

The main improvements in this paper with respect to Gordon (1994) are the presentation of five basic axioms—which could confidently be postulated in some other theorem prover—and the possibility of defining functions by iteration and primitive recursion on lambda-terms.

Acknowledgement Gordon is supported by a Royal Society University Research Fellowship. We are grateful to our colleagues at Cambridge and Glasgow and the anonymous referees for their comments.

References

Barendregt, H. P. (1984). *The Lambda Calculus: Its Syntax and Semantics* (Revised ed.), Volume 103 of *Studies in logic and the foundations of mathematics*. North-Holland.

Boulton, R., A. Gordon, M. Gordon, J. Harrison, J. Herbert, and J. Van Tassel (1992). Experience with embedding hardware description languages in HOL. In V. Stavridou, T. F. Melham, and R. T. Boute (Eds.), *Theorem Provers in Circuit Design: Theory, Practice and Experience: Proceedings of the IFIP TC10/WG 10.2 International Conference, Nijmegen, June 1992*, IFIP Transactions A-10, pp. 129–156. North-Holland.

Church, A. (1941). *The Calculi of Lambda-Conversion*. Princeton University Press.

Curry, H. B. and R. Feys (1958). *Combinatory Logic*, Volume 1. North-Holland.

de Bruijn, N. G. (1972). Lambda calculus notation with nameless dummies, a tool for automatic formula manipulation, with application to the Church-Rosser theorem. *Indagationes Mathematicae 34*, 381–392.

Despeyroux, J. and A. Hirschowitz (1994, July). Higher-order abstract syntax with induction in Coq. In F. Pfenning (Ed.), *Fifth International Conference on Logic Programming and Automated Reasoning (LPAR'94), Kiev*, Volume 882 of *LNAI*, pp. 159–173. Springer-Verlag.

Gordon, A. D. (1994). A mechanisation of name-carrying syntax up to alpha-conversion. In J. J. Joyce and C.-J. H. Seger (Eds.), *Higher Order Logic Theorem Proving and its Applications. Proceedings, 1993*, Number 780 in Lecture Notes in Computer Science, pp. 414–426. Springer-Verlag.

Gordon, M. J. C. and T. F. Melham (Eds.) (1993). *Introduction to HOL: A theorem-proving environment for higher-order logic.* Cambridge University Press.

Hindley, J. R. and J. P. Seldin (1986). *Introduction to Combinators and the λ-calculus.* Cambridge University Press.

Lambek, J. and P. J. Scott (1986). *Introduction to higher order categorical logic.* Cambridge University Press.

Landin, P. J. (1964, January). The mechanical evaluation of expressions. *Computer Journal 6*, 308–320.

Matthews, S. (1995, September). Implementing FS_0 in Isabelle: adding structure at the metalevel. In L. C. Paulson (Ed.), *Proceedings of the First Isabelle Users Workshop.* Available as Technical Report 379, University of Cambridge Computer Laboratory.

McKinna, J. and R. Pollack (1993). Pure Type Systems formalized. In *TLCA '93 International Conference on Typed Lambda Calculi and Applications, Utrecht, 16–18 March 1993*, Volume 664 of *Lecture Notes in Computer Science*, pp. 289–305. Springer-Verlag.

Melham, T. F. (1994). A mechanized theory of the π-calculus in HOL. *Nordic Journal of Computing 1*, 50–76.

Milner, R., J. Parrow, and D. Walker (1992). A calculus of mobile processes, parts I and II. *Information and Computation 100*, 1–40 and 41–77.

Nordström, B., K. Petersson, and J. M. Smith (1990). *Programming in Martin-Löf's Type Theory.* Clarendon Press, Oxford.

Owens, C. (1995, September). Coding binding and substitution explicitly in Isabelle. In L. C. Paulson (Ed.), *Proceedings of the First Isabelle Users Workshop.* Available as Technical Report 379, University of Cambridge Computer Laboratory.

Paulson, L. C. (1994). *Isabelle: A Generic Theorem Prover*, Volume 828 of *Lecture Notes in Computer Science.* Springer-Verlag.

Pfenning, F. and C. Elliott (1988, June). Higher-order abstract syntax. In *Proceedings of the ACM SIGPLAN '88 Symposium on Language Design and Implementation*, pp. 199–208.

Pollack, R. (1994). *The Theory of LEGO.* Ph. D. thesis, University of Edinburgh.

Stoughton, A. (1988). Substitution revisited. *Theoretical Computer Science 59*, 317–325.

Talcott, C. L. (1993). A theory of binding structures and applications to rewriting. *Theoretical Computer Science 112*, 99–143.

Set Theory, Higher Order Logic or Both?

Mike Gordon

University of Cambridge Computer Laboratory
New Museums Site
Pembroke Street
Cambridge CB2 3QG
U.K.

Abstract. The majority of general purpose mechanised proof assistants support versions of typed higher order logic, even though set theory is the standard foundation for mathematics. For many applications higher order logic works well and provides, for specification, the benefits of type-checking that are well-known in programming. However, there are areas where types get in the way or seem unmotivated. Furthermore, most people with a scientific or engineering background already know set theory, but not higher order logic. This paper discusses some approaches to getting the best of both worlds: the expressiveness and standardness of set theory with the efficient treatment of functions provided by typed higher order logic.

1 Introduction

Higher order logic is a successful and popular formalism for computer assisted reasoning. Proof systems based on higher order logic include ALF [18], Automath [20], Coq [9], EHDM [19], HOL [13], IMPS [10], LAMBDA [11], LEGO [17], Nuprl [6], PVS [22] and Veritas [14].

Set theory is the standard foundation for mathematics and for formal notations like Z [24], VDM [15] and TLA+ [16]. Several proof assistants for set theory exist, such as Mizar [23] and Isabelle/ZF [21].

Anecdotal evidence suggests that, for equivalent kinds of theorems, proof in higher order logic is usually easier and shorter than in set theory. Isabelle users liken set theory to machine code and type theory to a high-level language.

Functions are a pervasive concept in computer science and so taking them as primitive, as is done by (most forms of) higher order logic, is natural. Higher order logic is typed. Types are an accepted and effective method of structuring data and type-checking is a powerful technique for finding errors. Types can be used to index terms and formulae for efficient retrieval. General laws become simpler when typed.

Unfortunately, certain common mathematical constructions do not fit into the type disciplines associated with higher order logic. For example, the set $\{\emptyset, \{\emptyset\}, \{\emptyset, \{\emptyset\}\}, \{\emptyset, \{\emptyset\}, \{\emptyset, \{\emptyset\}\}\}, \ldots\}$ is traditionally used as the definition of the nateral numbers and lists are defined as the union of the infinite chain $\{\langle\rangle\} \cup (X \times \{\langle\rangle\}) \cup (X \times X \times \{\langle\rangle\}) \cup \ldots$. These sets are essentially untyped.

Furthermore, the traditional axiomatic method used in mathematics needs to be reformulated to fit into type theory [3].

There is no standard formulation of higher order logic. The various higher order logics/type theories differ widely both in the notation used and in their underlying philosophical conception of mathematical truth (e.g. intuitionistic or constructive versus classical). Automath is based on de Bruijn's own very general logic [20, A.2] (which anticipated many more recent developments). Coq and LEGO support different versions of the Calculus of Constructions. EHDM, PVS and Veritas each support different classical higher order logics with subtypes and/or dependent types. HOL and LAMBDA support similar polymorphic versions of the simple theory of types. IMPS supports monomorphic simple type theory with non-denoting terms and a theory interpretation mechanism. ALF and Nuprl support versions of Martin Löf type theory (a constructive logic with a very elaborate type system).

There is much less variation among set theories. The well known formulations are, for practical purposes, pretty much equivalent. They are all defined by axioms in predicate calculus. The only variations are whether proper classes are in the object or meta language and how many large cardinals are postulated to exist. The vast majority of mathematicians are happy with ZFC (Zermelo-Fraenkel set theory with the Axiom of Choice).

It would be wonderful if one could get the best of both worlds: the expressiveness and standardness of set theory with the efficient treatment of functions provided by typed higher order logic. In Section 2 an approach is outlined in which set theory is provided as a resource within higher order logic, and in Section 3 a reverse approach is sketched in which higher order logic is built on top of set theory. Both these approaches are explored in the context of the HOL system's version of higher order logic[1], but in the presentation I have tried to minimise the dependence on the details of the HOL logic. Some conclusions are discussed in Section 4.

2 Sets in Higher Order Logic

Set theory can be postulated inside higher order logic by declaring a type V and a constant $\in : V \times V \to bool$ (where $bool$ is the type of the two truthvalues) and then asserting the normal axioms of set theory. The resulting theory has a consistency strength stronger than ZF, because one can define inside it a semantic function from a concrete type representing first order formulae to V

---

[1] The HOL logic is just higher order predicate calculus with a type system, due to Milner, consisting of Church's simple theory of types [5] with type variables moved from the meta-language into the object language. In Church's system, a term with type variables is actually a meta-notation – a term-schema – denoting a family of terms, whereas in HOL it is a single polymorphic term. Other versions of mechanised simple type theory (e.g. IMPS, PVS) use uninterpreted type constants instead of type variables, and then permit these to be instantiated via a theory interpretation mechanism.

such that all the theorems of ZF can be proved.[2] However, a model for higher order logic plus V can be constructed in ZF with one inaccessible cardinal. Thus the strength of higher order logic augmented with ZF-like axioms for V is somewhere between ZF and ZF plus one inaccessible cardinal.[3]

An alternative approach to using some of the linguistic facilities of higher order logic, whilst remaining essentially first order, has been investigated by Francisco Corella. His PhD thesis [8] contains a very interesting discussion of the different roles type theory can have in the formalisation of set theory.

Defining set theory inside higher order logic is very smooth. For example the Axiom of Replacement is simply:

$$\forall f\ s.\ \exists t.\ \forall y.\ y \in t = \exists x.\ x \in s \land y = f(x) \qquad (1)$$

In traditional first-order formulations of ZF, the second-order quantification of f is not permitted, so a messy axiom scheme is needed. Another example of a useful second order quantification is the Axiom of Global Choice:[4]

$$\exists f.\ \forall s.\ \neg(s = \emptyset) \Rightarrow f(s) \in s \qquad (2)$$

Standard definitional methods allow all the usual set-theoretic notions to be defined and their properties established. Such notions include, for example, the empty set, numbers, Booleans, union, intersection, finite sets, powersets, ordered pairs, products, relations, functions etc.

When set theory is axiomatised in higher order logic, the Axiom of Separation interacts nicely with λ-notation to allow $\{x \in X \mid P(x)\}$ to be represented by Spec X ($\lambda x.\ P(x)$), for a suitably defined constant Spec.

More generally, $\{f(x_1, \ldots, x_n) \in X \mid P(x_1, \ldots, x_n)\}$ can be represented by Spec X ($\lambda x.\ \exists x_1 \ldots x_n.\ x = f(x_1, \ldots, x_n) \land P(x_1, \ldots, x_n))$.

In HOL, new types are defined by giving names to non-empty subsets of existing types. Each element s of type V determines a subtype of V whose characteristic predicate is $\lambda x.\ x \in s$ (i.e. the set of all members of set s). A type σ of HOL is *represented* by $s : V$ iff there is a one-to-one function of type $\sigma \to V$ onto the subtype of V determined by s. It is straightforward to find members of V corresponding to the built-in types of HOL, for example $\{\emptyset, \{\emptyset\}\}$ represents the type of Booleans, and $\{\emptyset, \{\emptyset\}, \{\emptyset,\{\emptyset\}\}, \{\emptyset,\{\emptyset\},\{\emptyset,\{\emptyset\}\}\}, \ldots \}$ represents the natural numbers.

Standard set-theoretic constructions can be used to mimic type operators. If s_1, s_2 represent types σ_1 and σ_2, respectively, then the Cartesian product of s_1 and s_2, which will be denoted here by $s_1 \times s_2$, represents the type $\sigma_1 \times \sigma_2$. The set of all total, single-valued relations between s_1 and s_2, denoted here by $s_1 \twoheadrightarrow s_2$,

---

[2] In HOL jargon, this is a deep embedding [4] of ZF in higher order logic plus V.
[3] These facts about consistency strength were provided by Ken Kunen (private communication).
[4] When type V is postulated in the HOL logic this is actually a theorem (because of Hilbert's ε-operator).

represents the type $\sigma_1 \to \sigma_2$.[5] Since there are lots of non-empty subsets of the class of sets, this provides a rich source of new types.

There are two ways this richness can be exploited: (i) to define types that could be defined without V in a slicker and more natural manner, and (ii) to define types that could not be defined at all.

An example of a construction that can be done in HOL without V, but is neater with it, is the definition of lists. In the current HOL system, lists of elements of type σ are represented as a subtype of the type $(num \to \sigma) \times num$, the idea being that a pair (f, n) represents the list $[f(0), f(1), \ldots, f(n-1)]$.[6] A more direct and natural approach uses $\langle x_1, \langle x_2, \ldots, \langle x_n, \text{True} \rangle \ldots \rangle\rangle$ to represent the list $[x_1, \ldots, x_n]$ (the empty list $\langle\rangle$ can be represented by an arbitary set). However, this is not 'well-typed' since tuples with different lengths have different types. Thus this approach cannot be used to define lists in higher order logic. However, the construction can easily be performed inside V, by defining (using primitive recursion):

$$\text{List}(X) = \{\langle\rangle\} \cup (X \times \{\langle\rangle\}) \cup (X \times X \times \{\langle\rangle\}) \cup \ldots \quad (3)$$

The required properties of lists are easily derived, such as the fixed-point property:

$$\forall X. \ \text{List}(X) = \{\langle\rangle\} \cup (X \times \text{List}(X)) \quad (4)$$

and structural induction:

$$\forall P \ X.
\begin{array}{l}
P(\langle\rangle) \wedge (\forall l \in \text{List}(X). \ P(l) \Rightarrow \forall x \in X. \ P(\langle x, l \rangle)) \\
\Rightarrow \\
\forall l \in \text{List}(X). \ P(l)
\end{array} \quad (5)$$

If $s:V$ represents a type σ, then $\text{List}(s)$ represents the type of finite lists of elements of type σ. Thus a type of lists of elements of type σ could be defined in HOL as the subtype of V determined by the predicate $\lambda x. \ x \in \text{List}(s)$. This illustrates how having a type of ZF-like sets can benefit developments in higher order logic.

An example of a construction using V that would be difficult or impossible without it, is Scott's classical model D_∞ of the λ-calculus. This is a subset of an infinite product $D_0 \times D_1 \times D_2 \ldots$ where D_0 is given and D_{i+1} is a subset of the set of functions from D_i to D_i. If the D_is are represented as sets inside V (they are actually sets equipped with a complete partial order structure) then Scott's classical inverse limit can be performed directly, as has been elegantly shown by Sten Agerholm [1]. However, there seems to be no way to do it within

---

[5] Details are in the technical report 'Merging HOL with Set Theory' [12] available at http://www.cl.cam.ac.uk/users/mjcg/papers/holst/index.html.

[6] To ensure that the pairs (f_1, x_1) and (f_2, x_2) are equal if and only if the corresponding lists are equal, it is required that pairs (f, n) representing lists have the property that $f \ m$ equals some canonical value when m is greater than or equal to the length n of the list. The subtype consisting of such pairs (f, n) is used to define lists.

higher order logics based on simple type theory (HOL, PVS, IMPS), though it could be done in type theories with dependent products.[7]

The set-theoretic construction of lists outlined above only works if the type of the list's elements has already been represented as a set. In type systems with polymorphism, like the HOL logic, a type-operator can be defined that uniformly constructs a type $(\alpha)list$ of lists over an arbitrary type α (α is a type variable). Ching-Tsun Chou[8] has proposed a method of defining such polymorphic operations via set theory. He suggests that instead of having just a type V, one could have a type operator that constructed a universe of sets, $(\alpha)V$ say, over an arbitrary type α of atoms ('urelements'). The type $(\alpha)list$ could then be defined as a suitable subtype of $(\alpha)V$, for an arbitrary α, using a similar construction to the one given above.

To make this work, a set theory with atoms needs to be axiomatised – that is, members of $(\alpha)V$ are either atoms – i.e. non-sets – or sets (but not both). The subtype of atoms is specified to be in bijection with α. Such atoms or 'urelements' have a long history in set-theory. For example, before the technique of forcing was developed, they were used for finding models in which the Axiom of Choice doesn't hold. The axioms of set theory need to be tweaked to cope with atoms; in particular, the Axiom of Extensionality needs to be restricted to apply only to sets.[9]

I have not tried working out the details of Chou's proposal, but it strikes me as very promising. Not only does it appear to enable generic constructions to be done over an arbitrary type, by it is also philsophically satisfying in that it seems to correspond to a certain common mathematical practice in which general (e.g. categorical) developments are done directly in a kind of informal type theory, and set-theoretic constructions are restricted to where they are needed.

Adding a type V of sets (or a type operator $(\alpha)V$) to higher order logic can be compared to the already successful mechanisations of first-order set theory provided by Isabelle/ZF and Mizar. It is not clear whether axiomatising ZF in higher order logic is really better than using first order logic.

Sten Agerholm has compared Isabelle/ZF with HOL + V for the construction of D_∞ [2]. The results of his study were somewhat inconclusive. He found that using higher order logic can simplify formulation and proof, but one is faced with a "difficult question" of "which parts of the formalisation should be done in set theory and which parts in higher order logic". Chains (sequences) of sets illustrate the issue: with higher order logic chains can be represented as logical functions from the type of numbers to V, i.e. as elements of type $num \to V$, but with first order logic the user is not burdened with this decision as chains have to be represented inside set theory as set-functions.[10] However, if chains are

---

[7] Define types D_i (for $i = 0, 1, 2, \ldots$), then D_∞ is a subtype of the product $\prod_{n=0}^{\infty} D_i$.
[8] Private communication.
[9] Ken Kunen provided me with information (private communication) on the use of urelements in set theory.
[10] Isabelle does have a polymorphic higher order metalogic but, as Agerholm puts it:

represented as set-functions, then certain things that correspond to type checking in higher order logic need to be done using theorem proving in Isabelle/ZF. Agerholm found that Isabelle's superior theorem proving infrastructure ensured that this was not much of a burden.

3 Higher Order Logic on top of Set Theory

In the previous section set theory – in the guise of the type V – was provided as a resource within typed higher order logic. An alternative approach is to turn this upside down and to take set theory as primary and then to 'build' higher order logic on top of it. The idea is to provide the notations of higher order logic as derived forms on top of set theory and then to use set-theoretic principles to derive the axioms and rules of higher order logic. A key component of this scheme would be the development of special purpose decision procedures that correspond to typechecking.

The set-theoretic interpretation of the HOL logic is straightforward [13, Chapter 15] and can be used to provide a shallow embedding [4] of it in set theory. Each type constant corresponds to a (non-empty) set and each n-ary type operator op corresponds to an operation, $|op|$ say, for combining n sets into a set. In HOL + V such an operation on sets can be represented as a function with type $V \to V \to \cdots \to V$. For example, $|\times|$ is the Cartesian product operation $\times$ and $|\to|$ takes sets X and Y to the set $X \to Y$.

The set, $[\![\sigma]\!]$ say, corresponding to an arbitrary type σ is defined inductively on the structure of σ. If σ is a a type constant c, then $[\![\sigma]\!]$ is just $|c|$. Type variables can be simply interpreted as ordinary variables ranging over V. A compound type $(\sigma_1,\ldots,\sigma_n)op_n$ is interpreted as the application of the logical function $|op_n|$ to the types corresponding to $\sigma_1,\ldots,\sigma_n$ – i.e. $[\![(\sigma_1,\ldots,\sigma_n)op_n]\!] = |op_n|\,[\![\sigma_1]\!]\ldots[\![\sigma_n]\!]$.

The interpretation of HOL constants in set theory is complicated by polymorphism, because the interpretation of a polymorphic constant depends on the sets corresponding to the type variables it contains. For example, the identity function in HOL is a polymorphic constant $\mathsf{I} : \alpha \to \alpha$. For any type α, I is the identity on α. The interpretation of I in set theory, $|\mathsf{I}|$ say, is the identity set-function on some set A – where the set-valued variable A corresponds to the type variable α. Thus $|\mathsf{I}|$ takes a set A and returns the identity set-function on A (so is a logical function of type $V \to V$). If c is monomorphic (has a type containing no type variables), then $|c|$ will have type V. If the type of c contains n distinct type variables, then $|c|$ will be a (curried) function taking n arguments of type V and returning a result of type V.

The fact that the type parameterisation of functions like I is hidden makes the HOL logic clean and uncluttered compared with set theory. One of the challenges in supporting higher order logic on top of set theory is to gracefully manage the correspondence between implicit type variables and explicit set-valued variables.

"The metalogic is meant for expressing and reasonining in logic instantiations ... not for formalising concepts in object logics".

The embedding of terms (i.e. the simply-typed λ-calculus) in set theory requires set-theoretic counterparts of function application and λ-abstraction. The application of a set-function f to an argument x is the unique y such that the pair $\langle x, y \rangle$ is a member of f (necessarily unique if f is a set-function – i.e. a total and single-valued relation). Let us write this set-theoretic application as $f \diamond x$, which is neatly defined using Hilbert's ε-operator by:

$$f \diamond x \ = \ \varepsilon y. \ \langle x, y \rangle \in f \tag{6}$$

The set-theoretic counterpart to λ-abstraction is Fn $x \in X. \ t[x]$, where $t[x]$ is a set-valued term containing a set variable x. The definition of this notation is:

$$\text{Fn } x \in X. \ t[x] \ = \ \{\langle x, y \rangle \in X \times \text{Image}(\lambda x. t[x]) X \ | \ y = t[x]\} \tag{7}$$

where Image $\mathcal{F} \ X$ is the image of set X under a logical function $\mathcal{F}$ (which exists by the Axiom of Replacement).

Each HOL term t is translated to a term $[\![t]\!]$ of type V as follows:

$$\begin{aligned}
[\![x : \sigma]\!] &= x : V & \text{(variables)} \\
[\![c : \sigma[\sigma_1, \ldots, \sigma_n]]\!] &= |c| \ [\![\sigma_1]\!] \ \ldots \ [\![\sigma_n]\!] & \text{(constants)} \\
[\![\lambda x : \sigma. \ t]\!] &= \text{Fn } x \in [\![\sigma]\!]. \ [\![t]\!] & \text{(abstractions)} \\
[\![t_1 \ t_2]\!] &= [\![t_1]\!] \diamond [\![t_2]\!] & \text{(applications)}
\end{aligned} \tag{8}$$

Notice that $[\![t]\!]$ lies in (monomorphic) simple type theory using just the type V. Applying this translation to the term $\forall m \ n. \ m + n \ = \ n + m$ results is:

$$\begin{aligned}
&(|\forall| \ |num|) \diamond \\
&(\text{Fn } m \in |num|. \\
&\quad (|\forall| \ |num|) \diamond \\
&\quad (\text{Fn } n \in |num|. \\
&\quad\quad ((|=| \ |num|) \diamond ((|+| \diamond m) \diamond n)) \diamond ((|+| \diamond n) \diamond m)))
\end{aligned} \tag{9}$$

This is a set-denoting term – i.e. a term of type V – the logical constants $\forall$ and $=$ have been 'internalised' into set-functions $|\forall|$ and $|=|$, respectively.

In HOL, Boolean terms can play the role of *formulae* which denote 'true' or 'false' – i.e. are judgements. Terms play this role when they are postulated as axioms or definitions or occur in theorems. When embedding higher order logic in set theory, formulae of the former should be translated to formulae of the latter. The translation of a Boolean term via (8) can be made into a logical formula of set theory by equating it to the set representing 'true' inside V, $|\text{T}|$ say. Thus the formula corresponding to $\forall m \ n. \ m + n \ = \ n + m$ is obtained by equating (9) to $|\text{T}|$. Using suitable definitions of the internalised constants, the resulting formula will be equivalent to:

$$\forall m \ n \in |num|. \ (|+| \diamond m) \diamond n = (|+| \diamond n) \diamond m \tag{10}$$

Free variables in formulae are interpreted as implicitly universally quantified. Thus the set-theoretic formula corresponding to $m + n = n + m$ should be equivalent to (10). If we want higher order logic formulae with free variables to translate to set-theoretic formulae with the same free variables, then the universal quantifier in (10) can be stripped off – but the restrictions that m and n be in $|bool|$ must be retained, i.e.:

$$m \in |num| \wedge n \in |num| \Rightarrow ((|+| \diamond m) \diamond n = (|+| \diamond n) \diamond m) \quad (11)$$

Thus when translating formulae, the typing of variables in higher order logic has to be converted into explicit set membership conditions in set theory.[11]

Using the scheme just described, a term tm of higher order logic is interpreted, depending on context, as the set-denoting term $[\![tm]\!]$ or the formula $x_1 \in [\![\sigma_1]\!] \wedge \ldots \wedge x_n \in [\![\sigma_n]\!] \Rightarrow [\![tm]\!] = |\mathsf{T}|$ (where the free variables in tm are $x_1{:}\sigma_1, \ldots, x_n{:}\sigma_n$). These interpretations could be handled by a parser and pretty printer (i.e. implemented as a shallow embedding).

In the HOL logic there are two primitive types $bool$ and ind and three primitive constants $\Rightarrow$, $=$ and ε. The internalised primitive types $|bool|$ and $|ind|$ are the set of two truthvalues and some arbitrary infinite set. The internalised primitive constants $|\Rightarrow|$, $|=|$ and $|\varepsilon|$ are easily defined – $|\Rightarrow|$ by explicitly writing down the set representing the appropriate set-function (details omitted), and the other two by:

$$\begin{aligned} |=|\ X &= \{\langle x,y \rangle \in X \mathbin{\mathbf{x}} X \mid x = y\} \\ |\varepsilon|\ X &= \mathsf{Fn}\ f.\ \mathsf{Choose}\{x \in X \mid f \diamond x = |\mathsf{T}|\} \end{aligned} \quad (12)$$

where Choose is a suitable choice operator (of logical type $V \to V$) legitimated by the Axiom of Choice.

If the shallow embedding described here is to provide the user with higher order logic, then the axioms and rules must be derived. An example of an axiom of the HOL logic is the Law of Excluded Middle: $\forall t.\ t=\mathsf{T} \vee t=\mathsf{F}$. Using (8) this is interpreted as the formula:

$$\begin{aligned} &(|\forall|\ |bool|) \diamond \\ &\quad(\mathsf{Fn}\ t \in |bool|. \\ &\qquad (|\vee| \diamond (((|=|\ |bool|) \diamond t) \diamond |\mathsf{T}|)) \diamond (((|=|\ |bool|) \diamond t) \diamond |\mathsf{F}|)) \\ &= |\mathsf{T}| \end{aligned} \quad (13)$$

which, with suitable definitions of $|\vee|$ and $|=|$ will be equivalent to:

$$\forall t \in |bool|.\ t = |\mathsf{T}| \vee t = |\mathsf{F}| \quad (14)$$

which can be proved in set theory if $|bool| = \{|\mathsf{T}|, |\mathsf{F}|\}$.

An example of a rule of inference in higher order logic is β-conversion. A β-redex $(\lambda x : \sigma.\ t_1[x])t_2$ translates to: $(\mathsf{Fn}\ x \in [\![\sigma]\!].\ [\![t_1[x]]\!]) [\![t_2]\!]$. Now it is a theorem of set theory that:

$$y \in X \Rightarrow (\mathsf{Fn}\ x \in X.\ t(x)) \diamond y = t(y) \quad (15)$$

---

[11] In HOL, different variables can have the same name as long as they have different types, so if a HOL formula contains two distinct variables with the same name, then these variables will need to be separated on translation (e.g. by priming one of them).

and hence (using higher-order matching etc.) β-conversion can be derived. Notice, however, that to apply (15) an instance of the the explicit set membership condition $x \in X$ has to be proved. In higher order logic this happens automatically via typechecking. In set theory, a special 'typechecking' theorem prover can be implemented [12, Section 6.6] using theorems such as:

$$f \in (X \twoheadrightarrow Y) \wedge x \in X \Rightarrow f \diamond x \in Y$$
$$(\forall x.\ x \in X \Rightarrow t[x] \in Y) \Rightarrow (\text{Fn } x.\ t[x]) \in (X \twoheadrightarrow Y)$$
(16)

The set-theoretic versions of the non-primitive types and constants could be defined by interpreting the HOL definitions in set theory. However, this leads to a potential confusion between certain standard set-theoretic constructions, and the versions obtained by translating HOL definitions. In particular, the translation of the HOL definition of ordered pairs (product types) via (8) does not result in the familiar model of pairing used in set theory. This is an area needing further thought. Probably the best strategy is to support on top of set theory a higher order logic with more primitives than the HOL logic (e.g. with pairing built-in) – and then to augment the translation (8) to interpret the additional constructs as their natural set-theoretic counterparts.

A potentially useful feature of having set theory as the underlying logical platform is that theories in set theory can be encoded as single (large) theorems in a way that can't be done for theories in some versions of higher order logic (e.g. HOL). A theory in set theory can be regarded as an implication with the antecedents being the axioms. Constants declared in the theory will be monomorphic and can just be treated as free variables. This doesn't work in HOL because polymorphic constants can occur at different type instances of their declared (i.e. generic) type in different axioms and theorems of a theory, but a polymorphic variable must have the same type at all its occurrences in an individual theorem. Relaxing this restriction is known to make the HOL logic inconsistent [7].

Being able to code up theories as theorems could enable 'abstract theories' to be naturally supported, since theory interpretation then becomes just ordinary instantiation.

4 Discussion and Conclusions

In Section 2 it was shown how by postulating V (or, better, $(\alpha) V$) it was possible to increase the power of higher order logic, whilst still retaining its attractive features. This idea has been explored in some detail by Sten Agerholm and seems a success.

In Section 3 a more radical idea is outlined in which higher order logic is 'mounted' on top of set theory as a derived language (via a shallow embedding). If this can be made to work – and it is not yet clear whether it can – then it would seem to offer the best of the two worlds. Users could choose to work entirely within higher order logic, but they could also choose to stray into the rich pastures of set theory. Furthermore, users could add additional constructs themselves without having to modify the core system. Thus, for example, the record and dependent subtypes of PVS could be added (type correctness conditions just being handled by normal theorem proving). However, this is currently all fantasy: it still remains to see whether it is possible to get an efficient and well-engineered type theory via a shallow embedding into set theory.

In conclusion, my answer to the question posed as the title of this paper is that both set theory and higher order logic are needed. In the short term useful things can be done by adding a type of sets to higher order logic, but building higher order logic on top of set theory is an exciting research challenge that promises a bigger payoff.

Acknowledgements

Sten Agerholm, Ching-Tsun Chou, Francisco Corella, John Harrison, Tom Melham, Larry Paulson and Andy Pitts provided various kinds of help in the development of the ideas described here.

References

1. S. Agerholm. Formalising a model of the λ-calculus in HOL-ST. Technical Report 354, University of Cambridge Computer Laboratory, 1994.
2. S. Agerholm and M.J.C. Gordon. Experiments with ZF Set Theory in HOL and Isabelle. In E. T. Schubert, P. J. Windley, and J. Alves-Foss, editors, *Higher Order Logic Theorem Proving and Its Applications: 8th International Workshop*, volume 971 of *Lecture Notes in Computer Science*, pages 32–45. Springer-Verlag, September 1995.
3. Jackson Paul B. Exploring abstract algebra in constructive type theory. In A. Bundy, editor, *12th Conference on Automated Deduction*, Lecture Notes in Artifical Intelligence. Springer, June 1994.
4. R. J. Boulton, A. D. Gordon, M. J. C. Gordon, J. R. Harrison, J. M. J. Herbert, and J. Van Tassel. Experience with embedding hardware description languages in HOL. In V. Stavridou, T. F. Melham, and R. T. Boute, editors, *Theorem Provers in Circuit Design: Theory, Practice and Experience: Proceedings of the IFIP TC10/WG 10.2 International Conference*, IFIP Transactions A-10, pages 129–156. North-Holland, June 1992.
5. A. Church. A formulation of the simple theory of types. *The Journal of Symbolic Logic*, 5:56–68, 1940.
6. R. L. Constable et al. *Implementing Mathematics with the Nuprl Proof Development System*. Prentice-Hall, 1986.
7. Thierry Coquand. An analysis of Girard's paradox. In *Proceedings, Symposium on Logic in Computer Science*, pages 227–236, Cambridge, Massachusetts, 16–18 June 1986. IEEE Computer Society.
8. Francisco Corella. Mechanizing set theory. Technical Report 232, University of Cambridge Computer Laboratory, August 1991.

9. G. Dowek, A. Felty, H. Herbelin, G. Huet, C. Murthy, C. Parent, C. Paulin-Mohring, and B. Werner. The Coq proof assistant user's guide - version 5.8. Technical Report 154, INRIA-Rocquencourt, 1993.
10. W. M. Farmer, J. D. Guttman, and F. Javier Thayer. IMPS: An interactive mathematical proof system. *Journal of Automated Reasoning*, 11(2):213–248, 1993.
11. S. Finn and M. P. Fourman. *L2 – The LAMBDA Logic*. Abstract Hardware Limited, September 1993. In LAMBDA 4.3 Reference Manuals.
12. M. J. C. Gordon. Merging HOL with set theory. Technical Report 353, University of Cambridge Computer Laboratory, November 1994.
13. M. J. C. Gordon and T. F. Melham, editors. *Introduction to HOL: A Theorem-proving Environment for Higher-Order Logic*. Cambridge University Press, 1993.
14. F. K. Hanna, N. Daeche, and M. Longley. Veritas+: a specification language based on type theory. In M. Leeser and G. Brown, editors, *Hardware specification, verification and synthesis: mathematical aspects*, volume 408 of *Lecture Notes in Computer Science*, pages 358–379. Springer-Verlag, 1989.
15. C. B. Jones. *Systematic Software Development using VDM*. Prentice Hall International, 1990.
16. L. Lamport and S. Merz. Specifying and verifying fault-tolerant systems. In *Proceedings of FTRTFT'94*, Lecture Notes in Computer Science. Springer-Verlag, 1994. See also: http://www.research.digital.com/SRC/tla/papers.html#TLA+.
17. Z. Luo and R. Pollack. LEGO proof development system: User's manual. Technical Report ECS-LFCS-92-211, University of Edinburgh, LFCS, Computer Science Department, University of Edinburgh, The King's Buildings, Edinburgh, EH9 3JZ, May 1992.
18. L. Magnusson and B. Nordström. The ALF proof editor and its proof engine. In *Types for Proofs and Programs: International Workshop TYPES '93*, pages 213–237. Springer, published 1994. LNCS 806.
19. P. M. Melliar-Smith and John Rushby. The enhanced HDM system for specification and verification. In *Proc. Verkshop III*, volume 10 of *ACM Software Engineering Notes*, pages 41–43. Springer-Verlag, 1985.
20. R. P. Nederpelt, J. H. Geuvers, and R. C. De Vrijer, editors. *Selected Papers on Automath*, volume 133 of *Studies in Logic and The Foundations of Mathematics*. North Holland, 1994.
21. L. C. Paulson. *Isabelle: A Generic Theorem Prover*, volume 828 of *Lecture Notes in Computer Science*. Springer-Verlag, 1994.
22. PVS Web page. http://www.csl.sri.com/pvs/overview.html.
23. Piotr Rudnicki. *An Overview of the MIZAR Project*. Unpublished manuscript; but available by anonymous FTP from menaik.cs.ualberta.ca in the directory pub/Mizar/Mizar_Over.tar.Z, 1992.
24. J. M. Spivey. *The Z Notation: A Reference Manual*. Prentice Hall International Series in Computer Science, 2nd edition, 1992.

A Mizar Mode for HOL

John Harrison

Åbo Akademi University, Department of Computer Science
Lemminkäisenkatu 14a, 20520 Turku, Finland

Abstract. The HOL theorem prover is implemented in the LCF manner. All inference is ultimately reduced to a collection of very simple (forward) primitive inference rules, but by programming it is possible to build alternative means of proving theorems on top, while preserving security. Existing HOL proofs styles are, however, very different from those used in textbooks. Here we describe the addition of another style, inspired by Mizar. We believe the resulting system combines the secure extensibility and interactivity of HOL with Mizar's readability and lack of logical prescriptiveness. Part of our work involves adding new facilities to HOL for first order automation, since this allows HOL to be more flexible, as Mizar is, over the precise logical connection between steps.

1 HOL

The HOL theorem prover [13] is descended from Edinburgh LCF. While adding features of its own like stress on definitional extension as a reliable means of theory development, it remains true to the basic idea of the LCF project. This is to reduce all reasoning to a few simple primitive (usually forward) inference rules, but to allow a full programming language to automate higher level 'derived rules', broken down into these primitives. For example, HOL includes derived rules for linear arithmetic, tautologies and inductive definitions. Ordinary users can simply invoke them without understanding their implementation, but because they do ultimately decompose to simple primitives, can feel confident in their correctness. Should they need other, perhaps application-specific, proof procedures in the course of their work, they can write them using the same methodology.

This combination of reliability and flexibility is the outstanding feature of LCF systems, and there is usually not a serious loss of efficiency in derived rules [17]. Some of these derived rules may present the user with a quite different view of theorem proving from that implemented in the logical core. Even in the original LCF publication [14] we find the following:

> The emphasis of the present project has been on discovering how to exploit the flexibility of the metalanguage to organise and structure the performance of proofs. The separation of the logic from its metalanguage is a crucial feature of this; different methodologies for performing proofs in the logic correspond to different programming styles in the metalanguage. Since our current research concerns experiments with proof methodologies – for example, forward proof versus goal-directed proof – it is essential that the system does not commit us to any fixed style.

To some extent, this theoretical flexibility is already a practical reality in HOL. In addition to the basic 'machine code' of forward primitive rules, there are several supported proof styles, all of which fit together smoothly:

- There are numerous more complicated forward proof rules, which can make the business of theorem proving much more palatable than it would be using the primitives. However, before each inference rule is applied, it's necessary to muster all the required hypotheses exactly, and either include their proofs verbatim, or bind them to names and use those. It's very hard to do proofs in this way unless the exact structure of the proof is already planned before starting to type.
- Backward, tactical proof was one of the most influential ideas in the LCF project. Most large HOL proofs are done in this way, perhaps because the required hypotheses appear naturally and determine the proof structure automatically. It also allows more convenient use of local assumptions and choosing variables. This flexibility is further increased if the tactic mechanism allows 'metavariables' whose instantiations can be delayed [32, 26].
- Equational reasoning is one of the most widely used parts of the HOL system, largely thanks to an elegant and flexible implementation [25]. Depth conversions and rewriting tools allow the convenient iterated instantiation and use of equations. There are also straightforward means of handling associative and commutative operators.
- Window inference [29] is a methodology for organizing localized proof efforts. Users may focus on a particular subterm or subformula and transform it, exploiting contextual information that comes from its position in the whole formula. For example, when transforming ψ into an equivalent formula ψ' in the expression $\phi \wedge \psi$, we may assume ϕ. Grundy [15] both mechanized window inference in HOL and generalized it to arbitrary preorder relations, such as implication and the refinement relation on programs.
- Prasetya [27] has written a package to support two features of textbook proofs: the use of a series of lemmas, and the use of iterated equations (we shall have more to say about this latter issue later).
- Specialized decision procedures for various particular domains such as linear arithmetic [6] are also available, as well as a number of derived definitional mechanisms [23, 24].

Most of these styles suffer from being rather low-level, making explicit too many details that are normally elided. More precisely, they are too *logically prescriptive*, demanding that even the most obvious steps be mediated by the exactly appropriate logical rule(s). This isn't just a problem because doing it is tedious. A beginner might well simply *not be able to drive the system well enough* to get it to do the requisite steps. For example, many HOL users find manipulation of assumptions difficult. Decision procedures, on the other hand, are perhaps too high-level, compressing into one line substantial mathematical detail.

Whether too high-level or too low-level, all the proof styles suffer from one common failing: the proofs are expressed using complicated combinations of arcane higher order functions in a computer programming language. Though it's easy to guess what many of them do, a HOL script looks nothing like a textbook proof. Even HOL experts cannot really read a typical HOL proof without replaying it in a session. Annotating proofs with intermediate theorems, as done by Paul Jackson in Nuprl, certainly makes them more readable. However it also causes them to expand substantially, and gives a rather artificial separation between the proof instructions to the machine and the parts that are intended for human consumption.

2 The Mizar Proof Script Language

The Mizar theorem prover,[1] developed by a team in the Białystok branch of Warsaw University under the leadership of Andrzej Trybulec, is quite different from HOL in almost every respect. It was designed primarily for the formalization of mainstream mathematical proofs rather than for verification applications; it is based on Tarski-Grothendieck set theory rather than simple type theory, and the proof checker is built on entirely different lines. However we believe that its proof script language provides many interesting ideas and lessons. We do not claim it is ideal for all applications; standard HOL styles may for example be better in many verification tasks, or where complex decision procedures are to be used. But for its original purpose of proofs in pure mathematics, it has a lot to recommend it — the enormous amount of mathematics that has been proof checked in the Mizar system stands as a testament to that.

2.1 Presenting natural deduction proofs

Systems of Natural Deduction seem to provide quite a direct rendering of typical mathematical reasoning, including common idioms such as reasoning from assumptions, performing case splits, etc. The actual format of natural deduction proofs as usually presented is, however, rather different from that of textbook proofs. The Mizar proof language improves things by associating deduction steps with English constructs that can be put together into a fairly conventional mathematical proof. For example:

- 'let x be α; $<$*proof of* $\phi[x]$ $>$' is a proof of $\forall x : \alpha. \phi[x]$.
- 'assume ϕ; $<$*proof of* ψ $>$' is a proof of $\phi \Rightarrow \psi$.
- 'take a; $<$*proof of* $\phi[a]$ $>$' is a proof of $\exists x. \phi[x]$.

These and other similar constructs define the 'proof skeleton', i.e. the overall logical structure of the proof.

2.2 Stepping beyond natural deduction

Though natural deduction captures many mathematical idioms, it is not ideal for every application. For example, equality reasoning is usually done using certain obvious techniques like substitution and rewriting, rather than by explicitly stringing together axioms for the equivalence and congruence properties of equality via natural deduction rules. And at the formula level, it is sometimes more attractive to reason directly

---

[1] See the Mizar Web page: 'http://web.cs.ualberta.ca:80/~piotr/Mizar/'.

with logical equivalence rather than decompose it to two implications [37].[2] In fact in HOL there is already a great emphasis on the use of equivalence: it is just equality on booleans, so all the powerful equational proof techniques like rewriting are available to exploit it.

In fact, one often wants to make very simple logical steps that do not correspond to individual natural deduction rules. Notwithstanding their theoretical interest, natural deduction rules are not sacrosanct. For example passing from $A \vee B$ and $\neg A$ to B (an instance of resolution) is at least as 'natural' as natural deduction $\vee$-elimination, let alone the sequence of ND steps needed for the above inference. In general we might be completely uninterested in the exact series of inferences, e.g. we might wish to pass from $a = 0x$ to $a = 0$ using the theorem

$$\forall x\ y.\ xy = 0 \equiv x = 0 \vee y = 0$$

without writing out a full natural deduction derivation. All this suggests beefing up natural deduction with the ability to make rather simple 'obvious' jumps, and this is precisely what the Mizar system does. The user may write 'ϕ by $A_1, \ldots, A_n$', meaning that ϕ is considered an obvious consequence of the theorems $A_1, \ldots, A_n$ (these are either preproved theorems or labelled steps in the present deduction).

The body of a Mizar proof contains a list of steps justified with 'by'; these are usually just formulas, with or without labels, but sometimes skeleton constructs also use 'by' for their justification. For example 'consider x such that $P[x]$' performs an $\exists$-elimination step; in subsequent steps, $P[x]$ may be assumed. However this requires justification for $\exists x.\ P[x]$, and enough theorems must be provided for this to be deduced. To avoid a proliferation of labels, the previous step may be implicitly assumed by prefixing a step with 'then'.[3] Finally, certain formulas are prefixed with 'thus' or 'hence' (the latter equivalent to 'then thus'); these are 'conclusions'. The set of conclusions collected in a list of steps should always be sufficient to justify the current objective or *thesis*. For example, if the thesis is $\phi \wedge \psi$, then one might have two conclusion steps containing ϕ and ψ; if the thesis is $\phi \equiv \psi$, the conclusions might be $\phi \Rightarrow \psi$ and $\psi \Rightarrow \phi$. To achieve a kind of bracketing of sets of conclusions, which could otherwise be ambiguous, an individual step can be justified not using 'by', but rather by a whole nested proof enclosed between 'proof' and 'end'.

The thesis is tracked automatically by the system as the proof script is processed, starting from the initial goal. For example, if the current thesis is $\forall x : A.P[x]$, then after processing a step 'let x be A' the thesis becomes $P[x]$. For one or two constructs, knowledge of the thesis is necessary; it cannot be constructed from the proof. For example 'take m' followed by a proof of $m \leq m$ could be a proof of $\exists x.\ x \leq m$ or of $\exists x.\ x \leq x$, among others. Apart from providing the system with additional information in such cases, the thesis is a useful sanity check, since the skeleton structure should

---

[2] Moreover, when doing exploratory interactive work, it is convenient that all equational steps are reversible, so one can feel confident that a provable subgoal is not being replaced by an unprovable one.

[3] The Mizar system makes formulas introduced using certain constructs like 'assume' available by default without labelling; our version makes this optional, and always allows labelling.

correspond to the thesis.[4] Moreover, it allows one to use the special word 'thesis' rather than repeatedly quote the formula; this is convenient since if several case splits are performed, there will typically be many conclusions 'thus thesis'. The system attempts to modify the thesis intelligently given a conclusion step. For example if the thesis is $\phi \equiv \psi$ and a conclusion step proves $\phi \Rightarrow \psi$, then the thesis becomes $\psi \Rightarrow \phi$. Mizar allows a nested proof within 'now ... end', which unlike the nested proofs within 'proof ... end', does not make the thesis known at the outset. In our HOL implementation we disallow this, since it does not fit very tidily with our reduction to tactics. It could easily be implemented, but would require the entire nested proof to be processed separately.

A few other mathematical idioms are admitted, in particular the kind of iterated equality reasoning whose usefulness we have already noted; our HOL version can handle other binary relations like numeric inequalities too, à la Grundy. In the HOL version, using three dots as the left-hand argument of a binary operator is a shorthand for the previous left-hand argument that was given explicitly, and there is also an implicit 'then' to link the previous step. So for example one may write:

```
    a = b by Th1,Th2;
    ... = c by Th3;
    ... = d by Th4,Th5,Th6;
```

which serves as a proof of $a = d$. The reader familiar with the typical calculational style of proof [12] will see that the use of iterated equality in the Mizar language is almost identical to that, each step in the transitivity chain being justified by a hint, albeit of a rather uniform kind.

In summary, Mizar scripts admit a division into the 'proof skeleton', which uses the special keywords to set out the basic structure of the proof, and the individual steps within the proof, mostly using 'by' and its relatives. There is thus an attractive combination of a clearly structured natural deduction proof together with flexibility over the individual inferences. Apart from simply making things easier, this might also appeal to the many mathematicians who are uninterested in, or actively dislike, logic and foundations.[5]

2.3 Stepping beyond Mizar

We have found it useful to add a few other features to the language beyond those included in Mizar itself (conversely of course, we have left out some features of the Mizar language such as its special type coercing functions that don't have a natural HOL counterpart). These new features do not by any means exhaust the possibilities; on the contrary a careful study of existing mathematical textbooks and papers, concentrating more on their form than their content, might reveal many other useful additions.

First, we allow the idiom (when trying to prove X, say): 'suffices to show X'', usually accompanied by a justification using 'by'. This construct allows backward

---
[4] This is decidable and in Mizar is checked by a separate pass.
[5] Probably its emphasis on practical usability rather than foundational questions is partly responsible for the amount of real mathematics done in Mizar.

proof. Sometimes the steps are quite trivial, corresponding to typical mathematical steps like 'by [induction] we need only prove $P[0]$ and $\forall n.\ P[n] \Rightarrow P[n+1]$', or 'by [symmetry in m and n] we may assume that $m \leq n$'; but they could be more substantial. Mathematics books often contain an admixture of backward proof, and some suggest that the proportion could profitably be increased.

Second, we allow the use of arbitrary HOL rules in justifications. As well as just a list of theorems, the 'by' command takes an optional identifier for a HOL inference rule. There is a standard default rule, of which more below, but users may use their own in special situations. By contrast Mizar has no facilities for extension with abbreviations for complicated proof idioms (e.g. repeated rewrites), not even a simple macro language as in PVS. So we can see that the traffic of ideas is not all in one direction: here we use HOL to address a weakness of Mizar.

3 Mizar Proofs in HOL

Our initial experiments involved taking a complete Mizar-style proof script and translating it to HOL primitive inferences. However in this way the Mizar proof style becomes decoupled from the others in HOL, whereas one of the attractions of the existing system is that say, forward and backward proof can be intermixed freely. In addition, we believe that another advantage HOL has over Mizar is that its style of interaction is less batch-oriented. In Mizar, the typical style of user interaction is an edit-compile cycle rather like the use of a programming language compiler, possibly processing a fairly large file each time, whereas with HOL one can try out proof steps, see their effect, and either press on or back up and try something else.

3.1 Mizar tactics

Therefore we now attempt to integrate Mizar-style proofs with HOL tactic proofs. If we think of the 'state' of a Mizar proof, that is, the current thesis and the list of facts derived and labelled so far, as the conclusion and hypotheses of a HOL goal, then there is a close relationship between most Mizar skeleton constructs and certain HOL tactics. Roughly speaking, the relationship is as follows:[6]

Mizar construct	HOL tactic	(Reversed) ND rule
assume	DISCH_TAC	$\Rightarrow$ intro
let	X_GEN_TAC	$\forall$ intro
take	EXISTS_TAC	$\exists$ intro
consider	X_CHOOSE_TAC	$\exists$ elim
given	DISCH_THEN o X_CHOOSE_TAC	$\Rightarrow$ intro and $\exists$ elim
suffices to show	MATCH_MP_TAC	$\Rightarrow$ elim
set	ABBREV_TAC	abbreviation

---

[6] ABBREV_TAC is not part of the HOL standard tactic collection, but is much used by the present author; it can be found for example in the code for the reals library.

For our purposes, it is desirable to extend HOL's tactic mechanism with the ability to label assumptions using chosen names. In this way we can associate assumptions in the goal with the appropriate Mizar labels. Such an extension is, we feel, desirable in any case. Manipulation of assumptions is a perennial problem in HOL, since both numbering and explicit term quotation can be sensitive to quite small changes in the proof, necessitating more sophisticated techniques [5]. The change to the HOL sources took only half an hour, mostly just inserting 'snd' or 'map snd' in various tactics. The types of goals and tactics change, but these are normally wrapped in aliases anyway when used at a higher level, so there seems little danger of proofs being broken by the change. Given this slight enhancement of the tactic mechanism, we are now quite close to an interpretation of Mizar steps as tactics. Note that the head of the assumption list is considered the 'previous step' in the Mizar sense, and is selected for 'then' linkage. Unless it is labelled, the next step deletes it from the assumption list, so that just as with Mizar, the previous result only exists ephemerally.

We actually define special 'Mizar tactics' which are very similar to their HOL analogs in the above table. For example, 'MIZAR_ASSUME_TAC' is just like HOL's 'DISCH_TAC' except that it checks that the term being discharged is the same as the one given as an argument, and attaches any specified label to that new assumption. If we were attempting to emulate Mizar's ability to process proofs without an explicit thesis, then it would be necessary to make these tactics work even in the event of a mismatch with the thesis; all that matters is that the subsequent proof reconstruction works as intended.

In order to reduce the load of user type annotation, all the Mizar tactics accept preterms rather than terms.[7] The Mizar tactics then typecheck them in the context of variable typings in the current goal. However if there are variables in the goal with the same name but different types, these are excluded, since an arbitrary choice could leave the user stymied. In that case, some annotation may be needed.

We define special constants that are expanded during Mizar's preterm to term translation. These are 'thesis' of type ':bool', which is expanded to the current thesis, and '...', of polymorphic type, which is expanded to the left hand of the previous step. We provide one that Mizar itself does not: 'antecedant' refers to the antecedent of the current goal. We quite like the idea of adding others, such as the first and second conjunct of a conjunction, and plan to experiment with this.

3.2 Case splitting constructs

It is necessary to deal with the constructs that can split a goal into several subgoals, namely nested subproofs and 'per cases'. Nested proofs are dealt with simply by using HOL's standard 'SUBGOAL_THEN' tactic, which sets up two subgoals: the lemma itself, and the original goal with the lemma as an extra assumption. However Mizar's case-splitting construct requires more care. Like 'DISJ_CASES_TAC', HOL's case-splitting tactic, it performs a natural deduction ∨-elimination step. It is used as follows:

---

[7] Readers unfamiliar with HOL preterms can think of them as untyped syntax trees that become terms only after typechecking.

```
    per cases by <justification>
      suppose X1
      ...
      ...
      hence thesis

    ...

      suppose Xn
      ...
      ...
      hence thesis
    end
```

The justification is supposed to be able to prove $X1 \vee \ldots \vee Xn$. Now, as compared with HOL's case-splitting construct, the above does not make explicit at the start how many subgoals will be generated, nor what the eventual disjunctive theorem to justify is. Therefore a direct translation into HOL's corresponding constructs would require processing of the complete construct, and as we've already said, we are keen to allow expansion of every stage of the proof interactively. Accordingly we proceed as follows.[8]

The 'per cases' is translated into a HOL tactic that simply yields the same subgoal, but with a justification phase which, on receiving a theorem with an additional assumption, tries to prove and so discharge it using the stated justification. Then each 'suppose X' causes a split into two subgoals, one with assumption 'X', one identical to the original. The assumption is that the second subgoal will in fact produce a theorem with some additional assumption; the justification stage of this tactic performs a HOL 'DISJ_CASES' step, i.e. $\vee$-elimination. Finally, the 'end' construct simply proves the goal under an assumption of falsity; this is trivially disposed of by the eventual disjunction justification.

The above scheme works rather well, and allows a direct step-by-step exploration of the Mizar proof using HOL's subgoal package. Note that because of the way 'per cases' is dealt with, a string of Mizar tactics has itself no structure, and needs to be applied repeatedly to the head of a current list of goals. Interactively, this is done by the goal stack anyway. However to compose a sequence of Mizar tactics, one must not use 'THEN', which applies its second argument tactic to all subgoals. Nevertheless it is easy to define a variant of 'THEN' that will package up a sequence of Mizar tactics into a single tactic.

3.3 Parsing

Writing proofs directly using the Mizar tactics is not a big improvement on the readability of standard HOL tactics, even though the structures into which they are organized may be more natural. Instead of that we have a special parser for Mizar texts.

---

[8] Note that some of the tactics mentioned below are 'invalid' in the LCF sense, i.e. they may not be able to reconstruct the goal from the subgoals.

Within this, we still use the HOL notation for terms, rather than Mizar's more readable but more verbose alternative. This could be changed if desired. There are actually one or two syntactic ambiguities arising from lumping arbitrary HOL terms together with Mizar keywords. For example 'let' could either introduce a Mizar step, or be the start of a HOL term; likewise 'L:A' could either be a labelled term 'A' or a term 'L' with type 'A' These could be cleared up without difficulty if they become troublesome.

In fact, we usually install as the default quotation parser a function that parses Mizar steps and reduces them to a tactic. The only problem with this is that we want to be able to refer not only to labels in the existing derivation, but also to pre-proved theorems. These theorems are all bound to ML identifiers, so to use them inside quotations, it is necessary to use antiquotation, i.e. precede each name by a carat. Since Slind's system for antiquotation [31] is polymorphic, this presents no problems in principle. However since our experiments are being conducted in a version of HOL without antiquotation,[9] we have adopted the temporary solution of pushing all required external lemmas onto the assumption list with appropriate labels. Similarly, we use a global list binding inference rules to names, which allows various inference rules to be used in the same quotation.

4 Proof support

We have dealt with parsing the proof, but have only discussed the processing of the skeleton constructs into HOL inferences. It remains to translate the individual 'obvious' proof steps in the same way. Since we allow essentially arbitrary ML code in a 'by' statement, any HOL rules can be used. However the existing HOL rules are not really capable of emulating Mizar's recognition of logically obvious steps. To maximize the benefits, something of similar scope is required. Mizar incorporates an automatic theorem prover for first order logic which, while not very powerful, has evolved over time to become extremely quick at checking 'obvious' inferences. For Mizar, speed is essential, since as we have already said, the system is normally used in batch mode. For us, high speed is less important, and is unlikely to be achievable anyway. So we can afford to err on the side of making the checker more powerful. We don't attempt to emulate Mizar's own prover, but start from scratch using fairly standard techniques for automated theorem proving. We provide two alternative provers, one based on tableaux, the other on model elimination. Before giving more details, we will make some general comments and look at how formulas are normalized for input to the provers.

Only first order logic?

Mizar's logic is almost first order, but it supports free second order variables, making axiom schemas much more civilized to deal with. (This logic is often facetiously referred to as 1.01^{th}-order logic.) The automated theorem prover that 'by' invokes is only for first order reasoning. When one wants to instantiate a second order variable,

---

[9] This version of HOL is implemented in CAML Light. See [19] for the starting point of this work.

e.g. in induction or set comprehension schemes, a separate command 'from' is invoked, together with an explicit instantiation for that variable.

In HOL, although higher order *features* are constantly used, many of the *proofs* are 'essentially first order'. We reduce higher order to first order logic in a well-known mechanical way: introduce a single binary function symbol a to represent 'application', and translate HOL's $f\ x$ into $a(f, x)$, etc.[10] Then it is often the case that when a theorem is provable in higher order logic, the corresponding first order assertion is also provable.

Proofs that cannot be done in the first order reduction are those that require the *instantiation* of higher order variables, i.e. the invention of lambda-abstractions. For example, when trying to prove $\forall n.\, n+0 = n$ by induction, the induction theorem needs to be specialized to the relation $\lambda n.\, n+0 = n$, or equivalently, to the set $\{n \mid n+0 = n\}$. There are techniques, mostly based on higher order unification [20], for finding higher order instantiations automatically — for example the TPS system [1] works in this way. Alternatively, it's possible to write down the combinator axioms in first order logic, so that in principle, lambda-abstractions (in combinator form) can be discovered using standard first-order proof search.[11] This was first proposed by Robinson [28], but his system appears to us unsound: since it does not respect types, the Russell paradox could apparently be derived quite easily by applying a fixpoint combinator to the negation operation. Dowek [11] gives a precise treatment, discussing how the type system can be used too.

We elected to accept the fact that certain higher order proofs cannot be found automatically (after all, other HOL rules can be used in 'by' if necessary). Note however that if the appropriate term is already bound to some function, then just throwing in the definition is enough; the lambda-term is then expressible in a first order way. Effectively this function works like the appropriate combinatory expression. For example, given the theorems:

$$\forall x.\, x \in Ins(s,y) \equiv x \in s \vee x = y$$

and

$$\forall x.\, x \in Del(s,y) \equiv x \in s \wedge x \neq y$$

as well as the defining property of the empty set, then our first order provers are quite capable of deducing the right instantiations to prove:

$$\forall s.\, s = \emptyset \vee \exists x, t.\, s = Del(t,x) \wedge x \notin t$$

Moreover there is no difficulty with *using* lambda-abstractions; we transform each term $P[\lambda x.\, t[x]]$ into $\forall f.\, (\forall x.\, f(x) = t[x]) \Rightarrow P[f]$ automatically.

---

[10] Constants (necessarily nullary) and variables can translated directly into first order logic constants and variables, and the logical connectives (at least when used in the standard way, using no higher order tricks) can be directly translated. Actually we optimize the above somewhat by using function and predicate symbols directly provided they are always used consistently in a first order way. This is in fact usually the case.

[11] This idea also lies behind the popularity in the first order ATP community of the finite NBG axiomatization of set theory [7] — in exactly the same way a finite set of building blocks replaces an infinite comprehension schema.

Another flaw in our system is that we do not preserve type information when translating to first order logic, which may lead to expansion of the search space with type-incorrect unifications, or could even result in proofs that fail when translated back to HOL inferences. In practice, however, this works surprisingly well in the domains we have tried. Much of the reasoning involves one or two types, which are implicitly encoded anyway in most formulas (note that we treat different instances of polymorphic constants, such as equality, as distinct). Providing better higher order and type-correct automation is an interesting research project. In any case, superior theorem-proving tools developed later may easily be hooked into our Mizar system.

Rather than work directly on the HOL term representation, our provers first translate into their own internal representation of first order logic, which is used during proof search. When a proof is found, it's then translated back into HOL. Systems like Isabelle which feature unification of metavariables in the tactic mechanism, can implement these rules very easily and directly. By contrast our approach looks a bit artificial. However it keeps the proof search fast, and this is the speed-critical part of the automated provers. The eventual proof is usually short and can be translated into HOL very quickly. Such an approach has already been used to implement provers rather similar to ours in HOL [21]. We elected to start from scratch rather than use their work, to make it easier for us to experiment with different ideas for proof automation, e.g. the incorporation of equality.

Preprocessing

When the system needs to prove that ϕ follows from assumptions $\psi_1, \ldots, \psi_n$, it begins, as does Mizar and as do most automated theorem provers, by forming the conjunction $\neg \phi \wedge \psi_1 \wedge \ldots \wedge \psi_n$ and attempting to refute it. The first stage is to convert it to negation normal form (i.e. a form where negations are applied only to atoms) and Skolemize it. Skolemization is done by a one-way process, specializing universal variables and introducing ε-terms for existential variables [4]. For example if the initial formula ψ is $\forall x\ y.\ \exists z.\ \phi[x, z]$, then we proceed through $\psi \vdash \exists z.\ \phi[x, z]$ to $\psi \vdash \phi[x, \varepsilon z.\ \phi[x, z]]$, introduce the local assignment $f = \lambda x.\ \varepsilon z.\ \phi[x, z]$ (it can easily be eliminated after refutation), and so get $\psi, f = \lambda x.\ \varepsilon z.\ \phi[x, z] \vdash \phi[x, f(x)]$. The preprocessing phase attempts to split formulas up into separate units as much as possible — in order to refute $\phi \vee \psi$, the disjuncts can be refuted separately. Conjunction is distributed over disjunction in an attempt to maximize this splitting (though this is disabled after a limit is reached, otherwise large tautologies lead to an exponential number of subtasks). Moreover, the expansion of bi-implications as either

$$(p \equiv q) \to (p \wedge q) \vee (\neg p \wedge \neg q)$$
$$\neg(p \equiv q) \to (p \wedge \neg q) \vee (\neg p \wedge q)$$

or

$$(p \equiv q) \to (p \vee \neg q) \wedge (\neg p \vee q)$$
$$\neg(p \equiv q) \to (p \vee q) \wedge (\neg p \vee \neg q)$$

is chosen to maximize splittability, and thereafter (i.e. after passing a universal quantifier) is chosen to keep the conjunctive normal form short, since one of our provers below uses CNF. (We do not use sophisticated 'definitional' techniques [9], which can give refutation equivalent CNF by introducing variables for all subexpressions, conjoining their definitions and forming the CNF of that.)

Splitting is most useful for proving equivalences: they are decomposed into two implications for the main prover to handle. In some contrived examples the improvement can be dramatic. For example, 'Andrews' Challenge':

$$((\exists x. \forall y.\ Px \equiv Py) \equiv ((\exists x.\ Qx) \equiv (\forall y.\ Qy)))$$
$$\equiv ((\exists x. \forall y.\ Qx \equiv Qy) \equiv ((\exists x.\ Px) \equiv (\forall y.\ Py)))$$

gets split into 32 independent subgoals, each of which is fairly easy. The problem as a whole, however, is a real challenge for CNF-based systems like the model elimination prover we describe below. On the other hand, our tableaux prover does this kind of splitting as part of the proof process anyway, so the gains from splitting are marginal.

A tableaux prover

Our first automatic prover is a simple tableaux prover, which is essentially a copy of leanT^{A}P [4]. It is extremely simple, but quite fast for moderately simple tasks. The idea of tableau provers is simply to perform backward search for a cut-free sequent proof, discovering variable instantiations by (first order) unification with Prolog-like backtracking. Beyond the limitations on search space already imposed by the underlying sequent calculus, the formulas are processed in a strictly round-robin manner. This means that universal assumptions can be instantiated n times, but only after all others have been tried at least $n - 1$ times. The Mizar notion of an 'obvious' inference [30], is that universal formulas are only instantiated *once*. So what we do is quite similar, but we just have a *bias* against re-using formulas, rather than a strict prohibition. Just as with Mizar, one can force multiple use of an assumption by listing it several times in the 'by' statement. Though from one point of view an artificial hack, this has some resemblance to a mathematical proof where one says for example 'using transitivity twice we get...'. Indeed, the equality-free part of Mizar is not unlike a tableau prover: it reduces the problem to disjunctive normal form (like the splitting of a tableau into separate branches) and then successively instantiates universal formulas until a refutation can be reached by unification with the negation of another formula [36].

The main extension over leanT^{A}P is a simple system for equality handling, which is necessary for many mathematical proofs. (Mizar includes its own equality-handling techniques.) Simply throwing in equality axioms is too inefficient given such undirected usage of assumptions. But dealing with equality in tableau provers is a hot research topic, especially since the key question of simultaneous rigid E-unification has recently been proved undecidable [10]. We chose a rather ad hoc method which nevertheless works quite well in practice. When a literal $P(s_1, \ldots, s_n)$ is processed given a complementary literal $\neg P(t_1, \ldots, t_n)$, we do not merely attempt to unify each (s_i, t_i) pair, but take each inequation $s_i \neq t_i$ and add it to the tableau branch, resulting in n new branches. And each time an inequation is the currently processed formula and there are at least some equations on the relevant branch, the equality-handling rules kick in. These simply search for a proof in equational logic, but cut down on redundancy by imposing strict canonicality requirements on the proof, e.g. that transitivity is applied after all congruence rules and is always chained right-associated, and that symmetry is only applied to axioms or assumptions.

A model elimination prover

As a more heavyweight and powerful alternative to the tableaux prover, we also developed a model elimination (MESON) prover, based on the Prolog Technology Theorem Prover [33]. Such systems work by reducing to clausal form and then further to a set of pseudo-Horn clauses that can be used for Prolog-style backward search. The default search mode is one of our own invention — see [18] for more details and a comparison with other techniques. The MESON prover is slower than tableaux for simple problems, because of the greater overhead of preprocessing into clauses. But on bigger examples, it usually outperforms tableaux. In particular it has a measure of goal-direction, which makes it practical in problems where large numbers of assumptions (even hundreds) are involved. These would almost certainly fail using tableaux, at least based on such a simple-minded round robin instantiation strategy. In this prover we deal with equality simply by throwing in all the equality axioms; though not dazzlingly efficient, it turns out to be satisfactory in most cases because of MESON's goal-direction. It is not necessary to include congruence axioms for Skolem functions [22].[12]

Because it is more powerful, we usually use MESON, together with the equality axioms, as the default prover. It seems quite a good choice for filling in obvious steps, the criticism being if anything that it is too powerful. (Actually, Tarver [35] also discusses using MESON in a supporting capacity within an interactive prover.) To avoid long delays where a theorem isn't actually provable (e.g. because the user has not supplied all the required assumptions), we place quite a strict limit on the number of inferences performed internally during search. However this isn't nearly as quick as the Mizar prover at detecting impossible goals.

5 Examples

We will now give a couple of examples of proofs in our Mizar format. Both of these just take a few seconds to process. The first is a rather cute predicate calculus fact due to Łoś. In fact, MESON is capable of proving this completely automatically, so a 1-step Mizar proof 'thus thesis' is sufficient. However the proof search takes rather a long time, and in any case it's more illuminating to see the reasoning involved. The thesis to be established is:

$$(\forall x\ y\ z.\ P(x,y) \land P(y,z) \Rightarrow P(x,z)) \land$$
$$(\forall x\ y\ z.\ Q(x,y) \land Q(y,z) \Rightarrow Q(x,z)) \land$$
$$(\forall x\ y.\ Q(x,y) \Rightarrow Q(y,x)) \land$$
$$(\forall x\ y.\ P(x,y) \lor Q(x,y))$$
$$\Rightarrow (\forall x\ y.\ P(x,y)) \lor (\forall x\ y.\ Q(x,y))$$

---

[12] Thanks to Geoff Sutcliffe for pointing out this piece of ATP folklore.

And the Mizar proof, verbatim, is as follows. Note that no type annotations are needed, as they are all derivable from the initial thesis (which gives all the first order variables type 'A'). The default quotation parser 'X' is set to generate Mizar tactics from the script; hence to set up the goal we locally reassert its usual definition. To emphasize the high level of interactivity, and the complete integration with the tactic mechanism, we show how the proofs can actually be entered in a HOL session, single-stepped through using the standard tactic expansion function 'e'.

```
let X = parse_term in
g '(!x y z. P x y /\ P y z ==> P x z) /\
  (!x y z. Q x y /\ Q y z ==> Q x z) /\
  (!x y. Q x y ==> Q y x) /\
  (!(x:A) y. P x y \/ Q x y)
  ==> (!x y. P x y) \/ (!x y. Q x y)';;

e 'assume L: antecedant';;
e 'Ptrans: !x y z. P x y /\ P y z ==> P x z by L';;
e 'Qtrans: !x y z. Q x y /\ Q y z ==> Q x z by L';;
e 'Qsym: !x y. Q x y ==> Q y x by L';;
e 'PorQ: !x y. P x y \/ Q x y by L';;
e 'per cases';;
e '  suppose !x y. P x y';;
e '  hence thesis';;

e '  suppose ?x y. ~P x y';;
e '  then consider a,b such that L1: ~P a b';;
e '  then L2: Q a b by PorQ';;
e '  per cases';;
e '    suppose !x. Q a x';;
e '    hence thesis by Qtrans,Qsym';;

e '    suppose ?x. ~Q a x';;
e '    then consider c such that L3: ~Q a c';;
e '    then L4: P a c by PorQ';;
e '    per cases by PorQ';;
e '      suppose P c b';;
e '      then P a b by Ptrans,L4';;
e '      hence thesis by L1';;

e '      suppose Q c b';;
e '      then Q a c by Qtrans,Qsym,L2';;
e '      hence thesis by L3';;
e '    end';;
e '  end';;
e 'end';;
```

Our second example is the fact that a group where the group operation is idempotent must in fact be Abelian.

$$(\forall x.\ xx = 1) \wedge$$
$$(\forall x\ y\ z.\ x(yz) = (xy)z) \wedge$$
$$(\forall x.\ 1x = x) \wedge$$
$$(\forall x.\ x1 = x)$$
$$\Rightarrow \forall a\ b.\ ab = ba$$

In the HOL version we use the symbol '#' for the group operation, after declaring it infix. This time we show the steps all folded together in a single quotation.

```
let X = parse_term in
g '(!x:A. x # x = i) /\
   (!x y z. x # (y # z) = (x # y) # z) /\
   (!x. i # x = x) /\
   (!x. x # i = x)
   ==> !a b. a # b = b # a';;

e 'assume L: antecedant;
   Idemp: !x. x # x = i by L;
   Assoc: !x y z. x # (y # z) = (x # y) # z by L;
   Ident: !x. i # x = x by L;
   Ident': !x. x # i = x by L;
   let a,b be A;
   (a # b) # (b # a) = a # (b # b) # a by Assoc;
                ... = a # i # a by Idemp;
                ... = a # a by Ident;
                ... = i by Idemp;
   then (a # b) = (a # b) # (a # b) # (b # a) by Ident';
                ... = ((a # b) # (a # b)) # (b # a) by Assoc;
                ... = i # (b # a) by Idemp;
   hence thesis by Ident';;
```

Conclusions

We have shown how another proof style can be added to the HOL system. The resulting system can be argued to combine the best features of HOL's and Mizar's theorem-proving technology. As the examples show, one can produce quite readable proof scripts and have HOL manage the internal decomposition to primitive inferences automatically. In fact, our work fully bears out the remark that 'transforming proofs that are capable of being validated with MIZAR's basic checker into formal natural deduction proofs would be straightforward' [36]. This is another indication of the flexibility and potential of the LCF approach.

We address two weaknesses of HOL: the unreadability of its tactic scripts, and its logical prescriptiveness. At the same time we provide a version of Mizar's proof language which is more interactive and allows secure extensibility. For one computer theorem prover to take ideas from others is in the spirit of the QED project [2], though we do not link actual systems as that project envisages. Experimentation with various proof styles, and experience with other systems generally, would be valuable. Probably there is no unique best style for all application areas, which makes it all the more attractive to allow the intermixing of different styles as we do here.

As well as the initial ease of construction and readability, an important consideration for formal proofs is their maintainability and modifiability [8]. It is interesting to enquire whether Mizar proofs are likely to be better in this respect. Since they are more readable and less sensitive to the precise choreographing of logical steps, they seem better; on the other hand they involve more extensive quotation of terms, and so could break more easily if these terms change. However by being explicit rather than implicit, they may be easier to change simply by a semi-automatic editing process.

Future work should probably focus on more powerful automation, integrating the type system and higher order instantiations in a more elegant way. For example, we could implement some of Andrews's techniques as HOL derived rules. It would also be worth experimenting with additions to the proof language. Another interesting idea is automatic proof presentation. For example, some readers might find the 'obvious' steps to be unobvious, but it would be possible to record the proof that the machine finds, and incorporate it into the proof script. Perhaps the processed script could be organized into a hypertext format to allow different readers to browse it at different levels of detail [16]. This could be integrated with more general work on producing a readable summary of machine proofs.

Finally, the theorems that get shipped to the automated prover might provide an interesting set of test cases for automated theorem proving. They have the merit of being realistic problems that arise in real proofs, whereas, for example, Andrews's challenge and its ilk are specifically developed with a view to providing problems for current technology. If they are too easy, then one could arrange for intermediate steps in the Mizar proof to be automatically excised until the proofs reach some given level of difficulty.

Acknowledgements

This work was inspired by the Mizar system; I'm very grateful to Andrzej Trybulec and others in Białystok who helped me to understand the system. I owe a great deal indirectly to Bob Boyer, whose energy and enthusiasm for the QED project has helped to bring users of different proof systems together. Comments on an earlier presentation from Ralph Back, Philipp Heuberger and Jockum von Wright were extremely helpful. I have also profited from discussions with Donald Syme. The automated theorem proving part of the work was inspired by Larry Paulson's implementation of MESON for Isabelle [26]. My work was very generously funded by the European Commission under the Human Capital and Mobility Programme.

References

1. P. B. Andrews, M. Bishop, S. Issar, D. Nesmith, F. Pfenning, and H. Xi. TPS: A theorem proving system for classical type theory. Research report 94-166, Department of Mathematics, Carnegie-Mellon University, 1994.

2. Anonymous. The QED Manifesto. In A. Bundy, editor, *12th International Conference on Automated Deduction*, volume 814 of *Lecture Notes in Computer Science*, pages 238–251, Nancy, France, 1994. Springer-Verlag.
3. M. Archer, J. J. Joyce, K. N. Levitt, and P. J. Windley, editors. *Proceedings of the 1991 International Workshop on the HOL theorem proving system and its Applications*, University of California at Davis, Davis CA, USA, 1991. IEEE Computer Society Press.
4. B. Beckert and J. Posegga. leanT^AP: Lean, tableau-based deduction. *Journal of Automated Reasoning*, 15:339–358, 1995. Available on the Web from ftp://sonja.ira.uka.de/pub/posegga/LeanTaP.ps.Z.
5. P. E. Black and P. J. Windley. Automatically synthesized term denotation predicates: A proof aid. In P. J. Windley, T. Schubert, and J. Alves-Foss, editors, *Higher Order Logic Theorem Proving and Its Applications: Proceedings of the 8th International Workshop*, volume 971 of *Lecture Notes in Computer Science*, pages 46–57, Aspen Grove, Utah, 1995. Springer-Verlag.
6. R. J. Boulton. Efficiency in a fully-expansive theorem prover. Technical Report 337, University of Cambridge Computer Laboratory, New Museums Site, Pembroke Street, Cambridge, CB2 3QG, UK, 1993. Author's PhD thesis.
7. R. S. Boyer, E. Lusk, W. McCune, R. Overbeek, M. Stickel, and L. Wos. Set theory in first order logic: Clauses for Goedel's axioms. *Journal of Automated Reasoning*, 2:287–327, 1986.
8. P. Curzon. Tracking design changes with formal machine-checked proof. *The Computer Journal*, 38:91–100, 1995.
9. T. B. de la Tour. Minimizing the number of clauses by renaming. In Stickel [34], pages 558–572.
10. A. Degtyarev and A. Voronkov. Simultaneous rigid E-unification is undecidable. Technical report 105, Computing Science Department, Uppsala University, Box 311, S-751 05 Uppsala, Sweden, 1995. Also available on the Web as ftp://ftp.csd.uu.se/pub/papers/reports/0105.ps.gz.
11. G. Dowek. Collections, sets and types. Technical report 2708, INRIA Roquencourt, 1995.
12. A. J. M. van Gasteren. *On the shape of mathematical arguments*, volume 445 of *Lecture Notes in Computer Science*. Springer-Verlag, 1990. Foreword by E. W. Dijkstra.
13. M. J. C. Gordon and T. F. Melham. *Introduction to HOL: a theorem proving environment for higher order logic*. Cambridge University Press, 1993.
14. M. J. C. Gordon, R. Milner, and C. P. Wadsworth. *Edinburgh LCF: A Mechanised Logic of Computation*, volume 78 of *Lecture Notes in Computer Science*. Springer-Verlag, 1979.
15. J. Grundy. Window inference in the HOL system. In Archer et al. [3], pages 177–189.
16. J. Grundy. A browsable format for proof presentation. In C. Gefwert, P. Orponen, and J. Seppänen, editors, *Proceedings of the Finnish Artificial Intelligence Society Symposium: Logic, Mathematics and the Computer*, volume 14 of *Suomen Tekoälyseuran julkaisuja*, pages 171–178. Finnish Artificial Intelligence Society, 1996.
17. J. Harrison. Metatheory and reflection in theorem proving: A survey and critique. Technical Report CRC-053, SRI Cambridge, Millers Yard, Cambridge, UK, 1995. Available on the Web as http://www.cl.cam.ac.uk/users/jrh/papers/reflect.dvi.gz.
18. J. Harrison. Optimizing proof search in model elimination. To appear in the proceedings of the 13th International Conference on Automated Deduction (CADE 13), Springer Lecture Notes in Computer Science, 1996.
19. J. Harrison and K. Slind. A reference version of HOL. Presented in poster session of 1994 HOL Users Meeting and only published in participants' supplementary proceedings. Available on the Web from http://www.dcs.glasgow.ac.uk/~hug94/sproc.html, 1994.

20. G. Huet. A unification algorithm for typed λ-calculus. *Theoretical Computer Science*, 1:27–57, 1975.
21. R. Kumar, T. Kropf, and K. Schneider. Integrating a first-order automatic prover in the HOL environment. In Archer et al. [3], pages 170–176.
22. W. McCune. Equality in automated deduction. In T. Dietterich and W. Swartout, editors, *Proceedings of the 8th National Conference on Artificial Intelligence*, pages 246–252, Boston, MA, 1990. MIT Press.
23. T. F. Melham. Automating recursive type definitions in higher order logic. In G. Birtwistle and P. A. Subrahmanyam, editors, *Current Trends in Hardware Verification and Automated Theorem Proving*, pages 341–386. Springer-Verlag, 1989.
24. T. F. Melham. A package for inductive relation definitions in HOL. In Archer et al. [3], pages 350–357.
25. L. C. Paulson. A higher-order implementation of rewriting. *Science of Computer Programming*, 3:119–149, 1983.
26. L. C. Paulson. *Isabelle: a generic theorem prover*, volume 828 of *Lecture Notes in Computer Science*. Springer-Verlag, 1994. With contributions by Tobias Nipkow.
27. I. S. W. B. Prasetya. On the style of mechanical proving. In J. J. Joyce and C. Seger, editors, *Proceedings of the 1993 International Workshop on the HOL theorem proving system and its applications*, volume 780 of *Lecture Notes in Computer Science*, pages 475–488, UBC, Vancouver, Canada, 1993. Springer-Verlag.
28. J. A. Robinson. A note on mechanizing higher order logic. In B. Meltzer and D. Michie, editors, *Machine Intelligence 5*, pages 123–133. Edinburgh University Press, 1969.
29. P. J. Robinson and J. Staples. Formalizing a hierarchical structure of practical mathematical reasoning. *Journal of Logic and Computation*, 3:47–61, 1993.
30. P. Rudnicki. Obvious inferences. *Journal of Automated Reasoning*, 3:383–393, 1987.
31. K. Slind. Object language embedding in standard ml of new jersey. Technical Report 91-454-38, University of Calgary Computer Science Department, 2500 University Drive N. W., Calgary, Alberta, Canada, TN2 1N4, 1991. Also appeared in Proceedings of 2nd ML Workshop.
32. S. Sokolowski. A note on tactics in LCF. Technical Report CSR-140-83, University of Edinburgh, Department of Computer Science, 1983.
33. M. E. Stickel. A Prolog Technology Theorem Prover: Implementation by an extended Prolog compiler. *Journal of Automated Reasoning*, 4:353–380, 1988.
34. M. E. Stickel, editor. *10th International Conference on Automated Deduction*, volume 449 of *Lecture Notes in Computer Science*, Kaiserslautern, Federal Republic of Germany, 1990. Springer-Verlag.
35. M. Tarver. An examination of the Prolog Technology Theorem-Prover. In Stickel [34], pages 322–335.
36. A. Trybulec and H. A. Blair. Computer aided reasoning. In R. Parikh, editor, *Logics of Programs*, volume 193 of *Lecture Notes in Computer Science*, pages 406–412, Brooklyn, 1985. Springer-Verlag.
37. J. G. Wiltink. A deficiency of natural deduction. *Information Processing Letters*, 25:233–234, 1987.

Stålmarck's Algorithm as a HOL Derived Rule

John Harrison

Åbo Akademi University, Department of Computer Science
Lemminkäisenkatu 14a, 20520 Turku, Finland

Abstract. Stålmarck's algorithm is a patented technique for tautology-checking which has been used successfully for industrial-scale problems. Here we describe the algorithm and explore its implementation as a HOL derived rule.

1 Introduction

To test whether a Boolean expression is a tautology, i.e. is true for all truth assignments of the variables it involves, there is no shortage of available methods. For example, one can construct its truth-table and make sure each row evaluates to 'true', convert it to conjunctive normal form and check that each conjunct contains some variable disjoined with its negation, or even translate it into an integer programming problem and solve it using a variety of standard techniques [9]. Moreover, many standard proof search procedures for first order logic yield a decision procedure for propositional logic as a special case, e.g. the Davis-Putnam procedure, resolution, model elimination and tableaux, including the more efficient variant of tableaux studied by d'Agostino [5].

Whatever their particular strengths and weaknesses, all known methods take time exponential in the size of the input formula, in the worst case. Since Cook [4] showed the dual problem of testing Boolean satisfiability to be NP complete, tautology checking is co-NP complete, and so it seems quite likely that *any* algorithm will have exponential complexity characteristics. However this still leaves open the possibility that some algorithms are acceptably efficient in a large number of important cases. Sometimes the traditional methods can shine in the right problem domain [17]. But in practical applications, the two most promising techniques seem to be binary decision diagrams [2], and a recent patented algorithm due to Stålmarck [14].

Binary decision diagrams have been most successful in hardware verification, though other applications have been explored — Bryant [3] gives a survey. Stålmarck's algorithm has also been applied to hardware verification, in a large number of real industrial situations. Some early examples [12] include reverse flushing control in a nuclear power station's emergency cooling system, and landing gear control logic for Saab military aircraft. More recently, the algorithm has also been applied to verification of, inter alia, car engine management systems, special service non-interaction in telephone networks, programmable logic controllers (PLCs) in various process industries (wood, steel, food etc.), and railway interlocking systems. This last application generally requires checking of tautologies involving something like 10^5 variables. This is too much for many traditional methods, including BDDs, though by using careful problem partitioning, e.g. a hiding technique proposed by Groote [7], they might still be tractable. Stålmarck's algorithm seems less sensitive to the number of variables than other methods, with runtimes depending more on certain proof-theoretic properties of the formula.

This paper discusses the implementation of Stålmarck's algorithm as a HOL derived rule. Part of our motivation was that it is an interesting experiment in LCF-style reduction of standard decision procedures, potentially yielding another valuable data point. Moreover, the algorithm itself is not yet widely known, and it's interesting to investigate how it performs, with or without integration as a HOL derived rule. But there are also deeper reasons.

The present author [8] has implemented the BDD algorithm as a HOL derived rule. This turned out to be possible with only a constant factor slowdown as compared with a direct ML implementation, but the constant factor was significant (40-50), and the implementation needed quite sophisticated tricks.[1] Stålmarck's algorithm has two features that make it seem a more promising target for HOL implementation. First, it is conceptually closer to a traditional natural deduction system of the kind HOL implements, so steps in the algorithm may be translated quite directly into HOL inferences. Second, the algorithm permits a significant separation of proof search from proof checking. It is well-known that this is advantageous for HOL implementation. The proof search stage need not produce HOL inferences; it need only record the proof eventually found, which may then be 'checked', i.e. translated to primitive inferences. By contrast, the HOL implementation of BDDs needs to track all stages of BDD construction by inference rules. (On the other hand the matter of deciding on a suitable variable ordering, which is of considerable significance for the efficiency of the BDD technique, *can* be separated.)

Another motivation is that Stålmarck's company Logikkonsult NP AB[2] is becoming increasingly interested in proof logging and proof checking for its implementation. Indeed, some of its customers are requesting (or insisting on) some kind of proof that can be independently checked. This phenomenon is likely to become increasingly common in applications of formal methods, as suggested by the U.K. Ministry of Defence [11]:

> 32.3.1 [...] It is [...] possible to remove the reliance on the correctness of the theorem proving assistant from the case for correctness of an application by arranging that a version of the final proof (omitting all history of its construction) is passed from the theorem proving assistant to a proof checker. For reasonable languages, such a proof checker could be a very simple program (perhaps ten pages in a functional programming language) that could be developed to the highest level of assurance.

Implementation as a HOL rule with the 2-pass structure that we have outlined *is* a form of proof checking. Indeed, it might be argued that it offers a rather high level of reliability. But even if HOL checkability doesn't coincide with the proof-checking interests of Logikkonsult or its customers, many of the same issues arise with other proof-checking arrangements, so some of our experiments here have a wider significance.

---

[1] In fact, not all of these are necessary given certain assumptions about equality testing and sharing inside terms.

[2] Swedenborgsgatan 2, S-118 48 Stockholm, Sweden.

2   Stålmarck's Algorithm

Stalmarck's method can deal with formulas involving all the usual logical connectives, $\neg$ (negation), $\wedge$ (conjunction), $\vee$ (disjunction), $\Rightarrow$ (implication) and $\Leftrightarrow$ (logical equivalence). However it avoids duplication later if some initial canonicalization is applied. The exact method chosen is not particularly important, though it should avoid the blowup in formula size that results from splitting logical equivalences. An early version of the algorithm [15] systematically pulled negations *up* the formula, leaving a body involving only conjunction, disjunction, implication and logical equivalence applied to unnegated propositional variables. This has the appealing feature that if the resulting formula is negated, it is immediately clear that it cannot be a tautology (set all propositional variables to 'true'); therefore in the main part of the algorithm, negation is avoided completely. However, we follow a more recent version of the algorithm in reducing the formula so that it uses only negation, conjunction and logical equivalence. Specifically, we apply the following transformations in a single bottom-up sweep, applying the first in the list where more than one is possible:

$$\neg\neg p \longrightarrow p$$
$$\neg p \vee \neg q \longrightarrow \neg(p \wedge q)$$
$$\neg p \vee q \longrightarrow \neg(p \wedge \neg q)$$
$$p \vee \neg q \longrightarrow \neg(\neg p \wedge q)$$
$$p \vee q \longrightarrow \neg(\neg p \wedge \neg q)$$
$$\neg p \Rightarrow \neg q \longrightarrow \neg(\neg p \wedge q)$$
$$\neg p \Rightarrow q \longrightarrow \neg(\neg p \wedge \neg q)$$
$$p \Rightarrow \neg q \longrightarrow \neg(p \wedge q)$$
$$p \Rightarrow q \longrightarrow \neg(p \wedge \neg q)$$
$$\neg \top \longrightarrow \bot$$
$$\neg \bot \longrightarrow \top$$
$$p \wedge \top \longrightarrow p$$
$$\top \wedge p \longrightarrow p$$
$$p \wedge \bot \longrightarrow \bot$$
$$\bot \wedge p \longrightarrow \bot$$
$$p \vee \top \longrightarrow \top$$
$$\top \vee p \longrightarrow \top$$
$$p \vee \bot \longrightarrow p$$
$$\bot \vee p \longrightarrow p$$
$$p \Rightarrow \top \longrightarrow \top$$
$$\top \Rightarrow p \longrightarrow p$$
$$p \Rightarrow \bot \longrightarrow \neg p$$
$$\bot \Rightarrow p \longrightarrow \top$$

$$p \Leftrightarrow \top \longrightarrow p$$
$$\top \Leftrightarrow p \longrightarrow p$$
$$p \Leftrightarrow \bot \longrightarrow \neg p$$
$$\bot \Leftrightarrow p \longrightarrow \neg p$$

If the formula is thus reduced to the logical constant $\top$ or $\bot$, then we are finished: in the first case the formula is a tautology; in the second it is not; indeed it is unsatisfiable. Otherwise we may suppose that the formula now involves only negation, conjunction and logical equivalence applied to propositional variables.

Reduction to triplets

Next, we imagine introducing a new propositional variable to represent each subformula that is either a conjunction or an equivalence, and assuming a set of logical equivalences representing their 'definitions'. For example given the formula:

$$\neg((a \Leftrightarrow b \wedge c) \wedge (b \Leftrightarrow \neg c) \wedge a)$$

we introduce new variables $v_1, \ldots, v_5$ defined by:

$$v_1 \Leftrightarrow b \wedge c$$
$$v_2 \Leftrightarrow a \Leftrightarrow v_1$$
$$v_3 \Leftrightarrow b \Leftrightarrow \neg c$$
$$v_4 \Leftrightarrow v_3 \wedge a$$
$$v_5 \Leftrightarrow v_2 \wedge v_4$$

Under these assumptions, the whole formula is equivalent to $\neg v_5$. Therefore to prove that formula it suffices to derive a contradiction from the above definitions together with the additional assumption v_5.

We said 'imagine' doing the above because it is, from a logical point of view, unnecessary. The intention is only to provide convenient short identifiers for subformulas, which is important since the algorithm works by assigning values to, and recording equivalences between, these subformulas. But the final proof discovered can perfectly well be written out with the subformulas used directly as their own names. This is what happens in the HOL version of the algorithm which we describe later. The above reduction may therefore be regarded as completely metalogical, a mere implementation convenience.

We have reduced the original formula to a set of 'triplets' of the form $p \Leftrightarrow q \otimes r$ where $\otimes$ is either conjunction or equivalence, p is a propositional variable (real or imaginary) and q and r are literals, i.e. either variables or their negations. The algorithm works by using these triplets to make logical inferences. All facts used and deduced by the algorithm can be considered as equations between literals,[3] if for the sake of regularity we treat $\top$ as another variable. That is, rather than v and $\neg v$, we use $v = \top$

---
[3] Or bi-implications, if the reader prefers to think of them that way.

and $v = \neg \mathsf{T}$ for actual truth-assignments. The starting point is the single equation $v^* = \neg \mathsf{T}$ where v^* represents the whole formula. In our example we start with $\neg v_5 = \neg \mathsf{T}$, or equivalently, $v_5 = \mathsf{T}$. The objective is to reach a contradictory equation of the form $v = \neg v$.

Simple rules

The basic means for deriving new equations from old is a set of 'simple rules'. These simply use the triplets together with the existing equations to deduce new equations via some obvious deductions. First there are rules for conjunctive triplets $p \Leftrightarrow q \wedge r$:

- if $p = \neg q$ then $q = \mathsf{T}$ and $r = \neg \mathsf{T}$
- if $p = \neg r$ then $q = \neg \mathsf{T}$ and $r = \mathsf{T}$
- if $q = r$ then $p = r$
- if $q = \neg r$ then $p = \neg \mathsf{T}$
- if $p = \mathsf{T}$ then $q = \mathsf{T}$ and $r = \mathsf{T}$
- if $q = \mathsf{T}$ then $p = r$
- if $q = \neg \mathsf{T}$ then $p = \neg \mathsf{T}$
- if $r = \mathsf{T}$ then $p = q$
- if $r = \neg \mathsf{T}$ then $p = \neg \mathsf{T}$

And there is also a similar set of rules for equivalential triplets $p \Leftrightarrow q \Leftrightarrow r$:

- if $p = q$ then $r = \mathsf{T}$
- if $p = \neg q$ then $r = \neg \mathsf{T}$
- if $p = r$ then $q = \mathsf{T}$
- if $p = \neg r$ then $q = \neg \mathsf{T}$
- if $q = r$ then $p = \mathsf{T}$
- if $q = \neg r$ then $p = \neg \mathsf{T}$
- if $p = \mathsf{T}$ then $q = r$
- if $p = \neg \mathsf{T}$ then $q = \neg r$
- if $q = \mathsf{T}$ then $p = r$
- if $q = \neg \mathsf{T}$ then $p = \neg r$
- if $r = \mathsf{T}$ then $p = q$
- if $r = \neg \mathsf{T}$ then $p = \neg q$

If we forget about the metalogical assignment of local variables and the breakdown into triplets, we can see all these as quite straightforward logical rules. For example the rule for conjunctive triplets 'if $p = \neg q$ then $q = \mathsf{T}$ and $r = \neg \mathsf{T}$' can be seen as:

$$\frac{\neg q \Leftrightarrow (q \wedge r)}{q \quad \neg r}$$

while 'if $p = \mathsf{T}$ then $q = \mathsf{T}$ and $r = \mathsf{T}$' looks even more like a standard natural deduction rule:

$$\frac{q \wedge r}{q \quad r}$$

0-saturation

Given a set of equations, the process of 0-saturation simply means deducing as many other equations as possible using only the simple rules. We also assume the use of symmetry and transitivity of equality together with the involution property of negation, e.g. going from $p = q$ and $\neg p = r$ to $q = \neg r$. From a theoretical point of view we can imagine deriving all possible equations using these extra properties, though of course in the actual implementation we do not derive such a heavily redundant set.

Let us see how our example can already be proved by 0-saturation. We start with just the equation $\neg v_5 = \neg \mathsf{T}$, i.e. $v_5 = \mathsf{T}$, and proceed as follows:

- By $v_5 = \mathsf{T}$ and triplet 5, $v_2 = \mathsf{T}$ and $v_4 = \mathsf{T}$
- By $v_4 = \mathsf{T}$ and triplet 4, $v_3 = \mathsf{T}$ and $a = \mathsf{T}$
- By $v_3 = \mathsf{T}$ and triplet 3, $b = \neg c$
- By $b = \neg c$ and triplet 1, $v_1 = \neg \mathsf{T}$
- By $v_2 = \mathsf{T}, a = \mathsf{T}$ and triplet 2, $v_1 = \mathsf{T}$
- By $v_1 = \mathsf{T}$ and $v_1 = \neg \mathsf{T}$, $v_1 = \neg v_1$, a contradiction.

Indeed, note that in general if the original formula uses local variable assignments in the following style:

$$\bigwedge_i (v_i = E_i) \Rightarrow \phi[v_1, \ldots, v_n]$$

then all the equivalences $v_i = E_i$ will be discovered by 0-saturation.

The Dilemma Rule

In general, 0-saturation alone is not sufficient to prove formulas. If after 0-saturation, no contradictory assignment has been reached, the next step is to use the so-called *dilemma rule*. This involves a case-split over a variable, though one of a rather sophisticated kind. Suppose that 0-saturation has yielded a set of equations Σ, and that we choose v as the variable to split over. Then we 0-saturate the sets $\Sigma \cup \{v = \mathsf{T}\}$ and $\Sigma \cup \{v = \neg \mathsf{T}\}$ to produce new sets of equations Δ_T and $\Delta_\perp$ respectively. Even if we have not gained a contradictory assignment in both Δ_T and $\Delta_\perp$, the case split may still yield new information. Set $\Delta = \Delta_\mathsf{T} \cap \Delta_\perp$ (if a set of equations contains a contradictory assignment, we think of it as contaning all possible equations between pairs of literals; so for example if Δ_T contains a contradiction then $\Delta = \Delta_\perp$). It is clear that we may now assume the set of equations Δ, since they hold whatever the value of v may be. We certainly have $\Sigma \subseteq \Delta$, and if Δ is a *proper* superset of Σ, new information has been obtained by the case split.

The process of 1-saturation means applying the dilemma rule to each variable (real or imaginary) in turn, repeating for all variables as long as new equations are obtained. If 1-saturation does not solve the problem, then 2-saturation is attempted. This is like 1-saturation, but case splits are tried over *pairs* of variables simultaneously. That is, for each pair of variables v and w, one 0-saturates the sets $\Sigma \cup \{v = \mathsf{T}, w = \mathsf{T}\}$, $\Sigma \cup \{v = \mathsf{T}, w = \neg \mathsf{T}\}$, $\Sigma \cup \{v = \neg \mathsf{T}, w = \mathsf{T}\}$ and $\Sigma \cup \{v = \neg \mathsf{T}, w = \neg \mathsf{T}\}$, and takes the intersection of the results, again repeating as often as new information is obtained. Similarly, one can n-saturate for any natural number n, case-splitting over n-tuples of variables simultaneously.

Stålmarck's method differs from naive methods using case splits in two important respects. First, it is possible to case-split not just over the primitive propositional variables, but over the imaginary ones too, i.e. over nontrivial subformulas. This may cause truth-assignments to propagate both up *and* down the formula's 'syntax tree'. For example if the formula contains a subformula $p \wedge (q \wedge r)$, then the assignment $q \wedge r = \neg\mathsf{T}$ will propagate up to yield $p \wedge (q \wedge r) = \neg\mathsf{T}$, while the assignment $q \wedge r = \mathsf{T}$ will propagate down to give $q = \mathsf{T}$ and $r = \mathsf{T}$. Second, rather than case-split over everincreasing numbers of variables until a contradiction is reached directly, the number of simultaneous case splits is kept as low as possible, with all new information arising carefully garnered and fed back into the next iteration. This avoids an immediate exponential blowup.

The algorithm gives rise to a new and interesting classification of tautologies according to their hardness. A tautology is said to be n-easy if it can be proved by n-saturation, and n-hard if it cannot be solved by $(n - 1)$-saturation. Obviously, the practicality of the algorithm for a large problem is likely to depend on the problem's being n-easy for reasonably small n. In fact, Stålmarck [13] shows that if a tautology A with size (number of connectives) $|A|$ is n-easy, then there is a proof of it with size $\leq |A|^{n+1}$. Moreover, the running time of his n-saturation algorithm, which is guaranteed to find such a proof, is bounded by $O(|A|^{2n+1})$.[4] For large formulas, even 2-saturation can be impractical, so the homophony of '2-hard' and 'too hard' is quite apt. Fortunately it turns out that many practical formulas arising in verification are in fact 1-easy. So often, in fact, that Logikkonsult's tools perform 1-saturation first, and if that fails, try to find a falsifying assignment, since experience shows that a formula is either 1-easy or not a tautology at all!

In particular, notice that 1-saturation will always discover common subformulas, modulo the symmetry of conjunction and equivalence, so assuming the formula is at least 1-hard, there is not much advantage in clever structure sharing when forming the triplets. If two identical formulas $p \otimes q$ appear in different places, giving rise to triplets $r_1 = p \otimes q$ and $r_2 = p \otimes q$, then splitting over either p or q will yield the equation $r_1 = r_2$. By the iteration of 1-saturation, this effect propagates up arbitrary common subformulas.

Patent

Note that the above algorithm is patented for commercial use.

---

[4] The algorithm described there and in other written presentations is more limited than the current version as we describe it here: rather than finding and using arbitrary equations between literals, it only uses true/false assignments. This makes it simpler, and also much closer to a standard natural deduction presentation. However the extension with arbitrary equations between variables is significantly more powerful, and is used in Logikkonsult's industrial practice.

3 Implementation

We shall first describe a direct implementation, since only minor modifications are required to produce a HOL derived rule. The algorithm is fairly straightforward to implement, the main difficulty being to represent efficiently the potentially quadratic number of equations derived.

Storing equations

Each (real or imaginary) variable is allocated a positive integer as an identifier. Logical negation is represented by numerically negating the corresponding integer. This means of course that we can't use 0. Moreover, we allocate 1 to the pseudo-variable $\top$. Now out of all the equations between variables $\pm v = \pm w$, we store only canonical ones in the form $v = \pm w$ where $w < v$ and w is the minimum possible value. So in particular, if a variable is assigned to $\top$ or $\bot$, then that assignment is the one stored. From these, other equations between variables may be created when required by plugging the stored equations together via transitivity and symmetry: if $x = y$ is derivable, then we must have $x = \pm v$ and $y = \pm v$ for the same canonical assignment v.

These canonical variable assignments are stored in assignment arrays. These have one cell per variable, which contains either its canonical assignment, if it has one, or otherwise the list of variables (with their signs) which are assigned to *it* (this list may be empty — initially all lists are empty).

When a new assignment $x = y$ is deduced, first of all the canonical assignments of x and y are taken from the assignment arrays, giving an equation $x' = y'$ between unassigned literals. Moreover, we orient it so that it is of the form $v = \pm w$ where $w < v$. Now this assignment for v is entered in the table, and moreover, each element in the list already assigned to v now becomes assigned to w (with the appropriate sign). The list of variables assigned to w has appended to it the list of those which were previously assigned to v.

Main loop

The main part of the algorithm is now quite simple. We repeatedly make the new assignments indicated then attempt to derive new facts using the simple rules, repeating until no new information can be derived. To avoid too much redundant searching of the simple rules, we initially produce lists of the triplets involving each variable. Then after a variable assignment to $v_1, \ldots, v_n$ (after canonicalization), only the triplets involving $v_1, \ldots, v_n$ need be examined for new information. Variables are tried in order of the number of triplets in which they appear, so that more apparently important variables are favoured.

During n-saturation, it is necessary to form intersections of the assignment arrays arising from the different settings of the variables. This is done by repeated pairwise intersection. Pairwise intersection, in the case where neither array is contradictory (this is indicated by assigning variable 0, which is otherwise unused), proceeds as follows. For each variable v assigned in both arrays (lists of assigned variables are maintained, again to avoid excessive searching), we find the respective canonical assignments w_1

and w_2. The assigned-to lists of these variables are examined, and the least common element, if any, extracted. This is now assigned to v in the result. When at least one array contains a contradiction, then it is merely necessary to copy the other. To avoid copying arrays, we proceed as follows. Three assignment arrays are allocated once and for all. The pointers to them are rotated if necessary by the intersection function, i.e. rather than copy one array to another, the pointers are merely exchanged. No other storage is needed.

After a saturation step resulting in new information, the triplets are rewritten with the assignments and all the assignment arrays reset. However, whatever internal assignments are derived during saturation, the triplets are not changed then.

As a derived rule

To implement the algorithm as a HOL derived rule requires remarkably modest changes to the overall structure of the code. The guiding principle is that each equation $v = \pm w$ is paired with a theorem (in general, with assumptions) justifying the equation between the corresponding subterms. The algorithm can be performed more or less as above, using symmetry and transitivity to justify certain procedures, like allocating the variables previously assigned to v to w when a new equation $v = w$ is derived. The simple rules also produce theorems, which are produced by matching against one of 30 pre-proved tautologies. The process of intersection becomes slightly more complex. After finding the least assignment common to both arrays, the corresponding theorems are combined using DISJ_CASES. For example, in 1-saturation, one theorem will have an assumption $v = \top$, the other will have $v = \neg\top$. The result will have an assumption $(v = \top) \vee (v = \neg\top)$. This assumption is itself a simple tautology and can therefore be eliminated quite easily. Where one assignment is contradictory, we first use *ex falso quodlibet* to derive identical conclusions. For example, from $\Gamma, v = \top \vdash x = y$ and $\Gamma, v = \neg\top \vdash \bot$, we first derive $\Gamma, v = \neg\top \vdash x = y$ then proceed as before. It would be much more efficient to retain a separate list of those assignments which gave a contradiction and eliminate them only at the end, but this is more tedious to implement.

Two passes

During the search for case splits which yield new information, there is no need to perform inference. Accordingly, we proceed as follows. The version which does not perform inference is run to discover the sequence of case splits which yields the result. Only these are then run performing inference. With more effort, the amount of work done in the inference phase could be cut back further. For example, once a branch of the case-split has derived all the equations that are known to hold in the final intersection, there is no need to proceed to others, since these will merely be thrown away. But the present organization seems reasonably satisfactory, and the proof log is of a very simple form: just a list of variable tuples to split over.

4 Results

In the long term, we hope to give results from Logikkonsult's own industrial examples. But some practical obstacles need to be cleared first. In any case, it is well not to let our implementation loose on really big examples until it has been improved. For the present, we use tautologies taken from three main sources.

1. An IFIP International Workshop on Applied Formal Methods For Correct VLSI Design 13-16 November 1989, IMEC Leuven, Belgium included a set of examples arising from hardware verification. These all involve verifying that two different combinational logic circuits implement the same function (in a couple of cases, modulo a few 'don't care' states).
2. The Second DIMACS Challenge[5] included a large number of Boolean satisfiability test cases. Those purported to be unsatisfiable were negated and used as test cases for tautology checking. They come from a range of applications; some are encodings of problems like graph-colouring, the Towers of Hanoi or the pigeonhole problem, some arise from circuit fault analysis, others are generated randomly. All are (when negated for satisfiability testing) in conjunctive normal form.
3. The validity of formulas $\forall x_1, \ldots, x_n . \exists y_1, \ldots, y_n . P[x_1, \ldots, x_n, y_1, \ldots, y_n]$, where no function symbols are involved, can be reduced to tautology checking [1]. We took most problems from the TPTP problem library [16] that fell into this subset (i.e. that when negated and reduced to clausal form for refutation, involved no non-nullary function symbols)[6] and generated the corresponding tautologies. Again, they are all in conjunctive normal form for refutation. Jeroslow [10] suggests more efficient translation techniques, e.g. exploiting 'type' information or checking smaller disjunctions first. But since we are interested in tautology checking per se, an inefficient translation is quite welcome! Nevertheless, a few examples, notably 'SAM's Lemma', were excluded because the resulting tautology was rather big (many millions of connectives).

All results are in seconds of user CPU on a Sparc 10. Note that this is running in interpreted CAML Light. It would probably run several times faster in a compiled version of ML. Moreover the inference-free 'oracle' could be rewritten in C and using a more efficient imperative style. We estimate that such an implementation would run at about 50 times the speed, even without improving the basic implementation structure explained above. We will first give results for some problems which are 1-easy. We list the problem name and the source (from the above three categories). Then we give the number of (primitive) propositional variables[7] and the number of connectives (conjunction and equivalence) they contain. The problems are sorted according to this latter figure, more or less the 'size' of the formula. Then runtimes are given to find the

---

[5] See http://mat.gsia.cmu.edu/challenge.html.
[6] Other first order logic problems, e.g. those involving only monadic predicates, could be reduced to this class [6].
[7] The IMEC examples use local variable assignments, so this is not an accurate indicator. Perhaps the number of variables remaining after 0-saturation would be a better one.

variable assignment sequence without performing inference, and to translate this into HOL inferences.

Problem	Source	variables	connectives	search time	proof time
syn323_1	TPTP	2	7	0.00	0.10
syn029_1	TPTP	3	9	0.00	0.11
syn052_1	TPTP	2	9	0.00	0.13
syn051_1	TPTP	3	11	0.00	0.13
syn044_1	TPTP	3	12	0.01	0.25
syn011_1	TPTP	7	16	0.01	0.18
syn032_1	TPTP	6	16	0.00	0.36
ex2_be	Imec	7	18	0.00	0.40
syn030_1	TPTP	5	21	0.03	0.31
transp_be	Imec	8	21	0.01	0.65
syn054_1	TPTP	8	23	0.01	0.23
gra001_1	TPTP	5	31	0.08	2.30
syn321_1	TPTP	10	31	0.00	0.20
rip02_be	Imec	9	41	0.20	2.96
puz014_1	TPTP	13	48	0.03	0.31
mul03_be	Imec	54	179	1.83	22.75
puz030_2	TPTP	10	213	4.55	67.23
puz030_1	TPTP	25	221	0.73	11.98
dk27_be	Imec	56	227	1.75	25.58
syn071_1	TPTP	16	233	0.61	1.30
aim_50_1_6_no_3	Dimacs	50	239	3.18	12.15
aim_50_1_6_no_4	Dimacs	50	239	1.98	3.88
hostint1_be	Imec	10	247	0.86	17.06
aim_50_2_0_no_4	Dimacs	50	298	6.91	22.21
aim_50_2_0_no_1	Dimacs	50	299	3.00	17.50
aim_50_2_0_no_2	Dimacs	50	299	6.93	21.81
aim_50_2_0_no_3	Dimacs	50	299	2.91	10.95
mul_be	Imec	14	324	2.73	31.20
dk17_be	Imec	63	327	18.23	92.11
risc_be	Imec	81	337	3.13	48.11
msc006_1	TPTP	32	449	1.98	2.91
syn072_1	TPTP	30	518	2.65	5.60
aim_100_2_0_no_1	Dimacs	100	599	7.71	5.23
aim_100_2_0_no_2	Dimacs	100	599	8.01	6.58
prv001_1	TPTP	115	600	3.00	2.18
ssa0432_003	Dimacs	435	2363	261.40	228.65
jnh211	Dimacs	100	3887	310.78	1451.21

The results are quite variable, but it is clear that to some extent the slowness of performing inference is compensated for by the fact that proof search does not need to be performed. In most cases the proof time is greater, but seldom by a dramatic margin. In some cases it is even shorter. Now let us look at some figures for 2-hard problems.

Problem	Source	variables	connectives	search time	proof time
rip04_be	Imec	19	97	4.02	6.62
ztwaalf2_be	Imec	13	107	9.11	36.75
ztwaalf1_be	Imec	15	111	57.17	32.01
z4_be	Imec	31	145	3.70	16.13
rip06_be	Imec	29	153	20.38	14.52
add1_be	Imec	59	173	11.43	9.95
rip08_be	Imec	39	209	70.08	20.43
aim_50_1_6_no_1	Dimacs	50	238	4.91	3.28
aim_50_1_6_no_2	Dimacs	50	239	8.88	5.25
vg2_be	Imec	77	277	30.10	26.38
misg_be	Imec	108	279	68.36	10.58
x1dn_be	Imec	79	279	261.70	61.36
counter_be	Imec	18	290	182.75	97.22
sqn_be	Imec	66	317	52.43	183.28
add2_be	Imec	144	407	38.11	27.05
dc2_be	Imec	87	409	90.43	231.28
f51m_be	Imec	101	449	371.25	207.10
aim_100_1_6_no_3	Dimacs	100	479	117.10	18.18
dubois20	Dimacs	60	479	335.65	41.31
add3_be	Imec	260	693	270.50	31.03
add4_be	Imec	376	1110	1267.03	85.18

Clearly the balance has shifted: here the proof search stage usually takes longer, significantly so in some cases. This is as expected, since the number of possible 2-saturations increases quadratically with the size of the formula, leading to much more expensive search. It is an interesting question whether very large 1-easy problems exhibit similar characteristics.

5 Conclusions

We have seen that Stålmarck's algorithm can be implemented reasonably straightforwardly as a HOL rule. This results in a substantial slowdown. However this slowdown (which in any case could no doubt be reduced by more efficient inference patterns) need not affect the proof-search phase. For problems where this dominates, particularly problems which are 2-hard, performance as a derived rule is quite good.

At present, neither the inference-free version nor the derived rule has been implemented especially efficiently. Future work will be to remedy this, and see how the balance of the above tables is modified. Rather than expend energy on a highly efficient C implementation of the algorithm, we hope to use Logikkonsult's own implementation to provide a proof log. It will be interesting to see how the HOL version can cope with truly large problems.

Acknowledgements

Thanks are due to Graeme Parkin and Tony Mansfield of the National Physical Laboratory for helping me to understand Stalmarck's algorithm, and most of all to Gunnar Stålmarck himself for explaining the algorithm in detail and providing stimulating discussions about this and many other topics. My work was conducted while I was a member of the Programming Methodology Group of Åbo Akademi University, supported by the European Commission under the Human Capital and Mobility scheme.

References

1. P. Bernays and M. Schönfinkel. Zum Entscheidungsproblem der mathematischen Logik. *Mathematische Annalen*, 99:401–419, 1928.
2. R. E. Bryant. Graph-based algorithms for Boolean function manipulation. *IEEE Transactions on Computers*, C-35:677–691, 1986.
3. R. E. Bryant. Symbolic Boolean manipulation with ordered binary-decision diagrams. *ACM Computing Surveys*, 24:293–318, 1992.
4. S. A. Cook. The complexity of theorem-proving procedures. In *Proceedings of the 3rd ACM Symposium on the Theory of Computing*, pages 151–158, 1971.
5. M. D'Agostino. Investigations into the complexity of some propositional calculi. Technical Monograph PRG-88, Oxford University Computing Laboratory, Programming Research Group, 11 Keble Road, Oxford, OX1 3QD, 1990. Author's PhD thesis.
6. M. Di Manzo, E. Giunchiglia, A. Armando, and P. Pecchiari. Proving formulas through reduction to decidable classes. In P. Torasso, editor, *Proceedings of the 3rd Congress of the Italian Association for Artificial Intelligence, AI*IA '93*, volume 728 of *Lecture Notes in Computer Science*, pages 1–10. Springer-Verlag, 1993.
7. J. F. Groote. Hiding propositional constants in BDDs. *Formal Methods in System Design*, 8:91–96, 1996.
8. J. Harrison. Binary decision diagrams as a HOL derived rule. *The Computer Journal*, 38:162–170, 1995.
9. J. N. Hooker. A quantitative approach to logical inference. *Decision Support Systems*, 4:45–69, 1988.
10. R. G. Jereslow. Computation-oriented reductions of predicate to propositional logic. *Decision Support Systems*, 4:183–197, 1988.
11. U. K. Ministry of Defence. The procurement of safety critical software in defence equipment. Interim Defence Standard 00-55, MOD Directorate of Standardization, Kentigern House, 65 Brown Street, Glasgow G2 8EX, UK, 1991.
12. M. Säflund. Modelling and formally verifying systems and software in industrial applications. Unpublished; available from the National Physical Laboratory, Teddington, Middlesex, TW11 0LW, UK, 1994.
13. G. Stålmarck. A proof theoretic concept of tautological hardness. Unpublished manuscript, 1994.
14. G. Stålmarck. System for determining propositional logic theorems by applying values and rules to triplets that are generated from Boolean formula. United States Patent number 5,276,897; see also Swedish Patent 467 076, 1994.
15. G. Stålmarck and M. Säflund. Modeling and verifying systems and software in propositional logic. In B. K. Daniels, editor, *Safety of Computer Control Systems, 1990 (SAFECOMP '90)*, pages 31–36, Gatwick, UK, 1990. Pergamon Press.

16. C. B. Suttner and G. Sutcliffe. The TPTP problem library. Technical Report AR-95-03, Institut für Infomatik, TU München, Germany, 1995. Also available as TR 95/6 from Dept. Computer Science, James Cook University, Australia, and on the Web.
17. T. Uribe and M. E. Stickel. Ordered Binary Decision Diagrams and the Davis-Putnam procedure. In J.-P. Jouannaud, editor, 1st *International Conference on Constraints in Computational Logics*, volume 845 of *Lecture Notes in Computer Science*, pages 34–49, Munich, 1994. Springer-Verlag.

Towards Applying the Composition Principle to Verify a Microkernel Operating System *

Mark R. Heckman, Cui Zhang, Brian R. Becker,
Dave Peticolas, Karl N. Levitt and Ron A. Olsson

Department of Computer Science, University of California,
Davis, CA 95616, USA
email: {heckman, zhang, beckerb, peticola, levitt, olsson}@cs.ucdavis.edu

Abstract. A compositional proof method allows the components of a system to be specified and verified independently, instead of having to verify the entire system as a monolithic unit. This paper describes how the composition principle of Abadi and Lamport can be applied to specify and compose systems that consist of both safety and progress properties, using the HOL theorem proving system. We discuss the translation of the composition principle into HOL and the resulting proof obligations, and introduce an example system, modeled after a microkernel operating system, that we composed using the method.

1 Introduction

A problem in specifying and verifying large systems is how to modularize a system and then prove that the components working together satisfy the overall system specification. This problem is compounded when a system is reactive and concurrent, as in a distributed operating system. Abadi and Lamport have provided a composition principle and proof rule for composing modular specifications that consist of both safety and progress properties [2]. Their composition method is based on the *transition-axiom* specification method [9] and a refinement mapping method of proving that one specification implements another [1]. In this paper we describe a translation of their specification, composition, and refinement mapping methods into HOL [7], and suggest how the composition principle could eventually be applied to specify and verify a microkernel based, message-passing operating system.

Abadi and Lamport developed a semantic model for the composition principle and did not tie it to any particular specification language or logic (although their model was influenced by Lamport's *temporal logic of actions* (TLA) [10]). The principle has been applied to specifications written in TLA [3] and an extension has been made to UNITY to allow the composition principle to be applied within the UNITY reasoning framework [6]. An application of the principle was also

---

* This work was sponsored by DARPA under contract USN N00014-93-1-1322 with the Office of Naval Research and by the National Security Agency's UR Program.

made to the composition of a secure system using the FDM theorem prover, but this application used only safety properties [8].

Our interest in a composition method stems from our current project to verify a system that includes a concurrent programming language and compiler, distributed and secure operating system, and reliable network [11, 12, 13]. CLI verified a "short stack"—a system that includes a microprocessor, compiler, assembler, and linker—using a vertical-layer proof method [5]. Their method was also applied to the proof of an operating system kernel [4]. CLI's system, however, was not distributed, and their approach was not intended to handle horizontal composition.

An advantage of composing modular specifications is that each module can be specified and proved independently, instead of having to prove the entire system as a single, large unit. Another advantage is that the system can be incrementally verified by composing a new module with a previously verified composition of modules. To realize these advantages, a system must be decomposable into well defined, relatively simple modules. The interfaces between the modules must also be well defined so that the effect of the behavior of one module on the behavior of another can be completely, and formally, specified. These concerns have led us to choose the microkernel operating system design, used in systems such as Mach, for our system.[2]

Our goal is to be able to independently verify each of the components of the operating system—the microkernel and operating system processes, called "server processes"—down to the code level, and then compose them into the complete system. An alternative method that we could use to verify our system would be to verify the composed system at the code level using conventional approaches (e.g., Hoare semantics for message passing). We believe, however, that a compositional proof method, like that of Abadi and Lamport, better structures the proof and will better allow us to deal with evolution of the system. This paper is a description of our first effort to apply Abadi and Lamport's method.

We have chosen to translate the semantic model into HOL for several reasons:

- As we expand our system into a stack like that of CLI's, we can make use of microprocessor specifications written in HOL.
- We intend to verify the operating system down to the code level, and we can embed a programming language in HOL.
- HOL also will allow us to eventually formalize the composition method and derive the proof rule, rather than simply using it as an axiom.

Section 2 summarizes the basic concepts of the transition-axiom method and the composition principle, section 3 discusses our translation of the method into HOL, and section 4 briefly touches on our example system. In the conclusion we explain the future direction of our work.

---

[2] The popularity of Mach-like operating systems has recently declined due to concerns about performance. Our primary interest here, however, is in improving verification performance, not run-time performance.

2 Semantic Model

In this section we summarize some of the main terms and concepts behind the transition-axiom, refinement mapping and composition methods of Abadi and Lamport.

2.1 Behaviors

In the transition-axiom method, systems are specified as infinite sequences of atomic state transitions, called *behaviors*, $s_0 \stackrel{\alpha_1}{\to} s_1 \stackrel{\alpha_2}{\to} \ldots$, where each s_k is a state (s_0 is the initial state) and each α_k is an *agent* that caused the state transition. States are an assignment of values to a set of state variables, and represent only the visible interface of a system, not internal variables.

State transitions occur due to actions taken by agents. In a nondeterministic, concurrent system, there is more than one agent, potentially many possible interleavings of events and, therefore, many possible behaviors. For example, a system with two agents running concurrently, α_1 and α_2, each with a single possible atomic action, a_1 and a_2, has two different possible interleavings of events, a_1, a_2 and a_2, a_1. The concurrency of the system is represented by the different possible event orderings.

A *stuttering step* is a transition from a state s to the same state s (i.e., the state doesn't change) [2]. If we consider non-stuttering transitions to be changes to the observable system state, then stuttering steps would correspond to invisible changes to the internal state of the system. We can take a behavior and "collapse" any consecutive sequence of stuttering steps into a single state (i.e., eliminating the stuttering steps) without changing the sequence of states and state transitions in the behavior. Any two behaviors that are equal after their stuttering steps are removed in this way are considered to be *stuttering equivalent*.

2.2 Properties

A set of behaviors where, for each behavior in the set, all the stuttering equivalent behaviors are also in the set is called a *property*. There are two types of properties: safety properties and progress properties. Intuitively, safety properties of a system define the acceptable initial states and the allowable state transitions. Progress properties assert that specific state transitions eventually occur.

The specification of a system consists of the conjunction of various safety and progress properties. If, for example, I is a state predicate that determines the set of valid initial states (interpreted as the property consisting of all behaviors whose initial state satisfies the predicate), $T(N)$ is a property that consists of all behaviors whose every state transition is either a stuttering step or else is allowed by a next-state relation N, and L is a progress property, then the specification for a system could be defined as the property $I \cap T(N) \cap L$, which means the set of behaviors whose initial state satisfies I, whose every state transition is a stuttering step or satisfies the next-state relation N, and that satisfies the progress requirement on the scheduling of transitions.

For example, consider the simple concurrent system described in section 2.1, with two agents α_1 and α_2, each with a single possible action, a_1 and a_2. We said that it has two different possible interleavings of events: a_1, a_2 and a_2, a_1. To specify this, we could write a property where I is always true (i.e., there are no constraints on the initial state), where N allows only the transitions a_1 and a_2 and allows them to happen only once each, and where the progress property asserts that a_1 and a_2 will eventually happen. Because of stuttering steps, an infinite number of possible behaviors satisfy the property, but they all have the general form $[x^*, a_1, x^*, a_2, x^*]$ and $[x^*, a_2, x^*, a_1, x^*]$, where x^* means zero or more stuttering steps by either or both agents. If we didn't have the progress property, our specification would also allow the additional behaviors $\emptyset$ (i.e., nothing happens), $[x^*]$ (nothing but stuttering steps happen), $[x^*, a_1, x^*]$ or $[x^*, a_2, x^*]$ (only one of the transitions happen).

2.3 System versus Environment

The inputs to a system come from the system's *environment*. For example, the environment for a server process in an operating system consists of the kernel and all other processes—both other servers and user processes.

A specification for a system could be written that accounts for the behavior of the system for all possible inputs. It is often simpler, however, to specify a system assuming that the inputs satisfy some definition of correctness. The server processes in an operating system, for example, cooperate to implement the overall operating system specification. It would be simpler to specify one server assuming that the other servers also conform to their specifications rather than specifying the behavior of the server in all possible conditions. A specification written this way must take into account that the system is only expected to work correctly if the environment works correctly. The specification of a system M and the assumptions about its environment E is the property $E \Rightarrow M$, which includes all behaviors where the system satisfies its specification, or the environment doesn't.

2.4 Composition Proof Rule

A specification for a system guarantees a property M assuming that the system's environment satisfies a property E. The environment of each component in the system consists of the overall system environment E together with all the other components of the system. Thus, a component guarantees a property M_i assuming E_i, but E_i holds only when $E \wedge \bigwedge M_j \Rightarrow E_i$, where $\bigwedge M_j$ is the conjunction of the properties guaranteed by all other components in the system. The problem with this idea is that the argument is circular. To prove that every E_i holds you have to assume that every M_i holds, but M_i holds only assuming that E_i holds. Despite the apparent circularity, Abadi and Lamport show that the rule is valid when the environment assumptions are safety properties. (They also describe how to manipulate specifications so that this requirement can always be met.)

Informally, this is Abadi and Lamport's composition proof rule: Given three specifications $E \Rightarrow M$ (the overall system specification), $E_1 \Rightarrow M_1$ and $E_2 \Rightarrow M_2$ (two component specifications), if

$$E \wedge \overline{M_1} \wedge \overline{M_2} \Rightarrow E_1 \wedge E_2,$$

then

$$E \wedge (E_1 \Rightarrow M_1) \wedge (E_2 \Rightarrow M_2) \Rightarrow (E \Rightarrow M_1 \wedge M_2),$$

where $\overline{M}$ is the smallest safety property that contains M.

This means that, if we prove that the conjunction of the overall system environment E and the safety properties of the two component systems implies the environments of the two component systems, then we can compose the two component system specifications into the form $E \Rightarrow M_1 \wedge M_2$.

The rule extends to n systems by requiring that the conjunction of the overall system environment and the safety properties of the n component systems implies the environments of all n component systems.

Strictly speaking, the consequent of the rule applies only to the "realizable" part of each specification, i.e., that part that has a correct implementation. We believe, but have not proven, that our approach to specifying a system using Abadi and Lamport's method ensures that our specifications are realizable.

2.5 Refinement Mappings

A refinement mapping is a function from states in one specification to states in another specification. Once component specifications M_1 and M_2 have been composed using the composition proof rule, the remaining step to complete the proof that the composition of M_1 and M_2 implements M is to find a refinement mapping from $M_1 \wedge M_2$ to M. In other words, the composition rule allows us to derive $E \Rightarrow M_1 \wedge M_2$ and the refinement step allows us to show that $M_1 \wedge M_2 \Rightarrow M$. From now on, we refer to $M_1 \wedge M_2$ as the "composed system" and M as the "overall system."

Abadi and Lamport describe four conditions that must be satisfied by a refinement mapping [2]:

1. f must preserve the external state so that the external state is the same for both the composed system and the overall system.
2. f must map initial states in the composed system to initial states in the overall system.
3. f must map state transitions in the composed system to state transitions (possibly stuttering steps) in the overall system.
4. f maps behaviors that satisfy the composed system specification to behaviors that satisfy the overall system's progress properties.

After applying the composition rule to create a composed system specification and proving that our refinement mapping from the composed system to the overall system is correct (i.e., satisfies the conditions), the composition is complete.

3 Adaptation to HOL

In this section we describe our translation of Abadi and Lamport's semantic model into HOL. At this point in our work we have not derived Abadi and Lamport's results, but simply use them as axioms.

3.1 Traces

In our adaptation of Abadi and Lamport's method to HOL we call behaviors "traces" (as in "execution traces"). Traces are defined as functions from numbers to $\langle agent, state \rangle$ pairs, called "trace elements."

For some trace *trace* and a number i, $trace(i)$ is the trace element that corresponds to agent α_i and state s_i in the behavior represented by *trace*. The state in $trace(0)$ (the first trace element) is the initial system state. The agent in the first trace element is undefined because there is no preceding state from which to transition to the initial state.

Abadi and Lamport define properties as sets of behaviors that are closed under stuttering. This could be represented in HOL (using the "sets" library) as

$$\{(trace : num \rightarrow trace\_element) \mid P(trace)\},$$

where P is a predicate that also allows stuttering steps. Practically speaking, however, we never need to use the set notation in applying the composition principle, but use only the predicate.

Following Abadi and Lamport, we distinguish safety properties from progress properties, and safety properties on the initial state from safety properties on state transitions.

3.2 Internal State

The state in a behavior is intended to completely describe the interface between the system and the environment. The system, however, may also have internal state that is not modifiable by the environment. Abadi and Lamport "hide" internal state using existential quantification.

The meaning of the existential operator in Abadi and Lamport's approach is that there exists an entire sequence of values for the internal state—call it the "internal trace"—that corresponds to a behavior (and therefore allows stuttering). Their approach also requires an assertion that the environment does not modify the internal state. A benefit of their approach is that all references to the internal state, including the assertion that the environment doesn't touch it, are encapsulated in the system property. No reference to the internal system state appears in the externally visible (interface) state or in the environment property. A HOL implementation of some property Q of a system that had internal state would look like the following:

$$\forall trace. \ \exists \ itrace. \ Q(trace, itrace),$$

where *trace* is a function from numbers to interface state, *itrace* is a function from numbers to internal state, and Q includes the assertion that any transitions in the trace by non-system agents leave the internal state unchanged.

In our example we tried another approach to implementing internal state. This approach extends the definition of system state in a trace element to include, but maintain separately, internal as well as external state. In other words, where the existential quantifier is used to define a sequence of internal state transitions that is parallel to the sequence of external state transitions in the trace, the second approach merges the internal state into each trace element.

We initially hoped that this method would eliminate the need for explicit assertions that the environment does not modify the internal state, but this did not turn out to be the case. The assertions must still be present but they become part of the safety properties of the environment instead of the system. One drawback to this approach that we found is that, when we compose systems and prove that the systems satisfy each other's environment assertions, we must explicitly prove that the component systems do not modify each other's internal state (see section 3.4). This proof would have been unnecessary had we specified the internal state using the existential operator. Another drawback of this approach is that it makes incremental composition more difficult, because the internal state for every component being composed must be added to the definition of a trace element.

3.3 Initial State Safety Properties

The general form of a predicate for an initial state property is

```
(∀ trace . initial_state_property trace = (I (trace 0)))
```

where I is a predicate on the initial state. Note that this definition implicitly allows stuttering steps because it cares only about the initial state, and so is a true property.

There is an initial state safety property for the external state, the "environment initial state property", and another initial state safety property that applies to the internal system state, the "system initial state property".

3.4 State Transition Safety Properties

The general form of a predicate for a state transition safety property is

```
(∀ trace . safe_transitions trace =
 (∀ i .
  let ss1 = (get_trace_state(trace i))
  and ss2 = (get_trace_state(trace(i+1)))
  in ((get_trace_agent ss2 = AGENT) ⇒
     ((ss1 = ss2) ∨ (ss2 = transition_function_1 ss1) ∨
     ... ∨ (transition_relation_n ss1 ss2)))))
```

Intuitively, this predicate says that, for all adjacent trace elements in the trace where the agent responsible for the state transition is AGENT, the states will be either identical (i.e., a stuttering step) or will satisfy the "next state" relation. The next state relation is implemented by the disjunction of the definitions of valid atomic state transitions, which are in the form of relations or functions from states to states.

State Transition Functions and Relations. There are two approaches to specifying how a system is allowed to transition from state to state. One approach emphasizes what a system is permitted to do; anything not specifically permitted is forbidden. The other approach emphasizes what a system may not do; anything not explicitly forbidden is permitted. We use both approaches in our specifications.

State transition relations are generally in the form of predicates on pairs of states (e.g., P $s1$ $s2$), where a relation is satisfied if the second state represents a valid transition from the first state. We define a special case of a transition relation, called a state transition "function," that has the following general form:

```
(∀ ss : system_state . transition_n ss =
   ¬(precondition ss) ⇒ ss | (next_state_n ss))
```

where next_state_n ss is the actual new state function that modifies the external state or the system's internal state. If the transition's precondition isn't met, then the state transition function implements a stuttering step.

Representing a state transition as a function has the advantage that portions of the state not explicitly changed by the function are implicitly left unchanged. This simplifies the transition specification because we do not have to exhaustively assert how every state variable changes or not. State transition functions, however, emphasize what the transition is permitted to do. In cases where a large number of possible state transitions must be defined, a specification may be more succinct if it instead emphasizes what a system may not do.

We have found that state transition functions are useful in describing the system state transitions, but that more general relations are more useful for specifying environment state transitions. This is because, when we compose two systems, we must prove that the safety properties of each system satisfy the environment assumption of the other. If we only specify state transition functions for the environment of a system, then we would have to anticipate all the functions of other systems with which we want to compose the system that we are specifying. A general state transition relation for the environment, however, specifies that *any* transition by the environment that isn't forbidden by the relation is allowed.

For example, consider a process that reads messages from the head of an input mailbox queue and then forwards the message to one of a set of other mailboxes, depending on the contents of the message. The safety properties for the environment constrain the environment to insert messages into the end of the input queue, but the behavior of the environment with regard to the other mailboxes is left unconstrained because they are irrelevant to the behavior of the process. The behavior of the environment could be described by a single relation that allowed the environment to insert zero or more messages into the end of the input queue and to arbitrarily modify the other mailboxes. Such a relation would be very simple, and limited only to specifying that the messages already present in the input queue remain unchanged.

In our example we made the internal state part of each trace element (see section 3.2). An effect of this approach was that the environment transition relations (not the environment transition functions, which we also used) needed to explicitly assert that the environment left the internal state unchanged. These extra assertions increased the complexity of the composition proof because we had to prove that each of the transition functions of the systems that we were composing satisfied the environment of the other systems, which means that we had to explicitly prove that each component system did not modify the internal state of each of the other components. Though not necessarily difficult, these proofs would have been unnecessary had we used the existential operator to hide the internal state.

3.5 Progress Properties

We specify progress properties using the following general form:

```
(∀ trace . progress_property_n trace =
  (∀ i . ∃ j .
    i≤j ∧ (get_trace_agent(trace(j+1)) = AGENT) ∧
    (get_trace_state(trace(j+1)) =
       transition_n(get_trace_state(trace j))) ∧
    (∀ (k : num) . i≤k ∧ k<j ⇒
       ¬(get_trace_agent(trace(k+1)) = AGENT) ∨
       ¬(get_trace_state(trace(k+1)) =
           transition_n(get_trace_state(trace k)))) ∨
       ((get_trace_state(trace(k+1))) =
         (get_trace_state(trace k)))))
```

where transition_n is a state transition function defined for the state transition safety property.

The progress property asserts that, at any point i in a trace, there is some point j, either now or in the future, where the transition transition_n will be made by agent AGENT. Furthermore, at all intermediate points k between i and j, the transitions will either not be made by AGENT, not be transitions of type transition_n, or will be stuttering steps. In other words, starting at any point in a trace, the transition will eventually occur after zero or more different transitions made by the same agent, transitions made by other agents, or stuttering steps.

The complete progress property for a system is the conjunction of the individual state transition progress properties. The complete progress property asserts that each transition will occur infinitely often. Note that a state transition defaults to a stuttering step when the transition's preconditions are not met, so it is possible that the state change defined by the transition will never occur unless the precondition becomes true at some point and remains true until after the transition has occurred. Our progress properties, therefore, implement a "weak fairness" scheduling policy on transitions.

We chose this form for progress properties for several reasons. First, a progress property in this form asserts no new safety properties. Abadi and Lamport use the term *machine-closed* to described specifications where the progress properties have this characteristic. A machine-closed specification should make it simple, for some system M, to create $\overline{M}$, the smallest safety property that contains M, as required by the antecedent of the composition proof rule. We merely take out the progress properties, leaving the initial state and state transition properties. Because the derivation of $\overline{M}$ is a proof obligation of composition, we will eventually need to prove formally that our definition of progress properties allows us to do this.

Another reason for using this general form for progress properties is that we hoped that the relative simplicity of the form would simplify the mapping of progress properties. This is certainly the case for any transition of the composed system that never maps to stuttering steps in the overall system, but maps only to a particular transition. In such a case it is easy to prove that, because the composed system transition always eventually happens, the overall system transition also always eventually happens. Transitions that always map to stuttering steps, of course, are even easier to map.

When, however, composed system transitions map to both stuttering steps and transitions of the overall system, additional proofs are required. To prove that a particular overall system transition eventually happens we must also prove that the composed system will eventually reach a state where a composed system transition will map to that particular transition, and that the composed system will remain in that state until the transition occurs. Note that this is a stronger requirement than just proving that the composed system transition's preconditions will eventually become and remain true, although that is part of it. We found that the clause in a progress property about intermediate transitions being different was helpful in proving these additional requirements. It may be possible, however, to derive this clause from the rest of the progress property definition (if it happens eventually then it must happen a first time), but we haven't tried to prove it yet.

The simplicity of this type of progress property, while it may simplify the proof of the refinement mapping, is not expressive enough to cover the types of progress properties that a system designer may wish to specify. Consider, for example, a specification of a queue that has two kinds of state transitions, data arriving into the queue from a source and data moving out of the queue to some destination. One progress property asserts that data will eventually arrive in the queue, while another asserts that any data in the queue will eventually move out. We might, however, want a single progress property for the entire system, that the queue works like a pipe and that data will eventually flow from the source to the destination. Clearly, though, the simple progress properties can be used to derive the more abstract property. Our simple progress properties on state transitions can generally be used to derive more abstract properties in this way.

3.6 Composition

Having specified the properties of each of the component systems, the next step in composing the system is to prove the antecedent of Abadi and Lamport's proof rule (summarized in section 2.4).

The antecedent of the rule for composing n components becomes the following HOL goal:

```
(∀ trace .
  ((env_init trace) ∧ (env_transition trace) ∧
   (sys1_init trace) ∧ (sys1_transition trace) ∧ ... ∧
   (sysn_init trace) ∧ (sysn_transition trace)) ⇒
  ((env1_init trace) ∧ (env1_transition trace) ∧ ... ∧
   (envn_init trace) ∧ (envn_transition trace)))
```

where env_init and env_transition are the overall environment initial state and state transition properties, sysn_init and sysn_transition are the initial internal state and state transition safety properties for component system n, and envn_init and envn_transition are the environment initial state and state transition properties for component system n,

Proof Complexity. The composition proof can rapidly grow in complexity as a function of the number of state transitions on the left-hand side of the implication and the number of systems that are being composed. For each system state transition property $sys\_transition\_x$ on the left side of the implication and environment state transition property $env\_transition\_y$ on the right side of the implication we must prove that every transition in $sys\_transition\_x$ implies at least one of the transitions in $env\_transition\_y$ (except in the case of $x = y$, because the agent sets of a system and its environment don't intersect).

In general, if we are composing n modules and, on average, each of the modules has m state transitions, then we must prove that each of the nm state transitions satisfy $n - 1$ environments. In the worst case we have to do $O(n^2 m)$

proofs. In practice, however, there may be ways to reduce the number of proofs. It may be the case, for example, that a composed system has fewer state transitions than the sum of the transitions of the modules that were composed. This can occur when external state is used exclusively by the component systems and not by the overall environment. The state becomes internal state once the system has been composed, and any system state transitions that modify only that state become stuttering steps. In the best case, incremental composition of components with this characteristic might lead to only a linear increase in the number of proofs.

3.7 Refinement

We define a refinement mapping f as a function that corresponds to Abadi and Lamport's f^* formula [2]. The function takes as input a system state of the form (t, y), where t is the external state and y is the internal state of the composed system, and returns as output a system state in the form (s, x), where s is the external state of the overall system and x is the internal state of the overall system. Because in our example we have merged the external and internal state, our refinement mappings are simply HOL functions of type $composed\_system\_state \to overall\_system\_state$.

Note that our definition of a refinement mapping does not satisfy Abadi and Lamport's first condition that the mapping preserves the external state (see section 2.5). This condition seems overly strict to us because, had we not relaxed the rule in our example, we would not have been able to do the mapping without otherwise unnecessary revisions to the specifications (see section 4.2).

Because we map the external state, we cannot assume that the environment state transitions will have the same behavior for the overall system as for the composed system. For this reason we must extend the third condition, which requires us to map system state transitions, to also map environment state transitions. In cases where the behavior of the environment is unchanged by the mapping, as in our example, mapping the external state seems safe, but we need to formalize this argument.

Mapping the Initial State. The goal for proving that the initial state in the composed system maps to the initial state in the overall system is

$$(\forall \text{ ss . I ss} \Rightarrow \text{I' (f ss)})$$

where f is the refinement mapping function from ss, a system state of the composed system, to a system state of the overall system; I is the conjunction of the environment and individual system initial state predicates of the composed system; and I' is the conjunction of the environment and system initial state predicates for the overall system.

Mapping the State Transitions. The next step in the refinement is to prove the mapping of each of the state transitions in the composed system to either a state transition or a stuttering step in the overall system. Because the specification of state transitions may implicitly assume that the system state is always in some kind of consistent state, it may not be the case that the composed system transitions map up to the overall system transitions for all possible values in the range of the state variables. Invariants are used to eliminate unreachable states from the proof of the mapping. The proof that the state transitions map up may, therefore, require the proof of the invariants as an obligation.

A goal to prove an invariant P has the form

```
(∀ trace i .
  ((env_init trace) ∧ (env_transition trace) ∧
   (sys_init trace) ∧ (sys_transition trace)) ⇒ P(trace i)
```

which means that P holds for all states in a trace assuming the initial states and state transitions defined for the environment and component systems. The invariant becomes a theorem that is used in mapping the state transitions.

The HOL goal for mapping state transitions is

```
(∀ trace i .
  ((env_init trace) ∧ (env_transition trace) ∧
   (sys_init trace) ∧ (sys_transition trace)) ⇒
     (N' (f'(trace i)) (f'(trace(i+1))) ∨
     ((f'(trace i)) = (f'(trace(i+1))))))
```

where N' is the next state relation for the overall system and f' maps trace elements. (The agent in a trace element, however, is unchanged by the mapping.) This means that the safety properties of the composed system must map up for every state transition to either a state transition or stuttering step in the overall system. The antecedents of this goal imply the invariants, which can then be used in the proof.

Mapping the Progress Properties. The goal to prove that the progress properties of the composed system imply the progress properties of the overall system has this form:

```
(∀ trace .
  (I trace ∧ TE trace ∧ TM trace ∧ L trace) ⇒ L' (f'' trace))
```

where I is the initial state property for the composed system, TE is the state transition property for the environment, TM is the state transition property for the composed system (which consists of the conjunction of the individual component system state transition properties), L is the progress property for the composed system, L' is the progress property for the overall system, and f'' applies the refinement mapping function to every element in a trace.

4 An Example Proof

Our interest in eventually applying the composition principle to the proof of a microkernel operating system guided our choice of an example. The system that we composed consists of two processes, running on a microkernel, that implement system calls. In this section we give a very high level overview of the system and discuss why it was necessary to apply a refinement function to the external state.[3]

4.1 System Description

In the specifications for the system and environment we assume the existence of a kernel, similar to the KIT kernel [4], that implements message passing and process separation. The kernel provides the processes with the basic operations for sending and receiving messages through mailboxes. Each process has its own mailbox from which it alone can read messages, but processes can send messages to any other process's mailbox. For simplicity in the specification, mailboxes have unlimited length.

The environment for our two-process system consists of all the other processes running on the kernel. All communication between the system processes and the environment processes occurs through message passing, so the external state of the system consists of all the mailboxes. An environment process makes a system call by sending a "request" message to one of the mailboxes of the system processes. A system process responds to a system call by sending a "response" message to the mailbox of the process that sent the "request" message.

The overall system specification specifies two system calls, F and FG. System call F accepts an input value x in a request message and returns the value $f(x)$. System call FG accepts an input value y in a request message and returns the value $f(g(y))$. System call F is implemented by a single server process, also called F, that applies the function f to the input value. System call FG is implemented by a server process, also called FG, that accepts an input value y in a request message and then sends the value $g(y)$ in a message to the F server. The F server, in accordance with its specification for all request messages, returns the value $f(g(y))$ to the FG server, which forwards the result to the process that originally sent y. This system is depicted in figure 1.

4.2 Mapping the External State.

In the composed system, the FG server sends messages to, and receives them from, the F server. These messages appear in the F and FG mailboxes, which are part of the external state. In the overall system specification, however, there are no server processes, only the two mailboxes and a "monolithic" system, and there are no state transitions that could account for the appearance of the messages.

---

[3] Due to space limitations we cannot give a detailed description of the example proof here, but will soon make one available in a forthcoming technical report.

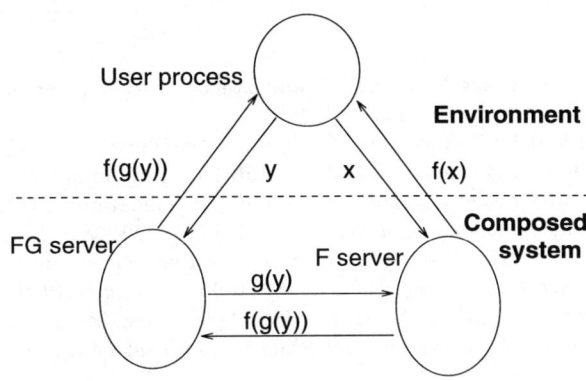

Fig. 1. Composition of F and FG servers

Our refinement mapping function must filter out the messages that are passed between the two servers, leaving all other messages and their ordering alone. The mailboxes have unlimited length, and any preconditions on the ability of the environment to put messages in the mailboxes are independent of the contents of the mailboxes. The proof that the behavior of the environment is unaffected by removing these messages from the system mailboxes should, therefore, be straightforward, although we have yet to formally prove this.

5 Conclusion

In this paper we described our first efforts to translate Abadi and Lamport's composition method into HOL and apply it to the proof of a two-process system. The proof that we did of our simple system was, to us, surprisingly complicated, and generated all sorts of unexpected subsidiary proof obligations. This was especially true of the proofs of the progress properties. Our experience was encouraging enough, however, that we intend to develop further the method in HOL. In the same way that we created templates for safety and progress properties, we will try to develop templates for proofs so that discovery of proof obligations will be less "ad-hoc" than in our first attempt.

Although the complexity of a composition proof can increase dramatically as the number of components being composed increases, the ability to incrementally compose components, and thereby keep the complexity of each incremental composition roughly constant, suggests that the method could reasonably scale to larger systems. In the next phase of our work we will attempt to demonstrate this by composing our example with additional processes. Longer term, we hope to be able to verify the composition of the kernel and server processes of a microkernel operating system using the composition principle.

References

1. Martín Abadi and Leslie Lamport. The existence of refinement mappings. *Theoretical Computer Science*, 82:253–284, 1991.
2. Martín Abadi and Leslie Lamport. Composing specifications. *ACM Transactions on Programming Languages and Systems*, 15(3):73–132, January 1993.
3. Martín Abadi and Leslie Lamport. Conjoining specifications. *ACM Transactions on Programming Languages and Systems*, 17(3):507–534, May 1995.
4. Willam R. Bevier. Kit: A study in operating system verification. *IEEE Transactions on Software Engineering*, 15(11):1382–1396, November 1989.
5. William R. Bevier, Warren A. Hunt, Jr., J. Strother Moore, and William D. Young. An approach to systems verification. *Journal of Automated Reasoning*, 5:411–428, 1989.
6. Pierre Collette. Application of the composition principle to unity-like specifications. In M.-C. Gaudel and J.-P. Jouannaud, editors, *TAPSOFT '93: Theory and Practice of Software Development*, number 668 in Lecture Notes in Computer Science, pages 230–242. Springer-Verlag, April 1993.
7. M. J. C. Gordon and T. F. Melham. *Introduction to HOL: A Theorem Proving Environment for Higer Order Logic*. Cambridge University Press, 1993.
8. Judith A. Hemenway and Dr. Jonathan Fellows. Applying the Abadi-Lamport composition theorem in real-world secure system integration environments. In *Proceedings of the Tenth Annual Computer Security Applications Conference*, pages 44–53, December 1994.
9. Leslie Lamport. A simple approach to specifying concurrent systems. *Communications of the ACM*, 32(1):32–45, January 1989.
10. Leslie Lamport. The temporal logic of actions. *ACM Transactions on Programming Languages and Systems*, 16(3):872–923, May 1994.
11. C. Zhang, R. Shaw, M. R. Heckman, G. D. Benson, M. Archer, K. Levitt, and R. A. Olsson. Towards a formal verification of a secure distributed system and its applications. In *Proceedings of the 17th National Computer Security Conference*, Baltimore, October 1994.
12. C. Zhang, R. Shaw, R. Olsson, K. Levitt, M. Archer, M. Heckman, and G. Benson. Mechanizing a programming logic for the concurrent programming language microSR in HOL. In *Proceedings of the International Higher-Order-Logic Theorem Proving Workshop*, pages 31–44, Vancouver, B.C., August 1993.
13. Cui Zhang, Brian R. Becker, Mark R. Heckman, Karl Levitt, and Ron A. Olsson. A hierarchical method for reasoning about distributed programming languages. In E. Thomas Schubert, Phillip J. Windley, and James Alves-Foss, editors, *Higher-Order-Logic Theorem Proving and Its Applications: 8th International Workshop*, number 668 in Lecture Notes in Computer Science, pages 385–400, Aspen Grove, Utah, September 1995. Springer-Verlag.

A Modular Coding of UNITY in COQ

Barbara Heyd[1] and Pierre Crégut[2]

[1] CRIN/CNRS-INRIA Lorraine, Nancy, France
[2] CNET - France Telecom, Lannion, France

Abstract. We present a modular embedding of UNITY in COQ. Special care has been put on the representation of UNITY programs and on the logic used. To keep elimination of invariants and composability of safety properties, we introduce a notion of context. The definition of progress is strengthened so that we can keep progress properties when programs are composed. This is a generalization of the ad'hoc notion of "conditional progress" properties. We present fully mechanized proofs of consistency and completeness for this new system.

1 Introduction

UNITY [CM89] is the union of a programming model based on asynchronous automata and a related and simple proof system expressing temporal properties of the described programs.

Its efficient set of derived rules can be used to establish complex theorems by combining simple safety properties. Moreover, contrary to TLA [Lam90], there are clear separations between the logic and programs. Applying the rules does not require complex syntactic manipulations.

Therefore the embedding of the logic in a logical framework can be shallow and a lot of theorem provers have provided an implementation of this theory [APP93, Gol90, Che95, BM93], but up to now not COQ. The main improvements of our implementation of UNITY in COQ over previous works are the following:

- Because verification is expensive, proof reuse is necessary. Most of the effort has been put on providing the modularity of the logical system. Mixing the previous works of Misra [Mis94], Sanders [San91], and Prasetya [Pra95], we provide a general but light notion of context that represents all the programs with which the system under study can cooperate. It solves the soundness problem of the so-called substitution axiom with composition of programs.
- Rules for combining safety properties but also *progress* properties are given.
- We use all the power of COQ type system to represent typed automata and the part of the state space they operate on. COQ section mechanism has been used to parameterize developments.
- We give a mechanized proof of consistency and completeness, with respect to classical operational definitions, of the logic used. The proof of completeness follows the principles given in [Päp95].

The paper is divided as follows. We give a brief introduction to UNITY and the base of its coding in COQ. Then we present our solution for composing of

UNITY safety properties and extend it to progress properties. We show how contexts are used in practice. Finally, we sketch the formal proofs of consistency and completeness of this new version of UNITY.

2 Unity

UNITY is constituted of:

- *a programming language:* programs are nondeterministic automata made of a finite set of unguarded transitions modifying a state.
- *a fragment of linear temporal logic expressing properties of programs:* it defines safety and progress operators and provides a set of derived rules to combine them.

Progress operators can be defined in terms of slightly enhanced safety properties because of the fairness property enforced by the programming model.

2.1 The Programming Language

Programs are divided into three parts :

- declaration of variables: a state is a mapping from a set of variable identifiers to values constrained by types.
- initialisation: a predicate $initial_P$ defines the set of correct initial states;
- assignments: a finite set of unguarded transitions; transitions are total deterministic functions from states to states acting on a finite part of the state.

A history is a sequence of states from a correct initial state obtained by applying successive transitions. An execution is the corresponding sequence of transitions. The following condition (weak fairness) must apply:

Each transition is chosen infinitly often.

2.2 The Temporal Logic

Operators are defined with *Hoare triples*: $\{p\}t\{q\}$, where t is an elementary action of the UNITY program, p and q are predicates over states. The triple is a valid proposition iff for each state s satisfying p, $q(t\ s)$ holds. We will study the exact scope of the quantification over s in the next sections.

SAFETY OPERATORS: The logic defines a basic safety operator, **unless**, stating that a property p will remain true unless another property q holds. [3]

$$p\ \mathbf{unless}_P\ q\ \equiv\ \langle \forall t.\ t \in P \Rightarrow \{p \wedge \neg q\}t\{p \vee q\}\rangle$$

---

[3] Misra recently proposed to use the more primitive operator **co** which is defined as:
$$p\ \mathbf{co}_P\ q\ \equiv\ \langle \forall t.t \in P \Rightarrow \{p\}t\{q\}\rangle$$

From which we can define two derived forms **stable** and **invariant**.

$$\textbf{stable}_P\ p\ \equiv\ p\ \textbf{unless}_P\ false$$

$$\textbf{invariant}_P\ p\ \equiv\ (\forall s.initial_P\ s \Rightarrow p\ s) \wedge \textbf{stable}_P\ p$$

PROGRESS OPERATORS: the first step is to define an elementary progress operator, **ensures**, similar to unless but that also requires the existence of a transition transforming any state satisfying p in a state satisfying q.

$$p\ \textbf{ensures}_P\ q\ \equiv\ p\ \textbf{unless}_P\ q \wedge \langle \exists t.\ t \in P \wedge \{p \wedge \neg q\}t\{q\}\rangle$$

The operator **leadsto**, written also $\leadsto$, is then defined as the transitive and (infinite) disjunctive closure of **ensures**:

$$\frac{p\ \textbf{ensures}_P\ q}{p \leadsto_P q} \qquad \frac{p \leadsto_P r \quad r \leadsto_P q}{p \leadsto_P q} \qquad \frac{\langle \forall m.m \in E \Rightarrow p_m \leadsto_P q\rangle}{\langle \exists m.m \in E \wedge p_m\rangle \leadsto_P q}$$

DERIVED RULES: combine temporal formulas to derive new ones. [CM89] gives a basic set of derived rules that are sufficient for most purposes. The following is a powerful induction principle for **leadsto** that will be used later on. $(W, \prec)$ is assumed to be well-founded. W indexes predicates whereas the version in [CM89] partitions the set of states.

$$\textbf{Leads\_to\_wf}\ :\ \frac{\langle \forall m : W.\ p_m \leadsto_P \langle \exists n : W.\ n \prec m \wedge p_n\rangle \vee q\rangle}{\langle \exists n : W.\ p_n\rangle \leadsto_P q}$$

3 Unity in Coq

3.1 An Overview of Coq

COQ [Coq94, CH88] is an interactive theorem prover based on the Calculus of Inductive Constructions. It is a higher order typed lambda-calculus with primitive mechanisms to define inductive types with their constructors, destructors and elimination theorems.

A theorem in COQ is a term of type Prop and a proof is an inhabitant of this type. Proofs are built by applying tactics to the current goal. A tactic builds a partial term whose type is the goal. The types of the holes are the subgoals generated by the tactic.

We have embedded a description of UNITY programs in COQ, coded the core logic and then proved the derived rules from this coding.

To make notations more readable, we have slightly modified the syntax of COQ in the presentation: we use ∀x:T.B and λx:T.B for (x:T) B and [x:T] B respectively, f a ≡ b for Definition f = [a:?] b. We use the usual symbols for the operators of the predicate calculus and we omit types when COQ can synthesize them (the "?" notation of COQ). Finally we use ML style pattern notation in the Case construction.

3.2 Coding the Programming Language

Closed Systems Variable identifiers are defined as an inductive type so that we can get separation for free. The domain of a variable is a dependent type. A state is a regular function from variable identifiers to their domain and a transition, a function from states to states. Note that in COQ, dependent product and universal quantification are the same notion.

<u>Inductive</u> variable : Set := x : variable | b: variable.
domain v ≡ <Set> <u>Case</u> v <u>of</u> x ⇒ nat | b ⇒ bool <u>end</u>.
state ≡ ∀ v:variable. (domain v).
transition ≡ state → state.
example_transition s v ≡ <domain> <u>Case</u> v <u>of</u>
   x ⇒ (Ifb nat (s b) ((s x) + 1) (s x)) | b ⇒ (s b)
<u>end</u>.

example_transition is the coding of the following transition

x := x + 1 if b

Modular Solution The definition of the state is shared by every automata present in the system. To develop independent libraries of verified programs, we must be able to specify and work on partial views of the state.

We use a COQ section (module) to state hypotheses (variables) over the state, they will be discharged in the objects (theorems) defined, when the section is closed. Theorems will be specialized to the actual definition of the state when the library is used.

variable and its visible members are declared as COQ variables. domain is defined, as a function, abstractly with bijections for each variable v between abstract values in (domain v) and actual values. The only useful property about variables is that they are distinct. This is captured by the existence of a constant case following the same reduction rules as the built-in operator Case.

<u>Variable</u> variable: Set, domain: variable → Set.
<u>Variable</u> x,b: variable.
<u>Variable</u> ext_x: (domain x) → nat, inj_x : nat → (domain x).
<u>Hypothesis</u> b1_x: ∀val:nat. (ext_x (inj_x val)) = val.
<u>Hypothesis</u> b2_x: ∀val:(domain x). (inj_x (ext_x val)) = val.
...
<u>Variable</u> case: ∀v:variable. (domain x) → (domain b) → (domain v).
<u>Hypothesis</u> case_x : ∀fx,fb. (case x fx fb) = fx.
example_transition s v ≡
(case v (inj_x (Ifb nat (ext_b (s b)) (((ext_x (s x)) + 1)
                                       (ext_x (s x))))
       (s b))
...

The definitions for closed systems are evident realizations of those hypotheses. [4]

3.3 Coding the Logic

Predicates are applied to states. We lift the definitions of logical operators from Prop to prop (a star distinguishes the lifted operators).

prop ≡ state → Prop.
p ∨* q ≡ λ s:state. (p s) ∨ (q s).

The translation of the definitions in terms of Hoare triples is direct. The definition of **reachable** will be given in Section 4.

unless p q P ≡
 ∀ s. (reachable P s) → ∀ t. (transition_of t P) →
  ((p ∧* ¬*q) s) → ((p ∨* q) (t s)).

The definition of **ensures** is similar. **leadsto** is coded by an inductive product:

lub E s ≡ ∃ p:prop. (E p) ∧ (p s).

<u>Inductive</u> leadsto [P:program] : prop → prop → Prop :=
  ens_leads   : ∀p,q:prop.(ensures p q P) → (leadsto P p q)
 | leads_trans : ∀ p,r:prop.(leadsto P p r) →
          ∀q:prop.(leadsto P r q) → (leadsto P p q)
 | leads_disj : ∀ p,q:prop. ∀ E:prop -> Prop.
          (∀ r:prop. (E r) → (leadsto P r q)) →
          (p ↔* (lub E)) → (leadsto P p q).

lub defines the least upper bound of a family of predicates.

3.4 An Example of Proof of a Derived Rule : Leads_to_wf

Theorem Leads_to_wf:
∀W:Set. ∀R:W→W→Prop. ∀P:program.
∀q:prop. ∀ p: W → prop. (has_a_transition P) →
(∀ m. (leadsto P (p m) (λ s.(∃ n.(R n m) ∧ (p n s)) ∨ (q s))))
→ (leadsto P λ s.(∃ n.(p n s)) q).

To prove the derived rule above, we need the well-founded definition. In CoQ, well founded sets are characterized by the accessibility of their elements [Coq94]:

<u>Inductive</u> Acc [A: Set; R: A → A → Prop] : A → Prop
  := Acc_intro: ∀ x. (∀ y. (R y x) → (Acc A R y)) → (Acc A R x).
well_founded A R ≡ ∀ a:A. (Acc A R a).

---
[4] To make instantiation practical, we use macros to implement a very crude mechanism of functor application (in the sense of SML modules). We need more support for abstract theories from the system especially for pretty printing.

Applying `leads_disj` simplifies the goal in $\forall n : W.\ p_n \leadsto q$. Now we can use the induction principle verified by well founded sets on the conclusion:

```
∀A: Set. ∀R: A → A → Prop. (well_founded A R) →
  ∀P:A→Prop. (∀x. (∀y. (R y x) → (P y))→(P x)) → ∀a. (P a)
```

We use the transitivity of `leadsto`, the disjunction of `leadsto` (a special case of `leads_disj` with two arguments) and an instance of the general `leads_disj` to prove the theorem.

4 Context

4.1 Definitions

The context of a program is the set of transitions that can be used by any programs it is composed with. This is the dual of the classical notion of interface which imposes restrictions from the context on the programs it calls.

A COQ-UNITY program is coded as a triple made of initial conditions, transitions and context. So `program` is an inductive type with one constructor:

```
set (T:Set) ≡ T → Prop.
Triple: (set state) → (set transition) → (set transition) → program.
```

`initial_of`, `transition_of` and `context_of` are defined as the three projections and the operator `lift` merges the context and the transitions of a program. If $P \equiv (I, T, C)$ then:

$$\text{lift } P \equiv (I, T \cup C, \emptyset)$$

We define composability of programs as the mutual inclusion of their sets of transitions in the context of the other one. If $P_i \equiv (I_i, T_i, C_i)$ then:

$$\text{composable } P_1\ P_2 \equiv T_1 \subset C_2 \wedge T_2 \subset C_1$$

`compose`, written also $[\![$, merges two programs:

$$P_1\ [\!]\ P_2 \equiv (I_1 \cap I_2, T_1 \cup T_2, C_1 \cap C_2)$$

Now we can inductively characterize states accessible during the execution of the program as either states satisfying the initial condition, or states resulting from the application of a transition from the program or the context to an accessible state.

```
Inductive reachable  [P:program]: state → Prop :=
  reach_init : ∀ s:state. (initial_of P s) → (reachable P s)
| reach_trans: ∀ s:state, t:transition.
       (transition_of P t) → (reachable P s) → (reachable P (t s))
| reach_ctxt: ∀ s:state, t:transition.
       (context_of P t) → (reachable P s) → (reachable P (t s)).
```

4.2 Modular Automata

The following paragraph is intended to give a more intuitive idea of the use of contexts. Here are some practical definitions:

- a context to express that some variables are local and cannot be accessed by the other programs:
$$\lambda t.\forall s.((t\ s)\ v_1) = (s\ v_1) \wedge \ldots \wedge ((t\ s)\ v_n) = (s\ v_n)$$

- a context to restrict the access of other programs to some interface functions $f_1 \ldots f_n$:
$$\lambda t.\forall s.(t\ s)\ \vec{v} = s\ \vec{v} \vee \exists \vec{x}_1.(t\ s)\ \vec{v} = f_1(\vec{x}_1, s) \vee \ldots \vee \exists \vec{x}_n.(t\ s)\ \vec{v} = f_n(\vec{x}_n, s)$$

We augment the syntax of UNITY programs used in [CM89] by introducing contexts (a new predicate inserted after the declarations) and the recursive definition of subprograms inside programs. Variables declared in a subprogram are local to this program and its subprograms.

5 Compositionality of Safety Properties

UNITY as defined by Chandy and Misra leaves the notion of "every states" in the definition of Hoare triples rather vague because the substitution principle and composition properties impose contradictory requirements on this quantification. Using context we develop a consistent combination of both notions.

5.1 Substitution "Axiom"

Invariants state a property that is valid in any state belonging to a correct execution (we call *reachable* such a state). So invariants are equivalent to *True* for reachable states.

We would like to substitute freely invariants in temporal formula. This substitution axiom is necessary to make the UNITY logic complete as shown by Sanders in [San91].

It is valid only if temporal operators are defined on states reachable by transitions of the program, not on every states. But invariance can also be defined by the stability of the property on the lifted program:

```
invariant i P ≡
    (∀ s. (initial_of P s) → (i s)) ∧ (stable i (lift P)).
```

This definition is equivalent to:

```
invariant' i P ≡ ∀ s:state. (reachable P s) → (i s).
```

and elimination theorems as the following hold:

```
∀ P, p,q,i. (has_at_least_one_transition P) → (invariant i P) →
    (leadsto P (p∧*i) q)  ↔  (leadsto P p q)
```

5.2 Program Composition

Composing programs over a given state space is obtained by merging their set of transitions and taking the conjunction of their initial properties. We would like to keep safety properties for combined programs if they were true for each component. [CM89] gives the following rule:

$$p \text{ safe}_P \ q \to p \text{ unless}_Q \ q \to p \text{ safe}_{P \| Q} \ q$$

where **safe** is either **unless** or **ensures**. It is valid if the quantifications over states in Hoare triples coincide for both programs. This is trivially true if there is no restriction on the quantifications and (usually) false if we restrict them to reachable states. But the theorem still holds if programs are compatible:

∀ P,Q. (composable P Q) →
∀ p,q. (safe p q P) → (unless p q Q) → (safe p q (P ∥ Q)).

The proof relies on the following theorem that does not hold with the usual notion of reachable used in [CM89]:

∀ P,Q. (composable P Q) →
∀ s. (reachable (P ∥ Q) s) → (reachable Q s).

6 How to Keep Leadsto Properties While Composing Programs

The previous section presented theorems to keep safety and **ensures** properties. We would also like to keep **leadsto** properties: i.e. if a program reaches a given property when used alone, it can also do it when it is a component of a system.

We can enforce a programming discipline like weak decoupling [Rao95]. A progress property is then always preserved but the class of expressible programs is very restricted.

Another solution is to restrict composition to well behaved programs. Here again contexts can capture those properties. A stronger notion of **ensures** is developed. It requires some stability properties from the context. **Leadsto** is then defined as the transitive and disjunctive closure of this operator. Any **ensures** property used in the proof holds in the composition. So the **leadsto** property still holds.

To be able to express a **leadsto** property of the program, we must strengthen its context and so restrict composable programs to a well-behaved class.

Section 5.2 states a way to keep **ensures** properties by composition: any transition from the context must satisfy the corresponding safety property. This leads us to the following definitions:

$$p \text{ unless}_P^c \ q = p \text{ unless}_{\text{lift}(P)} \ q$$
$$p \text{ ensures}_P^c \ q = p \text{ unless}_P^c \ q \wedge \langle \exists t : t \in P \wedge \{p \wedge \neg q\} t \{q\} \rangle$$

$$\frac{p \text{ ensures}_P^c\ q}{p \leadsto_P^c q} \qquad \frac{p \leadsto_P^c r \quad r \leadsto_P^c q}{p \leadsto_P^c q} \qquad \frac{\langle \forall m.m \in E \Rightarrow p_m \leadsto_P^c q \rangle}{\langle \exists m.m \in E \land p_m \rangle \leadsto_P^c q}$$

Then we can prove the following theorem:

$$\text{composable}(F, G) \to p \leadsto_F^c q \to p \leadsto_{F \| G}^c q$$

7 Consistency and Completeness

7.1 Operational Definitions of Temporal Operators

Operational definitions describe properties satisfied by any behaviour of the program. We need more formal characterizations of histories and executions than the one given in 2.1.

Executions and Histories

Executions. An execution is characterized by the program, its elements belong to:

```
execution ≡ nat → transition.
execution_of P e ≡ ∀ i. (transition_of P (e i)).
ext_exec_of P ≡ execution_of (lift P). ⁵
```

and by the fairness property verified. Note that we never require fairness for transitions of the context.

```
fair_execution P e ≡
   ∀ t,i. (transition_of P t) → ∃j.j ≥ i ∧ t = (e j).
fair_execution_of P e ≡ (execution_of P e)∧(fair_execution P e).
fair_ext_exec_of P e ≡ (ext_exec_of P e) ∧ (fair_execution P e).
```

Histories. A history can be described by an initial state and an execution. Histories are characterized by the properties of their related execution.

```
history ≡ nat → state.
Fixpoint is_history_of e is n : state := <state>Case n of
   0 ⇒ is | (S n') ⇒ (e n' (history_of e is n'))
end.
related p h e ≡ (p e) ∧ ∀ i.(h i)=(is_history_of e (h 0) i).
weak_history_of P t ≡ ∃e.(related (execution_of P) t e)
history_of P t ≡
   (initial_of P (t 0)) ∧ ∃e.(related (fair_execution_of P) t e).
ext_history_of P t ≡
   (initial_of P (t 0)) ∧ ∃e.(related (fair_ext_exec_of P) t e).
```

---
[5] ext_exec_of stands for extended execution of

Definitions There are two parts in a history h considered in the definition of a temporal operator. The first part leads to a state h_i through transitions from the program or the context. We assume that a property holds on h_i. The execution of the second part (after h_i) must lie within P and be fair (in fact fairness is imposed on the global history). At some later state of the history, a consequence must hold.

Operational Unless. The assumption is $p \wedge^* \neg q$ and the consequence is $p \vee^* q$ which must hold at h_{i+1}.

```
shift A i (s: nat → A) n ≡ (s (i+n)).
op_unless p q P ≡
   ∀ h. (ext_history_of P h) →
   ∀ i. (weak_history_of P (shift state i h)) →
   (p (h i)) ∧ ¬(q (h i)) → (p (h (i+1))) ∨ (q (h (i+1))).
```

Operational Leadsto. The assumption is p and the consequence is q which must hold later on.

```
op_leadsto p q P ≡
   ∀ h. (ext_history_of P h)→
   ∀ i. (weak_history_of P (shift state i h)) →
   (p (h i)) → ∃j. j ≥ i ∧ (q (h j))).
```

Operational C_leadsto. It is a simplified version of `op_leadsto`. The hypothesis about the secand part of the execution is not taken into account.

```
op_c_leadsto p q P ≡
   ∀ h. (ext_history_of P h)→
   ∀ i. (p (h i)) → ∃j. j ≥ i ∧ (q (h j))).
```

7.2 Proofs of Consistency

Consistency of Unless

∀ P,p,q.(unless p q P) → (op_unless p q P).

This theorem is fairly easy to prove and relies on the following lemma.

∀ i,h,P. (ext_history_of P h) → (reachable P (h i))

Consistency of Leadsto and C_leadsto.

∀ P,p,q.(progress p q P) → (op_progress p q P).

where `progress` is either `leadsto` or `c_leadsto`. Both proofs use the same argument. The proof is by induction on the algebraic definition. The non trivial step is the ensures case. Let h be a fair history such that $p(h_i)$. We look for j such that $q(h_j)$. We distinguish the following cases [6]:

---

[6] numbers between parenthesis refer to figure 1

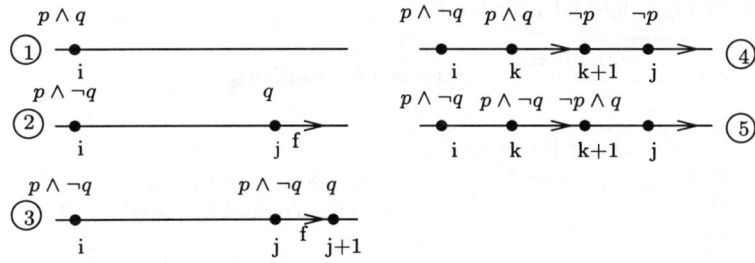

Fig. 1. Differents cases of traces for the consistency of leadsto

- q is satisfied at h_i and i is a solution (1).
- otherwise, there is a transition f that automatically yields a state satisfying q from a state satisfying $p \wedge \neg q$. Let j be a step such that the next transition is f. It exists because of the fairness assumption:
  - if $q(h_j)$, then j is a solution (2),
  - if $p(h_j) \wedge \neg q(h_j)$, then $q(h_{j+1})$ (3),
  - $\neg p(h_j)$. Then there is a k between i and j such that $p(h_k)$ and $\neg p(h_{k+1})$.
    * if $q(h_k)$ holds then k is a solution (4),
    * otherwise by the **unless** property implied by **p ensures q**, $p \vee^* q(h_{k+1})$ is satisfied, so $q(h_{k+1})$ holds (5).

The proof (induction on $\mathtt{i \leq j}$) of the next lemma (used in the previous proof) requires the decidability of P.

∀ P:nat → Prop; i,j. i ≤ j → (P i) → ¬(P j) →
∃k. i ≤ k ∧ k < j ∧ (P k) ∧ ¬(P (k+1)).

From now on, we assume that P is finite and is enumerated by $GenP$. There is a bijection between sequences over $\{1 \ldots N\}$ and executions.

7.3 Completeness of Unless

∀ P,p,q.(op_unless p q P) → (unless p q P).

First we show that a state is reachable if there is a fair history from an initial state to that state.

∀ P,e. (fair_execution_of P e) → ∀ s. (reachable P s) →
∃h,i. (ext_history_of P h) ∧ s = (h i).

The proof is by induction on the definition of **reachable**. We need to assume the existence of a fair execution of the program to solve the first case. $loop : n \mapsto GenP(n \bmod N)$ is such an execution.

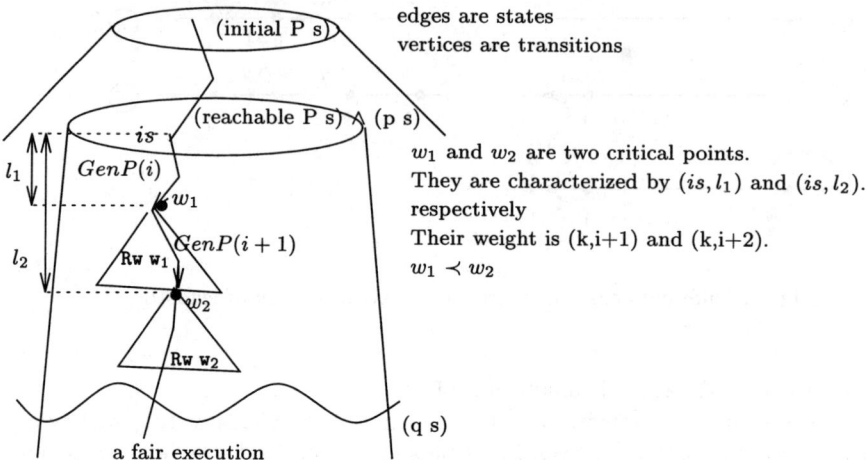

Fig. 2. Overview of the completeness proof for leadsto

The proof of the two other cases are similar: we modify one transition in the execution obtained from the induction hypothesis.

Then the proof of completeness is tedious but easy. We study each reachable state s and each transition t. Again let e be a fair execution of P and e' an execution going from an initial state reaching s. The execution e'' required to apply the definition of op_unless is defined by $e''_j = e'_j$ if $j < i$, $e''_i = t$ and $e''_j = e_j$ if $j > i$.

7.4 Completeness of Leadsto

∀ P,p,q.(op_leadsto p q P) → (leadsto p q P).

Our proof follows the principles given in [Päp95]. First, we reformulate the problem and hide the problem of reachability:

∀s.(reachable P s) → (p s) →
∀e.(fair_execution_of P e) → ∃ i.(q (is_history_of e s i))).

The main point is finding a way to use Leads_to_wf that captures progression.

Fragment of History. It is the pair of a state and a finite sequence of transitions. A history fragment (s, e) is a prefix of (s', e') if $s = s'$ and $e_i = e'_i$ for each valid index i of e. We write $(s, e) \prec (s', e')$. We consider all the fragments of fair histories beginning when p holds and such that q does not hold on their states (we call before_q this last property).

Characterization of W. We look for a family W of history fragments h and of associated predicates R_h, such that $(W, \prec)$ verifies the conditions of Leads_to_wf.

- $R_h(s')$ will hold if there is a history fragment h' such that h is the biggest prefix of h' belonging to W.
- We choose W so that the existence of an infinite chain for $\prec$ would imply the existence of an infinite *fair* history (limit of the history fragments). On this history q would never hold and this would contradict our assumptions. The abscence of infinite ascending chains implies that $(W, \prec)$ is well-founded [7]. The following function extracts an infinite fair execution chain from an infinite chain f (suposssing ascending). nth d n l gives back the n-ieth element of l if it exists or d. rev reverses a list and exc_of w is the execution part of w.

```
underlying_exec (f:nat → W) n ≡
   (nth 0 n (rev (exc_of (f (n+1)))))
```

Weight. To each history fragment, we attribute a weight [8] associating a pair of integers to a list of integers (coding a fragment of execution of P in reverse order):

- $dir\ [\ ] = (0,0)$
- $dir\ (n :: l) = (m+1, 0)$ if $dir\ l = (m, n)$ and $n = N - 1$;
- $dir\ (n :: l) = (m, n+1)$ if $dir\ l = (m, n)$ and $n \neq N - 1$;
- $dir\ (p :: l) = dir\ l$ if $dir\ l = (m, n)$ and $p \neq n$.

The second integer n of the result circles from 0 to $N - 1$ and states that we are waiting to see a n transition. The first integer codes the number of full iteration of this process already done.

A history fragment is *critical* if its weight is greater than the weight of its immediate prefix. W also is defined as the set of critical points before q.

```
is_critical l ≡ <Prop>Case l of
     [] ⇒ True | (_::l') ⇒ ¬(dir l)=(dir l')
   end.
```

Formal Definition of W. W is a dependent sum where the last four arguments are proofs of properties of the history fragment coded by the first two. bound requires that the fragment is a partial execution of P. execute_list s l is the result of applying the transitions of l on s. There are many more elements than expected (as many as the products of proofs and history fragments) but this does not influence the properties of the elements and the fact that the set is well founded.

---

[7] Note that this theorem cannot be proved in CoQ without the axiom of choice
[8] Päppinghaus uses a slightly different definition for dir. Ours is more deterministic in its evolution. This is useful to compute an index when we check the fairness property.

```
Inductive W : Set :=
  cw:∀ is:state, l: (list nat).
        (val_is: (p is) ∧ (reachable P is))
        (bound: (forall nat (λ m.m < N) l))
        (bel: (before_q is l))
        (crit: (is_critical l))
        W.

Rw w s ≡
  <prop>Case w of
    (cw is l _ _ _ _) ⇒
       ∃ l'. s = (execute_list is (l'@l)) ∧ (dir l)=(dir (l'@l)) ∧
             (before_q is (l'@l)) ∧ (forall nat (λm.m < N) l')).
  end.

orderW w w' ≡
  <Prop>Case w of
    (cw is l _ _ _ _) ⇒ <Prop>Case w' of
       (cw is' l' _ _ _ _) ⇒   is=is' ∧ (strict_suffix_of l l')
    end end.
```

Proving UNITY *Properties about* W Now we can check the basic **ensures** property

∀ w.(ensures (Rw w) λ s.((∃ u.(ordreW u w) ∧ (Rw u s)) ∨ (q s)) P).

We deduce a leadsto property from it and conclude by applying Leads_to_wf.

7.5 Completeness of C_leadsto

∀ P,p,q.(c_leadsto p q P) → (op_c_leadsto p q P).

The proof is similar to the previous one. We replace the lists of integers coding executions with lists of abs_trans where:

```
Inductive abs_trans : Set :=
  prog_trans: nat → abs_trans | ctxt_trans: transition → abs_trans.
```

8 Conclusion

8.1 Comparison with Other Works

Logics: Other closely related formalism for composition of Unity programs have been proposed.

- Sanders suggests to annotate properties with a predicate restricting the states considered. The only requirement is that it is an invariant of the program. But there is no reason to restrict the state space to something that does not correspond to the reachability induced by a group of programs.

- Misra presents the notion of closure of a program with respect to a set of programs. The set used plays the same role as our set of transitions but groups are kept separate. One can imagine to impose more restrictions on the program composed than simple safety properties but this increased power of expression has the cost of an intermediate level of quantification.

  The individual progress of a program relies only on safety properties satisfied by the other member of the group. If one wants to consider the global progress of the system, then there is no reason to study each program individually. But safety properties are predicates universally quantified over the set of transitions. Then there is no reason to keep the composable programs separate and our notion of context is a simplification of Misra's closure that captures all its power.

  We can use lift to access the context. There is nothing new to prove about temporal operators to state properties of the context. Note that if a context is an infinite UNITY program, this is not a problem as far as safety is concerned.
- On the other hand Prasetya suggests to keep information (read/write) on the use of variables. These are specific kinds of safety properties. We prefer to let the user give a more general notions of interface (for example by publishing functions (transitions)).

Implementations: We will not compare our implementation with implementations in first-order logic as they usually support only a subset of UNITY (for example leadsto_disj is restricted to finite disjunction).

Proof scripts of the core logic represents 1900 lines to which one must add 2000 lines for the consistency and completeness proofs (which are the first mechanical ones to our knowledge). This is a very concise kernel when compared to HOL implementations [APP93, Pra95]. The main reason is that all the power of COQ is used, that is the inductive types and the powerful tactics provided by COQ as shown by the definition of W. On the other hand, it is not always possible to share scripts among theorems. For example most of the 750 lines of completeness proofs for c_leadsto are specializations of the script for leadsto. The macro mechanism is not adapted for this purpose.

We also think that taking sets (predicates on the support type) for representing the body of programs simplified our proofs. For example it would not be possible to represent contexts as lists.

8.2 New Directions

We have used our tool on toy size examples: from [CM89], but also from [APP93] and a part of [Sta88] which showed some interesting weakness of the informal specification related to the fact that programs are implicit. We are now investigating a real size example, from telecommunication area.

The translator from UNITY programs to COQ still has to be improved. In particular the inclusion of external programs (what we have called libraries) should be automated and there should be more automated proof support.

Another direction could be to try to take a high level specification, to refine it to basic **ensures** properties and safety properties and then, for each **ensures**, to extract[9] a transition fulfilling the progress requirement and satisfying the invariants specified by the safety properties.

We have presented a completeness proof for our formalization of UNITY. Apart from its own theoretical interest, it can also be a tool for extending UNITY program relying on bisimulation methods.

References

[APP93] F. Andersen, K.D. Petersen, and J.S. Petterson. Program verification using HOL-UNITY. In J.J. Joyce and C.-J.H. Seger, editors, *International Workshop on Higher Order Logic Theorem Proving and its Applications*, pages 1–16, Vancouver, Canada, August 1993. University of British Columbia, Springer Verlag, Lecture Notes in Computer Science, No. 780 ,published 1994.

[BM93] Naïma Brown and Dominique Mery. A proof environment for concurrent programs. In *FME'93: Industrial-Strength Formal Methods*, number 670 in LNCS, pages 196–215, 1993.

[CH88] Th. Coquand and G. Huet. The calculus of constructions. *Information and Computation*, (76):95–120, 1988.

[Che95] Boutheina Chetali. Formal verification of concurrent programs: How to specify UNITY using the Larch Prover. Technical Report 2475, INRIA Lorraine, 1995.

[CM89] K.M. Chandy and J. Misra. *Parallel Program Design*. Addison-Wesley, Austin, Texas, May 1989.

[Coq94] Projet Coq. *The Coq Proof Assistant Reference Manual*. INRIA Rocquencourt and ENS Lyon, version 5.10 edition, 1994.

[Gol90] D.M. Goldschlag. Mechanically verifying concurrent programs with the Boyer-Moore prover. *IEEE Transactions on Software Engineering*, 16(9):1005–1022, September 1990.

[Lam90] Leslie Lamport. A temporal logic of actions. Technical Report SRC-57, Digital Equipment Corporation, 1990.

[Mis94] Jayadev Misra. Closure properties. unpublished manuscript on a new version of Unity, electronic version available under http://www.cs.utexas.edu/users/psp/newunity.html, 1994.

[Päp95] Päppinghaus. On the logic of UNITY. *Theoretical Computer Science*, 139, 1995.

[Pra95] I.S.W.B. Prasetya. *Mechanically Suported Design of Self-stabilizing Algorithms*. PhD thesis, University of Utrecht, october 1995.

[Rao95] J. R. Rao. *Extensions of the UNITY Methodology*. Number 908 in LNCS. Springer Verlag, 1995.

[San91] Beverly Sanders. Eliminating the substitution axiom from UNITY logic. *Formal Aspects of Computing*, 3(2):189–205, 1991.

[Sta88] Mark G. Staskaukas. The formal specification and design of a distributed electronic funds-transfer system. *IEEE Trans. on Computers*, 37(12):1515–1528, December 1988.

---

[9] in the sense of program extraction from COQ specification

Importing Mathematics from HOL into Nuprl

Douglas J. Howe

Bell Labs, Lucent Technologies
700 Mountain Ave., Room 2B-438
Murray Hill, NJ 07974, USA.
howe@bell-labs.com

Abstract. Nuprl and HOL are both tactic-based interactive theorem provers for higher-order logic, and both have been used in many substantial applications over the last decade. However, the HOL community has accumulated a much larger collection of formalized mathematics of the kind useful for hardware and software verification. Nuprl's relative lack impedes its application to verification problems of real practical interest. This paper describes a connection we have implemented between HOL and Nuprl that gives Nuprl effective access to mathematics formalized in HOL. In designing this connection, we had to overcome a number of problems related to differences in the logics, logical infrastructures and stylistic conventions of Nuprl and HOL.

1 Introduction

Nuprl [2] and HOL [4] are general-purpose interactive theorem proving systems with a number of similarities: their logics are higher-order type theories, their automated reasoning facilities are based on the tactic mechanism of LCF [3], and their main application has been to formal reasoning about computation. Both systems have been the focus of a great deal of research over the last decade. However, the overall thrust of the two research communities has been rather different.

Much of the work of the Nuprl project has involved the core of the system. There has been work on Nuprl's constructive type theory, on the proof editors, and on the basic architecture of Nuprl's automated reasoning support. In contrast, the core of the HOL system has remained stable for many years. The system has attracted a large number of users, and a great deal of the effort in the HOL community has gone into building the libraries of formal mathematics that are needed for verifying hardware and software of practical interest. There have been a number of substantial applications of Nuprl (see [7] for a recent example), but there has been nothing like the HOL community's sustained effort to formalize mathematics useful for verification.

Nuprl's relative lack of formalized mathematics impedes its application to verification problems of real practical interest. The goal of the present work is to redress this by importing the required mathematics from HOL.

Because of the similarities of the two systems, one might think that this is an easy thing to do. In one sense, it is. Just about any theorem-prover can embed the logic of any other simply by formalizing the syntax of proofs. However, the goal here is a practical one: to be able to effectively use HOL mathematics in Nuprl proofs. Not just any embedding will do. We need a strong connection between the mathematics developed in Nuprl and the mathematics imported from HOL, so that HOL facts will be applicable in Nuprl proofs. Furthermore, Nuprl's automated reasoning programs must be able to incorporate the HOL mathematics.

It turns out, perhaps surprisingly, that we have had to deal with a number of substantial problems. Some of the difficulties stem from the fact that the HOL and Nuprl type theories are, in fact, very different in some practically important ways. Nuprl has a constructive type theory, based on a type theory of Martin-Löf[9]. The theory contains an untyped programming language, and all objects have a computational interpretation. Programs are reasoned about directly in the logic, and the constructivity of the theory means that programs can be synthesised from proofs. On the other hand, HOL's theory is classical, and the way mathematics is encoded is similar to the way ordinary mathematics is done in ZF set theory. Functions are built in, but other objects, such as integers and lists, are given set-theory-like encodings with the aid of the "select operator" $@x \in T. \ P(x)$, which denotes some x of type T such that $P(x)$.

Aside from the logics, there are other differences between HOL and Nuprl that cause difficulties for importing HOL mathematics. In particular, there are substantial differences in:

- the logical infrastructure, including the definition and theory mechanisms,
- tactics,
- stylistic conventions for writing definitions and theorems, and
- implementation languages (HOL90 is SML, Nuprl is Common Lisp and "classic" ML).

Instead of reconciling the two logics, why not just replace Nuprl's type theory by HOL's? One answer is that Nuprl's type theory offers a number of advantages over HOL's. These include the following.

- *Expressive power of the type system.* Nuprl has subtypes and dependent function types. Also, through the use of universes and "sigma" types, one can express modules similar to the kind found in Standard ML [11].
- *Constructivity.* Experience with Nuprl has shown that for the mathematics of programs, constructivity comes at essentially no cost. Proofs about computationally meaningful objects such as hardware and software are naturally constructive, or can be made so with little effort. Thus one can, for example, build the same kinds of proofs as one does in HOL, with the additional benefit that programs can be extracted from the proofs.
- *Writing programs.* Nuprl includes a programming language which, while primitive, includes many of the features, such as function definition by general recursion, of a conventional functional programming language.

Most of these features have been advocated, in some form, as extensions to HOL. See, for example, [10, 8, 13]. Also, some of the type-theoretic features of Nuprl have been adopted in the PVS system [12]. Nuprl also has a large amount of automated support for making effective use of these features. Of course, one of the constraints on the embedding of HOL is that use of HOL mathematics does not prevent us from taking advantage of these features.

Another advantage of Nuprl is its user interface. Because of its separation of display and abstract syntax, Nuprl allows a great deal of flexibility in the use of notations in both the presentation and editing of mathematical syntax. A key component is Nuprl's novel structure editor. For details, see [6].

The main motivation for this work has been to make Nuprl more effective. However, another way of viewing the work is as the first case study of cooperation and result-sharing between different interactive theorem provers.

The body of the paper is divided into two parts. The first part gives a simple example illustrating the connection we have implemented between Nuprl and HOL90. The second part is a more detailed discussion of the problems we encountered and how we solved them. The hardest problem was one of semantics. To accommodate the embedding, we extended the Nuprl semantics to incorporate an operator similar to HOL's select. This extension is the subject of another paper [5], and will only be described briefly here. The other problems, though numerous, were not so difficult to deal with. This section also discusses some problems which we have not solved.

The work described in this paper is somewhat "in progress". The basic connection between the systems has been implemented, but it has not yet been tested in any substantial practical examples. We are currently working on a verification the SCI cache-coherency protocol [1] based on an importation of the HOL90 theory "List" (and all of its ancestors). A report on this work will be available soon via *http://www.research.att.com/howe*.

2 An Example

An HOL theory consists of some type and individual constants, some axioms, usually definitional, constraining the constants, and a set of theorems following from the axioms (and the axioms of ancestor theories). In contrast, the Nuprl logic is fixed; new constants and axioms are not introduced when building Nuprl's analogue of HOL theories. Formal mathematics in Nuprl is organized as a single list of *abstractions* and theorems. An example of an abstraction is

$$abs\_val(x) \;\; == \;\; if \; x < 0 \; then \; -x \; else \; x$$

which defines a new operator *abs_val* that takes one argument. This can be thought of as a one-argument term-macro. Abstractions in Nuprl are extralogical. The meaning of an expression containing abstractions is defined to be the meaning of the expressions obtained by expanding all abstractions (*i.e.* repeatedly replacing all occurrences by the corresponding right-hand sides). Nuprl also lets us define *display forms* for abstractions. For example, if we make the display-form definition

$$|\langle x \rangle| \;\; == \;\; abs\_val(\langle x \rangle)$$

then any subsequent occurrence of an expression of the form $abs\_val(e)$ will appear to the user as $|e|$. Because Nuprl's editors are structural, parsing/unparsing issues do not arise.

Consider now a truncated version of the HOL theory "bool" which introduces, among other things, some of the usual connectives of higher-order logic.

```
Parents:      min

Types:

Constants:    ?     :('a -> bool) -> bool
              T     :bool
              !     :('a -> bool) -> bool
              /\    :bool -> bool -> bool
              \/    :bool -> bool -> bool
              F     :bool
              [...]

Axioms:       BOOL_CASES_AX   |- !t. (t = T) \/ (t = F)
              IMP_ANTISYM_AX  |- !t1 t2. (t1 ==> t2) ==> (t2 ==> t1)
                                        ==> (t1 = t2)
              [...]

Definitions:  EXISTS_DEF  |- $? = (\P. P ($@ P))
              TRUTH       |- T = (\x. x) = (\x. x)
              FORALL_DEF  |- $! = (\P. P = (\x. T))
              AND_DEF     |- $/\ = (\t1 t2. !t. (t1 ==> t2 ==> t) ==> t)
              OR_DEF      |- $\/ = (\t1 t2. !t. (t1 ==> t) ==> (t2 ==> t)
                                              ==> t)
              [...]

Theorems:
```

The meaning of this theory is that for all values of the constants declared in the Constants section of the theory (and all ancestor theories), if the values satisfy the formulas in the Axioms and Definitions section of the theory (and all ancestor theories), then the formulas in the Theorems section are all true.

Suppose that we want to import this theory into Nuprl. We first invoke a program in HOL90 that writes the theory to a file in an intermediate form more suitable for Nuprl. This program also takes into account some hints supplied by the user. For example, for the theory bool some new names were supplied for some of the constants (most necessary renaming is done automatically). We then invoke a program in Nuprl that translates this file into the following Nuprl library fragment.

```
*C bool_begin    ************ BOOL ************
*A h_exists      ∃z_0:z_1. P1[z_0] ==    []
*A T             T ==    []
*A h_all         ∀z_0:z_1. P1[z_0] ==    []
```

```
*A h_and            P1 ∧ P2 == []
*A h_or             P1 ∨ P2 == []
*A F                F == []
[...]
#T h_exists_wf      ∀'a:S. ∀z_1:'a → o. (∃z_0:'a. z_1[z_0]) ∈ o
[...]
#T ...      ∀'a:S. (λz_0. ∃z_1:'a. z_0 z_1)
                    = (λP. P(@z_0:'a. P z_0))
#T ...      [T] ⟺ (λx. x) = (λx.x)
#T ...      ∀'a:S. (λz_0. ∀z_1:'a. z_0 z_1) = (λP. P = (λx. T))
#T ...      (λz_0,z_1. z_0 ∧ z_1)
                    = (λt1,t2. ∀t:o. (t1 ⇒ t2 ⇒ t) ⇒ t)
#T ...      (λz_0,z_1. z_0 ∨ z_1)
                    = (λt1,t2. ∀t:o. (t1 ⇒ t) ⇒ (t2 ⇒ t) ⇒ t)
#T ...      [F] ⟺ (∀t:o. [t])
```

Some of the object names have been elided for the sake of compactness. Also, Nuprl's type of booleans has been given the display form o.

Each line in the library fragment describes a single object. The second character in each line gives the kind of object: C for comment, A for abstraction and T for theorem. The first character is the status: * for complete and # for incomplete (e.g. when a theorem's proof has not been completed).

There are a number of apparent differences between this fragment and the HOL theory. We point out some of these here, but defer the explanations to the next section.

For each constant in the HOL theory, there are two Nuprl objects. The first is an abstraction with a right hand side which is initially a fixed constant. Corresponding to the constant /\ we have the abstraction named h_and. The left-hand side of the abstraction definition is actually h_and(P1;P2). The system displays it as P1 ∧ P2 because it is using a display form we have associated with the operator h_and. A Nuprl library may contain any number of such user-created display-form definitions. They have no logical significance; their only relevance is in display and editing of mathematical text.

Corresponding to the HOL existential operator ? is the abstraction h_exists. The left-hand side here is actually h_exists(z_1; z_0. P1[z_0]). This is a *second-order* abstraction: z_1 ranges over terms, but P1 ranges over terms with a distinguished free variable, and the expression P1[z_0] stands for the substitution of z_0 for the distinguished variable. As an example of the use of this operator, to express that there exists a natural number n equal to 0 one would write h_exists(N; n. n=0). On the right-hand side of the definition of h_exists one may apply $P1$ to any term. Note that the operator here takes two arguments, while the HOL constant takes only one.

The second object associated with a HOL constant is a *well-formedness theorem*, initially unproven. This essentially asserts that the abstraction has "same" type as the HOL constant. The fragment above shows only one such theorem, for h_exists. Note that we must explicitly quantify over the type 'a that is the first argument to h_exists. S represents the type of all HOL types. The expression z_1[z_0], after expanding abstractions, is just the function-application of z_1 to z_0.

Each definitional axiom of the HOL theory is mapped to a corresponding Nuprl theorem, initially unproven. Consider first the second theorem following the second [...]. This corresponds to the HOL definition TRUTH. Note that an HOL equality has been replaced by $\Longleftrightarrow$. In order to make translations of HOL theorems more directly applicable in Nuprl proofs, the main logical connectives of HOL have been translated into their Nuprl analogues. Thus HOL's "iff", which is equality over the type bool, maps to Nuprl's "iff". There is also an operator, denoted by [·], that coerces a value of type bool to a Nuprl proposition.

Note the difference between AND_DEF and the corresponding Nuprl theorem. "And" in HOL is just a function of two arguments. In Nuprl, it becomes a binary operator, and to make this into a function we need to explicitly abstract it.

If there were theorems in the HOL theory, then they would be translated in the same way as the axioms.

Given this library fragment, the user must now "instantiate" this Nuprl representation of an HOL theory with Nuprl objects. In particular, the user must do the following.

1. *Fill in the right hand sides of the abstraction definitions.* Usually this can be accomplished by cutting and pasting from the definitional axiom for the corresponding HOL constant. The key point here is that we will use computationally meaningful Nuprl objects to fill in these abstractions. For example, suppose bool (declared in the HOL theory "min") has been given the following definition in Nuprl as a subset of the integers.

$$\text{bool} == \{x:Z \mid x=0 \lor x=1\}.$$

We can then give a computable definition for, e.g., h_and:

$$P1 \land P2 == P1*P2.$$

There is a handful of HOL functions that cannot be given computable definitions. h_exists is such an example.

2. *Prove the well-formedness theorems.* These proofs are almost always done automatically by expanding the abstraction and running Nuprl's catch-all "autotactic".

3. *Prove the translations of the axioms.* In the case of definitional axioms, these proofs are usually trivial. They can, as in the case of type definitions, be quite non-trivial, though.

4. "Prove" the translations of the theorems. This is done automatically by using a "tactic" which works as follows. To prove a theorem ⊢ ϕ, the tactic first rewrites all the Nuprl logical operators to their HOL analogues. This reduces proving ϕ to proving [ψ] where ψ is an expression of type *bool*. Using the name of the theorem being proven, the tactic then looks up the theorem in Nuprl's internal image of the HOL theories loaded so far. It checks that the statement of this theorem is identical to ψ, then checks the completeness of the Nuprl import of the theory containing the theorem as well as the completeness of all ancestors. A Nuprl library fragment corresponding to a HOL theory is complete if all required axioms and and well-formedness theorems are present and completely proven. If the checks succeed, then Nuprl simply marks the theorem as proven.

Note that we are simply trusting HOL that its theorems are true. The correctness of our connection between HOL and Nuprl also relies on the correctness of the implementation of three programs: the SML program that preprocesses HOL theories and writes them to files, the Lisp/ML program that brings the files into Nuprl, and the ML program that marks imported facts as proven. In addition, soundness depends on the correctness of the semantic connection discussed in the next section.

3 Problems and Solutions

In this section we discuss some of the problems we encountered and how we solved them. Most of the problems relate to differences between the two type theories and the ways they are applied.

3.1 Semantics

HOL has a standard set theoretic semantics where propositions are booleans and function spaces contain all functions in a set-theoretic sense. In Nuprl, all semantic objects, including types themselves, are terms in an untyped programming language. One can think of this language as something like pure Lisp, or a variant of the untyped lambda-calculus. Thus, in the Nuprl semantics,

$$\lambda x.\ \textit{if } x{=}0 \textit{ then true else false}$$

is a member of the type $N \to \textit{bool}$. The reason is that if we evaluate the application of this expression to any natural number, the result is a boolean. In contrast, the corresponding semantic object in HOL is

$$\{(0, \textit{true}), (1, \textit{false}), (2, \textit{false}), \ldots\}.$$

In Nuprl, a term having a type is a semantic property. Something has a function type only if it has the right input-output behaviour. Thus we have the following typing, which has no direct translation into HOL.

$$(\textit{letrec } f(i) = \textit{if } i{>}100 \textit{ then } i{-}10 \textit{ else } f(f(i{+}11))) \in N \to N$$

Semantically, the function has this type simply because it terminates with a natural number on any natural-number input. In Nuprl, this typing requires a non-trivial proof, similar to what one would do to show termination informally (it requires a clever choice of well-founded ordering — try it!). This is just an instance of the general fact that in Nuprl, one can reason directly about functions defined by general recursion, but not in HOL.

Because the programming language is untyped, sensible terms can have "junk" subterms, as in

$$\text{if } 0 = 0 \text{ then } true \text{ else } 17 + \text{``foo''}$$

which has type *bool* since it evaluates to *true*.

We have reconciled these two semantics by extending the Nuprl semantics to include set-theoretic objects of the kind found in HOL. In particular, we add a collection of set-theoretic functions as new constants in Nuprl's programming language. We need to extend the notion of evaluation in the language. For example, we need to be able to evaluate applications like $\phi(e)$ where ϕ is one of the injected set theoretic functions and e is an arbitrary term. To do this, we introduce a notion of approximation. The assertion $\alpha \lhd e$ means that the set theoretic object α approximates the term e.

The evaluation relation of Nuprl's programming language is inductively defined by a set of rules. For example, the rule for ordinary function application is

$$\frac{f \Downarrow \lambda x.\, b \quad b[a/x] \Downarrow v}{f(a) \Downarrow v}$$

We add rules for $\lhd$, and add new evaluation rules for set theoretic objects. For example, we add two rules for set-theoretic functions ϕ:

$$\frac{f \Downarrow \phi \quad (\alpha, \beta) \in \phi \quad \alpha \lhd a}{f(a) \Downarrow \beta} \qquad \frac{\forall (\alpha, \beta) \in \phi.\ \beta \lhd b[\alpha/x]}{\phi \lhd \lambda x.\, b}$$

We illustrate this extension of evaluation with a few examples. Let $\phi = \{(0,4), (1,5)\}$, $\phi' = \{(0,2)\}$ and $\psi = \{(\phi, 17), (\phi', 18)\}$. We have

- $\phi(0+0) \Downarrow 4$ because $0 \lhd 0 + 0$.
- $\phi \lhd \lambda x.\, x+4$, but not $\phi' \lhd \lambda x.\, x+4$.
- $\psi(\lambda x.\, x+4) \Downarrow 17$.

Another set-theoretic notion we need to deal with is quotienting. Consider the rational numbers. In set theory, the rationals are represented as a set of pairs of integers, quotiented by the appropriate equality. The quotient in set theory is the set of all equivalence classes. In HOL, equivalence classes can be represented as predicates. This kind of representation of quotients is problematic computationally. For example, how does one do computations over rational numbers if a rational is represented as a function of type *int* × *int* → *bool*?

To deal with this, we add equivalence classes ξ to our language, and for computational purposes we also add a "polymorphic" equivalence class constructor. The polymorphic class $[\alpha]$ can be thought of as standing for any equivalence class ξ such that ξ has α as a member. We have the following evaluation/approximation rules for equivalence classes.

$$\frac{\alpha \in \xi \quad \alpha \lhd a}{\xi \lhd [a]} \quad \frac{a \Downarrow [e] \quad f(e) \Downarrow v}{f \cdot a \Downarrow v} \quad \frac{a \Downarrow \xi \quad \forall \alpha \in \xi. \ \beta \lhd f(\alpha)}{f \cdot a \Downarrow \beta}$$

The first rule captures the intuition given above for $[\cdot]$. The second and third rules describe how to compute with equivalence classes. To evaluate $f \cdot a$, where a evaluates to some equivalence class, we want to check that f returns the same value whenever it is applied to a member of the equivalence class, and then to return this value. When a evaluates to the polymorphic class $[e]$, there is no check to perform, so we just evaluate $f(e)$. In the case where e evaluates to a set-theoretic class ξ, we approximate the check by guessing some set-theoretic value β and checking that it approximates $f(\alpha)$ for every $\alpha \in \xi$. This introduces non-determinism.

Crucial for giving constructive implementations of HOL-defined types is Nuprl's quotient type. A simple example should suffice here. The quotient type $(x,y) : N//even(x - y)$ can be read as the quotient of N by the relation that equates numbers iff they have the same parity. Let $\xi_1 = \{0, 2, \ldots\}$ and $\xi_2 = \{1, 3, \ldots\}$. The type has as (canonical) members ξ_1, ξ_2 and $[\underline{n}]$ for $n \geq 0$. We have $\xi_1 \lhd [2]$ but not $\xi_2 \lhd [2]$. Also, if

$$f = \lambda x. \ \text{if } evenp(x) \ \text{then } 0 \ \text{else } 1$$

then $f \cdot [2] \Downarrow 0$ and $f \cdot \xi_2 \Downarrow 1$.

This semantics also justifies introduction of an analogue of HOL's select operator. We can extend Nuprl's programming language with an evaluation rule for select as follows. Note that non-emptiness of a type is taken to represent truth of the corresponding proposition.

$$\frac{T \Downarrow \gamma \quad \forall \alpha \in \gamma. \ P[\alpha/x] \Downarrow \gamma_\alpha \quad \alpha_0 \in \gamma \ \text{of minimum rank such that } \gamma_{\alpha_0} \neq \emptyset}{@x \in T. \ P \Downarrow \alpha_0}$$

Showing that a sensible semantics can be built based on the ideas described above is the subject of [5]. In this semantics, the Nuprl types are exactly the programs that evaluate to some set γ. The members of such a type are the members of γ together with all terms approximated by some $\alpha \in \gamma$.

3.2 Logic

In HOL logic is given a Tarskian semantics. A proposition is either true or false. In Nuprl, all propositions are represented as types. False propositions are empty types, and true propositions are types whose members represent the "computational content" of the proposition. For example, consider the proposition

$\forall x \in N.\ \exists y \in N.\ x < y\ \&\ prime(y)$. In HOL, the meaning of this is just the boolean *true*. In Nuprl, it is the type of all programs taking input $x \in N$ and returning as result a prime y with $x < y$.

We can give HOL propositions a direct interpretation (a "shallow" embedding in HOL parlance). The base logic has three constants

$$= : {}'a \to {}'a \to bool$$
$$==> : bool \to bool \to bool$$
$$@ : ({}'a \to bool) \to {}'a$$

for equality, implication, and "select", respectively. The $'a$ is a type variable. Given Nuprl's select operator, and the definition of *bool* from the previous section, it is trivial to give definitions in Nuprl for these constants, and to show that they have the required types.

This embedding by itself is not enough, however, since Nuprl's reasoning machinery is built around Nuprl's own logical operators. Fortunately, we can prove in Nuprl that the two different representations of logic are, in a sufficient sense, equivalent. For example, we can prove

$$\forall x, y \in bool.\ [x \Rightarrow y] \Leftrightarrow ([x] \Rightarrow [y])$$

where $\Rightarrow$ is overloaded notation, standing for HOL's version of implication on the left, and Nuprl's on the right, and where $[b]$ (not to be confused with the equivalence class constructor) is defined to be the proposition $b = true$. Also, we can prove

$$\forall A \in S.\ \forall P \in A \to bool.\ [\exists x \in A.\ P(x)] \Leftrightarrow (\exists x \in A.\ [P(x)]).$$

Putting a direct embedding of an HOL proposition into a form more suitable for Nuprl's tactics is thus just a matter of exhaustively applying rewrite rules such as the above.

3.3 Constructive vs Classical

In dealing with logic as we have above, we run the risk of losing the ability to extract programs from proofs. The program extracted is a member of the type representing the theorem proved. With our new semantics, we can always extract some "program", and it will have the right properties under evaluation, but the problem is that it might contain instances of the select operator, which is not computable. Nevertheless, we want to retain the ability to prove a theorem constructively and be assured that the extraction is computable.

The equivalence of the two representations of logic is highly non-constructive. In general, the object extracted from a proof of equivalence of two formulas will contain the select operator. Thus if we have tactics making unrestricted use of facts imported from HOL, it would be easy to unwittingly introduce a non-computable element.

The main reason we can solve this problem is because equalities in Nuprl have no computational content. So, for example, if a universally quantified equation is proved, then the program extracted from the proof is simply a constant function. This has two main consequences. First, if we are proving an equation (possibly under assumptions) in Nuprl, we can safely use any HOL theorem whatsoever. Second, no matter what we are proving, it is always safe to use HOL facts, such as universally quantified equations, that have no computational content. Fortunately, the vast majority of HOL theorems fit this category, and the vast majority of the work in proving any theorem about software involves computationally trivial facts (mostly equations and inequations). Most of the work in Nuprl proofs is done by term rewriting. All the programs that apply term rewriting can safely use any HOL theorem.

It is easy to modify the Nuprl system to ensure that non-computable "programs" are not inadvertently extracted from proofs. For example, we can add a bit to each proof node, where a true bit means that the extracted program of the subproof rooted at the node must not contain the select operator. The user sets the bit at the root, and the system computes the bit when the proof is extended by refinement, setting it to false when the node being refined has a conclusion which is computationally trivial, and simply propagating it otherwise. Inference steps may not mention the select operator, or use lemmas whose top bit is false, if the bit at the node being refined is true. This scheme for containing non-constructive reasoning has not yet been implemented, so it is currently up to the user to exercise appropriate care.

3.4 Constructivizing HOL Type Definitions

When new types are introduced in HOL, they are often given non-constructive implementations. Fortunately, when we translate into Nuprl we are not stuck with these implementations. This is because types in HOL are only defined up to isomorphism.

Consider the disjoint union $A + B$ of types A and B. In HOL, members of $A + B$ are represented as functions of type $bool \to A \to B \to bool$. For example, the injection of $a \in A$ into $A+B$ is represented as the function which returns true iff its first argument is *true* and its second argument is a. $A + B$ is axiomatized to be isomorphic to the collection of such representations by the following.

```
IS_SUM_REP
|- !f.
      IS_SUM_REP f =
      (?v1 v2.
        (f = (\b x y. (x = v1) /\ b)) \/
        (f = (\b x y. (y = v2) /\ ~b)))
sum_TY_DEF  |- ?rep. TYPE_DEFINITION IS_SUM_REP rep
```

IS_SUM_REP is thus defined to be predicate that picks out the members of $bool \to A \to B \to bool$ that serve as representations, and the second axiom states the existence of a bijection between $A + B$ and collection the objects satisfying IS_SUM_REP.

In contrast, Nuprl has a built-in type for disjoint union, with members $inl(a), inr(b)$. We use this as a definition of the disjoint union imported from HOL. We are then left with proof obligations to show that the above two axioms hold for this implementation, i.e. to show that Nuprl's disjoint union type and the HOL representation are in bijective correspondence. The Nuprl proof of this is non-constructive, but we will never need to refer to these axioms in other proofs.

3.5 Polymorphism and Type Inference

Although Nuprl's programming language is itself untyped, in practice the typing of programs follows a rather familiar type assignment style á la Curry. A difference is that types must be explicitly quantified in Nuprl. Since Nuprl has type universes, we can define a type S that contains all (small) types that are non-empty. The set S is sufficiently large to represent all HOL types. So, in HOL the polymorphic identity has the following typing

$$\lambda x.\ x \in\ 'a \to\ 'a,$$

whereas in Nuprl we would write

$$\forall\ 'a \in S.\ \lambda x.\ x \in\ 'a \to\ 'a.$$

Despite this similarity, there is still the crucial difference that type assignments can be statically determined in HOL. Given any expression, most-general types can be determined for the expression and all of its subexpressions. This is not possible in Nuprl. This gives rise to a slightly nasty problem with the select operator.

Consider the HOL expression $@x.P(x)$. Whenever this expression is used in some other expression, we can determine a type T for it, and the expression will denote some value in (the meaning of) T. In Nuprl, we cannot determine such a type in general, and the type must be passed in as an argument. Thus the expression is translated into Nuprl as $@x \in T.\ P(x)$.

Because of this, a definition whose right-hand side mentions the select operator may translate to an operator that passes type arguments. Fortunately, this happens relatively rarely. The default behaviour of the translator is to not pass type arguments. Exceptions must be indicated by the user through the "hint" mechanism.

3.6 Constants vs Operators

As indicated by the example in Section 2, definitions in Nuprl are typically operators with arguments. In HOL, defined objects that take arguments are represented as functions. This could also be done in Nuprl, by simply making each abstraction 0-ary and using λ-abstractions on the right-hand side. The reason this is not done is partly because of the undecidability of type-checking and type-inference in Nuprl. The details are somewhat technical, but it has turned out to be easier to organize facts related to typability (and a few other properties) around operators with arguments. This approach has also allowed us to incorporate second-order pattern matching in a straightforward way in type checking and rewriting.

Whether or not this difference is valuable, Nuprl's tactic collection relies on it heavily, so to incorporate HOL facts the constants need to be adjusted accordingly. For each constant in an HOL theory to be imported, an *arity* is computed. The arity is simply the number of (curried) arguments indicated by the type of the HOL constant. The user can also supply arities for cases when this default arity is undesirable. For example, the type of the operator o for composing two functions gives a default arity of 3, while the desired value is 2. Also, occasionally one will want the constant to map to a binding operator. This is the case, for example, for a constant declared to be a "binder" in HOL, e.g. h_exists in Section 2.

The arities are used when translating HOL expressions to Nuprl ones. When a constant of arity n is applied to at least n arguments, the n-ary application is replaced by an instance of the corresponding n-ary operator. If there are fewer than n arguments, η-expansions are first done to add a sufficient number of arguments.

3.7 Partial Functions

In Nuprl, as in HOL, function types are total: a function of type $A \to B$ produces a value of type B for every input of type A. However, most partial functions can be given convenient types in Nuprl because of Nuprl's subset type. For example, consider hd, the function that takes the head of a list. In Nuprl, $hd([])$ is undefined, but we can give hd the type

$$\{l \in N \ list \mid l \neq nil\} \to N,$$

for example.

In HOL, hd is defined on all lists: $hd([])$ is @$x \in$ 'a. *true*. This is a problem because when the theory containing hd is translated into Nuprl, we have to prove the well-formedness theorem for hd, and this requires showing that the Nuprl-defined hd is defined on the empty list. This forces us into giving the uncomputable HOL definition for hd in Nuprl. This kind of use of @ appears to be frowned on in the HOL community and does not seem to arise much.

Nevertheless, such cases do arise. Fortunately, in all the cases we have examined so far, we can work around the problem. Consider again the example of hd. We make the uncomputable definition for hd, but we also define a computable version, call it hd', and use this version in all subsequent definitions and in all theorems except for the definitional axiom for hd. This works because, although hd is defined on the empty list, this property is not taken advantage of. Consider a theorem asserting that the head of the list formed by consing x onto l is x.

The theorem that is directly justified by imported HOL theories is something like

$$\forall x \in A. \ \forall l \in A \ list. \ hd(x::l) = x.$$

But in Nuprl this formula can be proved equal to

$$\forall x \in A. \ \forall l \in A \ list. \ hd'(x::l) = x$$

by using the following lemma as a conditional rewrite rule:

$$\forall l \in A \ list. \ l \neq nil \Rightarrow hd(l) = hd'(l).$$

Note that this equivalence is within Nuprl, so it is irrelevant whether the HOL proof of the theorem relied on hd being defined on the empty list.

3.8 Theorems and Tactics

A number of tactics require additional information to make effective use of the theorems in Nuprl's library. For instance, some tactics require theorems of a certain kind to be annotated with special abstractions that have no logical significance but provide guidance to the tactic. More commonly, tactics require explicit indications, either by naming conventions or by explicit updates to reference variables via ML objects in the library, of relevant theorems and associated information. Currently, this must be dealt with by hand for imported theorems, just as with ordinary Nuprl theorems.

3.9 Unsolved Problems

One immediate problem we have not dealt with yet is HOL definitional packages. For example, there are packages that simulate various convenient forms of inductive definition and provide useful tactics for reasoning about the definition. While the theory objects generated by these packages can be readily imported using our scheme, the result is too low-level in some ways. The connection of the translated objects with the original higher-level definition is lost. It remains to be seen how effective Nuprl's tactics will be with such theories.

Other problems to be addressed in the future include abstract theories and making importation more incremental. Also, it might be interesting to try to import HOL tactics as well. The main obstacle to doing so is that HOL's tactic mechanism is incompatible with Nuprl's. In Nuprl, each time a tactic is applied by the user to refine a node in a proof tree, one is guaranteed by the system that the inference is sound. There is no such guarantee in HOL; soundness is guaranteed only for complete proofs. One way to fix this would be to redefine HOL's tactic type to be like Nuprl's, replacing the type *thm* with the type *proof* of (possibly incomplete) proofs. Making such a change would probably only affect the lowest levels of the system. Another way, incurring only a slight risk of unsoundness, is to take the approach of HOL's subgoal package.

References

1. *Part IIIA: SCI Coherence Overview, 1995.* Unapproved draft IEEE-P1596-05Nov90-doc197-iii.
2. R. L. Constable, et al. *Implementing Mathematics with the Nuprl Proof Development System.* Prentice-Hall, Englewood Cliffs, New Jersey, 1986.
3. M. J. Gordon, R. Milner, and C. P. Wadsworth. *Edinburgh LCF: A Mechanized Logic of Computation,* volume 78 of *Lecture Notes in Computer Science.* Springer-Verlag, 1979.
4. M. J. C. Gordon and T. F. Melham. *Introduction to HOL: A Theorem Proving Environment for Higher Order Logic.* Cambridge University Press, Cambridge, UK, 1993.
5. D. J. Howe. Semantics foundations for embedding hol in nuprl. *Proceedings of AMAST'96,* 1996. (to appear).
6. P. Jackson. *Nuprl 4.2 Reference Manual.* Cornell University, 1995. Available from ftp://cs.cornell.edu/pub/nuprl/doc.
7. P. B. Jackson. Exploring abstract algebra in constructive type theory. In A. Bundy, editor, *12th Conference on Automated Deduction,* Lecture Notes in Artifical Intelligence. Springer, June 1994.
8. B. Jacobs and T. Melham. Translating dependent type theory into higher order logic. In *Proceedings of the Second International Conference on Typed Lambda Calculi and Applications,* volume 664 of *Lecture Notes in Computer Science,* pages 209–229. Springer, 1993.
9. P. Martin-Löf. Constructive mathematics and computer programming. In *Sixth International Congress for Logic, Methodology, and Philosophy of Science,* pages 153–175, Amsterdam, 1982. North Holland.
10. T. Melham. The HOL logic extended with quantification over type variables. *Formal Methods in System Design,* 3(1–2):7–24, August 1993.
11. R. Milner, M. Tofte, and R. Harper. *The Definition of Standard ML.* MIT Press, 1990.
12. S. Owre, S. Rajan, J. Rushby, N. Shankar, and M. Srivas. PVS: Combining specification, proof checking, and model checking. In *Proceedings of CAV'96,* Lecture Note in Computer Science. Springer Verlag, 1996.
13. M. van der Voort. Introducing well-founded function definitions in HOL. In *Higher Order Logic Theorem Proving and Its Applications,* volume A-20 of *IFIP Transactions,* pages 117–131. North-Holland, 1993.

A Structure Preserving Encoding of Z in Isabelle/HOL[1]

Kolyang[*], T. Santen[‡], B. Wolff[*]

[*]Universität Bremen, FB3
P.O. Box 330440
D-28334 Bremen
{bu,kol}@informatik.uni-bremen.de

[‡]GMD FIRST Berlin
Rudower Chaussee 5
D-12489 Berlin
santen@first.gmd.de

Abstract. We present a semantic representation of the core concepts of the specification language Z in higher-order logic. Although it is a "shallow embedding" like the one presented by Bowen and Gordon, our representation preserves the structure of a Z specification and avoids expanding Z schemas. The representation is implemented in the higher-order logic instance of the generic theorem prover Isabelle. Its parser can convert the concrete syntax of Z schemas into their semantic representation and thus spare users from having to deal with the representation explicitly. Our representation essentially conforms with the latest draft of the Z standard and may give both a clearer understanding of Z schemas and inspire the development of proof calculi for Z.

1 Introduction

Implementations of proof support for Z [Spi 92, Nic 95] can roughly be divided into two categories. In *direct implementations*, the rules of the logic are directly represented by functions of the prover's implementation language. Since hundreds of rules are needed to reason about the mathematical toolkit of Z which defines relations, functions, data types, etc., these implementations are error-prone and tend to be difficult to modify. Moreover, they often lack implementations of advanced deduction techniques like, e.g., higher-order rewriting or resolution.

In contrast to direct implementation, one can choose to *semantically embed* Z in a logical framework. An implementation within a "tactical theorem prover" in the tradition of LCF like *HOL* [GM 93] or *Isabelle/HOL* [Pau 94] is particularly attractive: large libraries, e.g. for set theory, that are proven consistent with the kernel logic can be used to implement and enhance the mathematical toolkit of Z. The machinery of the Isabelle prover, in particular, supports deriving new rules and allows the systematic development of proof calculi for Z. Coming with an open system design going back to Milner, these systems allow for safe user-programmed extensions to support tedious proof tasks often arising in explicit (predicative) type checking or in transformational program development over specifications [KSW 96].

1.1 The Challenge

A lot of criticism from other scientific communities on Z is related to the fact that Z defines itself as a *notation* on the basis of set theory. The essential vehicle to

---

[1] This work has been supported by the BMBF projects **UniForM** and **ESPRESS**.

structure Z specifications at the level of concrete syntax is called a *schema*. Schemas have significantly more syntactic flavour than, for example, the notion of a "parameterised abstract data type" used in the algebraic specification community [EM 85].

A *purely* syntactical understanding of structuring would in fact neither be satisfying for theoretical nor for practical purposes. It would mean that structured specifications would have to be *expanded* (flattened) before any semantic treatment. As a consequence, the schema calculus and reasoning at the structural level would be impossible. The resulting formulas would get so large that the advantage of structuring specifications would disappear at the moment where reduction of complexity is of vital importance— namely when reasoning about specifications.

In our approach, schemas are represented as *boolean valued functions*. On the syntactic side, a schema declaration introduces an identifier of a particular class of types together with a particular signature that is used for parsing and pretty-printing purposes. On the semantic side, a schema declaration is a constant definition and may be unfolded during proofs at will. Since schemas are represented as first-class objects of the object logic, the schema calculus of Z can be represented and structured reasoning over specifications is possible.

The major contribution of this paper is to clarify the distinction between syntactic and semantic facets of schemas, at least from a HOL perspective. The practical consequences lie in the implementation of parsing- and pretty-printing machinery as well as in proof support at the level of schemas.

1.2 A Z-Journey Through Related Approaches

Since Z represents a particularly attractive goal for an encoding, several attempts to represent Z in logical frameworks and in particular in higher-order logic have been undertaken. Before we discuss some of them, let us briefly introduce the concepts of Z, crucial for such an embedding. We base the paper on the draft standard [Nic 95], which we will call "The Z Notation" (TZN) hereafter. An alternative would have been to refer to "The Z Reference Manual" (ZRM) [Spi 92], but we felt it more appropriate to follow the proposal for a future ISO standard for Z, although it may still evolve.

Central Concepts of Z. The semantics of Z given in TZN is based on Zermelo-Fränkel set theory (ZF) — which is untyped — but nevertheless Z is a strongly typed language. Each "given set" of a Z specification is associated with a primitive type. Type constructors for power-set types, product types, and schema types correspond to the respective set theoretic constructions. Type correctness with respect to these types is checked by Z type checkers like *fuzz* [Spi 92a]. In the following, we use the term "type" for Z types while "HOL-type" shall refer to types of higher-order logic. Types are not explicitly denoted in Z specifications. Instead, a declaration $x : S$ is a membership statement interpreted as $x \in S$ that implicitly declares x to have the — unique — type of the members of S.

The most important structuring concept of Z are schemas, like A, B and C below:

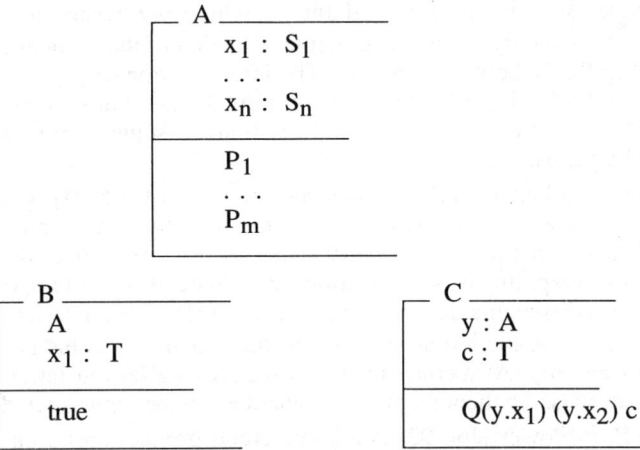

A schema consists of a list of declarations and a list of — semantically conjoined — predicates which are separated by a horizontal line in the concrete syntax.

A schema may be referenced by its name in subsequent schemas and schema-expressions. This may be *as import* like A in B (the declaration parts semantically are *sets of declarations*, hence their order and multiple occurrences are irrelevant — but S_1 and T must have the same type) or *as a set* like in C. Schemas can also serve as predicates like in $B \Rightarrow x_1 \in S_1 \cap T$. Moreover, it is possible to form new schemas in expressions of the schema calculus like $A \Rightarrow B$ or $\exists C \bullet B$ where $\Rightarrow$ is a schema connective and $\exists$ is a schema quantifier. It is important to note that the signature of a referenced schema is unified with the environment on the basis of *lexical* identity: in B, the explicitly declared x_1 is identified with the corresponding declaration in A.

Approaches to Mechanising Z. Bowen and Gordon's encoding "Z in HOL" [BG 94] represents A in the syntax of HOL88 by:

$$x_1 \text{ IN } S_1 \wedge \ldots \wedge x_n \text{ IN } S_n \wedge P_1 \wedge \ldots \wedge P_m$$

where S_i is a HOL term denoting a set (in HOL set theory) and $P_1 \ldots P_m$ are boolean terms constituting the schema's predicate. The S_i can have different HOL types. The mathematical toolkit of Z is represented by appropriate constant definitions based on HOL's set theory. Schema references as imports and sets, and schema expressions are represented by expansion into the representation above; hence their structure is "parsed away". Pretty-printing suppresses the printing of the expanded schemas — this helps the eye, but not the prover.

In Kraan and Baumann's representation "Z-in-Isabelle" [KB 95] a schema A is represented as a *theory* in Isabelle and the variables in the declaration part as *constants*:

```
AA = Toolkit +
consts     x1 :: "S1_t"
           ...
           xn :: "Sn_t"
translations
           "A" == " [ x1 : S1; ...; xn : Sn | P1 ...Pm]"
```

Here the $S_{i\_}t$ are the types of the S_i which are terms denoting a set in an independent set-theory based on an implementation of the sequent calculus LK that is contained in the Isabelle distribution. The interpretation of [..|..] is similar to the one of "Z in HOL": all parameters are conjoined. The consequence of making the signatures of schemas globally visible is that all schema expressions have to be expanded by the parser.

Compiling schemas to (PVS)-theories, the work of [ES 94] is conceptually rather close to "Z-in-Isabelle". However, due to the introduction of intermediate constants for predicative parts, it comes very close to our ideal of a structure preserving encoding — except that lists of equations have to be generated for variables stemming from different Z schema declarations (like x_1 in B). The compilation cannot support schema-as-sets. Deduction at the level of the schema-calculus is possible in a very restricted way only. As a consequence of the lacking re-translation (pretty-printing), intermediate stages in theorem proving cannot easily be reinterpreted as schemas.

ICL's ProofPower [Jon 92] is a commercial product based on a deep encoding. Schemas are represented as sets of bindings, and schema operations work on these sets. More information on this work can be found in [BG 94].

The theorem prover Ergo [RS 93] implements an untyped meta-logic in Prolog on top of which Zermelo-Fränkel set theory and a theory for Z are encoded. The basic proof mechanism of Ergo is window inference which is augmented by a tactic language. Because of the lack of an underlying type system, many "typing" subgoals arise during proofs. Automatically invoked tactics called *elves* are used to try and proof these subgoals. If the elves fail, the user has to tackle these subgoals interactively.

Like Ergo, the implementation of Z in EVES [Saa 92, MS 95] incorporates the mathematical toolkit of Z and uses theorems about it as rewrite rules. The basic proof commands of EVES are specialised on syntactic categories of formulas, e.g. "eliminate a quantifier". Proven theorems are automatically used by more complex rewriting and simplification procedures.

Jigsa$\mathcal{W}$ [Mar 94] is a deep encoding done in 2OBJ for $\mathcal{W}$, a logic proposed for Z [WB 92], which it faithfully encodes. Since there are no meta-variables in 2OBJ, general theorems about Z cannot be expressed schematically. Proof procedures that produce proofs for each instance of the schematic theorem have to be coded instead. Reasoning in a deep encoding in this way reduces performance considerably. Furthermore, Maharaj has encoded Z in type-theory using LEGO [Mah 90].

There are also direct implementations for dialects of Z, e.g. Balzac [Har 91, Jor 91] and CADiZ [TH 95] allowing at least for single step inferences. CADiZ is based on a sequent calculus that is applied using a proof procedure called "Gentzen". It also incorporates a decision procedure for integers, and is currently being extended with a tactic language.

1.3 Overview of this Paper

This paper proceeds as follows: TZN identifies the following hierarchy of syntactic categories: *expressions, predicates, schema-expressions, paragraphs* and *theories*. In Section 2, we present our encoding bottom-up, successively giving a semantic representation for each of these categories and demonstrate its consequences for Z. In Section 3, we discuss proof support for our encoding "Z in Isabelle/HOL".

2 Representing Z in Isabelle/HOL

2.1 Conformance with TZN

For any embedding of a logic, the question of a "faithful encoding of one calculus in another" has to be raised. The question is moot here, since a Z calculus is not available now: the draft definition of a calculus in Appendix F of TZN is very rudimentary. Our embedding *conforms* to Z in the sense of TZN (p. 2):

A specification conforms to the standard for the Z notation, if and only if the formal text is written in accordance with the syntax rules and is well-typed.

A deductive system for Z conforms to the standard, if and only if its rules are sound with respect to the semantics.

The core of TZN consists of the definition of two *partial* functions $\langle\ \rangle^\tau$ and $\langle\ \rangle^M$ that assign to an element of each syntactic category a *type* and a *value (meaning)*. Therefore a calculus conforms to the standard if it reflects the semantic function *where it is defined*. The semantic functions are interpreted in an untyped universe of Zermelo-Fränkel set theory. Thus, in the semantic universe, objects like $\{0, \{0\}\}$ may occur that are illegal in the typed set theory of HOL. This does not mean that $\{0, \{0\}\}$ is legal in Z; in fact, one of the major objectives of $\langle\ \rangle^\tau$ is to rule out such expressions.

We are confident that our encoding conforms to the draft standard. In the following, we emphasise the strong similarities between our encoding and the intuition about Z gained from TZN showing the conceptual correspondence to our encoding.

2.2 Syntactic Issues

Previous Z encodings cope differently with the question of Z syntax. Most of them try to encompass the LaTeX presentation of the Z syntax, e.g. [BG 94, KB 95]. Since lexical issues are very important for both presentation and readability, we decided to remain close to the Z syntax as presented in TZN. The draft standard proposes several ASCII-based representations of Z. The interchange format is based on SGML, while the email format is primarily conceived as a human readable lightweight interchange format, rather than for processing by tools. In the email format the character % is used to flag special strings, to disambiguate them from others.

Concerning the lexical representation, our encoding supports the email format. The following table gives an impression of how our representation relates to the various lexical representations:

Z operator	Email Format	Z in Isa/HOL	Meaning
;	%;	%;	compose
∘	%o	%o	functional compose
◁	<:	<:	domain restrict
↣	>-->>	>-->>	bijection

Below we present the exceptions made necessary in order to cope with Isabelle/HOL types:

Z operator	Email Format	Z in Isa/HOL	Meaning
:	:	::	type membership
∈	%e	:	set containment

Parsing is done at two different levels, interleaved by a conversion phase. For the e-mail format level, we introduce a theory file Zproto.thy providing the necessary syntax and translation rules. The second level is the semantic representation level. ML-functions convert the abstract syntax parsed by the e-mail format level to the latter. A short excerpt of Zproto.thy is shown below:

```
Zproto = HOL+
...
syntax
...
"_sch3"::[Name, DeclParts] => 'a Zschema          ("+---_/---_/---")
"_sch4"::[Name, DeclParts, Predicate] => 'a Zschema   ("+---_/---_/|--_---")
...
"_sch7"::[Name,Formals,DeclParts,Predicate]=>'a Zschema  ("+---_[_]/---_/|--_---")
translations
"+-- N ---   D ---"              == " N = Schema D True"
"+-- N ---   D |-- P ---"        == "(N = (Schema D P))"
"+-- N [M] ---   D |-- P ---"    == "(N = (%M. Schema D P))"
```

In Isabelle, syntactic sorts like Name or DeclParts can be introduced that are treated as non-terminals of a grammar (adapted from TZN), while constant declarations like _sch3 can have the character of a grammar rule, with the pragma ("+--_/---_/---") introducing alternative mixfix syntax where the / informs the pretty-printer where to place optional linebreaks. Translation rules may successively transform such raw-syntax-trees into trees where only symbols occur for which semantic information is available. Schema is such a function that takes a declaration and a predicate and returns a Z schema (of the type Zschema).

According to Zproto.thy, the example of schema A can be presented as follows:

```
"+-- A ---
     x1:S1;
     ...;
     xn:Sn
 |--  P1 &
     ... &
     Pm
 ---"
```

All schemas are parsed according to the signature of Zproto.thy. After computing its signature, a schema is translated to the semantic level, where the right binders with the right scope are generated. The pretty-printing will, according to a user-controlled variable pretty_level, partially or completely perform the retranslation. In the sequel, we focus on the description of the semantic representation level.

2.3 Expressions

The predicate logic of Z uses the usual connectives and quantifiers with their standard semantics. We therefore model Z predicates as Boolean functions P::'a => bool. Consequently, we can directly map the logical connectives of Z to the ones provided by HOL. Representing the set theory of Z is similarly simple, since Z is strongly typed.

In particular, all elements of a set must have a common type. It follows that there is a universal set for each type that contains all elements of that type. This is why there is no need to explicitly refer to types in the Z language. From these observations, we conclude that the set theory of HOL is convenient to represent Z sets. The set constructor 'a set models the powerset operator P of Z at the level of types, and the set operations of Z directly translate to the corresponding operations of HOL.

Primitive Operations and the Mathematical Toolkit. The mathematical toolkit is introduced as an Isabelle theory ZMathTool. It is a conservative extension of some theories from the HOL library just as all constants of the toolkit in TZN are *defined* by the core language of Z:

```
ZMathTool = Finite + Nat + Arith +
types ('a,'b) "<=>" = "('a*'b) set"        (infixr 20)
```

A pivotal type in the mathematical toolkit is the *relation* which is defined as an infix type constructor <=>. This type will be used to shape all sorts of function spaces.

According to our lexical principles, we are now able to present the toolkit as a suite of constant definitions (the technique is equivalent to [BG 94]). On the right-hand side of the type definition some parsing information is given together with the binding values.

```
consts
...
rel              ::"['a set, 'b set] => ('a <=> 'b) set"   ("_ <-->_")    [5,4]4)
partial_func     ::"['a set,'b set] => ('a <=> 'b) set"    ("_ -|-> _")   [5,4]4)
total_func       ::"['a set,'b set] => ('a <=> 'b)"        ("_ ---> _")   [5,4]4)
partial_inj      ::"['a set,'b set] => ('a <=> 'b)"        ("_ >-|-> _")  [5,4]4)
total_inj        ::"['a set,'b set] => ('a <=> 'b)"        ("_ >--> _")   [5,4]4)
...

defs
...
rel_def           "A <--> B  == P (A >< B)"
partial_fun_def   "S -|-> R  == {f. f:(S<-->R) & (ALL x:S y1:R y2:R.
                                          (x,y1):f & (x,y2):f-->(y1=y2))}"
total_func_def    "S ---> R  == Union{f. f:(S -|-> R) & (dom f) = S }"
partial_inj_def   "S >-|-> R == Union{f. f:(S -|-> R) &
                                    (EX x1:dom f x2:dom f. ((f^^x1)=(f^^x2))
                                                      --> (x1 = x2))}"
total_inj_def     "S >--> R  == (S >-|-> R) Int (S ---> R)"

end
```

Conformance of the toolkit with TZN is easy to verify: Just compare the definitions in ZMathTool to the ones in TZN. Furthermore, the laws given in [Spi 92] can be derived as theorems from ZMathTool. Especially the theorem:

$$\bigcap_{x:\{\}} P(x) = \{y.\ true\}$$

holds, in contrast to ZF where the result of this intersection over the empty index set is defined equal to { } because there are no universal sets in this untyped theory. In *typed* set theories like in Z or in HOL, the complement of a set is always defined.

Set Expressions, Function Application and Abstraction: On the basis of ZMathTool, it is straight-forward to represent declarations by membership predicates, for example the declaration of a partial injective function f:

   f : N >-|-> N

Here ":" is the usual set membership operator of HOL set theory and N is the set of all natural numbers. The type of f, $P(Z \times Z)$ is modelled by

   f :: nat set <=> nat set

This type is automatically inferred from the predicate. As a consequence of modelling functions as binary relations, we need a new application operator to apply f

   "^^":: ['a set <=> 'b set, 'a] => 'b

which is defined by the Hilbert operator (as in [BG 94]):

   f ^^ x == (@y. (x,y) : f)

This treatment of partiality again conforms to TZN, where this extension of the semantic function is explicitly justified (p. 36):

> An example is the definition of application: for example, in function application, when the argument is outside the domain of the function, then no meaning is explicitly given. Different interpretations of Z can ascribe different meanings to an ill-formed function application.

There is a lively debate on the issue of partiality in the Z community and in the specification community in general. TZN's approach of using partial semantic functions allows designers of proof support systems to resolve issues like partiality the way it suits them best. Although counter-intuitive propositions like 1/0 = 1/0 are indeed provable by reflexivity in our encoding — in contrast, e.g., to the one of Kraan and Baumann [KB 95] — we believe that our model of function application considerably simplifies deduction while still conforming to the semantics of Z. Since the semantics of Z is partial, specifiers cannot rely on any semantics of function applications outside their domain. Reasoning about expressions like "1/0" cannot provide them with any information about their specification. On the other hand, modelling the partiality of the semantics function, e.g. by making "^^" partial, would only complicate deduction.

Abstraction is defined analogously[2]::

   consts   LAM :: ['a => bool, 'a => 'b] => 'a <=> 'b
   defs   LAM S E == { (x,y) . S x & y = E x }

The general type of schemas which are introduced in the next section is 'a => bool. Thus LAM takes a schema S and an expression E in the signature of S and returns the desired relation.

Schema Expressions. The core problem of any shallow embedding of Z is the treatment of the sets of declarations defined in a schema. This set is called the *signature* of a schema and is computed in TZN via the already mentioned partial function $\langle\ \rangle^\tau$. Our example B of schemas as imports demonstrates the delicate fusion

---

[2] Since functional abstraction is not defined in the current version of TZN, we use the definition of [Spi 92] to justify the conformity of our encoding.

of a schema's signature with the signature of its context: identifiers with equal *names* in the declaration part of the imported schema *A* and in the surrounding declarations of *B* are identified

Our approach suggests making this dependency explicit and considering a schema *S* as a boolean-valued function on the sub-signature of the context. It is the task of the parser to keep track internally of the signature of schemas and schema expressions. It must hence substitute a schema references *S* in a schema context by an application *S x*, where *x* represents the variables of the surrounding context that have to be identified with the ones declared in *S*. Note that the construction of signatures will always assure that the signature of the context will contain the elements of *x*.

However, the type-scheme

$$\tau = \alpha_1 \to ... \to \alpha_n \to \text{bool}$$

for the representation of a schema is inappropriate because, in some places, we need an Isabelle type-scheme that subsumes all types of schema representations. This is necessary to represent schemas-as-sets and to deal with expressions of the schema calculus, as well. For this reason, we use the uncurried version of the type scheme above:

$$\tau' = \alpha_1 \times ... \times \alpha_n \to \text{bool}$$

This type-scheme can be generalised to the type $\alpha \to \text{bool}$. This means that schemas are basically represented as predicates over tuples of variables in our encoding.

The fundamental mechanism to define representations of schemas are *schema binders* that are defined as follows[3]:

```
consts    SB0  :: "('a => bool) => ('a => bool)"           (binder "SB0" 10)
          SB   :: "(['a, 'b] => bool) => (('a * 'b) => bool)"   (binder "SB" 10)
defs      SB0_def  "SB0 P == P"
          SB_def   "SB P  == (% (x,y). (P x y))"
```

The pragma (binder "SB" 10) tells Isabelle to treat SB like a quantifier and provide it with additional syntax; the term:

SB x_1. ... SB x_{n-1}. SB0 x_n. P x_1 ... x_n

has the type τ' iff P has the type τ. We call such a term a *schema lifter* over P of *length* n.

Schema lifters are intended to be suppressed by the pretty-printer. The conversion-phase within the parser generates lifters that are lexically sorted because schemas with permutated declarations are semantically identical.

Simple Schemas. Schema binders only model the binding structure of schemas. Simple declarations in Z not only declare a new variable but also contain membership constraints. These are reflected by the schema declaration DECL.

```
consts    DECL :: [ bool, bool ] => bool    ("[ _ | _ ]" ...)
defs      DECL_def   "[P | Q ]  == P & Q"
```

---

[3]In our implementation, the situation is slightly more complex for purely syntactic reasons.

This enables us to represent a schema

$$
\begin{array}{|l}
\hline
S \\
\hline
x : \mathbb{N} \\
y : \mathbb{Z} \\
f : \mathbb{Z} \rightarrowtail \mathbb{Z} \\
\hline
P\ x\ f\ y \\
\hline
\end{array}
$$

in Isabelle as:

```
consts    S:: "int <=> int * int * int => bool".
def       S == SB f. SB x. SB0 y.[ x : N & y : Z & f: Z >-->Z | P x f y ]
```

Semantically, a simple schema is just the conjunction of the two parameters. The new constructor DECL is needed to reflect the structure of schemas in the representation. Thus, the pretty-printer can reproduce the concrete syntax of a schema. In particular, the *user-defined* ordering of the declarations in a schema and the distinction between membership propositions in declarations and predicates can be reconstructed. The schema binders are not printed at all, and the sorting of declarations remains internal. However, the parser has to store the lexical names of the variables in the signature so the pretty-printer can hide the effects of α-conversions on schema binders.

Generic Schemas. The representation of simple schemas naturally extends to the case of *generic* schemas by abstracting the body schema S in the form: $\lambda M :: \alpha$ *set*. S. The generic schema

$$
\begin{array}{|l}
\hline
P[R] \\
\hline
x : R \rightarrowtail \mathbb{Z} \\
\hline
T\ x \\
\hline
\end{array}
$$

is converted by the parser to

```
consts    P:: "'a set => ( 'a set <=> 'int set ) => bool"
defs      P == % R . SB0 x . [ x : R -|-> Z | T x ]
```

The polymorphism of P in 'a models exactly the type constraints of Z that are checked by type-checkers like *fuzz*. How the parameter set R is instantiated depends on the particular application context: if R is not instantiated explicitly, the universal set {x :: 'a. true} over the parameter *type* 'a can be used.

Schema-as-import, Schema-as-sets. We recall the structures B and C from the introduction:

$$
\begin{array}{|l}
\hline
B \\
\hline
A \\
x_1 : T \\
\hline
\text{true} \\
\hline
\end{array}
\qquad
\begin{array}{|l}
\hline
C \\
\hline
y : A \\
c : T \\
\hline
Q(y.x_1)\ (y.x_2)(c) \\
\hline
\end{array}
$$

hat are consequently represented by:

```
defs
    B == SB x₁. ... SB0 xₙ.[ A(x₁,...,xₙ) & x₁ : S₁ | true]
    C == SB y. SB0 c. [ y : {z. A z} & c : T | Q (fst y) ((fst o snd) y) (c)]
```

Schema Expressions. The expressions of the schema calculus closely resemble predicate logic formulas. There are the usual connectives as well as universal and existential quantification, and, as one expects, their intuitive semantics can be based on the understanding of predicate logic.

The construction of schema lifters mimics the construction of signatures of schemas in TZN. Hence, logical connectives and quantifiers can be lifted to the corresponding schema expressions using an appropriate schema lifter.

In the case of the schema connectives, the signature of schema expressions is the union of the signatures' operands if the variables in the intersection of the signatures have identical types. Otherwise, the schema expression is ill-formed. For example, the schema-conjunction $N \cong S \wedge R$, where the signature of S is (s, u) and the one of R is (t, u, v) is represented by the following lifter over the conjunction:

```
    defs "M == SB t. SB0 v. (ALL (s,u). (S (s,u)) --> (R (t,u,v)) )"
```

As with simple schemas, the pretty-printer can decide on the basis of the predicate following the schema lifters which schema expression is represented, and print it accordingly.

2.4 Z Paragraphs and Sections

The formal text of a Z specification consists of a sequence of paragraphs which gradually introduce the given types, global variables, and schemas of the specification. The paragraphs forming a specification can be organised as a collection of named sections. Each paragraph can augment the environment by declarations of constants or schemas.

In our encoding, Z paragraphs and sections are represented as Isabelle theories. The way new schemas are introduced, by the key word **defs**, ensures a conservative construction of these theories. This guarantees that the representation process of Z into Isabelle does not introduce inconsistencies.

3 Proof Support

Since our encoding preserves the structure of specifications and allows us to deal with the schema calculus, two possibilities for structured reasoning about specifications arise. First, we can combine theorems about single schemas in a controlled way, and second, we are able to lift theorems of predicate logic to the level of schema expressions, and thus come to a truly mechanised schema *calculus*.

3.1 Structured Theories about Z Specifications

Complex Z specifications are usually structured as follows: first, axiomatic declarations introduce the constants which the rest of the specification is based upon (the "model" of the specification). Afterwards, simple schemas provide the basic notions relevant for the system to specify. Finally these are combined, possibly in several steps, to schemas that provide "the" system specification. For sequential systems, for example, a common approach is to combine several schemas describing "substates" of the system to the final state schema, and to provide schemas specifying different aspects of an operation, e.g. various cases of normal behaviour and behaviour in error cases, that are combined, using the schema calculus, to a schema specifying "the" operation.

Since we can refer to schemas as first-class objects in our encoding, we have an elegant way to form "local theories" about single schemas and augment schema references with the information provided by these theories. A theorem P about schema A can be expressed as an implication of the Isabelle meta-logic

$$A\ (x_1,...,x_n) ==> P\ x_1\ ...\ x_n$$

This theorem can be used to extend the predicate of

$$B == SB\ x_1.\ ...\ SB0\ x_n.[\ A(x_1,...,x_n)\ \&\ x_1 : S_1\ |\ true]$$

Another possibility to reason about schemas is to successively transform them into equivalent representations by equational reasoning. One starts with an equation whose right-hand side is a meta-variable

$$A\ (x_1,...,x_n) = ?X$$

Using transitivity of equality and Isabelle's simplifier or other suitable tactics one can now gradually instantiate the metavariable ?X and prove

$$A\ (x_1,...,x_n) = Asimp$$

where Asimp is a simplified version of A. This equation can now be used — possibly again using the simplifier — to replace references to A in other schemas by Asimp.

Given the explicit representation of schema references, these transformations are technically simple. Still, they provide the possibility to develop the theory of a specification in "layers", starting at the layer of the mathematical toolkit and the axiomatic declarations, and ascending via simple schemas to the more complex schema expressions that finally make up a system description.

First experiences with reasoning about a non-trivial Z specification show the practical advantage of our structure preserving encoding. We have applied the technique described above to transform operation schemas of a specification of a simplified embedded controller into disjunctive normal form and used the simplifier — instantiated with rules about the mathematical toolkit — to eliminate unsatisfiable disjuncts from the normal form. This procedure is part of an approach to generate test cases from model-based specifications [DF 93].

The major practical problem here is the performance of the prover when faced with large subgoals. With encodings that do not preserve the structure of specifications, all schema references in the operation schema to transform are inevitably expanded. This is neither appropriate nor practically feasible: on one hand, one will want to control expansion of definitions to come to sensible test cases. On the other hand, each additional literal in the schema's predicate increases computation time and space requirements of the prover because the distributivity laws duplicate subformulas and context-dependent simplification is mandatory to find unsatisfiable disjuncts.

We computed the normal form of our example operation schema in a bottom-up fashion: first computing the normal form of the state schema and of the operation schema without expanding schema references, and second combining the results and computing the normal form of the combination. Simplification helps a lot to control combinatorial explosion: more than two thirds of the state schema's disjuncts can be reduced to *false*, and only 8 of 44 disjuncts of the combined schema are satisfiable and form the final outcome of the computation. Each disjunct consists of about 20 literals. Just expanding schema references and compute the normal form in one step is absolutely infeasible: the algorithm did not terminate in an over-night run!

3.2 Lifting Predicate Logic to Schema Expressions

A technically more demanding possibility arising from our encoding of schema expressions is to lift theorems of predicate logic to the corresponding schema expressions. The basic idea is to use Isabelle's lifting of meta-variables that occurs when resolving a theorem with the matrix of the (meta-)universal quantifier "!!".

Consider what happens when resolving the extensionality theorem ext

!! x. ?f x = ?g x ==> ?f = ?g

with a theorem of predicate logic, say and_commute

?A & ?B = ?B & ?A

Forward chaining (and_commute RS ext) results in

%x. ?A x & ?B x = %x. ?B x & ?A x

This means the meta-variables ?A and ?B of and_commute are lifted to *functions* in x. We can use this mechanism to lift predicate logic theorems to *schema binders* using the theorems SB0_ext and SB_ext:

(!!x. f x = g x) ==> (SBinder0 f) = (SBinder0 g)
(!!x. f x = g x) ==> (SBinder f) = (SBinder g)

Combining these two in an ML function lift

```
fun lift th 0 = th
  | lift th 1 = th RS SB0_ext
  | lift th n = (lift th (n-1)) RS SB_ext;
```

allows us to lift a predicate logic theorem to a theorem about schemas with signatures of arbitrary length, e.g.

```
lift and_commute 3;
val it =    "(SB x xa. SB0 xb. ?A x xa xb & ?B x xa xb) =
            (SB x xa. SB0 xb. ?B x xa xb & ?A x xa xb)" : thm
```

This lifting mechanism enables us to reason *directly* about schema expressions simply by reusing predicate logic theorems. In particular, it is not necessary to — implicitly or explicitly — expand the schemas of a particular expression and reason at the level of their (combined) predicates. Because the length of schemas can always be inferred by the parser it is possible to totally hide schema lifting from the user's view by providing versions of Isabelle's resolution functions like RS or rtac that implicitly lift predicates to schemas of appropriate length.

4 Conclusion

We have presented a shallow, TZN-conforming encoding of Z into higher-order logic that nevertheless preserves the structure of specifications. Moreover, our representation allows deductions at a structural level, i.e. in the schema calculus. Other approaches to proof assistants for Z either implement a proof tool from scratch and accept the disadvantages in terms of implementation work, lack of reuse, and error-proneness to come up with a tool tailor-made for Z, or they encode Z into a logical framework like HOL or ZF. The latter approaches usually choose a deep encoding if they want to deal with the schema calculus, or they sacrifice the structure of Z specifications and represent only "flattened" specifications where all schema references are expanded. Providing a shallow encoding that still allows us to deal with schemas at the logical level is the major contribution of our work.

There is a price to pay for a shallow embedding: we cannot represent all aspects of the semantics of Z in logical terms, some have to be dealt with at the level of syntax. We chose to put the dividing line between syntax and logic at exactly the point of the Z semantics where it gets a very "syntactic" flavour, i.e. where the signatures of schemas and the binding structure induced by schema references are concerned. The mechanisms needed here can easily be understood as manipulations of sets of identifiers, and can hence safely be implemented in a parser by introducing appropriate schema binders "SB".

An advantage of this approach is that we, unlike deep embeddings like ProofPower [Jon 92], need not deal with "syntactical" issues in the logic. This also includes the issue of type checking. Since Z types are handled by the Isabelle parser, we do not need to reason about them explicitly. Untyped provers like Ergo [RS 93] must provide specialised tactics to prove type constraints. A first experiment based on the ZF encoding of Isabelle has shown that the type constraints provided by HOL greatly enhance efficiency of deduction. Similar proofs of (simple) propositions about Z schemas as predicates need much more search in ZF than in HOL (we usually had to use **best_tac** instead of **fast_tac** to find a proof).

Depending on the context of their use, schemas can have three different interpretations in Z: as sets, in schema calculus, and as predicates. We picked the latter as the basis of our embedding because schemas are often used in predicative context and proof engines like Isabelle are tuned to deal with predicates most efficiently. This choice lead to the idea of schema lifters and consequently enabled us to come up with a representation of the schema calculus which, to our knowledge, is the first in a shallow embedding.

As did Bowen and Gordon [BG 94], we map the set theory of Z to the one of HOL. Z is strongly typed and the apparent similarities to the HOL set theory are much greater than to other set theories like, e.g., ZF. We believe this and the encoding of the schema calculus in higher-order abstract syntax justifies our claim that our embedding conforms to the Z draft standard.

4.1 Future Work

Many specifications using Z heavily depend on the mathematical toolkit, and consequently many subgoals arising while reasoning about Z specifications are basically propositions about the toolkit. Such "standard" subgoals may be provable automatically by specialised proof procedures that are based on an extensive collection of theorems about the toolkit. The generic proof infrastructure of Isabelle, notably the highly customisable simplifier, provides a means to implement such proof procedures. We are currently investigating what degree of automation can be reached, in particular for specifications stemming from specific application domains.

For the time being, we have concentrated on representing the most crucial features of Z — this subset should be augmented and combined with other formal methods as envisaged in [Kri+95]. The powerful pretty-printer of Isabelle provides much potential to support other syntactic representation like the *TZN interchange format.*, LaTeX or Tcl/Tk for the interactive vizualisation of mathematical, high-quality notation. In this way, the elaborate theorem proving facilities of Isabelle are made available to support practical work with Z such as analyses of specifications, transformational development, and test case generation and evaluation on the basis of Z. Refining the theorem proving support for testing based on formal specifications will be one major focus of activities in the project ESPRESS (see also [Jäh+95]).

Acknowledgement. We thank the referees for constructive comments which helped improving the presentation. Thanks also for pointing us to [ES 95].

References

[BG 94]   Bowen, J. P., Gordon, M. J. C.: Z and HOL. In Bowen, J.P. and Hall, J.A. (ed.): *Z Users Workshop*, Cambridge 1994, Workshops in Computing, pp. 141-167, Springer Verlag, 1994

[DF 93]   Dick, J., Faivre, A.: Automating the Generation and Sequencing of Test Cases from Model-Based Specifications. In Woodcock, Larsen (eds.), Proc. Formal Methods Europe, pp. 268-284, LNCS 670, Springer Verlag, 1993.

[EM 85]   Ehrig, H. Mahr, B.: *Fundamentals of Algebraic Specification: Volume 1: Equations and Initial Semantics*, Springer Verlag, 1985

[ES 95]   M. Engel, J.U.Skakkebæk: Applying PVS to Z. ProCoS II document [ID/DTU ME 3/1], Technical University of Denmark. 1995.

[GM 93]   Gordon, M.J.C., Melham, T.M.: *Introduction to HOL: a Theorem Proving Environment for Higher order Logics*, Cambridge University Press, 1993.

[Har 91] Harwood, W. T.: Proof rules for Balzac. Technical Report WTH/P7/001, Imperial Software Technology, Cambridge, UK, 1991.

[Jäh+95] S.Jähnichen (director): ESPRESS — Engineering of safety - critical embedded systems. Online information available via http://www.first.gmd.de/org/espres.html.

[Jon 92] Jones, R. B.: ICL ProofPrower. BCS FACS FACTS Series III, 1(1):10-13, Winter 1992.

[Jor 91] Jordan, L. E.: The Z Syntax Supported by Balzac II/1. Technical Report LEJ/S1/001. Imperial Software Technology, Cambridge, UK, 1991.

[KB 95]   Kraan, I., Baumann, P.: Implementing Z in Isabelle. In Bowen, Hinchey (eds.), ZUM '95: The Z Formal Specification Notation, pp. 355-373, LNCS 967, Springer Verlag, 1995.

[Kri+95] Krieg-Brückner, B., Peleska, J., Olderog, E.-R., Balzer, D., Baer, A.: Uniform Workbench — Universelle Entwicklungsumgebung für formale Methoden. Technischer Bericht 8/95, Universität Bremen, 1995. Also available online via http://www.informatik.uni-bremen.de/~uniform.

[KSW 96] Kolyang, Santen, T., Wolff, B: Correct and User-Friendly Implementations of Transformation Systems. Proc. Formal Methods Europe, Oxford. LNCS 1051, Springer Verlag, 1996.

[Mah 90] Maharaj, S.: Implementing Z in LEGO. Unpublished M.Sc.thesis. Departement of Computer Science, University of Edinburgh, September 1990.

[Mar 94] Martin, A.: Machine-Assisted Theorem-Proving for Software Engineering, Unpublished PhD Thesis, University of Oxford, 1994.

[MS 95] Meisels, I., Saaltink, M.Z.: The Z/EVES Reference Manual (draft). Technical report TR-95-5493-03, ORA Canada, December 1995

[Nic 95] Nicholls, J. (*ed.*, prepared by the members of the Z Standards Panel): Z - Notation. Version 1.2. ISO-Draft. Online: http://www.comlab.ox.ac.uk/oucl/users/andrew.martin/zstandard/.14th September 1995.

[Pau 94] Paulson, L. C.: *Isabelle - A Generic Theorem Prover.* LNCS 828, Springer Verlag, 1994.

[RS 93] Robinson, P.J., Staples, J.: Formalizing a Hierarchical Structure of Practical Mathematical Reasoning. *Journal of Logic and Computation* 3 (1), pp. 47-61, 1993

[Saa 92] Saaltink, M.Z.: Z and EVES. In Nicholls, J.E. (ed.) Z User Workshop, York 1991, Workshops in Computing, pages 223- 242. Springer Verlag 1992

[Spi 92] Spivey, J.M. : *The Z Notation: A Reference Manual* (2nd Edition). Prentice Hall, 1992.

[Spi 92a] Spivey, J.M.: The *fuzz* Manual, Computing Science Consultancy, 2 Willow Close, Garsington, Oxford OX9 9AN, UK 2nd edition, 1992

[TH 95] Toyn, I., Hall, J.: Proving Conjectures using CADiZ. York Software Engineering Ltd., September 1995.

[WB 92] Woodcock, J.C.P., Brien, S.M.: $\mathcal{W}$: A logic for Z. In Nicholls, J.E. (ed.) *Z User Workshop*, York 1991, Workshops in Computing, pp. 77-96. Springer Verlag 1992

Improving the Result of High-Level Synthesis Using Interactive Transformational Design

Mats Larsson *
Volvo Technological Development/6240
Chalmers Teknikpark, S-412 88 Göteborg, Sweden
Tel: +46 31 7724169
Email: mala@vtd.volvo.se

Abstract. This paper reports on work to alleviate the problem that high-level synthesis tools cannot in general produce results that are comparable in quality to manual design. We propose to use pre-synthesis transformations of algorithmic input specifications to improve the quality of the synthesis. By mechanising transformational reasoning about specifications as interactive theorem-proving in a proof system no errors are introduced during transformation, the designer can guide the tuning process and we can make behavioural transformations which extend the capabilities of the synthesis tool.

1 Introduction

Due to the ever increasing complexity of digital designs it is becoming increasingly necessary to design at higher levels of abstraction. High-level synthesis tools provide mechanical support for this task by producing an RTL-level implementation from an algorithmic input specification written in a hardware design language (HDL). A problem with high-level synthesis tools is the poor quality of the resulting circuits. Such tools cannot in general produce results that are comparable to manual design. One remedy is for the designer to manipulate the original HDL description into a tuned design that have the same external behaviour but will produce a better circuit when synthesised by a high-level synthesis tool. We call this process *transformational design* and each step in this process is called a *behavioural transformation*.

This paper reports on work done on how transformational design of design specifications can be rendered as theorem proving in the HOL proof assistant. The idea is that rather than rely on paper and pen calculations, the designer can use the HOL system to prove that the original and tuned designs in fact have the same meaning. This way we can ensure that no errors are introduced in the process since each behavioural transformation will have to be proved before it can be applied.

A design tool based on these principles would allow the designer to prune the design space for good solutions by incremental transformation from an initial design.

---
* This work was done while the author was at Linköping University, Sweden

1.1 Hardware Design Language

The hardware design language we use is a slightly extended subset of ADDL (Algorithmic Design Description Language), henceforth called MINI-ADDL. ADDL is an imperative digital design language that was designed to allow design specifications for the CAMAD synthesis system [9] to be written in a high-level language.

MINI-ADDL, is an extension of the While-language presented by Camilleri and Melham and distributed as an example of inductive definitions with the HOL system [2]. Our work is based on that work and part of our embedding is the same as theirs. The extensions we have made are to consider a slightly richer language that includes three new commands, parallelism, for loops and local variable declarations. In addition to that we have also added array types which means that the value of an expression can be either a number or an array of values. The abstract syntax for MINI-ADDL can be seen in Figure 1 where sequential composition of commands is denoted by ; and parallel composition is

$$
\begin{aligned}
C ::= \ &\textbf{skip} \\
| \ &L := E \\
| \ &C_1 \ ; \ C_2 \\
| \ &C_1 \ ! \ C_2 \\
| \ &\textbf{if } B \textbf{ then } C_1 \textbf{ else } C_2 \\
| \ &\textbf{while } B \textbf{ do } C \\
| \ &\textbf{for } I := E_1 \textbf{ to } E_2 \textbf{ do } C \\
| \ &\textbf{local } I \textbf{ in } C \\
L ::= \ &I \\
| \ &L[E] \\
E ::= \ &I \\
| \ &E_1[E_2] \\
| \ &E_1 \textbf{ op } E_2 \\
| \ &\underline{n} \\
B ::= \ &E_1 \textbf{ bop } E_2 \\
| \ &\underline{b}
\end{aligned}
$$

Fig. 1. Abstract syntax of MINI-ADDL

separated by !. The operations, **op**, are $*, +$ and $-$, and the boolean operations, **bop**, are $=, \geq, \leq, <$ and $>$.

Parallel composition of commands in MINI-ADDL specify that the two program branches may execute in parallel. This means in practice that the two branches may not read or write to the same set of variables since there are no mechanisms for process synchronisation and communication.

1.2 A Motivating Example

To motivate this research we start by giving a small example of the sort of program transformations we would like to be able to support and the effect of such transformations. We have chosen to use a small FIR filter design as example:

```
ack := 0 ;
for i := 1 to (n+n) do
   ack := ack + (x(i) + x(m-i)) * h(i)
```

Where x is an input vector, h is a vector of coefficients and m is the length of the input vector. Note that m should be equal to 4*n and that the index variable, i, is a global variable whose value is affected by assignments inside the loop body. We wish to transform the specification into the following specification:

```
ack := 0 ;
for i := 1 to (n+n) do
   j := i ;
   ((ack1 := x(j) + x(m-j) ;
     ack1 := ack1 * h(j) ;
     ack := ack + ack1) ;
   !
   (i := i + 1 ;
     ack2 := x(i) + x(m-i) ;
     ack2 := ack2 * h(i))) ;
   ack := ack + ack2
```

This design is behaviourally equivalent with the initial but it is tuned in the sense that the result of synthesising this description is a significantly faster design. Below we will experimentally verify this assertion.

Evaluation A link to high-level synthesis already exists since ADDL is the input language to the CAMAD high level synthesis system. We use this by synthesising both the initial and the tuned design description and evaluate the resulting implementations regarding speed.

A high-level synthesis tool such as CAMAD performs mainly three tasks; resource allocation, resource binding and operation scheduling. It does so under the control of an optimisation heuristic that tries to optimise a cost function while satisfying some design constraints. Such a cost function typically consists of the execution time or an estimate of the area needed to implement the design. A typical design constraint can be maximum delay, maximum area or maximum number of I/O pins.

We have run the CAMAD synthesis tool in the default mode. This means using execution time as cost function, using a standard library of components for implementation and the default parametrisation of the optimisation heuristic. It is fully possible that by tampering with some of these parameters we might get a different result but we believe that running the initial and the tuned designs through CAMAD in default mode gives a fair estimate of the sort of delay reductions we can expect.

Synthesising the original and the tuned specification with n set to 4 gives the following properties for the resulting implementations:

specification	timing (ns)	cost ($10^3 * \lambda^2$)
original	9600	1241
tuned	6240	1872

Thus we have reduced the total execution time from 9600 to 6240 ns which is a reduction by 35 percent. Note that we at the same time have increased the cost by 51 percent. This is because we achieve shorter execution time by allowing operations to be done in parallel which requires more resources. Note that these figures are only estimates of the final timing and area cost. It is possible that lower level tools, such as logic synthesis, will do significantly better on one of the designs than on the other.

2 Embedding Mini-ADDL in HOL

In the HOL community the embedding of languages is often classified into two categories: *deep* and *shallow* embeddings. In this paper we use a hybrid model where we have a deep embedding of commands and a shallow embedding of expressions. The idea is that we should always use the simplest possible adequate embedding. Use the power of a deep embedding only when it is motivated and use the flexibility of a shallow embedding for the rest of the system.

2.1 Modelling State and Values

To represent state we use a naive approach where the state is represented by a function from (the representation of) variables to (the type of) values. The limitation of this approach is that we cannot do type checking. Variables are represented as strings.

Our language contains arrays so the value of an expression can be either a simple type or a composite array type, where simple types are numbers. Arrays are our most important extension relative to the language considered by Camilleri and Melham. We represent values as a recursive type:

$$value ::= \texttt{MK\_NUM}\ num$$
$$|\ \texttt{MK\_ARRAY}\ (value)list$$

This type involves recursion beneath the list type constructor so we cannot use the standard define_type tool. Instead we have used Gunter's nested_rec tool [6]. We have also defined predicates and destructors for numbers and arrays as well as functions for updating an array, assigning into an array, updating a state and incrementing a number.

2.2 Modelling Syntax

We define the syntax of commands in higher order logic as a logical type comm using the built-in HOL tools for defining concrete recursive datatypes. As a result we get eight constructors representing the different syntactic classes of commands. Assignment, sequential and parallel composition are represented by the infix constructors (:=), (;;) and (!!) respectively:

   skip, (:=), (;;), (!!), if, while, for and   local

For expressions we introduce a semantic constant to interpret each form of syntax. Expressions can be of two types, *value expressions* and *boolean expressions* but only value expressions can be stored in the state. Boolean expressions are only used to control data flow. There are two interpretations of value expressions: as R-values and as L-values. An R-value denotes the contents of a storage location. whereas an L-value denotes the storage location itself. An expression on the left hand side of an assignment is interpreted as an L-value and an expression on the right hand side of an assignment is interpreted as an R-value.

For L-values we have two constructors, one for variable locations and one for array locations, where each constructor is a function from a state to a (string,num list) pair: LVAR and LSUB.

For R-values we have a constructor for variables, arrays, operators in the language and for value numerical constants. Each constructor is a function from a state to a value: VAR, SUB, INC, ADD, SUBT, MUL and VAL.

For boolean expressions too we have a constructor for each boolean operator in the language and one for boolean constants. They are functions from a state to a boolean: EQL, GEQ, LEQ, GE, LE and BVAL.

Compositional Semantics We must define a mapping from the abstract syntax of the source language into the syntax model presented above. We do this by defining a compositional semantics that define translation rules for mapping statements in the source language into terms of the HOL logic. The compositional semantics can be mechanised by writing a parser and a pretty-printer for MINI-ADDL.

2.3 Operational Semantics

The semantics of MINI-ADDL is encoded in a compiler to the ETPN (Extended Timed Petri Net) design representation used by the CAMAD system. This is a Petri-net based graph language with a semantics based on event structures.

From the MINI-ADDL to ETPN mapping and the semantics of ETPN we have manually derived an operational semantics for MINI-ADDL.

The operational semantics of MINI-ADDL commands is represented as a relation, EVAL, such that EVAL C s_1 s_2 holds when executing the command C in the initial state s_1 terminates in the final state s_2. The semantics is defined in HOL using the tool for defining inductive definitions [2]. For example, the operational rules for parallel composition are:

$$\frac{\text{EVAL } C_1\ s_1\ s_2 \quad \text{EVAL } C_2\ s_2\ s_3}{\text{EVAL } (C_1\ !!\ C_2)\ s_1\ s_3} \qquad \frac{\text{EVAL } C_2\ s_1\ s_2 \quad \text{EVAL } C_1\ s_2\ s_3}{\text{EVAL } (C_1\ !!\ C_2)\ s_1\ s_3}$$

The parallel rules say that either C_1 takes place before C_2 or the other way around. Thus we have a coarse grained definition of parallelism. Note that this rule introduces non-determinism into the language. The ETPN semantics states that there can be no interference between parallel commands. To capture this requirement we have made an inductive definition of the well-formed commands saying that $C_1\ !!\ C_2$ is well-formed iff C_1 and C_2 are themselves well-formed and a set of restrictions on variable accesses and assignments hold that ensures non-interference of C_1 and C_2. All other commands are well-formed if their component commands are well-formed.

The semantics of expressions is represented by the semantic constants introduced in Section 2.2. For example, the semantic constants for R-values return a function from a state to a value:

$$\text{VAR } i\ s \equiv s\ i$$
$$\text{VAL } n\ s \equiv \text{MK\_NUM } n$$

2.4 Formalising Data-dependencies

When reasoning about MINI-ADDL programs it is often necessary to be able to state conditions on data-dependencies, e.g. to know if a particular variable is assigned to by a command. Such conditions typically appear as proof obligations in transformations. For us to be able to state such transformations as theorems in HOL it is necessary to define what we mean by variable assignment and variable access and formalise these properties in HOL.

We can state variable assignment as a function, ASS, from commands to a set of variables. Thus ASS(C) denotes the set of variables assigned to by the command C. For reasons explained later we have chosen not to define the function ASS in HOL. Instead we define a predicate "C ASSIGNS V" intended to mean that command C may assign to the variables in the set V in some run, and we define this predicate inductively using the inductive definitions package, e.g. the rule for parallel composition is:

$$\frac{C_1 \text{ ASSIGNS } V_{C_1} \quad C_2 \text{ ASSIGNS } V_{C_2}}{(C_1\ !!\ C_2) \text{ ASSIGNS } (V_{C_1} \cup V_{C_2})}$$

For variable accesses we use the corresponding function ACC(C) and relation ASSIGNS.

3 Transformational Design

In Section 2 we mechanised a definition of MINI-ADDL in HOL. Using this mechanised definition we can prove the equivalence of the original and the tuned program in Section 1.2. But such a proof would be both difficult and tedious since we would have to prove everything from first principles. In addition, we want to support experimental discovery of a tuned program, by incremental transformation from an initial design. Thus, we need an infrastructure for incremental transformational reasoning.

In such an infrastructure the language embedding is just one component. We also need a catalogue of MINI-ADDL source-to-source transformations and pre-proved theorems representing the transformations as well as tools and methods to support interactive transformation.

We formalise the idea of a source-to-source transformation as a HOL sequent of the following form:

$$\Gamma \vdash [\![\phi]\!] \equiv [\![\psi]\!]$$

which can be interpreted as "the two MINI-ADDL phrases ϕ and ψ have the same meaning, given the hypothesis Γ". Below is a collection of transformations on for loops:

$$\vdash [\![\text{for } i := (m+1) \text{ to } m \text{ do } C]\!] \equiv [\![\text{skip}]\!] \quad (1)$$

$$\neg(m_2 < m_1) \vdash \begin{array}{l} [\![\text{for } i := m_1 \text{ to } m_2 \text{ do } C]\!] \equiv \\ [\![i := m_1 \; ; \; C \; ; \; \text{for } i := (i+1) \text{ to } m_2 \text{ do } C]\!] \end{array} \quad (2)$$

$$\begin{array}{l} \neg(i \in \text{ASS}(C_1)), \\ \neg(i \in \text{ASS}(C_2)), \vdash \begin{array}{l} [\![\text{for } i := m_1 \text{ to } m_2 \text{ do } C_1]\!] \equiv \\ [\![\text{for } i := m_1 \text{ to } m_2 \text{ do } C_2]\!] \end{array} \\ (C_1 = C_2) \end{array} \quad (3)$$

The first transformation says that a for loop terminates if it is out of bounds, the second shows how to unroll a loop from the bottom and the last is a congruence transformation saying that we can substitute the for loop body command for an equal command without changing the meaning of the for loop.

Each transformation is a rewrite rule with which a designer can manipulate a program; the conditions in the assumptions are proof obligations that must be satisfied for the transformation to be valid.

3.1 The Example Revisited

In Section 1.2 we presented an example of the sort of tuning of design descriptions that we would like to be able to support in a mechanised tool. Here we will step through the transformation of the tuned description from the initial description shown below:

```
ack := 0 ;
for i := 1 to (n+n) do
    ack := ack + (x(i) + x(m-i)) * h(i)
```

To speed-up the design we must transform the specification into a more parallel one. We start by splitting the loop body so that two consecutive assignments of the accumulated result are performed in one loop iteration:

```
ack := 0 ;
for i := 1 to (n+n) do
    ack := ack + (x(i) + x(m-i)) * h(i) ;
    i := i + 1 ;
    ack := ack + (x(i) + x(m-i)) * h(i)
```

We perform this step by instantiating the following transformation:

$$\neg(i \in \text{ASS}(C)) \vdash \begin{array}{l} [\![\textbf{for } i := 1 \textbf{ to } (n+n) \textbf{ do } C]\!] \equiv \\ [\![\textbf{for } i := 1 \textbf{ to } (n+n) \textbf{ do } C\ ;\ i := (i+1)\ ;\ C]\!] \end{array}$$

The next step is to 'localise' the consecutive assignments of ack so that we can later parallelise them. By localise we mean to make each assignment independent of the other. Remember that two programs can only be parallel in ADDL if they do not read or write the same variables. For this reason we must introduce the local variables, ack1 and ack2 to hold the partial results of ack. Furthermore we cannot use i as address variable in both branches since they need to have different values. For this reason we introduce the local variable j to hold one of the address values:

```
ack := 0 ;
for i := 1 to (n+n) do
    local j
        j := i ;
        local ack2
            local ack1
                ack1 := (x(j) + x(m-j)) * h(j) ;
                ack := ack + ack1 ;
            i := i + 1 ;
            ack2 := (x(i) + x(m-i)) * h(i) ;
            ack := ack + ack2
```

This process consists of multiple steps. For example, we apply the following transformation once for every local variable we introduce:

$$\begin{array}{l} \neg(i \in \text{ASS}(C)), \\ \neg(i \in \text{ACC}(C)) \end{array} \vdash [\![C]\!] \equiv [\![\textbf{local } i\ C]\!]$$

For the other steps we currently have no suitable transformation but we have to prove these steps using interactive theorem proving.

Having sorted this out it is time to parallelise the two assignments:

```
ack := 0 ;
for i := 1 to (n+n) do
  local j
    j := i ;
    local ack2
      (local ack1
        ack1 := (x(j) + x(m-j)) * h(j) ;
        ack := ack + ack1
      !
      (i := i + 1 ;
        ack2 := (x(i) + x(m-i)) * h(i))) ;
    ack := ack + ack2
```

Once again we can perform this step by instantiating an existing transformation:

$$\begin{array}{l} \mathrm{ASS}(C_1) \bigcap \mathrm{ACC}(C_2) = \\ \mathrm{ASS}(C_2) \bigcap \mathrm{ACC}(C_1) = \vdash [\![\, C_1 \,;\, C_2 \,]\!] \equiv [\![\, C_1 \,!\, C_2 \,]\!] \\ \mathrm{ASS}(C_1) \bigcap \mathrm{ASS}(C_2) = \phi \end{array}$$

We can still improve the final design since CAMAD is not powerful enough to always find the best match between two parallel paths. In this case the following translation helps CAMAD to produce a better circuit:

```
ack := 0 ;
for i := 1 to (n+n) do
  local j
    j := i ;
    local ack2
      (local ack1
        ack1 := (x(j) + x(m-j)) ;
        ack1 := ack1 * h(j) ;
      ack := ack + ack1 ;
      !
      (i := i + 1 ;
        ack2 := (x(i) + x(m-i)) ;
        ack2 := ack2 * h(i))) ;
    ack := ack + ack2
```

Also this final step can be performed by instantiating an existing transformation:

$$i \notin \mathrm{EACC}(E_2) \vdash [\![\, i := E_1 * E_2 \,]\!] \equiv [\![\, i := E_1; i := i * E_2 \,]\!]$$

Where $\mathrm{EACC}(E)$ denotes the set of variables accessed by the expression E.

3.2 Transformations as Theorems

Since we have a deep embedding of ADDL we can represent transformations directly as theorems in the HOL logic. This is in contrast with previous work (See Section 5) where sequents are mechanised in two stages. First, a theorem about the HOL semantic constants represents the essence of each transformation. Second, a conversion function represents each sequent in the sense that given a left-hand side, it proves the whole sequent by instantiating the general theorem.

The result from developing the proof infrastructure is a set of theorems about program equivalences. These can then be used for transformational reasoning about MINI-ADDL programs. We have proved such theorems about each of the MINI-ADDL constructs. For example, transformation (3) is represented by the following HOL theorem:

$$
\vdash (C_1 \text{ ASSIGNS } V_{C_1}) \land \neg(i \text{ IN } V_{C_1}) \land \qquad (4)
$$
$$
(C_2 \text{ ASSIGNS } V_{C_2}) \land \neg(i \text{ IN } V_{C_2}) \implies
$$
$$
(\text{EVAL } C_1 = \text{EVAL } C_2 \implies
$$
$$
(\text{EVAL (for } i \text{ (VAL 1) (VAL } m) \text{ } C_1) =
$$
$$
\text{EVAL (for } i \text{ (VAL 1) (VAL } m) \text{ } C_2)))
$$

4 Support for Interactive Transformation

Applying transformations to programs is modelled by rewriting the logical theorem that represent programs with the equational theorems that represent transformations. In the simplest case we can use the built-in rewriting tools, but in many cases this is not possible either because there are proof obligations associated with the transformation that do not follow immediately from the assumptions of the program theorem or because the command we want to transform is a sub-command of the program. We discuss each of these cases below but first we give an example of a program and explain our terminology. Assume we have the following program:

$$
\Gamma \vdash \text{EVAL (for } i \text{ (VAL 1) (VAL 3)} \qquad (5)
$$
$$
(\text{LVAR } "x" \text{ LSUB VAR } "i" := \text{VAR } "i"))
$$

Then Γ is the set of assumptions, for the top-level command and the array assignment LVAR $"x"$ LSUB VAR $"i"$:= VAR $"i"$ a sub-command of the program.

4.1 Applying Transformations to Top-level Commands

When applying transformations to top-level commands the potential difficulties are twofold. First, we need to instantiate our transformation so that it matches the top-level command; Second, we must handle possible proof obligations before we can apply the transformation. We discuss each of these separately below but first we introduce an example. Given the program above, assume we want to unroll, from the top, a single loop iteration. The transformation for doing this is represented by the following theorem:

$$
\vdash (C \text{ ASSIGNS } V_C) \land \neg(i \text{ IN } V_C) \implies \qquad (6)
$$
$$
(\text{EVAL (for } i \text{ (VAL 1) (VAL } (m+1)) \text{ } C) =
$$
$$
\text{EVAL (for } i \text{ (VAL 1) (VAL } m) \text{ } C \text{ ;;}
$$
$$
(\text{LVAR } i := \text{VAL } (m+1) \text{ ;; } C)))
$$

Instantiation of Transformations In this case instantiation is not trivial for two reasons. First, we have an expression, $m+1$, as the upper bound on the for loop index variable in the transformation; and Second, we do not know how to instantiate V_C.

To instantiate m we need to solve the equation $m+1 = 3$ and we do this with some arithmetic reasoning using the built-in tools of HOL. To be able to instantiate V_C we need to calculate the set of assigned variables and for this we have an ML function. Finally, we have an ML function, inst_for_UNROLL_TOP_COROLLARY, that encapsulate these two calculations, and, applied to Theorem (6), returns the following instantiated theorem:

$\vdash$ ((LVAR "x" LSUB VAR "i" := VAR "i") ASSIGNS $\{$"x"$\}) \wedge$
$\neg(i$ IN $\{$"x"$\}) \implies$
(EVAL (for i (VAL 1) (VAL 3))
(LVAR "x" LSUB VAR "i" := VAR "i")) =
EVAL (for i (VAL 1) (VAL 2)
(LVAR "x" LSUB VAR "i" := VAR "i") ;;
(LVAR i := VAL 3) ;;
(LVAR "x" LSUB VAR "i" := VAR "i")))

But still we cannot apply this theorem to our program since we do not know if the proof obligations are satisfied or not.

Handling Proof Obligations It is important that the process of proving the proof obligations is automated to the extent possible. This is for two reasons. First, a designer would like to focus on the task of transforming the design without getting distracted by possibly intricate questions of showing that proof obligations are satisfied; and Second, we want to hide as much theorem proving details as possible from the application engineer.

We have chosen to prove proof obligations before applying a transformation. For this reason we implement a proof tool for each type of proof obligation in our catalogue of transformations. An example is the ML function, prove_ASSIGNS, that proves proof obligations of the type $i \notin \text{ASS}(C)$. It uses repeated application of IN_CONV from the set library, with string_EQ_CONV as equality predicate, and a bespoke tactic, ass_TAC, where ass_TAC consecutively applies a bespoke tactic for each type of command in MINI-ADDL until one succeeds or fails if none of them do. Applying prove_ASSIGNS to the instantiated theorem above results in the following theorem:

$\vdash$ EVAL (for i (VAL 1) (VAL 3))
(LVAR "x" LSUB VAR "i" := VAR "i")) =
EVAL (for i (VAL 1) (VAL 2)
(LVAR "x" LSUB VAR "i" := VAR "i") ;;
(LVAR i := VAL 3) ;;
(LVAR "x" LSUB VAR "i" := VAR "i"))

This theorem can now be used to rewrite the design in Theorem 5.

Proof Tools To make transformational reasoning more convenient for the application engineer we design a bespoke proof tool for each transformation in the catalogue of transformations in the form of a conversion that encapsulate the instantiation and proof obligation requirements for a particular transformation. For Theorem (6) the conversion looks like this:

```
fun for_unroll_top_conv tm =
    prove_ASSIGNS(inst_for_UNROLL_TOP_COROLLARY tm);
```

From this we derive a top-level conversion, for_UNROLL_TOP_CONV, that applies for_unroll_top_conv to all subterms, an inference rule, UNROLL_TOP_RULE, and a tactic, for_UNROLL_TOP_TAC.

4.2 Applying Transformations to Sub-commands

The problem associated with the application of transformations to sub-commands is best explained by an example. Assume we have the following composite command:

$$\Gamma \vdash \text{EVAL (LVAR } "y" := \text{VAL 1 ;;} \qquad (7)$$
$$\text{for } i \text{ (VAL 1) (VAL 3)}$$
$$\text{(LVAR } "x" \text{ LSUB VAR } "i" := \text{VAR } "y"))$$

Then Γ is the set of assumptions, sequential composition the top-level command and there are three subcommands — the assignment of $"y"$, the for loop and the assignment of $"x"$ in the for loop body.

Assume now that we would like to unroll the for loop once from the top as in the previous section. The problem is that we cannot do that since the for loop is not the top-level command. Thus we need a method to "open up" the program structure so that we get access to subcommands. One method is to use congruence theorems.

The MINI-ADDL language is congruent. By this we mean that we can replace equals with equals anywhere in a MINI-ADDL program without changing the meaning of the program. To prove a language congruent we would prove congruence theorem(s) for each construct of the language. An example of a congruence theorem is Theorem (4).

Proof Tools We have identified two proof tools that we believe are useful to support the application of transformations to sub-commands. The first is a tool that applies the transformation to the first suitable sub-command(s) encountered in top-down order. We have written an ML program, apply_trans_conv, based on congruence theorems that implement this functionality and we have derived a top-level conversion, APPLY_TRANS_CONV, that applies apply_trans_conv to all subterms, an inference rule, APPLY_TRANS_RULE, and a tactic, APPLY_TRANS_TAC. Using these tools we can transform program (7) in the desired way by applying APPLY_TRANS_RULE parametrised with for_unroll_top_conv to it. This results in the following theorem:

$$\Gamma \vdash \text{EVAL } (\text{LVAR } "y" := \text{VAL } 1 \text{ ;;} \tag{8}$$
$$\text{for } i \text{ (VAL 1) (VAL 2)}$$
$$(\text{LVAR } "x" \text{ LSUB VAR } "i" := \text{VAR } "y") \text{ ;;}$$
$$\text{LVAR } i := \text{VAL } 3 \text{ ;;}$$
$$\text{LVAR } "x" \text{ LSUB VAR } "i" := \text{VAR } "y")$$

The reason we do not use for_UNROLL_TOP_CONV as parameter to APPLY_TRANS_RULE is that apply_trans_conv depends on that the supplied conversion fails if it cannot be applied to a term which for_UNROLL_TOP_CONV does not do since it is embedded in DEPTH_CONV.

The other tool which we believe is useful is a tool that applies the transformation to the first sub-command(s) encountered in a top-down order that match a given pattern. Thus we transform specific instead of all suitable sub-commands. This program, match_trans_conv, closely resembles the previous except for having an additional pattern parameter against which each sub-command is matched to see if the supplied conversion should be applied. The matching is done by the built-in match_term function that when applied to two terms attempts to find a set of type and term instantiations for the first term (only) to make it alpha-convertible to the second. Thus, by using variables in the pattern we can make the tool match more general sub-commands.

As before we derive a top-level conversion, MATCH_TRANS_CONV, that applies match_trans_conv to all subterms, an inference rule, MATCH_TRANS_RULE, and a tactic, MATCH_TRANS_TAC. With these tools we can transform the program represented by Theorem 7 in the desired way by applying the following command to it:

```
MATCH_TRANS_RULE for_unroll_top_conv
    (--'for i (VAL m) (VAL n) (LVAR a LSUB b := VAR c)'--)
```

This results in the following theorem that represents the transformed program:

$$\Gamma \vdash \text{EVAL } (\text{LVAR } "y" := \text{VAL } 1 \text{ ;;} \tag{9}$$
$$\text{for } i \text{ (VAL 1) (VAL 2)}$$
$$(\text{LVAR } "x" \text{ LSUB VAR } "i" := \text{VAR } "y") \text{ ;;}$$
$$\text{LVAR } i := \text{VAL } 3 \text{ ;;}$$
$$\text{LVAR } "x" \text{ LSUB VAR } "i" := \text{VAR } "y")$$

4.3 The Example Once More

In the previous sections of this chapter we have derived a rudimentary infrastructure for transformational reasoning about MINI-ADDL programs and in this section we demonstrate, in some detail, how selected parts of the derivation of the tuned program is done using this. The purpose is to give the reader an idea of how the tuning process is carried out using our set of tools. We assume that the initial program given in Section 1.2 is represented by the following theorem:

$$\Gamma \vdash \text{EVAL (LVAR } "ack" := \text{VAL } 0 \;;;$$
$$\text{for } "i" \text{ (VAL 1) (VAL } (n+n))$$
$$(\text{LVAR } "ack" := \text{VAR } "ack" \text{ ADD}$$
$$((\text{VAR } "x" \text{ SUB VAR } "i") \text{ ADD}$$
$$(\text{VAR } "x" \text{ SUB (VAL } m \text{ SUBT VAR } "i")))$$
$$\text{MUL (VAR } "h" \text{ SUB VAR } "i")) \tag{10}$$

As we saw in Section 3.1 the initial step is to split the loop body to increase the potential for parallelism. To justify this step we use the following theorem that we encapsulate in the function for_split_conv in analogy with the encapsulation of Theorem (6) by for_unroll_top_conv in Section 4.1:

$$\vdash (C \text{ ASSIGNS } V_C) \wedge \neg(i \text{ IN } V_C) \implies$$
$$(\text{EVAL (for } i \text{ (VAL 1) (VAL } (n+n)) \; C) =$$
$$\text{EVAL (for } i \text{ (VAL 1) (VAL } (n+n))$$
$$(C \;;; \text{LVAR } i := \text{VAR } i+1 \;;; C)))$$

To apply this theorem to the body of the for loop in the theorem representing the initial program we assert the following command:

`APPLY_TRANS_RULE for_split_conv`

The result of applying the above command to Theorem (10) is the following theorem that represents the transformed program:

$$\Gamma \vdash \text{EVAL (LVAR } "ack" := \text{VAL } 0 \;;;$$
$$\text{for } "i" \text{ (VAL 1) (VAL } (n+n)) \tag{11}$$
$$(\text{LVAR } "ack" := \text{VAR } "ack" \text{ ADD}$$
$$((\text{VAR } "x" \text{ SUB VAR } "i") \text{ ADD}$$
$$(\text{VAR } "x" \text{ SUB (VAL } m \text{ SUBT VAR } "i")))$$
$$\text{MUL (VAR } "h" \text{ SUB VAR } "i") \;;;$$
$$\text{LVAR } "i" := \text{VAR } "i" \text{ ADD VAL } 1 \;;;$$
$$\text{LVAR } "ack" := \text{VAR } "ack" \text{ ADD}$$
$$((\text{VAR } "x" \text{ SUB VAR } "i") \text{ ADD}$$
$$(\text{VAR } "x" \text{ SUB (VAL } m \text{ SUBT VAR } "i")))$$
$$\text{MUL (VAR } "h" \text{ SUB VAR } "i"))$$

The next step is to introduce local variables. We show here the introduction of the variable "j". The justification for this is the following theorem that we encapsulate in the function local_intro_conv:

$$\vdash (C \text{ ACCESSES } V_{ACC}) \wedge \neg(i \text{ IN } V_{ACC}) \wedge$$
$$(C \text{ ASSIGNS } V_{ASS}) \wedge \neg(i \text{ IN } V_{ASS}) \implies$$
$$(\text{EVAL } C = \text{EVAL } (\textbf{local } i \; C))$$

We apply it to theorem 11 with the following command:

`MATCH_TRANS_RULE local_intro_conv`
   `(--'(LVAR "ack" :=` E_1 `;; LVAR "i" :=` E_2 `;; LVAR "ack" :=` E_3`)'--)`

Note the use of variables, E_1, E_2 and E_3, in the pattern. The resulting theorem representing the new current program looks as follows:

$\Gamma \vdash$ EVAL (LVAR "ack" := VAL 0 ;;
    for "i" (VAL 1) (VAL $(n+n)$)
        (**local** " j"
            (LVAR "ack" := VAR "ack" ADD
                ((VAR "x" SUB VAR "i") ADD
                    (VAR "x" SUB (VAL m SUBT VAR "i")))
                MUL (VAR "h" SUB VAR "i") ;;
            LVAR "i" := VAR "i" ADD VAL 1 ;;
            LVAR "ack" := VAR "ack" ADD
                ((VAR "x" SUB VAR "i") ADD
                    (VAR "x" SUB (VAL m SUBT VAR "i")))
                MUL (VAR "h" SUB VAR "i")))

This process continues in the same fashion by applying theorems encapsulated into conversions that handle instantiation and proof obligations using either APPLY_TRANS_RULE or MATCH_TRANS_RULE until the theorem representing the final program in Section 3.1 is the current theorem.

5 Related work

McFarland and Parker [8] report the seminal work on proving hardware description language transformations correct in terms of behavioural semantics. They prove algorithmic transformations correct manually whereas we prove manually applied transformations correct mechanically.

In the HOL-SILAGE project the goal was to obtain an interactive system that under the guidance of the designer can prove that an initial and a tuned program have the same behaviour. The idea was to embed SILAGE programs as HOL terms, and represent transformational design as proof of equations in HOL. The main effort in this project was on the formal definition of SILAGE [5]. A digital filter example is given and a naive and a tuned implementation of it are proved equal. This example shows the verification of a given transformation but experimental discovery of a tuned program, by incremental transformation from an initial program is not supported.

Gordon have also worked on ELLA transformations [4] in the framework of the HOL-ELLA project [1]. Here Gordon built a prototypical infrastructure for transformational design using the HOL-ELLA system and a small example was proved. A set of non-trivial transformations was implemented by a combination of theorems and conversionals but no decision procedures for proof obligations was designed.

A major difference from our work is that both SILAGE and ELLA are functional languages and thus naturally have a denotational style semantics. In addition, both the HOL-SILAGE and the HOL-ELLA projects use a shallow embedding style. This means that transformations of the type we have given in this chapter cannot be represented directly as theorems in these systems.

There are mechanical transformation systems not based on a general-purpose theorem-prover, such as the DDD system [7]. The advantage of using an LCF-style theorem-prover such as HOL is that theorems are an abstract type distinguished from sentences of the logic, and so programming errors are unlikely to lead to a proof of a false transformation. Furthermore, a system based on a theorem-prover need not be based on a fixed set of transformations: any logically provable transformation can be made available by resorting to interactive theorem-proving.

6 Conclusions and Future work

In order to do the most effective transformations one really needs to analyse the results from CAMAD carefully. Some transformations, like the final one in our example, can have a surprisingly big effect. In order for our approach to work smoothly the operation of the synthesis tool must be transparent to the designer transforming the input specification.

Future extensions include adding simple (batch) I/O capabilities, a parser and pretty printer to mechanise the compositional semantics and deriving Hoare-style proof rules for our language. The latter would make it possible to prove an initial program correct according to a specification before applying optimising transformations to it.

This work has been done in cooperation with Andy Gordon and I gratefully acknowledge his contribution. I also thank the anonymous referees for their constructive comments.

References

1. R. Boulton. A HOL Semantics for a Subset of ELLA. Technical Report 254, University of Cambridge, Computer Laboratory, New Museums Site, Pembroke Street, Cambridge CB2 3QG, England, Apr. 1992.
2. J. Camilleri and T.Melham. Reasoning with Inductively Defined Relations in the HOL Theorem Prover. Technical Report 265, University of Cambridge, Computer Laboratory, Aug. 1992.
3. *Proceedings of the 1991 International Workshop on Formal Methods in VLSI Design*, Miami, Jan. 1991.
4. A. D. Gordon. Transformational Design using a Proof Assistant. Oct. 1992.
5. A. D. Gordon. A Mechanised Definition of Silage in HOL. Technical Report 287, University of Cambridge, Computer Laboratory, Feb. 1993. ESPRIT BRA 3215.
6. E. Gunter. The nested_rec contrib library. HOL 90.7 Documentation, 1994. Available with the HOL 90.7 release.
7. S. D. Johnson and B. Bose. DDD — A System for Mechanized Digital System Design Derivation. In Formal Methods in VLSI Design [3].
8. M. C. McFarland and A. C. Parker. An Abstract Model of Behavior for Hardware descriptions. *IEEE Transactions on Computers*, C-32(7):621–637, July 1983.
9. Z. Peng and K. Kuchcinski. Automated Transformation of Algorithms into Register-Transfer Level Implementations. *IEEE Transactions on Computer-Aided Design of Integrated Circuits and Systems*, 13(2):150–166, Feb. 1994.

Using Lattice Theory in Higher Order Logic

Linas Laibinis

Turku Center for Computer Science,
Lemminkäisenkatu 14, FIN-20520 Turku, Finland
llaibini@ra.abo.fi

Abstract. We describe an implementation of general (abstract) lattice theory in the HOL system and its use in transformational reasoning within concrete instances of lattices, using the window inference of HOL. The implementation is extensible; users can add new instances of lattices and all the existing transformation rules are then available for the added structures. As a particularly promising application we briefly describe how our system can be used as part of a tool for transformational reasoning about programs (program refinement).

1 Introduction

The HOL system (Higher Order Logic) [4] is a theorem prover which can be used to formalise theories and verify proofs of theorems within these theories. It is based on the LCF approach [5] with a secure datatype of theorems and a programmable user interface. The HOL system was first mainly used for formal specification and verification of hardware, but recently there has been a growing interest in other applications, e.g., the formalisation of mathematical theories and programming calculi.

Working with formalised concepts of program refinement [3, 14], we have encountered structures that are instances of lattices on different abstraction levels of the theory. In this paper we show how it is possible to create a single abstract theory of lattices in HOL, which can then be instantiated in different ways and used efficiently when reasoning within these structures. Thus we prove properties of lattices once and for all in the abstract theory. When we have an instance of a lattice (i.e., when we have shown that a certain structure satisfies the defining properties of lattices), then theorems are easily shown to hold for this instance also. The basic principles of abstract theories are taken from Gunter's work on abstract group theory in HOL [7].

We show how lattice properties can be used in derivations using window inference style of reasoning [6]. We extend window inference with new transformation and window rules for working with lattices. Transformation rules are special inference rules working with the current window stack. Transformation rules for lattices allow us to introduce and eliminate lattice constructs in the style of natural deduction. Window rules allow us to focus on some subcomponent and do local transformations using context monotonicity properties.

Furthermore, we show how our transformation rules and other infrastructure for transformational reasoning work together with a tool for program refinement that has previously been implemented [9]. This tool is built on top of HOL and the HOL window Library, with a HOL theory of program refinement and an X Window System interface for doing transformations in the style of window inference.

2 Formalising Lattice Theory in HOL

We start by considering what lattices are and how they can be formalized in HOL. The basic ideas of formalising lattice theory in HOL are described in more detail by J. von Wright[13].

2.1 Doing Algebra in HOL

Theories of algebra generally assume an underlying set of anonymous elements and operators that work on these elements. Examples of algebraic theories are the theories of groups and lattices. Group theory in HOL is described by Gunter [7], with the underlying set represented by its characteristic function and n-ary operators represented by n-ary (curried) functions. The same approach can be used for other theories of algebra as well.

Consider the theory of posets (partially ordered sets) as an example. Assume that $(A, \sqsubseteq)$ is a poset, where the elements of A belong to an unspecified type α. The partial order $\sqsubseteq$ is formalised as a function $po : \alpha \to \alpha \to bool$ with the following interpretation: if x and y are elements of A then $po\ x\ y$ holds if and only if $x \sqsubseteq y$. In HOL, we define the predicate POSET so that POSET(A,po) holds if and only if po satisfies the properties of reflexivity, antisymmetry and transitivity on A. In HOL the definitional theorem is as follows:

```
POSET_DEF =
 ⊢ POSET(A,po) =
    (∀x. A x ⇒ po x x) ∧
    (∀x y. A x ∧ A y ⇒ (po x y ∧ po y x ⇒ (x=y))) ∧
    (∀x y z. A x ∧ A y ∧ A z ⇒ (po x y ∧ po y z ⇒ po x z))
```

Theorems proved in the theory of posets will generally have an assumption stating that the set under consideration is a poset.

From now on we will consistently work with structures where the underlying set is a whole type, i.e. *universal set* $U = (\lambda x : \alpha.\ T)$. Everything we do could also be done with subsets of U, but this would add membership conditions to almost all definitions and theorems which would make them (even) harder to read.

2.2 What is a Lattice?

We shall now show how basic definitions and properties of lattices are formalised in HOL. We first recall some basic concepts of lattices. Then we show how to define lattices, meets, joins, etc. in HOL.

We assume that the reader is familiar with the concepts of partial orders, meets (greatest lower bounds) and joins (least upper bounds).

A partially ordered set $(A, \sqsubseteq)$ is a *lattice* if the meet (greatest lower bound) $x \sqcap y$ and the join (least upper bound) $x \sqcup y$ exist for arbitrary elements x and y in A. If every subset B of the set A has a meet $\sqcap B$ and a join $\sqcup B$ in A we say that $(A, \sqsubseteq)$ is a *complete lattice*. The least (bottom) element of a complete lattice is denoted $\bot$ and the greatest (top) element is denoted $\top$. If $(A, \sqsubseteq)$ is a complete lattice where the following conditions hold for arbitrary $x \in A$ and $B \subseteq A$:

$$x \sqcap (\sqcup B) = \sqcup \{x \sqcap y | y \in B\}$$
$$x \sqcup (\sqcap B) = \sqcap \{x \sqcup y | y \in B\}$$

we say that $(A, \sqsubseteq)$ is *infinitely distributive*. Finally, a *(complete) boolean lattice* is an infinitely distributive lattice where every element x has an unique inverse x^{-1} satisfying

$$x \sqcap x^{-1} = \bot \text{ and } x \sqcup x^{-1} = \top$$

Sometimes it is more convenient to use indexed notation for general meets and joins. We then write general meet as $(\sqcap v \in I.\ s)$ (where I is index set and s is an expression with variable v free) and general join as $(\sqcup v \in I.\ s)$.

2.3 Definitions

We define lattices in the same way as we defined posets above; the defining theorem states that in order to be a lattice, a set must be a poset and every pair of elements must have a greatest lower bound (meet) and a least upper bound (join):

```
LAT_DEF =
 ⊢ LAT(U,po) =
       POSET(U,po) ∧
       (∀a b. (∃m. po m a ∧ po m b ∧
                  (∀m'. po m' a ∧ po m' b ⇒ po m' m)) ∧
              (∃j. po a j ∧ po b j ∧
                  (∀j'. po a j' ∧ po b j' ⇒ po j j')))
```

The (binary) meet and join operators can now be defined using Hilbert's choice operator ε. The defining theorem for binary meet is as follows:

```
meet2_DEF =
 ⊢ meet2(U,po) a b =
       (εm. po m a ∧ po m b ∧
           (∀m'. po m' a ∧ po m' b ⇒ po m' m))
```

and binary join, join2, is defined dually. Note that the first argument of the constants POSET, LAT and meet2 is a pair (U,po) which indicates what lattice we are considering. Note also that universal set U in the definitions is redundant in the sense that it can be easily inferred from the type of the ordering po. We left it in the paper only for clarity reasons.

2.4 Basic Properties

The definitions of meet2 and join2 are hard to work with in practice, due to the occurrences of the choice operator ε. We want to reason about lattices in the ordinary mathematical way, relying on the characteristic properties of meets and joins: idempotency, commutativity, associativity and absorption, as well as on the identities relating meets and joins to the partial order. Once we have proved these properties, we can reason about lattices much as we do in ordinary mathematics.

The proofs of the characteristic properties of the meet and join operators are quite straightforward. We show as example the theorem about commutativity of a binary meet (other properties are defined similarly):

```
meet2_comm =
  LAT(U,po) ⊢ meet2(U,po) a b = meet2(U,po) b a
```

2.5 Complete and Boolean Lattices

We define complete lattices, infinitely distributive lattices and boolean lattices using the same principles as above. We show only definitions of complete lattices, arbitrary meets and bottom elements (arbitrary joins and top elements are defined dually):

```
CLAT_DEF =
  ⊢ CLAT(U,po) =
      LAT(U,po) ∧
      (∀B. (∃m. (∀x. B x ⇒ po m x) ∧
              (∀m'. (∀x. B x ⇒ po m' x) ⇒ po m' m)) ∧
          (∃j. (∀x. B x ⇒ po x j) ∧
              (∀j'. (∀x. B x ⇒ po x j') ⇒ po j j')))

meet_DEF =
  ⊢ meet(U,po) B =
      εm. (∀x. B x ⇒ po m x) ∧
         (∀m'. (∀x. B x ⇒ po m' x) ⇒ po m' m)

bot_DEF =
  ⊢ bot(U,po) = εb. ∀x. po b x
```

2.6 Fixpoints

Monotonicity of functions over complete lattice guarantees existence of (least and greatest) *fixpoints* which can be very useful for defining recursion and iteration. (Fixpoints of a function $f : \alpha \to \alpha$ are solutions of the equation $f\, x = x$).

The Knaster-Tarski theorem gives the following explicit constructions of the least(μf) and greatest(νf) fixpoints:

$$\mu f = (\sqcap x \in A \mid f\, x \sqsubseteq x.\, x)$$
$$\nu f = (\sqcup x \in A \mid x \sqsubseteq f\, x.\, x)$$

where $(A, \sqsubseteq)$ is a complete lattice and f is a monotonic function on A.
At first we define monotonicity property of functions over lattice:

```
Lmono_DEF =
 ⊢ Lmono(U,po) f =
     (∀x y. po x y ⇒ po (f x) (f y))
```

Then we define a least fixpoint using the characterisation given above (the definition of a greatest fixpoint is dual):

```
lfix_DEF =
 ⊢ lfix(U,po) f =
     meet (U,po) (λx. po (f x) x)
```

Using this definition we can prove basic properties of fixpoints. Each such theorem will have assumptions stating that the type under consideration is a complete lattice and the function is monotonic on this lattice.

2.7 Inference Rules for Lattices

Abstract lattices introduce special constructs to operate with - binary meets and joins, general meets and joins, tops and bottoms. It is convenient to work with them using the style of natural deduction, with special inference rules for introduction and elimination of different lattice constructs. Such rules are useful when the aim is to prove a theorem of the form $\vdash t \sqsubseteq t'$ by stepwise transformational reasoning, i.e., by proving first $\vdash t \sqsubseteq t_1$, then $\vdash t_1 \sqsubseteq t_2$ etc. up to $\vdash t_n \sqsubseteq t'$.

Our inference rules for introducing and eliminating lattice constructs differ slightly from the traditional rules for logical connectives. This is because we want to express the rules as properties of the lattice ordering. The names 'introduction' and 'elimination' here refer to the fact that the specific construct is introduced (or eliminated) when we move from the left to the right hand side of the ordering. The advantages of this aproach will become clear in Section 4 where we consider reasoning about lattices using the window Library of the HOL system.

Below we present inference rules for basic lattice constructs. For binary meet we have the following rules:

$$\frac{\Phi \vdash s \sqsubseteq t \quad \Phi' \vdash s \sqsubseteq t'}{\Phi \cup \Phi' \vdash s \sqsubseteq t \sqcap t'} \qquad (binary \sqcap introduction)$$

$$\vdash t \sqcap t' \sqsubseteq t \qquad \vdash t \sqcap t' \sqsubseteq t' \qquad (binary \sqcap elimination)$$

For general meet the rules are as follows:

$$\frac{\Phi, v \in I \vdash s \sqsubseteq t}{\Phi \vdash s \sqsubseteq (\sqcap v \in I. \, t)} \qquad (general \sqcap introduction)$$

- v not free in s, I or Φ

$$t' \in I \vdash (\sqcap v \in I.t) \sqsubseteq t[v := t'] \qquad (general \sqcap elimination)$$

- t' is free for v in t

The rules for joins are dual.

For least fixpoint we have the elimination rule (the rule for greatest fixpoint introduction is dual):

$$\frac{f\,x \sqsubseteq x}{\mu f \sqsubseteq x} \qquad (\mu\ elimination)$$

For bottom and top the rules are as follows:

$$\Phi \vdash \bot \sqsubseteq t \qquad (bottom\ elimination)$$

$$\Phi \vdash t \sqsubseteq \top \qquad (top\ introduction)$$

In addition to these rules, there are inference rules expressing monotonicity properties of lattice constructs. For example, the rule stating that binary meet is monotonic in its left argument is:

$$\frac{\Phi \vdash t \sqsubseteq t'}{\Phi \vdash t \sqcap s \sqsubseteq t' \sqcap s} \qquad (left\ monotonicity\ of\ binary\ \sqcap)$$

Similar rules exist for the right argument of the binary meet as well as for both arguments of the binary join. For general meets ($\sqcap v \in I.\ s$), we have monotonicity in the body s and antimonotonicity in the index set I of the argument (as a set, I belongs to a powerset lattice ordered by set inclusion). For general joins ($\sqcup v \in I.\ s$), we have monotonicity in both the body s and the index set I of the the argument.

All these rules are implemented in HOL as ML functions taking as arguments a term (representing left hand side of ordering in conclusion expression) and one or several theorems (hypotheses of the inference rule) and returning the theorem in the conclusion of the rule. In Section 4 we show how the rules are used in transformational reasoning.

3 Using Lattice Properties in Various Domains

We now consider concrete examples of lattices encountered in various domains and show how abstract lattices (as they were formalised in the preceding section) can be instantiated.

3.1 Concrete Instances of Lattices

We can encounter concrete lattices in various contexts. We start with the most commonplace domain in classical logic - truth values.

The truth values with implication as the ordering form a complete, boolean and totally ordered lattice. Other logical connectives can be treated as lattice operations as well (conjuction is binary meet, disjunction is binary join and negation is the complement operation). The constants F and T are bottom and top elements of the lattice respectively. The *bounded* universal quantification

($\forall v \in I.\ t$) stands for general meet, and *bounded* universal quantification ($\exists v \in I.\ t$) for general join.

Thus the truth values form very specific and restricted lattice structure with many useful properties. There is a direct connection between inference rules for lattices and inference rules for introducing and eliminating logical connectives.

Pointwise extension is a general method by which operations on a type α can be lifted to operations on functions from some type β to α. If we introduce an ordering in this new type $\beta \to \alpha$ by pointwise extension, i.e.,

$$f \sqsubseteq_{\beta \to \alpha} g\ =\ (\forall x : \beta.\ f\,x \sqsubseteq_\alpha g\,x)$$

then new type $\beta \to \alpha$ inherits property of being lattice, as well as completeness, distributivity, and being a boolean lattice. The lattice operations in $\beta \to \alpha$ are defined in terms of the corresponding operations on α by pointwise extension.

We know that *bool* is a complete boolean lattice. Using pointwise extension we get that $\alpha \to bool$ (predicates or subsets of α) is a complete boolean lattice, for arbitrary type α. The ordering in this lattice is defined by pointwise extension:

$$p\ implies\ q\ =\ (\forall x : \alpha.\ p\,x\ \Rightarrow\ q\,x).$$

Other lattice operations are also lifted to the new lattice by pointwise extension. For example, binary meet in the predicate lattice is defined as

$$p\ and\ q\ =\ (\lambda x : \alpha.\ p\,x\ \wedge\ q\,x).$$

Doing one more step, we get that relations (as functions $\beta \to \alpha \to bool$) form a complete boolean lattice as well.

3.2 Refinement Calculus Theory

Predicates are only one of many interesting lattice structures, others include imperative programs ordered by program refinement. In particular, we will consider lattice structures found in the Refinement Calculus theory [1, 2]. The Refinement Calculus is a calculus for development of programs that are (totally) correct by construction, using the stepwise refinement paradigm.

The focus of this theory is imperative state based programs. Structures considered include state predicates, state relations, state functions and predicate transformers. Predicate transformers are used to model *program statements*, modelling them as functions that map postconditions to preconditions.

The state is modelled as a type α (which for concrete programs can be instantiated in different ways). Above, we showed that state predicates (functions $\alpha \to bool$) and state relations (functions $\beta \to \alpha \to bool$) are complete boolean lattices. Using pointwise extension once more for state predicates we get the type $(\beta \to bool) \to (\alpha \to bool)$ (predicate transformers) which is also a complete boolean lattice. The ordering on this lattice is called *refinement ordering* and is defined as follows:

$$S\ ref\ T\ =\ (\forall q : \beta \to bool.\ S\,q\ implies\ T\,q).$$

Note that we have constructed a pointwise extension hierarchy of types starting from *bool*. At all levels of this hierarchy we have complete boolean lattices and we can use all properties of abstract lattices of this kind that were mentioned in Section 2.

3.3 Instantiation of the Abstract Lattice Theory

In Section 2 we showed how the HOL theory of abstract lattices can be created. Just above we present several concrete domains which turned out to be lattices of some kind. How can we use properties of abstract lattices in the concrete domains?

In order to instantiate lattice theory we need a partial order on a type, satisfying the defining property of lattices (complete lattices etc.). As our example we take predicates, defined semantically as boolean functions on an unspecified state space. Thus predicates have type $\alpha \to bool$.

The partial order on predicates is the implication order (lifted from the booleans), formalised by the infix constant `implies`:

```
implies_DEF =
  ⊢ p implies q = (∀s. p s ⇒ q s)
```

Connectives are also lifted as explained in Section 3.1 (we name them **and**, **or** and **not**).

We can now use HOL to prove that the predicates are in fact a lattice:

```
pred_lat =
  ⊢ LAT(U,implies)
```

By the definition of LAT, we must prove that `implies` is a partial order, and that a meet operator (which we call **and**) and a join operator (which we call **or**) both exist.

The predicates are, of course, also a boolean lattice. To prove this, we first define the general meet and join operators `glb` and `lub`:

```
glb_DEF =
  ⊢ glb P = (λs. ∀p. P p ⇒ p s)
```

```
lub_DEF =
  ⊢ lub P = (λs. ∃p. P p ∧ p s)
```

Note that both these operators take a set of predicates as their argument. Now we can prove that the predicates are a complete lattice, with `glb` as meet and `lub` as join operator:

```
pred_lat =
  ⊢ CLAT(U,implies)
```

The proof in HOL that the predicates satisfy the infinite distributivity property is tedious but not difficult. Finally, the predicates can be proved to be a boolean lattice with **not** as the inverse operator.

The next step would be instantiate and specialise inference rules and general theorems. But it is not necessary because these things are done automatically in the inference rules and special rewrite rules. In the next section we show how it can be done in examples.

4 Using Window Inference

In this section we show how the window inference style of reasoning can be used when working with lattices. Window inference is based on the idea of proofs by contextual transformation [11] and was extended and implemented as a HOL library by Grundy [6].

Window inference is a style of reasoning where the user may transform an expression by restricting attention to a subexpression and transforming it. In this way user can transform the subexpression without changing the remainder of the enclosing expression. Also, while transforming a subexpression, the user can use assumptions that are based on the context of the subexpression.

In the window inference style of proof, a user starts with an expression s and transforms it to t such that $s \; R \; t$ holds for some relation R, thus creating a proof of the theorem $\vdash s \; R \; t$. The relation must be preorder, i.e. reflexive and transitive.

Within the window inference system, reasoning is conducted with a stack of windows. Each window has a *focus* (the expression to be transformed), a set of formulae Γ that can be assumed true in the context of the window (the assumptions), and a relation R that must be preserved. It can also have a set of goals (conjectures or proof obligations).

4.1 Transformation Rules for Lattices

The ordering relation on lattices is a partial order and therefore also a preorder. Thus, we can use the lattice ordering as the relation to be preserved in window inference. By default, the window inference system supports three relations - equality, forward implication and backward implication, but we can add new relations to the system. New relations are added by providing theorems about their reflexivity and transitivity.

In order to transform the focus in the window inference system we must provide special ML functions called *transformation rules*. A transformation rule takes the focus s and the current relation R as arguments and returns the theorem that $s \; R \; t$ holds for some t that has been computed by the rule. The system then automatically transforms the focus s to t.

We have written a transformation rule for each introduction and elimination rule presented in Section 2.6. Recall that the conclusion parts of our inference rules are of the form $t \sqsubseteq t'$ where the relation is the lattice ordering. If the focus s can be matched to the left hand side of conclusion expression, then the conclusion of the inference rule is the required theorem for transformation of the focus.

If the conclusion of the inference rule has hypotheses, then they become proof obligations (if some hypothesis can be matched to the one of the contextual assumptions, then it is discharged automatically). All necessary instantiations of the abstract lattice inference rule are done automatically as well.

4.2 Window Rules

When opening a window on some subterm of the current focus, the system uses information that the focus is monotonic in the position where the subterm occurs. This information must be provided by general inference rules for monotonicity. These rules are called *window rules*, and the system keeps them in a database together with information about applicability conditions.

After opening the subwindow on a subterm of the focus, the system starts a subderivation transforming the subterm while preserving some relation (which can be different from the one of the main derivation). After finishing the subderivation (closing the subwindow), the system uses the monotonicity inference rule to prove the theorem for transforming the focus of the main window. Finally, the system transforms the focus according to the provided theorem.

Because lattice constructs are monotonic (or antimonotonic) in their arguments, we have written one window rule for each argument position of each lattice operator.

For example, a window rule for opening subwindow on the left argument of binary meet uses the fact that the binary meet is monotonic in its left argument (see the monotonicity rule for binary meet in Section 2.6). Similar rules exist for the right argument of the binary meet as well as for both arguments of the binary join.

Similarly, we have implemented two window rules for general meet ($\sqcap v \in I.\ s$) (one for the index set I and one for the body expression s) and also two rules for general join.

4.3 Basic Rewrites

Because equality is the smallest preorder relation, it is always possible to transform the current focus using equational theorems (provided the left hand side of the equation matches the current focus). For this purpose the command REWRITE_WIN is used. Its format and functionality is similar to the REWRITE_RULE of HOL but it operates on a window stack rather than a theorem.

It is convenient to have a similar rewrite rule for working with lattices as well. We have implemented such a rule (called LAT_REWRITE_WIN) which does all necessary instantiations in order to make an equational theorem expressing some basic property of the abstract lattice applicable to the current focus. LAT_REWRITE_WIN with an empty list as argument transforms the focus using a list of trivial properties of abstract lattices (such as $s \sqcup s = s$ and $s \sqcap \top = s$).

4.4 Example of a Derivation

Here we show a simple example of a derivation in the window inference system using general lattice properties. As concrete domain we take the predicate lattice instantiated in the way presented in Section 3.4 (we assume that the relation *implies* has been added as a relation that can be preserved in window inference system). We set up a window stack with p or $(q$ and $r)$ as focus, r implies p as assumption and *implies* as the relation to be preserved. The HOL window Library prints this stack as follows:

```
           ! r implies p
implies * p or (q and r)
```

Here the first line is a contextual assumption (marked with an exclamation mark) and the second line contains the relation to be preserved and the current focus (separated by the asterisk).

Let us now concentrate on the right subterm of the focus. We can do it because *or* (a binary join operator) is monotonic in its right argument and we have added a corresponding window rule for the abstract case to the system. All necessary instantiations to the concrete level of predicates are done automatically.

To open a window on a subterm in the window inference system we use the command OPEN_WIN which takes as argument a path describing the position of the desired subterm within the focus. A path is a list made up of the constructors RATOR, RAND and BODY.

```
- DO(OPEN_WIN [RAND]);
           ! r implies p
implies * q and r
```

In the same way we can open a window on r using the fact that binary meet is monotonic in its right argument as well. Then we can use the contextual assumption to transform the focus. This is done with the command TRANSFORM_WIN. This command takes a theorem in the form $s\ R\ t$ where R is transformation relation and if the left hand side expression matches the current focus, then the focus is transformed accordingly:

```
- DO(OPEN_WIN [RAND]);
- DO(TRANSFORM_WIN (ASSUME (--'r implies p'--)));
- DO(CLOSE_WIN);
           ! r implies p
implies * q and p
```

Next, we simplify the focus using the transformation rule for binary meet elimination $(a \sqcap b \sqsubseteq b)$:

```
- DO(MEET2_ELIM_WIN2);
           ! r implies p
implies * p
```

Closing the window with the command CLOSE_WIN we get a transformed version of our initial expression:

```
- DO(CLOSE_WIN);
      ! r implies p
  implies * p or p
```

In the final step the default rewrite applies the idempotence property of binary join ($a \sqcup a = a$):

```
- DO(LAT_REWRITE_WIN[]);
      ! r implies p
  implies * p
```

With the command WIN_THM we can now retrieve the theorem that has been proved in the derivation:

```
- WIN_THM();
val it =
   r implies p |- p or (q and r) implies p  : thm
```

Note that our steps in this example are independent of what concrete lattice we are using. If we had some other lattice instead of the predicate one, our actions would be absolutely the same, and the end result would have been another instance of the theorem $r \sqsubseteq p \vdash p \sqcup (q \sqcap r) \sqsubseteq p$.

5 Implementation Issues

In the previous section we showed how to use lattice properties in derivations working with Window Inference Library in HOL. However, working with subexpressions (selecting subexpression by path, regrouping expressions by explicit application of commutativity and associativity, etc.) is very tedious with the standard command line interface to HOL.

5.1 The Refinement Calculator

The Refinement Calculator tool [9] was developed in order to provide a user friendly interface for transformational reasoning. The interface permits the user to select subexpressions by pointing and clicking and to select transformations from menus.

The Refinement Calculator consists of a number of layers built on top of the HOL system and its Window Inference library. These layers include a theory of programs, transformation rules for program refinement and an X Window System based graphical user interface.

If the user selects some subexpression of the current focus with the mouse and presses a special window opening button, then the system computes the path to the subexpression and executes the appropriate window opening command and a subderivation starts. By closing a window the user ends the subderivation and the focus is transformed according to the appropriate monotonicity rule.

We have added general lattice transformations to the system and made them available in a separate menu. When a user chooses a menu alternative, the appropriate transformation rule is applied to the current focus. If the rule requires arguments, then a dialog window pops up and the user can type in them. It is also possible to indicate (by pointing and clicking) that a transformation should be applied to a specific subterm of the current focus, rather than to the whole focus.

It should be pointed out that the lattice transformations can also be used in ordinary proofs, since the booleans with forward or backward implication as the ordering are a complete boolean lattices. Thus transforming a boolean expression to truth while preserving backward implication is a special case of preserving a lattice ordering.

5.2 A Database of Concrete Lattices

The window inference system stores information about supported relations and window rules in a database. Similarly, our lattice tool has a small database with information about preproved lattices and lattice instantiations. By default, there are theorems that show (among others) that *bool*, predicates and predicate transformers are complete boolean lattices. In all situations, when properties of such concrete lattices are used, corresponding assumptions about being (boolean, complete) lattice are automatically discharged. The database also contains information about lattice instantiations, i.e., theorems of the form $< operator > = < lattice\ construct >$. For example,

$$\vdash \wedge\ =\ \mathtt{meet2}(\mathtt{U}, \Rightarrow)$$

is the binary meet instantiation for the lattice *bool* with the forward implication ordering.

Before a lattice transformation is applied to the focus, all concrete lattice operators (i.e., instantiations of meet, join, etc.) are rewritten into abstract form. After the transformation is done, the converse rewriting takes place. Thus we are not forced always to use the abstract form of lattice constructs in order to transform expressions according to some lattice property.

The user can add further lattice structures to the database, by supplying the appropriate theorems (i.e., theorems stating that the structure is a lattice, that a specific operator is the meet etc).

5.3 Pretty Printer

The Refinement Calculator contains a parser and pretty-printer in order to allow users to interact with the system using syntax that they are accustomed to. We plan to extend the parser and the pretty-printer to allow traditional mathematical syntax for general meets and joins: $(\sqcap v \in I.\ s)$ and $(\sqcup v \in I.\ s)$ rather than the the current HOL syntax which is hard to read. For example, the HOL syntax for general meet is

$$\mathtt{meet2}\ (\mathtt{U}, \mathtt{po})\ (\lambda \mathtt{x}.\ \exists \mathtt{v}.\ \mathtt{I}\ \mathtt{v}\ \wedge\ (\mathtt{x}\ =\ \mathtt{s}))$$

where **po** is the ordering.

With pretty-printing we can still open subwindows on subexpressions of the focus and transform them using applicable window rules. The pretty-printer automatically translates the path to the subexpression on the screen to the actual path in abstract HOL syntax.

5.4 Example of a Refinement

Let us try very simple refinement example using the Refinement Calculator. Suppose we are refining the program containing as a subcomponent the nondeterministic assignment statement "x := x'. x' = 0 ∨ x' = 1" (x is assigned either 0 or 1).

In the Refinement Calculus theory a nondeterministic assignement is defined as a predicate transformer in the form

$$(\lambda q : \beta \to bool. \lambda \sigma : \alpha. \forall \gamma : \beta. R\,\sigma\,\gamma \Rightarrow q\,\gamma)$$

where q is a postcondition, σ an initial state, γ a final state and R is a relation on initial and final states.

It can be proved that a nedeterministic assignment in general is equivalent to a general meet on predicate transformers of such form:

$$(\sqcap (\sigma, \gamma) \in R.\ (\lambda q.\ \lambda \sigma'.\ (\sigma' = \sigma) \Rightarrow q\,\gamma)\,).$$

Intuitively it means a general meet on all such statements (predicate transformers) which started in any state $\sigma \in Dom\,R$ reach the state γ such that $R\,\sigma\,\gamma$ holds.

Therefore, focusing on the body of our nondeterministic assigment

$$x := x'.\ x' = 0 \vee x' = 1$$

we in fact focus on the index set of a special form of general meet. Because a general meet is antimonotonic in its index set we get the following starting expression for the subderivation:

$$\lambda(x, x').\ x' = 0 \vee x' = 1$$

The relation we have to preserve is *implied_by* - the ordering reverse to the *implies* ordering on predicates defined earlier. The easiest way to refine such expression is to focus on the body of the lambda expression, thus moving from the predicate to the boolean level. In this case we have on the screen:

$$x' = 0 \vee x' = 1$$

The relation we have to preserve now is $\Leftarrow$ (backward implication). $(bool, \Leftarrow)$ is a complete boolean lattice dual to the $(bool, \Rightarrow)$ lattice. Because of duality ∨(disjunction) is binary meet in the lattice $(bool, \Leftarrow)$. Therefore, we can apply binary meet elimination rule to simplify the current expression to the left disjunct. Then we get

$$x' = 0$$

Closing windows we get the initial subcomponent refined:

$$x := x'.\ x' = 0.$$

Easy to show that such nondeterministic assignment is equivalent to the ordinary assignment x:=0. Applying the appropriate transformation rule we finish our derivation. The final result is shown as the theorem (in the pretty-printed form):

$$\vdash\ (x := x'.\, x' = 0 \lor x' = 1)\ \mathtt{ref}\ (x := 0).$$

This example shows how the nondeterministic assignment statement can be refined using lattice properties. The body of the nondeterministic assignment is very simple in the example but the same principles could be applied for more complex cases as well.

6 Conclusion

The work presented in this paper can be seen as an example of a novel way of implementing abstract theories in the HOL system. Rather than changing the logic or the system or adding extra layers of syntax, we have created a general theory of lattices and provided ways of instantiating this theory to concrete lattices. This means that the same underlying theory can be used for transformational reasoning in situations that on the surface seem to have very little in common, such as backward proof (transforming boolean terms under backward implication) and program development (transforming programs under a correctness preserving refinement relation).

As related works we should mention formalisations of abstract theories in HOL done by Gunter[8] and Windley[12]. Regensburger[10] formalised the theory of complete partial orders (cpo's) in Isabelle using type classes mechanism. However, Isabelle's type classes allow a type to be a lattice only in one way (with one ordering relation). Our approach is more flexible in this sense.

The system described here depends heavily on the Window Inference library of the HOL system. The Refinement Calculator tool provides a good interface and numerous instances of lattices, but it is also possible to use our system with only the standard HOL system. A moderate amount of pretty-printing and parsing is still needed to make the system easy to use for someone who is not familiar with the HOL system.

In the near future, we intend to extend our system with more facilities for handling fixpoint constructs. We also intend to add a way of handling monoid structures which interact with the lattice ordering (e.g., so that the composition operation is monotonic in both arguments with respect to the ordering).

Acknowledgements

We wish to thank Joakim von Wright for his help and advice preparing this paper and Ralph Back for his suggestion of writing rules for lattices in the introduction/elimination form. We are also grateful to anonymous referees for their suggestions.

References

1. R.J.R. Back. A calculus of refinements for program derivations. *Acta Informatica*, 25:593–624, 1988.
2. R.J.R. Back and J. von Wright. Duality in specification languages: a lattice-theoretical approach. *Acta Informatica*, 27:583–625, 1990.
3. R.J.R. Back and J. von Wright. Refinement concepts formalised in higher-order logic. *Formal Aspects of Computing*, 2:247–272, 1990.
4. M.J.C. Gordon. HOL: A proof generating system for higher-order logic. In *VLSI Specification, Verification and Synthesis*. Kluwer Academic Publishers, 1988.
5. M.J.C. Gordon, R. Milner and C. Wadsworth. Edinburgh LCF: A mechanised logic of computation. In *LNCS* 78. Springer–Verlag, 1979.
6. J. Grundy. Window Inference in the HOL System. In *Proc. of the Int. Workshop on the HOL Theorem Proving System and Its Applications*, Aug. 1991. IEEE Computer Society Press.
7. E. Gunter. Doing algebra in higher order logic. In the HOL system documentation, Cambridge, 1990.
8. E. Gunter. The Implementation and Use of Abstract Theories in HOL In *Proc. of the Third HOL Users Meeting*, Aarhus, Denmark, October 1990. Technical Report DAIMI PB – 340, 1990.
9. T. Långbacka, R. Rukšėnas and J. von Wright. TkWinHOL: A tool for doing window inference in HOL. In *LNCS* 971, 245–260. Springer–Verlag, 1995.
10. F. Regensburger. HOLCF: Higher Order Logic of Computable Functions. In *LNCS* 971, 293–307. Springer–Verlag, 1995.
11. P.J. Robinson and J. Staples. Formalising the hierarchical structure of practical mathematical reasoning. *Logic and Computation*, 1:47–61, 1993.
12. P.J. Windley. Abstract Theories in HOL. In *Proc. of the Int. Workshop on the HOL Theorem Proving System and Its Applications*, September 1992. IFIP Transactions A-20, 197–210.
13. J. von Wright. Doing Lattice Theory in Higher Order Logic. In Technical Report 136, Reports on Computer Science and Mathematics, Series A. Åbo Akademi, Turku, 1992.
14. J. von Wright. Program refinement by theorem prover. In *Proc. 6th Refinement Workshop*, London, January 1994. Springer–Verlag.

Formal Verification of Algorithm $\mathcal{W}$: The Monomorphic Case

Dieter Nazareth* and Tobias Nipkow

Technische Universität München**

Abstract. A formal verification of the soundness and completeness of Milner's type inference algorithm $\mathcal{W}$ for simply typed lambda-terms is presented. Particular attention is paid to the notorious issue of "new" variables. The proofs are carried out in Isabelle/HOL, the HOL instantiation of the generic theorem prover Isabelle.

1 Introduction

Type systems for programming languages are usually defined by type inference rules which inductively define the set of well-typed programs. Functional languages of the ML-tradition also come with a type inference algorithm which computes the (most general) type of a program. The inference algorithm needs to be sound and complete w.r.t. the rules: the rules and the algorithm must determine the same set of type correct programs.

The idea of type inference goes back to Hindley [7]. Milner [14] gave the first account of type inference for a simply-typed lambda-calculus with let, the core of ML [15]. In particular, he presented a type inference algorithm $\mathcal{W}$ based on unification of types. Although Milner's original article only proves soundness of $\mathcal{W}$ w.r.t. the rules, completeness has been settled in the mean time [4, 25]. This polymorphic type system forms the basis of most modern functional languages, usually in some extended or generalized form. Each extension of the type system requires a corresponding modification of algorithm $\mathcal{W}$, which again has to be proved sound and complete, e.g. [9, 17, 16, 22]. However, all these proofs have been carried out in a mathematical or informal way. We are not aware of any completely machine checked proof.

This paper presents a formalization of the original type system restricted to the monomorphic case. The soundness as well as the completeness property of the type inference algorithm $\mathcal{W}$ is formally established within the Isabelle/HOL system. The paper provides the complete definition of all concepts, the key lemmas and theorems, but no proofs. The complete development is accessible via http://www4.informatik.tu-muenchen.de/~nipkow/isabelle/HOL/MiniML/.

The rest of the paper is organized as follows: after a brief introduction to Isabelle/HOL in Section 2, Section 3 deals with the formalization of the type level.

---

* Research supported by ESPRIT BRA 6453, *Types*.
** Institut für Informatik, 80290 München, Germany.
 http://www4.informatik.tu-muenchen.de/~{nazareth,nipkow}/

It includes theories of substitution and unification and the treatment of "new" type variables. In Section 4 the type inference system is presented. Algorithm $\mathcal{W}$ is formalized in Section 5: we briefly show how to model exception handling in a functional style using monads and discuss the soundness and completeness proofs of $\mathcal{W}$. In Section 6 we compare our case study to related work. Section 7 concludes with some lessons learned and some future work.

2 Isabelle/HOL

Isabelle is an interactive theorem prover which can be instantiated with different object logics. One particularly well-developed instantiation is Isabelle/HOL, which supports Church's formulation of Higher Order Logic and is very close to Gordon's HOL system [6]. In the remainder of the paper HOL is short for Isabelle/HOL.

We present no proofs but merely definitions and theorems. Hence it suffices to introduce HOL's surface syntax. A detailed introduction to Isabelle and HOL can be found elsewhere [18]. We have intentionally refrained from recasting HOL's ASCII syntax in ordinary mathematical symbols to give the reader a bit of an idea what interacting with HOL looks like.

Terms and formulae The following table summarizes the correspondence between ASCII and mathematical symbols:

!, !!	?	%	-->, ==>	&	|	=, ==	~=	:	Un	UN	<=
$\forall$	$\exists$	λ	$\Longrightarrow$	$\wedge$	$\vee$	$=$	$\neq$	$\in$	$\cup$	$\bigcup$	$\leq, \subseteq$

The two universal quantifiers, implications and equalities stem from the object and meta-logic, respectively. The distinction can be ignored while reading this paper. The notation [| A_1;...;A_n |] ==> A is short for the nested implication $A_1 \Longrightarrow \ldots \Longrightarrow A_n \Longrightarrow A$. The predicate <= is overloaded and applies to natural numbers ($\leq$) and to sets ($\subseteq$).

Types follow the syntax for ML-types, except that the function arrow is => rather than ->. The notation [τ_1, ..., τ_n] => τ abbreviates τ_1 => ... => τ_n => τ. A term t is constrained to be of type τ by writing t::τ.

Isabelle also provides Haskell-like *type classes* [8], the details of which are explained as we go along. A type variable 'a is restricted to be of class c by writing 'a::c.

Theories introduce constants with the keyword **consts**, non-recursive definitions with **defs**, and primitive recursive definitions with the keyword **primrec**. For general axioms the keyword **rules** is used. Further constructs are explained as we encounter them.

Although we do not present any of the proofs, we usually indicate their complexity. If we do not state any complexity the proof is almost automatic. That means, it is either solved by rewriting or by the "classical reasoner", fast_tac in Isabelle parlance [19]. The latter provides a reasonable degree of automation for predicate calculus proofs. Note, however, that its success depends on the right selection of lemmas supplied as parameters.

3 Types and Substitutions

This section describes the language of object-level types used in our case study. They should not be confused with Isabelle's built-in meta-level type system described in the previous section. To emphasize this distinction we sometimes call the object-level types *type terms*.

3.1 Types

Type terms consist only of type variables and the function space constructor. They are expressed as an inductive data type with two constructors. Type variables are modeled by natural numbers.

```
datatype typ = TVar nat | "->" typ typ (infixr 70)
```

We do not need quantified types because our term language does not contain let-expressions.

3.2 Substitution

A substitution is a function mapping type variables to types. In HOL, all functions are total. The identity substitution is denoted by id_subst.

```
types    subst = nat => typ

consts   id_subst :: subst

defs     id_subst == (%n.TVar n)
```

Substitutions can be extended to type terms, lists of type terms, etc. Type classes, i.e. overloading, allow us to use the same notation in all of these cases.

```
classes  type_struct < term
```

introduces a new class type_struct as a subclass of term, the predefined class of all HOL types. Class type_struct is meant to encompass all meta-level types which substitutions can be applied to. This is expressed by declaring

```
consts   app_subst :: [subst, 'a::type_struct] => 'a ("$")
```

The purpose of app_subst is to apply substitutions to values of types in class type_struct. $ is syntactic sugar for app_subst. Because identifiers in Isabelle do not contain "$" we may write $s instead of $ s. The notation $s emphasizes that we regard $ as a modifier acting on the substitution s.

So far there is no definition of app_subst, but there will be several, one for each instance of 'a we are interested in. Hence app_subst will be overloaded. In Haskell, app_subst is a *member function* of class type_struct.

Now we want to turn typ into an element of class type_struct by extending substitutions from type variables to types in the usual fashion. This requires two steps. First we simply tell Isabelle that typ is an element of type_struct:

```
arities  typ :: type_struct
```

The general form of the arities declaration is $t :: (C_1, \ldots, C_n)C$, where t must be an n-ary type constructor. It expresses that $(\tau_1, \ldots, \tau_n)t$ is of class C provided each τ_i is of class C_i.

Then we define the appropriate instance of app_subst by primitive recursion over typ:

```
primrec app_subst typ
        $ s (TVar n) = s n
        $ s (t1 -> t2) = ($ s t1) -> ($ s t2)
```

In Haskell, both steps are combined into the instance construct.

In the same way we extend app_subst to lists:

```
arities   list :: (type_struct)type_struct
```

Hence (τ)list is of class type_struct provided the element type τ is. A substitution is applied to a list by mapping it over that list, where map is predefined:

```
defs   $ s == map ($ s)
```

Note that $ s on the left has type 'a list => 'a list and on the right type 'a => 'a, where 'a::type_struct.

In the sequel, a *type structure* is a type in class type_struct.

Now we can prove that the extension of the identity substitution to type terms and list of type terms again yields identity functions:

```
$ id_subst = (%t::typ.t)
$ id_subst = (%ts::typ list.ts)
```

For the composition of substitutions the following propositions hold:

```
$ g ($ f t::typ) = $ (%x. $ g (f x) ) t
$ g ($ f ts::typ list) = $ (%x. $ g (f x)) ts
```


3.3 Free Type Variables

The set of type variables occurring in a type structure is denoted by free_tv. Again, we overload this function by using class type_struct. The definitions below describe the usual behaviour of the typ and list instances, respectively:

```
consts  free_tv :: 'a::type_struct => nat set

primrec free_tv typ
        free_tv (TVar m) = m
        free_tv (t1 -> t2) = (free_tv t1) Un (free_tv t2)

primrec free_tv list
        free_tv [] = {}
        free_tv (x#xs) = free_tv x Un free_tv xs
```

Note that infix # :: ['a, 'a list] => 'a list is the list constructor adding an element to the front of a list.

These definitions enable us to show some interesting properties:

```
[| $ s1 (t::typ) = $ s2 t; n : free_tv t |] ==> s1 n = s2 n
(!n. n: free_tv t --> s1 n = s2 n) ==> $ s1 (t::typ) = $ s2 t
(t::typ) mem ts ==> free_tv t <= free_tv ts
```

The first one states that if applying two different substitutions to the same type term yields the same result, then the substitutions coincide on the free type variables occurring in the type term. The second one reverses this implication. The third one states that if a type term is an element (infix mem) of some list, then the set of free type variables of the type term is a subset of the set of free type variables of the list. The first two propositions have also been proved for lists of type terms.

Domain and codomain of a substitution are defined in the usual way:

```
consts  dom, cod :: subst => nat set

defs    dom s == {n. s n ~= TVar n}
        cod s == UN m:dom s. free_tv (s m)
```

The set of variables occurring either in the domain or the codomain of a substitution is called the set of free variables of a substitution. We want to use the identifier free_tv to denote this set. Hence, we must add type subst to class type_struct. Type subst, however, is only an abbreviation for the composed type nat => typ. Just like Haskell, Isabelle does not allow to add composed types to type classes directly. Instead, we must state the propagation of class membership for the type constructor =>. We already met this mechanism when adding the list types to class type_struct. In the same way we define an arity for the type constructor => which is the infix name for fun:

```
arities  fun::(term,type_struct)type_struct
```

Because nat::term and typ::type_struct, this implies subst::type_struct. Note, however, that it also implies further types to belong to class type_struct, for example bool => typ. This may seem a bit permissive but does no harm. Now we can define the subst instance of free_tv:

```
defs   free_tv s == (dom s) Un (cod s)
```

We do not define other possible instances of free_tv. Neither do we provide an instance for function app_subst at type subst.

The following lemmas capture important relationships between substitution application and free type variables:

```
free_tv ($ s (t::typ)) <= cod s Un free_tv t
[| v : free_tv(s n); v ~= n |] ==> v : cod s
free_tv (%n::nat. $ s1 (s2 n) :: typ) <= free_tv s1 Un free_tv s2
```

We need these propositions in the soundness and completeness proofs of the type inference algorithm.

3.4 New Type Variables

Algorithm $\mathcal{W}$ needs to generate new type variables. This mechanism is rarely formalized in the description of the algorithm. It is simply assumed that there always exists some type variable never used before. However, to perform formal proofs we have to completely formalize the algorithm. The obvious way of handling the generation of new type variables is as follows: the set of already used type variables is explicitly passed to $\mathcal{W}$, all new variables generated during the execution are added to this set, and the enlarged set is returned upon successful termination [17]. Because we use natural numbers for type variables we have a total ordering on these variables. Instead of passing all used type variables to $\mathcal{W}$, we only pass the successor of the greatest one used up to now. Each time algorithm $\mathcal{W}$ needs a new type variable it uses the counter and increments it by one. Predicate new_tv formalizes our notion of a new type variable. It takes a type variable and a type structure and determines whether the given variable is greater than any type variable occurring in the structure. Such a type variable is called *new w.r.t. the structure*.

```
consts   new_tv :: [nat,'a::type_struct] => bool

defs     new_tv n ts == ! m. m:free_tv ts --> m<n
```

This predicate is a necessary precondition for most propositions about algorithm $\mathcal{W}$. To prove these propositions we need some theorems about new_tv. The first one is quite simple: it states that all greater type variables are also new type variables. This holds not only for type terms, but also for lists of type terms and for substitutions.

```
[| n<=m; new_tv n (t::typ) |] ==> new_tv m t
```

This proposition is proved by induction on the structure of t. The next theorems show how substitutions and new_tv interact:

```
[| new_tv n s; new_tv n (t::typ) |] ==> new_tv n ($ s t)
[| new_tv n (s::subst); new_tv n r |] ==> new_tv n (($ r) o s)
new_tv n s = ((!m. n <= m --> s m = TVar m) &
              (! l. l < n --> new_tv n (s l) ))
```

The first two theorems tell us that the new type variable property is preserved by application and composition (o) of substitutions. The third one is more complex to express and prove: it requires about 15 proof steps.

3.5 Unification

The goal of unification is to unify two terms, i.e. to find a substitution of terms for variables which makes the two terms syntactically identical. A *unification algorithm* either computes a most general unifier of two terms or fails, if the two terms are not unifiable. The first in a long line of unification algorithms is due to Robinson [21]. Of course the correctness of type inference does not depend on any particular implementation of unification but merely on general properties of unification. Therefore we introduce a function *mgu* (*most general unifier*), specify its characteristic properties, but provide no implementation. This is the only point in the whole development where we introduce new axioms as opposed to consistency preserving definitions. Of course we know that a function mgu which satisfies the axioms exists. Alternatively, we could have made mgu a parameter of all functions using it and the axioms about mgu preconditions of the theorems about those functions. However, that is overkill because mgu is not intended as a parameter of individual functions but of the whole development. It requires some form of parameterized theories to express this.

Unification may fail. To model the distinction between a successful computation and a failure situation we define

```
datatype  'a maybe =  Ok 'a | Fail
```

Unification either terminates normally returning Ok(s) for some substitution s, or indicates a failure situation by returning Fail:

```
consts   mgu ::  [typ,typ] => subst maybe
```

A most general unifier should satisfy the following three axioms:

```
rules    mgu t1 t2 = Ok u ==> $u t1 = $u t2
         [| mgu t1 t2 = Ok u; $s t1 = $s t2 |] ==> ? r. s = $r o u
         $s t1 = $s t2 ==> ? u. mgu t1 t2 = Ok u
```

The first axiom requires the result of mgu to be a unifier of the given type terms. The second one states that the computed unifier is a most general one: each unifier can be obtained by composing the computed one with some substitution. The third one requires mgu to return an Ok result if the two types are unifiable. This prevents trivial implementations which satisfy the first two axioms by always returning Fail.

Most general unifiers are only unique up to consistent renaming of variables. Such a renaming may even introduce type variables not occurring in the type terms to unify. However, because we want to keep track of used variables we need one last axiom:

```
mgu t1 t2 = Ok u ==> free_tv u <= free_tv t1 Un free_tv t2           (0)
```

This ensures that the algorithm does not introduce new type variables. We can then show that unification preserves the new type variable property:

```
[| mgu t1 t2 = Ok u; new_tv n t1; new_tv n t2 |] ==> new_tv n u
```

4 Well-Typed Lambda Terms

Lambda terms are represented in de Bruijn notation [5] which can conveniently be expressed as an inductive data type with three constructors for variables, abstraction and application:

```
datatype   expr = Var nat | Abs expr | App expr expr
```

The index i in a subterm Var i indicates that, when moving upward, i abstractions must be traversed until the corresponding binder is found. For example, $\lambda x.x(\lambda y.y\ x)$ becomes Abs(App(Var 0)(Abs(App(Var 0)(Var 1)))). This shows that

- different occurrences of the same index may represent different bound variables, and
- different occurrences of the same bound variable may be represented by different indexes, depending on how far below the binding λ they are.

Free variables are those without enough enclosing applications, e.g. Var 0 on its own.

Note that we have restricted our language to pure lambda terms without let-construct. Because let-bound identifiers are the only source of polymorphism in ML-like languages we do not need quantified type terms.

The datatype expr represents all untyped lambda-terms. Now we need to define the subset of *well-typed* lambda-terms. To keep track of the types of bound variables we use a context assigning type terms to variables. Because we use de Bruijn notation this context is simply a list of type terms. The type of variable i can be found at the i-th list position. Well-typedness is a relative notion because it depends on the context. Therefore we introduce a relation between contexts, lambda-terms and type terms:

```
consts  has_type :: (typ list * expr * typ)set
```

The proposition (a,e,t):has_type should be read as "In context a expression e has type t". This allows a term to have more than one type in a given context. The following bit of syntactic sugar (which we will not explain in detail) allows us to read and write the more conventional a |- e :: t:

```
syntax "@has_type":: [typ list,expr,typ] => bool  ("_ |- _ :: _" 60)
translations  a |- e :: t == (a,e,t) : has_type
```

Warning: the delimiter :: is now used for type annotations both in the logic and in the object level lambda-terms (expr). The latter is easily distinguished by its leading |-.

Relation has_type is defined inductively, i.e. by a set of inference rules. Proposition a |- e :: t holds iff it can be derived from the inference rules. HOL provides a package for defining inductive sets. The following text defines has_type to be the least set closed under the given inference rules.

```
inductive has_type
    [| n < length a |] ==> a |- Var n :: nth n a
    [| t1#a |- e :: t2 |] ==> a |- Abs e :: t1 -> t2
    [| a |- e1 :: t2 -> t1; a |- e2 :: t2 |] ==> a |- App e1 e2 :: t1
```

Note that nth :: [nat, 'a list] => 'a selects the nth element of a list.

Modulo the fact that we use de Bruijn notation, these are the usual type inference rules for lambda terms. The reader not familiar with this type system is referred to [2]. Note that the simplicity of the Abs-rule is due to de Bruijn notation: the extended context t1#a takes care of the fact that when descending into an abstraction all references to variables bound outside shift by 1.

The following theorem shows that has_type is closed w.r.t. substitution:

a |- e :: t ==> $ s a |- e :: $ s t

It is proved by induction on the derivation of a |- e :: t. This leads to three subgoals (corresponding to the three inference rules given above), each of which is proved almost automatically.

5 Type Inference

The purpose of *type inference* (or *type reconstruction*) is to find the most general type t for a given term e in a given context a such that a |- e :: t. Interpreting the type inference rules as a Prolog program yields such a type inference algorithm. Using a functional implementation language, the computation of a most general type requires a separate algorithm which was first presented by Milner [14] who called the algorithm $\mathcal{W}$ (Well-typing).

5.1 Programming with Monads

Given a context a and a term e, there may be no type t such that a |- e :: t. In this case $\mathcal{W}$ should "fail" with some meaningful error message. The easiest way of handling error messages is to use so-called *impure* features, like side effects or exceptions. Wadler [23] introduced the idea that *monads* could be used as a practical method for modeling such impure features in a purely functional way. In this paper we only give a brief introduction into programming with monads. A good presentation of this topic can be found in [24].

A *monad* is a data type consisting of a type constructor M and two operations unit and bind with the following functionalities:
```
unit :: 'a => 'a M
bind :: ['a M, 'a => 'b M] => 'b M
```
A value of type 'a M represents a computation which is expected to produce a result of type 'a. Function unit turns a value into the computation that returns that value and does nothing else. Function bind provides a means of combining computations. It applies a function of type 'a => 'b M to a computation of type 'a M.

We already met a monad in this paper, in connection with unification, where failure is also an issue. Together with appropriate unit and bind functions, type constructor maybe forms a monad. Values of type 'a maybe represent computations that may raise an exception. Constructor Ok denotes the unit function. It turns a value a into Ok a. Because we only gave an abstract requirement specification for mgu we did not explain how to propagate failures. This is the task of the `bind` function defined in the following way:

```
consts   bind :: ['a maybe, 'a => 'b maybe] => 'b maybe (infixl 60)

defs     m bind f == case m of Ok r => f r | Fail => Fail
```

The call m `bind` f examines the result of the computation m: if it is a failure, it is propagated; otherwise the function f is applied to the value of the computation. In most cases f is expressed as a λ-term:

```
m bind (%x.c)
```

This can be read as follows: perform computation m, bind the resulting value to variable x, and then perform computation c. In an imperative language this would be expressed in the following way:

```
x := m; c
```

The powerful translation mechanism of Isabelle allows us to use exactly this notation for the `bind` function:

```
syntax "@bind" :: [pttrns,'a maybe,'b maybe] => 'b ("_ := _; _" 0)
translations P := E; F  ==  E bind (%P.F)
```

Now we can define algorithm $\mathcal{W}$ in an imperative style without using any impure feature.

Warning: do not confuse the monad ";" with the separator for hypotheses.

5.2 Algorithm $\mathcal{W}$

If $\mathcal{W}$ succeeds, it returns a substitution s and a type t such that $s a |- e::t. At certain points $\mathcal{W}$ requires new type variables. As already explained in Section 3.4, we handle the generation of new type variables by passing the successor of the greatest type variable used up to now. Thus, we need an additional result component for this counter. Altogether, $\mathcal{W}$ has the following type:

```
consts   W :: [expr, typ list, nat] => (subst * typ * nat)maybe
```

$\mathcal{W}$ is recursively defined on the term structure. Because the type inference rules given in Section 4 are syntax-directed, each case of function $\mathcal{W}$ corresponds to exactly one rule.

```
primrec W expr
  W (Var i) a n = (if i < length a then Ok(id_subst, nth i a, n)
                                   else Fail)
  W (Abs e) a n = ( (s,t,m) := W e ((TVar n)#a) (Suc n);
                    Ok(s, (s n) -> t, m) )
  W (App e1 e2) a n = ( (s1,t1,m1) := W e1 a n;
                        (s2,t2,m2) := W e2 ($s1 a) m1;
                        u := mgu ($s2 t1) (t2 -> (TVar m2));
                        Ok($u o $s2 o s1, $u (TVar m2), Suc m2) )
```

A call `W e a n` fails if a contains no entry for some free variable in e, or if a call of the unification algorithm `mgu` fails. The failure propagation is invisibly handled by our mixfix `bind` notation.

The main goal of this case study was to formally show the correctness of algorithm $\mathcal{W}$. Correctness is defined as soundness and completeness w.r.t. the type inference rules. We start with soundness of $\mathcal{W}$:

```
W e a n = Ok(s,t,m) ==> $s a |- e :: t
```

This proposition is shown by induction on the structure of e. The proof can be performed directly, without any additional auxiliary proposition. However, both the Abs and App case require explicit instantiation of variables. Therefore the user needs to be familiar with the details of the proof. An automatic proof which synthesizes these instantiations seems unlikely.

5.3 Completeness of $\mathcal{W}$

The proof of completeness of $\mathcal{W}$ w.r.t. the type inference rules is more complex. We have to prove some auxiliary lemmas first. All these lemmas deal with the problem of new type variables. In mathematical proofs about type inference algorithms this problem is simply ignored. The first lemma states that the counter for new type variables is never decreased:

```
W e a n   = Ok (s,t,m) ==> n<=m
```

It is proved by induction on the structure of e and in turn helps us to prove the following lemma:

```
[| new_tv n a; W e a n = Ok (s,t,m) |] ==> new_tv m s & new_tv m t
```

It says that the resulting type variable is new w.r.t. the computed substitution as well as the computed type term. This fact ensures that we can safely use the returned type variable as new type variable in subsequent computations. Again we use induction on the structure of e. The most difficult part of the proof is case App e1 e2. It takes about 30 proof steps; the degree of automation is quite low.

Now we can prove a proposition that seems to be quite obvious. Roughly speaking, it says that type variables free in either the computed substitution or the computed type term did not materialize out of the blue: they either occur in the given context or were taken from the set of new type variables. Formally, the proposition is expressed as follows:

```
[| W e a n = Ok (s,t,m); v : free_tv s | v : free_tv t; v<n |]
==> v : free_tv a
```

The proof is performed by induction on the structure of term e. The complexity of this proof is roughly the same as the complexity of the last proof.

With the help of these propositions we are able to show completeness of $\mathcal{W}$ w.r.t. the type inference rules: if a closed term e has type t' then $\mathcal{W}$ terminates sucessfully and returns a type which is more general than t', i.e. t' is an instance of that type.

```
[] |- e :: t' ==> ? s t. (? m. W e [] n = Ok(s,t,m)) &
                         (? r. t' = $r t))
```

This proposition needs to be generalized considerably before it is amenable to induction:

```
[| $s' a |- e :: t'; new_tv n a |]
==> ? s t. (? m. W e a n = Ok (s,t,m)) &
            (? r. $s' a = $r ($s a) & t' = $r t))
```

This theorem is the most difficult one to prove. Although the proof plan in [16] is quite detailed, translating it into Isabelle turned out to be hard work. The proof starts with an induction on the structure of term e. Again, case `App e1 e2` causes most of the problems. Proving this case requires about 90 proof steps. Isabelle is used mainly to keep track of the proof and to avoid foolish mistakes.

Let us have a brief look at the `App` case: the main problem is to show successful termination. The algorithm may fail during unification of `$s2 t1` and `t2 -> (TVar m2)`. Hence, we have to prove that these terms are indeed unifiable, i.e. that there exists a substitution u such that

```
$u ($s2 t1) = $u (t2 -> (TVar m2)).
```

In our proof we use the witness u given in [16], which differs slightly from the one used in much of the published literature (for example [9, 17]). We establish that this witness is indeed a unifier for the above type terms.

5.4 Algorithm $\mathcal{I}$

Milner [14] also presents a more efficient refinement of algorithm $\mathcal{W}$ called $\mathcal{I}$, where substitutions are extended incrementally instead of computing new substitutions and composing them later. We have also formalized $\mathcal{I}$ [1]

```
consts I :: [expr, typ list, nat, subst] => (subst * typ * nat)maybe

primrec I expr
  I (Var i) a n s = (if i < length a then Ok(s,nth i a,n) else Fail)
  I (Abs e) a n s = ( (s,t,m) := I e ((TVar n)#a) (Suc n) s;
                      Ok(s, TVar n -> t, m) )
  I (App e1 e2) a n s =
                    ( (s1,t1,m1) := I e1 a n s;
                      (s2,t2,m2) := I e2 a m1 s1;
                      u := mgu ($s2 t1) ($s2 t2 -> TVar m2);
                      Ok($u o s2, TVar m2, Suc m2) )
```

and shown that it correctly implements $\mathcal{W}$:

```
[| new_tv m a;  new_tv m s;  I e a m s = Ok(s',t,n) |]
==> ? r. W e ($s a) m = Ok(r, $s' t, n) & s' = ($r o s)

[| new_tv m a;  new_tv m s;  I e a m s = Fail |]
==> W e ($s a) m = Fail
```

For lack of space, the details cannot be presented here.

---

[1] Ideally, mgu should take the substitution s2 as a separate argument. For simplicity we have applied s2 explicitly to the two type arguments of mgu.

6 Comparison

The literature contains many accounts of type systems and type inference algorithms for lambda-terms which boil down to the rules and algorithm we used [14, 3, 4, 2, 25, 1]. However, ours seems to be the first formal verification of $\mathcal{W}$. There are three key differences between the existing literature and our formal proof:

1. We do not treat `let`. This is a regrettable omission but is justified by the complexity of the formal proof for the `let`-free system.
2. We treat the issue of "new" variables, which is almost universally ignored (an exception is [17]). Ironically, it is this very issue which really complicates the proof for us.
3. In the literature, completeness of $\mathcal{W}$ is always proved under the assumption that the unification algorithm returns idempotent substitutions. In contrast, we require axiom (0) (see the end of Section 3.5), which turns out to be weaker than idempotence.

We would like to concentrate on the last point for a moment, employing the usual mathematical notation from unification theory [10].

A substitution σ is *idempotent* if $\sigma \circ \sigma = \sigma$. Axiom (0) requires $V(\sigma) \subseteq V(s,t)$ for the mgu σ of two terms s and t, where $V(\sigma) = dom(\sigma) \cup cod(\sigma)$ and $V(s,t)$ is the set of variables in s and t.

Theorem 1. *Let σ be a most general unifier of s and t. If σ is idempotent, then $V(\sigma) \subseteq V(s,t)$.*

Proof. It is known that $V(\sigma_1) = V(\sigma_2)$ holds for any two idempotent mgus of $s = t$ [11, Prop. 4.11]. If s and t are unifiable, there exists an idempotent mgu σ_0 such that $V(\sigma_0) \subseteq V(s,t)$ (use any of the standard unification algorithms). Thus the claim follows for every idempotent mgu σ.

Idempotence is strictly stronger than the variable condition as the following example [11] shows: $\sigma = \{x \mapsto f(x), y \mapsto x\}$ is a most general unifier of $x = f(y)$ which satisfies the variable condition but is not idempotent.

Thus we have verified $\mathcal{W}$ under weaker assumptions than usual. However, this is merely a theoretical curiosity: practical unification algorithms do return idempotent substitutions. In fact, it is hard to imagine a unification algorithm computing σ in the example above.

The only other published formal verification of a type checking algorithm we are aware of is by Pollack [20]. His object-language (a subset of "Pure Type Systems") is much more powerful than ours but his terms already contain types. Hence he does not need a separate theory of substitutions and unification to support type inference. On the other hand his proofs involve a substantial amount of lambda-calculus theory, and he also faces the issue of new (term) variables [13]. A final difference is that he does not verify an explicit algorithm but performs a constructive proof which embodies the algorithm.

7 Conclusion and Future Work

The results of this case study can be summarized as follows:

Specification: Isabelle/HOL offers a mature specification environment with a flexible syntax (mixfix) and type system (classes).

Proof: Isabelle/HOL provides some automation on the predicate calculus level, but not nearly enough for our case study. Especially reasoning by transitivity and monotonicity needs to be improved. Decidable subtheories (e.g. fragments of nat and set) and better predicate calculus support would help, but they need to be interleaved with user interactions for providing key instantiations.

$\mathcal{W}$: The proof has confirmed our suspicion that the issue of "new" variables is nontrivial. Although it is true that the formalization of this aspect has not given us any deeper insight into the algorithm, it has helped us to elucidate some finer points like the non-requirement of idempotence. It also helps us to avoid mistakes. Although we are not aware of any incorrect published proofs of $\mathcal{W}$, they do occur in closely related areas: for example, the completeness statement and proof of SLD-resolution in [12] are incorrect because they ignore the "new" variable issue.

Despite all this, the proof has left us with the feeling that there should be a simpler way to treat variables and substitution. However, this is only brought to a head by the requirement for complete formalization and is a sentiment known to most people who have worked with substitutions.

The most important next step is to extend the object-language with a let-construct and polymorphic types. Although we do not expect any major new problems, it is likely to be a substantial piece of work.

Acknowledgements We thank Thomas Stauner for his help with the Isabelle proofs, Franz Baader for an email discussion on the intricacies of mgus, and two anonymous referees for their constructive comments.

References

1. L. Cardelli. Basic polymorphic typechecking. *Sci. Comp. Programming*, 8:147–172, 1987.
2. D. Clément, J. Despeyroux, T. Despeyroux, and G. Kahn. A simple applicative language: Mini-ML. In *Proc. ACM Conf. Lisp and Functional Programming*, pages 13–27, 1986.
3. L. Damas and R. Milner. Principal type schemes for functional programs. In *Proc. 9th ACM Symp. Principles of Programming Languages*, pages 207–212, 1982.
4. L. M. M. Damas. *Type Assignment in Programming Languages*. PhD thesis, Department of Computer Science, University of Edinburgh, 1985.
5. N. G. de Bruijn. Lambda calculus notation with nameless dummies, a tool for automatic formula manipulation, with application to the Church-Rosser theorem. *Indagationes Mathematicae*, 34:381–392, 1972.

6. M. Gordon and T. Melham. *Introduction to HOL: a theorem-proving environment for higher-order logic.* Cambridge University Press, 1993.
7. J. R. Hindley. The principal type-scheme of an object in combinatory logic. *Trans. Amer. Math. Soc.*, 146:29–60, 1969.
8. P. Hudak, S. Peyton Jones, and P. Wadler. Report on the programming language Haskell: A non-strict, purely functional language. *ACM SIGPLAN Notices*, 27(5), May 1992. Version 1.2.
9. M. P. Jones. Qualified Types: Theory and Practice. Technical Monograph PRG-106, Oxford University Computing Laboratory, Programming Research Group, July 1992.
10. J.-P. Jouannaud and C. Kirchner. Solving equations in abstract algebras: A rule-based survey of unification. In J.-L. Lassez and G. Plotkin, editors, *Computational Logic: Essays in Honor of Alan Robinson*, pages 257–321. MIT Press, 1991.
11. J.-L. Lassez, M. Maher, and K. Mariott. Unification revisited. In J. Minker, editor, *Foundations of Deductive Databases and Logic Programming*, pages 587–625. Morgan Kaufman, 1987.
12. J. W. Lloyd. *Foundations of Logic Programming.* Springer-Verlag, 1987.
13. J. McKinna and R. Pollack. Pure type systems formalized. In M. Bezem and J. Groote, editors, *Typed Lambda Calculi and Applications*, volume 664 of *Lect. Notes in Comp. Sci.*, pages 289–305. Springer-Verlag, 1993.
14. R. Milner. A Theory of Type Polymorphism in Programming. *Journal of Computer and System Sciences*, 17:348–375, 1978.
15. R. Milner, M. Tofte, and R. Harper. *The Definition of Standard ML.* MIT Press, 1990.
16. D. Nazareth. *A Polymorphic Sort System for Axiomatic Specification Languages.* PhD thesis, Technische Universität München, 1995. Technical Report TUM-I9515.
17. T. Nipkow and C. Prehofer. Type reconstruction for type classes. *J. Functional Programming*, 5(2):201–224, 1995.
18. L. C. Paulson. *Isabelle: A Generic Theorem Prover*, volume 828 of *Lect. Notes in Comp. Sci.* Springer-Verlag, 1994.
19. L. C. Paulson. Generic automatic proof tools. Technical Report 396, University of Cambridge, Computer Laboratory, 1996.
20. R. Pollack. A verified typechecker. In M. Dezani-Ciancaglini and G. Plotkin, editors, *Typed Lambda Calculi and Applications*, volume 902 of *Lect. Notes in Comp. Sci.* Springer-Verlag, 1995.
21. J. Robinson. A machine-oriented logic based on the resolution principle. *J. ACM*, 12:23–41, 1965.
22. M. Tofte. Type inference for polymorphic references. *Information and Computation*, 89:1–34, 1990.
23. P. Wadler. Comprehending monads. In *Conference on Lisp and Functional Programming*, pages 61–78, June 1990.
24. P. Wadler. The essence of functional programming. In *Proc. 19th ACM Symp. Principles of Programming Languages*, 1992.
25. M. Wand. A simple algorithm and proof for type inference. *Fundementa Informaticae*, 10:115–122, 1987.

Verification of Compiler Correctness for the WAM

Cornelia Pusch[*]

Fakultät für Informatik, Technische Universität München
80290 München, Germany

E-mail: pusch@informatik.tu-muenchen.de

Abstract. Relying on an derivation of the Warren Abstract Machine (WAM) by stepwise refinement of Prolog models by Börger and Rosenzweig we present a formalization of an operational semantics for Prolog. Then we develop four refinement steps towards the Warren Abstract Machine (WAM). The correctness and completeness proofs for each step have been elaborated with the theorem prover Isabelle using the logic HOL.

1 Introduction

In the area of logic programming, Prolog ranks among the most prominent programming languages. The development of efficient compilation techniques allows the application of logic programming even in large-scale software development. One of the main contributions to this field is due to D. Warren, having developed a sophisticated compilation concept known as the Warren Abstract Machine (WAM), which serves as basis for a large number of prolog implementations.

While Prolog benefits from a well-defined semantics derived from its logical roots, the development of the WAM is not based on theoretical investigations, correctness is in general "proved" by successful testing. For this reason, several approaches have been made to develop a formal verification for the WAM [Rus92], [BR94].

While in [Rus92], correctness is shown for a specific Prolog compiler translating Prolog programs into WAM code, [BR94] provides a correctness proof for a whole class of compilers by formulating general compiler assumptions. The specification of an operational semantics for Prolog is given in terms of *evolving algebras*, and a development of the WAM is given by stepwise refinement, outlining the proofs for correctness and completeness of each refinement step. However, the proofs are not complete, their description remains semi-formal.

A first attempt to check the proofs by machine was made with the theorem proving system KIV [Sch95]. This case study revealed that the formal proofs were significantly more involved than estimated by [BR94].

---

[*] Research supported by DFG grant Br 887/4-3, Deduktive Programmentwicklung

In our paper we present the formalization of the correctness proofs in the Isabelle system [Pau94]. In contrast to the KIV formalization we do not use the framework of evolving algebras. Our development starts from a slightly different operational semantics for Prolog [DM88]. The operational semantics and all the refinement steps towards the WAM considered so far are formalized in higher order logic. The reasons for refraining from embedding the formalism of evolving algebras are discussed in section 4.

The rest of this paper is structured as follows. Section 2 provides a short introduction to Isabelle. In section 3 formalizations of the syntax and operational semantics of Prolog are given. In section 4 the refinement steps towards the WAM are elaborated. In section 5 the proof principles are discussed, and section 6 summarizes the results of the case study and outlines future work.

2 Isabelle

Isabelle is a generic theorem prover, where new logics are introduced by specifying their syntax and rules of inference. Proof procedures can be expressed using tactics and tacticals. A detailed introduction to the Isabelle system can be found in [Pau94].

The formalization and proofs described in this paper are based on the instantiation of Isabelle for higher order logic, called Isabelle/HOL.

The new release of Isabelle/HOL comes along with a graphical interface, allowing the use of mathematical symbols like $\forall$ and $\exists$. Therefore, the presentation of the formalization in this paper corresponds to the Isabelle input (except the introduction of some abbreviations for better readability).

3 Prolog Syntax and Semantics

This section describes the syntactic categories of Prolog programs and their formalization in Isabelle/HOL. Then we give an operational semantics by means of inference rules. More detailed information about logic programming can be found in [Apt90] and [Llo87].

3.1 Syntax

Since we do not have to reason about the exact structure of terms and formulae, we start with the notions of predicates and atoms. Furthermore we do not deal with the explicit construction of atoms by a predicate symbol followed by a list of terms. We just assume the existence of types for predicates and atoms together with a function returning the predicate symbol of an atom. Therefore, this part of our formalization is not definitional:

$$\begin{aligned}\text{types}\quad &\text{Pred}\\ &\text{Atom}\\ \text{consts pname} &:: \text{Atom} \Rightarrow \text{Pred}\end{aligned}$$

In logic programming syntax, a positive literal is an atom, a negative literal is the negation of an atom, and a program clause is a set of literals, containing exactly one positive literal, which is called the head. The remaining negative literals are called the body of the clause. Finally, a logic program is a set of program clauses.

However, these definitions are not suitable for the discussion of Prolog implementations. In order to describe the computational behavior of Prolog programs, one usually considers specific depth-first search strategies (SLD-resolution) and the use of the cut symbol, which is an impure but central control facility of Prolog. Therefore, we have to redefine the notions of literals and program clauses as follows[2]:

$$\begin{aligned}\text{datatype Lit} &= \text{Cut} \\ &\mid \text{Atm Atom} \\ \text{types Clause} &= \text{Atom} \times (\text{Lit list})\end{aligned}$$

In our terminologies a literal is either the cut symbol or an atom. This covers the notion of a negative literal and introduces the cut symbol. A program clause is a pair consisting of an atom (the head) and a list of literals (the body). This definition ensures that the cut never occurs as the head of a clause. A logic program is again a list of program clauses and a goal is a list of literals:

$$\begin{aligned}\text{types Program} &= \text{Clause list} \\ \text{Goal} &= \text{Lit list}\end{aligned}$$

In the following we introduce the concepts of substitution, unification and renaming. Since we abstracted away from terms and the construction of atoms we cannot give complete definitions for these functions. However, this is not a severe drawback as these concepts are well understood. Therefore we just axiomatize some minimal properties essential to the further proofs.

Substitutions are represented by the type Subst coming along with the functions

$$\begin{aligned}\text{consts @subcomp} &:: \text{Subst} \Rightarrow \text{Subst} \Rightarrow \text{Subst} & (\text{"\_ o \_"}) \\ \text{@subapp} &:: \text{Subst} \Rightarrow \text{Atom} \Rightarrow \text{Atom} & (\text{"\$"})\end{aligned}$$

for composition and substitution application. The identifiers beginning with a @ declare the names to be used just for the internal representation in Isabelle. The user has to apply the names given in parentheses, where o is introduced as infix operator. The function

$$\text{consts mgu} :: \text{Atom} \Rightarrow \text{Atom} \Rightarrow \text{Result}$$

returns the most general unifier of two atoms, if there is one, and a fail value otherwise. This optional result value is modeled by defining an error monad:

$$\begin{aligned}\text{datatype 'a maybe} &= \text{Ok 'a} \\ &\mid \text{Fail} \\ \text{types Result} &= \text{Subst maybe}\end{aligned}$$

---

[2] The **datatype** construct generates axioms for free data types: injectiveness, distinctness and an induction rule. The **types** construct is used here to introduce type synonyms.

When selecting a clause for unification, all variables occuring in that clause should be renamed, such that the so called variant does not have a variable name in common with the original goal and the set of clauses already used in the derivation process. The relevance of this renaming is discussed for example in [Apt90]. As clauses are built up by an atom and a list of literals, it is convenient to define an overloaded renaming function working on atoms and lists of literals. Overloading of function symbols can be realized in Isabelle by introducing a new type class, which is a subclass of the predefined class term of higher order terms:

$$\text{classes renamecl} < \text{term}$$

We then make Atom and Lit elements of renamecl. Furthermore we have to ensure that application of the type constructor list to an element of class renamecl yields an element of the same class:

$$\begin{aligned}
\text{arities Atom} &:: \text{renamecl} \\
\text{Lit} &:: \text{renamecl} \\
\text{list} &:: (\text{renamecl})\text{renamecl}
\end{aligned}$$

The intention behind the renaming function is that it takes a value and a renaming index returning a value which is equal to the input up to the variable names. By that each instance of a variable is made unique. As we will see later the renaming index is increased by the interpreter function after each successfull unification:

$$\begin{aligned}
\text{types Rename} &= \text{nat} \\
\text{consts @rename} &:: ('a :: \text{renamecl}) \Rightarrow \text{Rename} \Rightarrow' a :: \text{renamecl} \quad ("\uparrow")
\end{aligned}$$

For the proofs carried out so far, we only needed the following basic properties of the functions for substitution, unification and renaming:

$$\begin{aligned}
\text{rules pname a1} &\neq \text{pname a2} \Longrightarrow \text{mgu a1 a2} = \text{Fail} \\
\text{pname } (\uparrow \text{ a vi}) &= \text{pname a} \\
\text{pname } (\$ \text{ sub a}) &= \text{pname a}
\end{aligned}$$

The first axiom states that unification fails for two atoms with different predicate symbols. The next two axioms express that renaming and substitution application do not affect the predicate symbol of an atom.

3.2 Operational Semantics

The semantics of Prolog programs is usually given in terms of the model theory of first order logic. Since this approach ignores the behavioral aspects of Prolog like sequential depth-first search and the cut control facility, S. Debray and P. Mishra developed a denotational as well as an equivalent operational semantics expressing the properties of interest [DM88].

Hereafter we will give an operational semantics by the definition of an interpreter, which is almost identical to the one described in [DM88] with just some slight modifications. In our approach we chose a different renaming function, according to the formalization of Börger and Rosenzweig [BR94]. Furthermore we considered just one single possible answer substitution (an extension to multiple answer substitutions is under construction). Another difference consists in the formalization of the interpreter function. In [DM88], the interpreter is defined by a set of recursive equations. Because computation does not necessarily terminate in Prolog programming, we have to deal with partial functions. Since in Isabelle/HOL functions are total, we have to model partiality by inductive relations.

The computation state of the interpreter is described by so called configurations. A configuration consists of a list of clauses describing the Prolog program, a computation stack storing different backtrack points, and a renaming index. Therefore, we define:

$$\text{types Config} = \text{Program} \times \text{CStack} \times \text{Rename}$$

Each element of the computation stack consists of the substitution computed so far, the goal still to be executed, and a list of candidate clauses that are yet to be tried in solving the leftmost literal of the corresponding goal.

To model the effect of cuts in Prolog the goal is not presented linearly but decomposed in a list of "decorated" subgoals: each subgoal has to maintain its own continuation information, which is just a part of the entire computation stack. If a cut is encountered while processing a subgoal, the tail of the current computation stack is replaced by the continuation stack stored along with the subgoal. This corresponds to the deletion of all those backtrack points set up by literals on the left of the cut as well as the backtrack point for the parent goal (i.e. the goal which caused the clause containing the cut to be activated), which is the usual Prolog semantics for cut. At the end of this section we will see an example for this. Now we define the computation stack as:

datatype CStack = Es
 | "##" (Subst × (Goal × CStack)list × Clause list) CStack
   (infixr 70)

Now, we define an interpreter relation for Prolog programs:

consts   interp0   :: (Config × Result)set
syntax   @interp0   :: Config ⇒ Result ⇒ bool   (" _ $\xrightarrow{i_0}$ _"[0,95]95)
translations config $\xrightarrow{i_0}$ res == (config, res) ∈ interp0

The interpreter relation interp0 is defined as a set of pairs. The syntax section introduces an infix operator $\xrightarrow{i_0}$ for this relation, for which a translation into the set representation is given. Note that the i_0 denotes the base interpreter. We will refer to the interpreter after n refinement steps as i_n.

We give an inductive definition of the interpreter $\xrightarrow{i_0}$ by means of inference rules, where multiple premises are stacked on top of each. Note that the predefined type of lists comes along with [] for the empty list and the infix operator # for list construction:

If the computation stack is empty, execution terminates returning a fail value:

$$\text{query\_failed} \quad \frac{}{(\text{db}, \text{Es}, \text{vi}) \xrightarrow{io} \text{Fail}}$$

If the current list of decorated subgoals is empty, execution terminates returning the current substitution:

$$\text{query\_success} \quad \frac{}{(\text{db}, (\text{sub}, [], \text{cll})\#\#\text{sl}, \text{vi}) \xrightarrow{io} \text{Ok sub}}$$

If we are interested in all possible answer substitutions, execution has to be continued by processing the tail sl of the stack.

If the first element of the current decorated subgoals list is empty, execution continues by processing the rest of the decorated subgoals list:

$$\text{goal\_success} \quad \frac{(\text{db}, (\text{sub}, \text{ds}, \text{db})\#\#\text{sl}, \text{vi}) \xrightarrow{io} \text{res}}{(\text{db}, (\text{sub}, ([], \text{ctp})\#\text{ds}, \text{cll})\#\#\text{sl}, \text{vi}) \xrightarrow{io} \text{res}}$$

If the first subgoal of the current decorated subgoals list begins with a cut, the tail of the computation stack is replaced by the continuation ctp of the current subgoal:

$$\text{cut} \quad \frac{(\text{db}, (\text{sub}, (\text{ls}, \text{ctp})\#\text{ds}, \text{db})\#\#\text{ctp}, \text{vi}) \xrightarrow{io} \text{res}}{(\text{db}, (\text{sub}, (\text{Cut}\#\text{ls}, \text{ctp})\#\text{ds}, \text{cll})\#\#\text{sl}, \text{vi}) \xrightarrow{io} \text{res}}$$

All remaining rules hold for configurations, where the current decorated subgoals list is not empty and the first subgoal does not start with a cut but with an atom.

If the current choice point contains no more candidate clauses, execution proceeds by backtracking to the most recent choice point, i.e. popping the current one from the computation stack:

$$\text{back} \quad \frac{(\text{db}, \text{sl}, \text{vi}) \xrightarrow{io} \text{res}}{(\text{db}, (\text{sub}, ((\text{Atm } x)\#\text{ls}, \text{ctp})\#\text{ds}, [])\#\#\text{sl}, \text{vi}) \xrightarrow{io} \text{res}}$$

If the list of clauses still to be tried contains at least one element, two different cases have to be considered:

If unification of the leftmost literal in the current subgoal with the head of the first candidate clause fails, the next candidate clause has to be tried:

$$\text{atm1} \quad \frac{\text{mgu}(\$ \text{ sub } x)(\uparrow h \text{ vi}) = \text{Fail} \quad (\text{db}, (\text{sub}, ((\text{Atm } x)\#\text{ls}, \text{ctp})\#\text{ds}, \text{cs})\#\#\text{sl}, \text{vi}) \xrightarrow{io} \text{res}}{(\text{db}, (\text{sub}, ((\text{Atm } x)\#\text{ls}, \text{ctp})\#\text{ds}, (h, b)\#\text{cs})\#\#\text{sl}, \text{vi}) \xrightarrow{io} \text{res}}$$

If unification succeeds with a substitution sub', execution proceeds by extending the computation stack with a new choice point chp and incrementing the renaming index:

$$\text{atm2} \quad \frac{\text{mgu}(\$ \text{ sub } x)(\uparrow h \text{ vi}) = \text{Ok sub}'}{(\text{db}, \text{chp}\#\#(\text{sub}, ((\text{Atm } x)\#\text{ls}, \text{ctp})\#\text{ds}, \text{cs})\#\#\text{sl}, \text{vi}+1) \xrightarrow{io} \text{res}}$$
$$\frac{}{(\text{db}, (\text{sub}, ((\text{Atm } x)\#\text{ls}, \text{ctp})\#\text{ds}, (h, b)\#\text{cs})\#\#\text{sl}, \text{vi}) \xrightarrow{io} \text{res}}$$

where $\text{chp} = (\text{sub}' \circ \text{sub}, (\uparrow b \text{ vi}, \text{sl})\#(\text{ls}, \text{ctp})\#\text{ds}, \text{db})$

The new choice point chp contains an updated substitution, a subgoal list where the unified atom is replaced by a new subgoal containing the body of the selected clause, and the whole program serving as candidate clauses for the new subgoal. The decoration of the new subgoal has to be set to the tail of the old computation stack. In the old choice point, the selected clause has to be removed from the list of untried clauses.

The formalization of inductive sets is supported in Isabelle/HOL by a special package, where all generated rules are automatically proved as theorems.

Example 1 In the following we give a little example, to see how the interpreter works on the computation stack. The most interesting point is to see how the list of decorated subgoals evolves. Therefore we omit in the representation the substitutions and candidate clauses:
Consider the Prolog program

$$\begin{array}{ll} o : - \; p, x. & q : - \; s. \\ p : - \; q, !, r. & q \; . \\ p \; . & x \; . \end{array}$$

and query o. Computation starts with an initial computation stack containing the query decorated with an empty stack:

$$[\langle [o], \text{Es} \rangle]$$

After some execution steps the computation stack looks as follows:

$$[\langle [s], c_2 \rangle, \langle [!, r], c_1 \rangle, \langle [x], \text{Es} \rangle, \langle [], \text{Es} \rangle]$$
$$[\langle [q, !, r], c_1 \rangle, \langle [x], \text{Es} \rangle, \langle [], \text{Es} \rangle]$$
$$[\langle [p, x], \text{Es} \rangle, \langle [], \text{Es} \rangle]$$
$$[\langle [o], \text{Es} \rangle]$$

where $c_1 = [\langle [o], \text{Es} \rangle]$ and $c_2 = \dfrac{[\langle [p, x], \text{Es} \rangle, \langle [], \text{Es} \rangle]}{[\langle [o], \text{Es} \rangle]}$

Since unification with s fails for all program clauses, the top element of the computation stack is popped. The next clause to be tried for q succeeds immediately, then we get:

$$[\langle [!, r], c_1 \rangle, \langle [x], \text{Es} \rangle, \langle [], \text{Es} \rangle]$$
$$[\langle [p, x], \text{Es} \rangle, \langle [], \text{Es} \rangle]$$
$$[\langle [o], \text{Es} \rangle]$$

Now a cut is encountered in the current subgoal. As described, the tail of the computation stack is replaced by c_1, which yields:

$$\frac{[\langle[r], c_1\rangle, \langle[x], \mathsf{Es}\rangle, \langle[], \mathsf{Es}\rangle]}{[\langle[o], \mathsf{Es}\rangle]}$$

Unification with r fails for all program clauses, therefore backtracking has to be executed. We see now that the remaining clause for p is no more considered, according to the meaning of the cut.

4 Towards the WAM

The first Prolog compiler was developed at the University of Edinburgh by D.H. Warren in 1977. The Warren Abstract Machine (WAM) is a refinement of this system. Roughly speaking, the WAM is an abstract machine consisting of a memory architecture and an instruction set tailored to Prolog. It is based on the concept of a virtual machine in order to achieve portability to a wide range of hardware configurations [Boi93]. In this paper, we do not describe the details of the WAM, since we are just doing some refinement steps towards the WAM, starting from our operational semantics presented in the previous section. For more information the reader might refer to [AK91], which gives a more detailed introduction to the WAM, rather than the original paper [War83].

In [BR94] Börger and Rosenzweig developed a methodical derivation of the WAM starting from an operational semantics for Prolog. They provide a correctness proof for a whole class of compilers by formulating general compiler assumptions. The specification of an operational semantics for Prolog is given in terms of *evolving algebras* [Gur95]. Their development of the WAM is partitioned into 12 refinement steps, each of which introduces a new aspect of the WAM. For each step, they outline the proofs for correctness and completeness. However, these proofs are not complete, their description remains semi-formal.

A first attempt to check the proofs by machine was made with the theorem proving system KIV [Sch95]. This case study revealed that the formal proofs were significantly more involved than estimated by [BR94]. For example, the correctness proof of the first refinement step used an invariant property which covered an entire page. Studying this proof, we got the impression that this complexity is caused to a large extent by the formalism of evolving algebras. For instance, the manipulation of inductive data structures seems to be quite tedious. However, this concept turned out to be central to the formalization of this case study. On the other hand higher order logic offers advanced features for the treatment of inductive data structures. Therefore, we coded the operational semantics directly in HOL as presented above and refrained from embedding the formalism of evolving algebras.

Nevertheless, we could adopt the structure of the refinement steps developed by Börger and Rosenzweig in [BR94] which turned out to be very suitable for the realization of the proof task.

We now outline the steps of our development and present the formal definition of the refined interpreter after the fourth step.

4.1 Introduction of pointers

Copying parts of the computation stack into the decorated subgoals list is very inefficient. Therefore, the first improvement consists of replacing the copies by pointers to the original stack, called cut points. Hence, we define:

$$\text{types Index} = \text{nat}$$
$$\text{CArray} = (\text{Subst} \times (\text{Goal} \times \text{Index})\text{list} \times \text{Clause list})\text{list}$$

The type CArray replaces CStack in the configuration. Since the new stack definition no longer contains nested recursion, the definition of CArray can be based on the predefined type of lists. We do not need to introduce a new data type. To allow the deletion of choice points on the stack up to a given index, we provide a function

$$\text{consts ntail} :: \text{Index} \Rightarrow \text{CArray} \Rightarrow \text{CArray}$$

which takes an index i and a computation stack, and returns the back end of the stack of length i.

4.2 Optimizing the list of candidate clauses

Up to now, the list of candidate clauses for a new subgoal consists of the entire program. However, it is clear that only some of the clauses are likely to match the selected atom. Therefore, in a second step we restrict the set of candidate clauses by a preselection depending on the currently selected atom. This is done by a function

$$\text{consts pdef} :: \text{Atom} \Rightarrow \text{Program} \Rightarrow \text{Clause list}$$

which filters out those clauses from a given program whose heads consist of the same predicate symbol as the currently selected atom. Additionally, the configuration is extended by a new component which describes a triple of registers holding the current values for substitution, decorated subgoals list and candidate clause list. The computation stack is left to maintain the remaining backtracking points.

4.3 Reusing choice points

During execution, it is often the case that information is popped from the stack into the registers, and in a later stage, almost identical information is pushed back onto the stack. This information transfer can be optimized by leaving the formerly popped choice point on the stack and just changing part of its contents. This is related to the well-known peephole optimization in compiler construction.

4.4 Deleting useless choice points

The next optimization step consists of deleting trivial choice points. This means a choice points including an empty candidate clause list is no longer pushed onto the stack: whenever execution returned to this point, it would be immediately popped by backtracking.

4.5 The optimized interpreter model

After these four refinement steps, the formalization of the interpreter has undergone the following changes:

The configuration of our interpreter has been extended by two components. The first one describes different modes of the computation. We distinguish four modes: In call mode execution proceeds until an atom is encountered in the current subgoal or computation terminates. In try mode a choice point is pushed onto the computation stack. In enter mode unification is attempted, and in retry mode reuse of choice points is done. The second extension is an index to the computation stack, which stores the value of the current cut point. This cut point register will be needed for the further development. We therefore define:

$$\text{datatype Mode} = \text{Call} \mid \text{Try} \mid \text{Enter} \mid \text{Retry}$$
$$\text{types} \quad \text{Regs} = (\text{Subst} \times (\text{Goal} \times \text{Index})\text{list} \times \text{Clause list})$$
$$\text{Config} = \text{Program} \times \text{CArray} \times \text{Regs} \times \text{Rename} \times \text{Mode} \times \text{Index}$$

The inductive definition of the interpreter $\xrightarrow{i_4}$ is as follows:

If the decorated goals register is empty, the query was successful, returning the content of the substitution register as result:

$$\text{query\_success} \quad \frac{}{(\text{db}, \text{arr}, (\text{sreg}, [], \text{creg}), \text{vi}, \text{Call}, \text{ct}) \xrightarrow{i_4} \text{Ok sreg}}$$

If the first subgoal in the decorated subgoals register is empty, execution proceeds the rest of the decorated goals list:

$$\text{goal\_succes} \quad \frac{(\text{db}, \text{arr}, (\text{sreg}, \text{ds}, \text{creg}), \text{vi}, \text{Call}, \text{ct}) \xrightarrow{i_4} \text{res}}{(\text{db}, \text{arr}, (\text{sreg}, ([], \text{ctp})\#\text{ds}, \text{creg}), \text{vi}, \text{Call}, \text{ct}) \xrightarrow{i_4} \text{res}}$$

If a cut is encountered, the backtracking stack is shortened upto the cut point of the current subgoal:

$$\text{cut} \quad \frac{(\text{db}, \text{ntail ctp arr}, (\text{sreg}, (\text{ls}, \text{ctp})\#\text{ds}, \text{creg}), \text{vi}, \text{Call}, \text{ct}) \xrightarrow{i_4} \text{res}}{(\text{db}, \text{arr}, (\text{sreg}, (\text{Cut}\#\text{ls}, \text{ctp})\#\text{ds}, \text{creg}), \text{vi}, \text{Call}, \text{ct}) \xrightarrow{i_4} \text{res}}$$

The following two rules hold for configurations, where the current subgoal begins with an atom, but the predicate of the current atom is not defined. This is the case, if the current atom does not occur in any head of a program clause.

If the backtracking stack is empty, computation fails:

$$\text{call1} \quad \frac{\text{pdef x db} = []}{(\text{db}, [], (\text{sreg}, ((\text{Atm x})\#\text{ls}, \text{ctp})\#\text{ds}, \text{creg}), \text{vi}, \text{Call}, \text{ct}) \xrightarrow{i_4} \text{Fail}}$$

If the backtracking stack is not empty, execution is processed in Retry mode:

$$\text{call2} \quad \frac{\text{pdef x db} = [] \quad (\text{db}, \text{x}\#\text{xs}, (\text{sreg}, ((\text{Atm x})\#\text{ls}, \text{ctp})\#\text{ds}, \text{creg}), \text{vi}, \text{Retry}, \text{ct}) \xrightarrow{i_4} \text{res}}{(\text{db}, \text{x}\#\text{xs}, (\text{sreg}, ((\text{Atm x})\#\text{ls}, \text{ctp})\#\text{ds}, \text{creg}), \text{vi}, \text{Call}, \text{ct}) \xrightarrow{i_4} \text{res}}$$

If the definition of the current atom contains at least one clause, computation continues with mode set to Try and the candidate clauses and cut point registers updated:

$$\text{call3} \quad \frac{\text{pdef x db} = \text{c\#cs} \quad (\text{db}, \text{arr}, (\text{sreg}, ((\text{Atm x})\#\text{ls}, \text{ctp})\#\text{ds}, \text{c\#cs}), \text{vi}, \text{Try}, \text{length arr}) \xrightarrow{i_4} \text{res}}{(\text{db}, \text{arr}, (\text{sreg}, ((\text{Atm x})\#\text{ls}, \text{ctp})\#\text{ds}, \text{creg}), \text{vi}, \text{Call}, \text{ct}) \xrightarrow{i_4} \text{res}}$$

If computation is in Try mode, two different cases have to be distinguished.

In the first case, the candidate clauses register contains at least two clauses, one to be tried immediately and at least one to be pushed onto the stack. Then execution proceeds in Enter mode with a new choice point pushed onto the stack:

$$\text{try1} \quad \frac{(\text{db}, (\text{sreg}, \text{dreg}, \text{c2\#cs})\#\text{arr}, (\text{sreg}, \text{dreg}, \text{c1\#c2\#cs}), \text{vi}, \text{Enter}, \text{ct}) \xrightarrow{i_4} \text{res}}{(\text{db}, \text{arr}, (\text{sreg}, \text{dreg}, \text{c1\#c2\#cs}), \text{vi}, \text{Try}, \text{ct}) \xrightarrow{i_4} \text{res}}$$

If there is only one candidate clause to be tried, no additional choice point has to be stored on the stack:

$$\text{try2} \quad \frac{(\text{db}, \text{arr}, (\text{sreg}, \text{dreg}, [\text{c}]), \text{vi}, \text{Enter}, \text{ct}) \xrightarrow{i_4} \text{res}}{(\text{db}, \text{arr}, (\text{sreg}, \text{dreg}, [\text{c}]), \text{vi}, \text{Try}, \text{ct}) \xrightarrow{i_4} \text{res}}$$

If unification fails in Enter mode, the result of the computation depends on the contents of the backtracking stack.

If there are no more backtracking points, computation terminates returning a fail value:

$$\text{enter1} \quad \frac{\text{mgu}(\$ \text{ sreg x})(\uparrow \text{ h vi}) = \text{Fail}}{(\text{db}, [], (\text{sreg}, ((\text{Atm x})\#\text{ls}, \text{ctp})\#\text{ds}, (\text{h}, \text{b})\#\text{cs}), \text{vi}, \text{Enter}, \text{ct}) \xrightarrow{i_4} \text{Fail}}$$

If the backtracking contains at least one element, computation is continued in Retry mode:

$$\text{enter2} \quad \frac{\text{mgu}(\$ \text{ sreg x})(\uparrow \text{ h vi}) = \text{Fail} \quad (\text{db}, \text{x\#xs}, (\text{sreg}, ((\text{Atm x})\#\text{ls}, \text{ctp})\#\text{ds}, (\text{h}, \text{b})\#\text{cs}), \text{vi}, \text{Retry}, \text{ct}) \xrightarrow{i_4} \text{res}}{(\text{db}, \text{x\#xs}, (\text{sreg}, ((\text{Atm x})\#\text{ls}, \text{ctp})\#\text{ds}, (\text{h}, \text{b})\#\text{cs}), \text{vi}, \text{Enter}, \text{ct}) \xrightarrow{i_4} \text{res}}$$

If unification succeeds, execution proceeds in Call mode after updating the registers:

$$\text{enter3} \quad \frac{\text{mgu}(\$ \text{ sreg x})(\uparrow \text{ h vi}) = \text{Ok sub}' \quad (\text{db}, \text{arr}, \text{regs}, \text{vi}+1, \text{Call}, \text{ct}) \xrightarrow{i_4} \text{res}}{(\text{db}, \text{arr}, (\text{sreg}, ((\text{Atm x})\#\text{ls}, \text{ctp})\#\text{ds}, (\text{h}, \text{b})\#\text{cs}), \text{vi}, \text{Enter}, \text{ct}) \xrightarrow{i_4} \text{res}}$$

where regs = $(\text{sub}' \circ \text{sreg}, (\uparrow \text{b vi}, \text{sl})\#(\text{ls}, \text{ctp})\#\text{ds}, (\text{h}, \text{b})\#\text{cs})$

In Retry mode, the information of the top level backtracking point is reused, where two different cases have to be considered:

In the first case, the top level element contains more than one candidate clauses. Then the backtrack information is loaded into the registers while the candidate clauses list is updated in the top level stack element:

$$\text{retry1} \quad \frac{(\text{db}, (\text{sub}, \text{dcl}, \text{c1}\#\text{cs})\#\text{xs}, (\text{sub}, \text{dcl}, \text{c}\#\text{c1}\#\text{cs}), \text{vi}, \text{Enter}, \text{length xs}) \xrightarrow{i_4} \text{res}}{(\text{db}, (\text{sub}, \text{dcl}, \text{c}\#\text{c1}\#\text{cs})\#\text{xs}, (\text{sreg}, \text{dreg}, \text{creg}), \text{vi}, \text{Retry}, \text{ct}) \xrightarrow{i_4} \text{res}}$$

If the backtracking point contains just one single clause still to be tried, the backtrack information is loaded into the registers and the current top level stack element is deleted:

$$\text{retry2} \quad \frac{(\text{db}, \text{xs}, (\text{sub}, \text{dcl}, [\text{c}]), \text{vi}, \text{Enter}, \text{length xs}) \xrightarrow{i_4} \text{res}}{(\text{db}, (\text{sub}, \text{dcl}, [\text{c}])\#\text{xs}, (\text{sreg}, \text{dreg}, \text{creg}), \text{vi}, \text{Retry}, \text{ct}) \xrightarrow{i_4} \text{res}}$$

Example 2 Now we will see how computation has changed in our example: In addition to the computation stack there is now a register containing the current decorated subgoal list. In the initial state, the query is stored in the register and the stack is empty. After some execution steps these components look as follows:

$$[\langle [\text{s}], 1 \rangle, \langle [!, \text{r}], 0 \rangle, \langle [\text{x}], 0 \rangle, \langle [], 0 \rangle]$$

$$\begin{array}{|c|} \hline [\langle [\text{q},!,\text{r}], 0 \rangle, \langle [\text{x}], 0 \rangle, \langle [], 0 \rangle] \\ \hline [\langle [\text{p},\text{x}], 0 \rangle, \langle [], 0 \rangle] \\ \hline \end{array}$$

You may notice that the bottom stack element of the example in 3.2 does no longer occur in this computation. This results of the fact that there exists just one single program clause for o. After having tried it, it would be useless to return to this point since the list of candidate clauses would be empty.

Since there is no program clause for s the information of the top level backtracking point is reused. There is just one clause still to be tried, therefore the backtracking point is popped from the stack into the register:

$$[\langle [!,\text{r}], 0 \rangle, \langle [\text{x}], 0 \rangle, \langle [], 0 \rangle] \qquad [\langle [\text{p},\text{x}], 0 \rangle, \langle [], 0 \rangle]$$

Now a cut is encountered in the current subgoal which causes the backtracking stack to be set to the empty stack:

$$[\langle [\text{r}], 0 \rangle, \langle [\text{x}], 0 \rangle, \langle [], 0 \rangle] \qquad []$$

Since there is no program clause for r, computation terminates returning Fail.

5 Proof principles

In a refinement step, a more concrete interpreter model is developed from an abstract model. To establish a relationship between two different levels, we have to define an abstraction function F, translating configurations of the concrete interpreter to configurations of the abstract one.

We call an interpreter $\xrightarrow{i_1}$ a correct refinement of the interpreter $\xrightarrow{i_0}$, if every computation of $\xrightarrow{i_1}$ starting with an initial configuration terminates returning a result res provided the computation of $\xrightarrow{i_0}$ returns res starting with an equivalent initial configuration. The notion of initial configuration is explained below.

A configuration of $\xrightarrow{i_0}$ is a triple consisting of the Prolog program, a computation stack, and a renaming index. In an initial configuration, the computation stack contains exactly one choicepoint, consisting of a substitution which is typically set to the identity map, a decorated subgoals list containing the goal to be solved decorated by the empty stack, and a list of candidate clauses which is typically set to the whole program. The initial configuration for $\xrightarrow{i_1}$ just differs in the decoration of the goal, where now a pointer to the empty stack is held. Application of the abstraction function F to the initial configuration of $\xrightarrow{i_1}$ returns the equivalent initial configuration of $\xrightarrow{i_0}$. The correctness theorem is then formalized as follows:

$$\text{correctness} \quad \frac{((\text{db}, [(\text{subst}, [(\text{goal}, 0)], \text{cll})], 0) \xrightarrow{i_1} \text{res})}{(\mathsf{F}(\text{db}, [(\text{subst}, [(\text{goal}, 0)], \text{cll})], 0) \xrightarrow{i_0} \text{res})}$$

Since this assertion cannot be proved directly, we have to show the validity of a more general theorem, holding for any given configuration. The following theorem can be proved by rule induction:

$$\text{i1\_implies\_i0} \quad \frac{\text{config\_ok config} \quad \text{config} \xrightarrow{i_1} \text{res}}{(\mathsf{F}\ \text{config}) \xrightarrow{i_0} \text{res}}$$

Here, we had to introduce an additional assumption. The predicate config_ok restricts config to admissible configurations. One of the central proof tasks is to find the right restrictions. For each refinement step, several attempts were necessary to find the final solution.

Proving correctness is not sufficient to assure a really useful implementation. We could implement $\xrightarrow{i_1}$ by a never-halting function fulfilling the correctness property. Therefore, we have to verify the completeness of the development step as well, which assures that every solution computed by $\xrightarrow{i_0}$ can be found by $\xrightarrow{i_1}$:

$$\text{completeness} \quad \frac{(\mathsf{F}(\text{db}, [(\text{subst}, [(\text{goal}, 0)], \text{cll})], 0) \xrightarrow{i_0} \text{res})}{((\text{db}, [(\text{subst}, [(\text{goal}, 0)], \text{cll})], 0) \xrightarrow{i_1} \text{res})}$$

Here again, a generalization of the theorem has to be proved:

$$\text{i0\_implies\_i1} \quad \frac{\text{config\_ok config}' \quad \mathsf{F}\text{config}' \xrightarrow{i_0} \text{res}}{\text{config}' \xrightarrow{i_1} \text{res}}$$

This technique of defining an abstraction function F and inductively proving correctness and completeness by finding suitable restrictions was common to all refinement steps considered so far.

6 Results and Future Work

The formalization and implementation of the proofs for four development steps took seven months in total. The formalization in Isabelle comprises about 900 lines, the proofs for correctness and completeness consist of approximately 3500 user interactions. Although Isabelle offers a certain degree of automation, significant parts of the proofs have to be guided by user interaction. Better proof support by the system would facilitate the realization of complex case studies like the present one. This concerns in particular an improvement of error messages returned by the system.

As described, we decided to refrain from embedding the formalism of evolving algebras and coded the different refinement steps of an Prolog interpreter directly in higher order logic. Because of this, we were able to make intensive use of Isabelle's features concerning the treatment of inductive data structures and recursive concepts. The type class mechanism was profitably used for overloading. It is our opinion that the adaption of the formalization to higher order logic simplified the complexity of the proof invariants to a large extent. Due to that, we were able to conduct a large-scale case study like the present one: as far as we know this is one of the biggest mechanized proofs concerning operational semantics. In general this cannot be realized without careful decomposition of the proof task. Here the adaption of the refinement steps developed by Börger and Rosenzweig was essential to reduce the complexity of each step to a manageable size.

Our next steps consist in extending our formalization to the computation of multiple answer substitutions, which corresponds closer to a real Prolog interpreter. However, we do not think that proofs will become more complicated by that.

Furthermore, the development steps towards the WAM not yet considered remain to be done. The next refinement step introduces parts of the WAM instruction set: the list of clauses defining a predicate is now translated by an abstract compiler into a sequence of instructions that achieves the indexing of the clauses together with its backtracking management [Boi93]. The proofs for this step are presumed to be even more complex than the presented ones due to the formalization of suitable compiler assumptions.

Acknowledgements I wish to thank Tobias Nipkow and Franz Regensburger for helpful discussions and constructive criticism.

References

[AK91] Hassan Aït-Kaci. *Warren's Abstract Machine, A Tutorial Reconstruction.* MIT Press, Cambridge, Massachusetts, 1991.

[Apt90] Krzysztof R. Apt. Logic programming. In J. van Leeuwen, editor, *Handbook of Theoretical Computer Science*, chapter 10, pages 495–574. Elsevier Science Publishers B.V., 1990.

[Boi93] Patrice Boizumault. *The Implementation of Prolog*. Princeton Series in Computer Science. Princeton University Press, Princeton, New Jersey, 1993.

[BR94] E. Börger and D. Rosenzweig. The WAM - Definition and Compiler Correctness. In C. Beierle and L. Plümer, editors, *Logic Programming: Formal Methods and Practical Applications*. Elsevier, 1994.

[DM88] Saumya K. Debray and Prateek Mishra. Denotational and Operational Semantics for Prolog. *J. Logic Programming*, (5):61–91, 1988.

[Gur95] Yuri Gurevich. Evolving Algebras 1993: Lipari Guide. In E. Börger, editor, *Specification and Validation Methods*, pages 9–36. Oxford University Press, 1995.

[Llo87] J. W. Lloyd. *Foundations of Logic Programming*. Springer, 1987.

[Pau94] L.C. Paulson. *Isabelle: A Generic Theorem Prover*, volume 828 of *LNCS*. Springer, 1994.

[Rus92] David M. Russinoff. A Verified Prolog Compiler for the Warren Abstract Machine. *J. Logic Programming*, (13):367–412, 1992.

[Sch95] G. Schellhorn. Von PROLOG zur WAM - Compilerverifikation mit KIV. Talk at the annual meeting of the GI section "Logic in Computer Science", Karlsruhe, Juni 1995.

[War83] D. H. Warren. An Abstract Prolog Instruction Set. Technical Report 309, SRI International, 1083.

Synthetic Domain Theory in Type Theory :
Another Logic of Computable Functions

Bernhard Reus

Ludwig-Maximilians-Universität
Munich, GERMANY
reus@informatik.uni-muenchen.de

Abstract. We will present a Logic of Computable Functions based on the idea of Synthetic Domain Theory such that all functions are *automatically* continuous. Its implementation in the LEGO proof-checker – the logic is formalized on top of the Extended Calculus of Constructions – has two main advantages. First, one gets machine checked proofs verifying that the chosen logical presentation of Synthetic Domain Theory is correct. Second, it gives rise to a LCF-like theory for verification of functional programs where continuity proofs are obsolete. Because of the powerful type theory even modular programs and specifications can be coded such that one gets a prototype setting for modular software verification and development.

1 Introduction

There exist several theorem provers and proof checkers supporting a logic of domains like the LCF system [Pau87], and higher-order versions like HOLCF [Reg94] or HOL-CPO [Age94]. All these systems provide a (higher-order) logic for classical domain theory and differ substantially in the way they treat domains or cpo-s as types. In HOL-CPO a cpo is described as a carrier set together with an order relation on it, so there is no proper type of cpo-s. HOLCF uses type classes, such that cpo-s and domains form a class. However, there is no proper *type of domains* in any approach. Of course, only fixpoints of *continuous functions* can be built.

Synthetic Domain Theory provides a setting for denotational semantics in which *all functions are continuous*. This is due to Dana Scott's slogan "domains as sets". Several approaches can be found in the literature [FMRS92, Hyl91, Tay91, LS95]. These approaches make heavy use of category and topos theory without consequently the internal language or they simply work in a PER-model. By contrast, in [RS93a, Reu95] we presented a *model-free* axiomatization of the complete ExPERs, called Σ-cpo-s, in a higher-order intuitionistic logic with additional axioms. This gives rise to a Σ-cpo-theory which can be extended to Σ-domains (Σ-cpos with least element). Domain constructors like $\longrightarrow$, $\longrightarrow_\bot$, $\times$, $(\_)_\bot$, $+$, $\otimes$, $\oplus$ can be defined as functors on the category of Σ-domains and strict maps. These functors must be automatically locally continuous. Additionally, one can prove that recursive domain equations given by internal mixed-variant functors can be solved in the category of Σ-domains with strict maps.

By contrast to LCF, admissibility can be expressed inside the logic, so we have all the necessary tools for program verification.

Through this logical approach we also gain access to formalization. The whole theory can be implemented in an appropriate interpreter for type theory, LEGO [LP92], where one can also build a *type of all domains*. In this paper we shall explain how this can be done. As we will have dependent products and sums, it is even possible to build modules, i.e. program modules as well as specification modules. For type theoretical specifications the *deliverables*-approach [BM92, RS93b] is particularly useful, so the resulting language provides a good playground for deriving modular programs together with formal correctness proofs.

More details as well as all proofs omitted in this paper can be found in [Reu95].

The paper is organized as follows. Section 2 introduces the ECC and the extension by an additional impredicative universe Set that contains the propositions. In the next section we add several non-logical axioms on top of ECC. The ideas of Synthetic Domain Theory are briefly discussed in Section 4 followed by the SDT-Axioms (Sect. 5). Then we browse through the core theory: Σ-posets, Σ-cpos and Σ-domains (Sect. 6). We will discuss the solution of recursive domain equations in Sect. 7. The last section is devoted to a short review of a sample correctness proof. Finally, the conclusions will point out some loose ends.

2 Extending the Extended Calculus of Constructions

2.1 Extended Calculus of Constructions (ECC)

The type theory we use is the Extended Calculus of Constructions (ECC) [Luo94] which combines an impredicative universe with predicative Martin-Löf type theory.

Informally, the hierarchy of predicative universes Type_j is ordered by a subtype relation $\preceq$, that is appropriately extended on Π and Σ-types. Moreover, any Type_j is an element of Type_{j+1}. The impredicative universe Prop of propositions is an element of Type_0 and also a subtype of Type_0. The subtype relation is transitive and closed under conversion, i.e. if $A \simeq B$ then A is a subtype of B and vice versa. In the next section this is extended by a new universe.

Naive set theoretic models exist neither for System F nor for CC because of impredicativity. Fortunately, the partial equivalence relations (PERs) provide an adequate semantics for System F, CC, and ECC (e.g. [Str91, Luo94]). A PER-model for the "extended" ECC will be discussed briefly in Section 2.3.

2.2 Adding new universes to ECC

The systems CC and ECC provide just one impredicative universe Prop. For SDT, however, it is convenient to have a second universe Set of sets. Therefore, we have to extend ECC by an additional impredicative universe. One must be careful, since Coquand has shown that adding impredicative universes can lead to inconsistencies [Coq86]. This is, however, only true for cumulative hierarchies of impredicative universes. Since Prop is *not* an element of our new universe Set, in our case there is no danger of inconsistency. This can be proved by providing a (realizability) model. Henceforth the extension of ECC by Set will be called ECC*.

Recall the following type formation rules in ECC. Note that we use a more informal calculus (á la Tarski) where one does not distinguish whether a type is considered as a type or an object. A more accurate description following Streicher [Str91] can be found in [Reu95].

$$\overline{\vdash \mathsf{Prop}:\mathsf{Type}_0} \qquad \overline{\vdash \mathsf{Type}_j:\mathsf{Type}_{j+1}}$$

$$\frac{\Gamma,x:A\vdash P:\mathsf{Prop}}{\Gamma\vdash \Pi x:A.P:\mathsf{Prop}} \qquad \frac{\Gamma\vdash A:\mathsf{Type}_j \quad \Gamma,x:A\vdash B:\mathsf{Type}_j}{\Gamma\vdash \Pi x:A.B:\mathsf{Type}_j}$$

$$\frac{\Gamma\vdash A:\mathsf{Type}_j \quad \Gamma,x:A\vdash B:\mathsf{Type}_j}{\Gamma\vdash \sum x:A.B:\mathsf{Type}_j} \qquad \frac{\Gamma\vdash M:A \quad \Gamma\vdash A':\mathsf{Type}_j}{\Gamma\vdash M:A'} \;(A\preceq A')$$

Moreover, $\mathsf{Prop} \preceq \mathsf{Type}_0 \preceq \mathsf{Type}_1 \ldots$ as described above.

DEFINITION 2.1 In order to get ECC* we add some similar rules for Set:

$$\overline{\vdash \mathsf{Set}:\mathsf{Type}_0} \qquad \mathsf{Prop} \preceq \mathsf{Set} \preceq \mathsf{Type}_0$$

$$\frac{\Gamma\vdash A:\mathsf{Type}_j \quad \Gamma,x:A\vdash B:\mathsf{Set}}{\Gamma\vdash \Pi x:A.B:\mathsf{Set}} \qquad \frac{\Gamma\vdash A:\mathsf{Set} \quad \Gamma,x:A\vdash B:\mathsf{Set}}{\Gamma\vdash \sum x:A.B:\mathsf{Set}}$$

So Set is an impredicative universe being an element and a subset of Type_0, closed under set-indexed sums.

These rules together with the rules of ECC form the extended system ECC*. We have coded them in the LEGO-interpreter. The new resulting system, called SDT-LEGO, is the one we will use in the following. But first we argue that this extension is logically sound by providing a model.

2.3 Extending the realizability-model for ECC

We change and extend the realizability model for ECC [Luo94, Str91].

DEFINITION 2.2 The interpretation of the type universe Type remains unchanged, namely

$$[\![\Gamma \vdash \mathsf{Type}_j : \mathsf{Type}_{j+1}]\!](\gamma) = \nabla\,(\omega\text{-}\mathbf{Set}(j)^{\mathrm{obj}}),$$

where ω-$\mathbf{Set}(j)$ is the category of ω-sets whose carrier sets are in the universe V_{κ_j} of the cumulative hierarchy of sets [Luo94] and ∇M denotes the ω-$\mathbf{Set}$ with carrier M and full (i.e. trivial) realizability relation.

In order to give a nice interpretation to Set we must change the interpretation of Prop. In fact, Prop will become proof-irrelevant:

$$[\![\Gamma \vdash \mathsf{Prop} : \mathsf{Type}_0]\!](\gamma) = \nabla\ (\mathbf{PER}_1^{\mathrm{obj}}),$$

where $\mathbf{PER}_1$ is the full subcategory of ω-**Set** which is isomorphic to the category of partial equivalence relations with at most one equivalence class. It is easy to see that $\mathbf{PER}_1$ is closed under arbitrary products.

Now we can use the category of partial equivalence relations to interpret Set:

$$[\![\Gamma \vdash \mathsf{Set} : \mathsf{Type}_0]\!](\gamma) = \nabla\ (\mathbf{PER}^{\mathrm{obj}}),$$

where $\mathbf{PER}$ is the full subcategory of ω-**Set** which is isomorphic to the category of partial equivalence relations. In [Luo94, Str91] it is shown that $\nabla\ (\mathbf{PER}^{\mathrm{obj}})$ has the required closure properties and that it lives in Type_0 since there it is used to interpret Prop.

It is obvious that $\mathbf{PER}_1^{\mathrm{obj}} \subseteq \mathbf{PER}^{\mathrm{obj}}$.

3 The logic

The logic we use is the one we get from ECC by the Curry-Howard-Isomorphism, i.e. higher-order intuitionistic logic.[1] In order to mimic a topos logic one needs some more principles. Sometimes we shall present axioms and definitions also in LEGO-syntax to give a "look-and-feel" of the theory. For the reader not familiar with LEGO we refer to [LP92] but shortly recall the most important tokens: curly brackets denote Π-types or $\forall$ (in case of propositions), -> denotes (non-dependent) function space or implication (in case of propositions), angle brackets denote $\sum$-types, square brackets denote extension of the current context, x == t denotes the definition of a macro x for t, [x:A]t denotes $\lambda x{:}A.\,t$. Our notion of equality is Leibniz equality Q. The abbreviations Ex, ExU, and, or, neg, iff stand for the logical connectives $\exists$, $\exists!$, $\wedge$, $\vee$, $\neg$, $\Leftrightarrow$, respectively. In function or type definitions we sometimes write x|A instead of x:A which means that the corresponding argument can be left out in applications. Tuples are written in round brackets and first and second projections are written x.1 and x.2, respectively. Application is denoted f x or f(x) and sometimes we also use x.f. We will write Type for Type(0) – anyway LEGO will compute the correct level of the universe.

We assume that extensionality for functions and the Axiom of Unique Choice hold. Moreover, we assume to have natural numbers $\mathbb{N}$ (in LEGO: N) and Booleans $\mathbb{B}$ (B) as inductive types with correspsonding induction principles. Note that the Axiom of Unique Choice is formulated with a sum rather than an existential quantifier in the conclusion, such that one gets a function by first projection i.e.

```
[ ACu_dep : {A : Type}{C : A->Set}{P : {a:A}(C a)->Prop}
            ( {x:A} ExU (P x) ) -> < f:{a:A}C a > {a:A} P a (f a)  ];
```

---
[1] SDT is not consistent with classical logic.

3.1 Subset types

There is no standard higher-order intuitionistic logic with subset types. We therefore model subset types – as usual – by sums, e.g. $\{x \in A \,|\, p(x)\}$ as $\sum x{:}A.\,p(x)$, in LEGO we write <x:A>P x. Because of the coding we must use coercion maps, i.e. if $y \in \sum x{:}A.\,p(x)$, then $\pi_1(y) \in A$. Thus $\sum x{:}A.\,p(x)$ must live in Set again. Since we know that Set is closed under dependent sums (of families of sets indexed by sets) and that Prop $\preceq$ Set, the "Σ-coded subsets" of a type $A \in$ Set indeed live again in Set.

Any mono $m : X \rightarrowtail Y$ describes a subset via $\{y \in Y \,|\, \exists x{:}X.\,m(x) = y\}$. The mono m is called $\neg\neg$-closed if $\exists x{:}X.\,m(x) = y$ is $\neg\neg$-closed[2] for all y, i.e. $\neg\neg(\exists x{:}X.\,m(x) = y) \Rightarrow (\exists x{:}X.\,m(x) = y)$. Note that for $\neg\neg$-closed propositions P the proof rule $\frac{A \Rightarrow P \quad \neg\neg A}{P}$ is valid. This rule is often used with $A \equiv \neg B \vee B$ which does not hold intuitionistically, but $\neg\neg(\neg B \vee B)$ holds for any B. In order to prove $P \equiv \exists x{:}X.\,m(x) = y$ (i.e. "$y \in X$") using the rule above, one needs that $m : X \rightarrow Y$ is $\neg\neg$-closed. The predicate "$x \in X \subseteq Y$" is mirrored by image x m as outlined below. Here the mono m:X->Y codes the set X as a subset of Y. Consequently, one can define what a $\neg\neg$-closed map (mapDnclo) and a $\neg\neg$-closed mono (dnclo_mono) is.

```
dnclo     == [p: Prop] (not(not(p))) -> p;
image     == [X,Y|Type][f:X->Y] [y:Y] Ex [x:X]  Q (f x) y;
mapDnclo  == [X,Y|Type][f:X->Y] {y:Y} dnclo (image f y);
mono      == [X,Y|Type][m:X->Y] {x,y:X} (Q (m x)(m y)) -> Q x y;
dnclo_mono == [X,Y|Type][m:X->Y] and (mono m) (mapDnclo m);
```

The equality on a subset should, of course, coincide with the equality on the superset. This can be achieved by stipulating the follwoing two axioms:

```
[ proof_irrelevance: {P|Prop}{p,q:P} Q p q ];
[ surj_pair: {X|Type}{A|X->Type}{u:<x:X>A x} Q ( u.1, u.2: <x:X>A x ) u ]
```

The first axioms says that all proofs of one and the same proposition P are equal. The second is necessary, since in LEGO the sums are not inductively defined, but built-in in a somehow *ad hoc* way.

The coding for subsets sometimes gets clumsy so it would be much more convenient to work with a system that supports subtypes in a nice and easy fashion. Up to now, unfortunately, there is no such system available.

4 Synthetic Domain Theory – Ideas and Motivation

In this section we will briefly present the ideas of SDT. The analytical method in domain theory is well-known. It describes domains as ideal completions of some bases. Compound domains are constructed usually by patching together

---

[2] A proposition ϕ is $\neg\neg$-closed if $\neg\neg\phi \Rightarrow \phi$, a predicate $p \in X \rightarrow$ Prop is $\neg\neg$-closed if $p(x)$ is $\neg\neg$-closed for any $x \in X$.

partial orders [Pau87] or using Scott's neighbourhood systems. The synthetic approach treats *domains as sets with special properties*. Compound domains can be put together by set constructions. Of course, one must prove that the "special properties" are *preserved* by these constructions. This axiomatic setting is formally analogous to SDG, Synthetic Differential Geometry (cf. [Koc81]), where the name "synthetic" stems from.

So the starting point of Synthetic Domain Theory is to assume a distinguished domain Σ (the simplest non-trivial one) which is described axiomatically and to associate with an arbitrary set X its "natural topology" by defining the open sets of X as the functions from X to Σ. The computational intuition behind these "open sets" is that they correspond to *semi-decidable predicates* which constitute the most general form of experiment which can be applied to a computational object. The objects $\top$ and $\bot$ of Σ correspond to the propositions expressing termination and nontermination, respectively. Thus Σ is considered as the subset of the set Prop of propositions that intuitively corresponds to Σ_1^0-sentences. So we will have to stipulate that Σ is closed under conjunction, disjunction, and existential quantification over $\mathbb{N}$.

It is known that a function is Scott-continuous, if (and only if) the inverse image of a Scott-open set is Scott-open. It is even simpler: Scott-continuity already follows from the fact that the "open sets" of the form $X \longrightarrow \Sigma$ (or shorter Σ^X) are Scott-open[3]. So one has to assure that any "open set" P satisfies

$$x \in P \wedge x \sqsubseteq y \Rightarrow y \in P$$

and for any ascending chain $(x_n)_{n \in \mathbb{N}}$

$$\sup_n x_n \in P \Rightarrow \exists n{:}\mathbb{N}.\,(x_n \in P).$$

The first condition suggests to define $x \sqsubseteq y$ as $\forall P{:}D \to \Sigma.\, x \in P \Rightarrow y \in P$ (cf. Definition 6.1). To satisfy the second condition we simply define sup by the condition $\sup_n x_n \in P \iff \exists n{:}\mathbb{N}.\,(x_n \in P)$ (cf. Definition 6.2). Without further requirements this supremum is not necessarily unique unless any object of a domain is determined by the results of all possible experiments applied to it (i.e. its observational behaviour). This will be ensured by the definition 6.3 of Σ-posets.

By definition we get that all functions $f : D \longrightarrow E$ are monotone and continuous (provided unique suprema exist).

5 The SDT-Axioms

We shortly discuss the SDT-axioms and refer to [Reu95] for an exact treatment.

---

[3] In [Pho90] this is proved in the PER-model.

5.1 The set of r.e. propositions Σ

DEFINITION 5.1 Let $\Sigma \in $ Set be a distinguished set with the following properties:

- $\Sigma \subseteq$ Prop
- $\top, \bot \in \Sigma$ with $\neg(\bot = \top)$
- If $p, q \in \Sigma$ then $p \wedge q,\ p \vee q \in \Sigma$
- If $f \in \mathbb{N} \longrightarrow \Sigma$ then $\exists n{:}\mathbb{N}. fn \in \Sigma$
- $\forall x, y{:}\Sigma. ((x = \top) \Leftrightarrow (y = \top)) \Leftrightarrow x = y.$

This means that Σ is a Set having the closure properties of r.e. propositions. In LEGO the above requirements are expressed as follows:

```
[Sig : Set]
[top,bot : Sig] ;
def == [x : Sig] Q x top ;        (* embedding Sig->Prop *)
[ Prf_botF : not (def bot) ] ;    (*      bot <> top        *)
[ extSig : {p,q : Sig} iff ( iff (def p)(def q) ) (Q p q) ];

[Or,And : Sig->Sig->Sig] [Join : (N->Sig) -> Sig] ;
[Or_pr   : {x,y : Sig} iff (def (Or x y))  (or (def x) (def y))]   ;
[And_pr  : {x,y : Sig} iff (def (And x y)) (and (def x) (def y))]  ;
[Join_pr : {p : N->Sig} iff (def (Join p)) (Ex ([n:N] def (p n)))] ;
```

Remark: One has to use the mono def in order to represent the subobject $\Sigma \subseteq$ Prop. Note that in the premiss of the Axiom extSig we use equivalence rather than equality, since we do not require that equivalent propositions are equal. Actually, we cannot claim that because we do not know whether there exist non-trivial impredicative universes in toposes. So there is a rather subtle difference between type theory and the internal language of a topos: in type theory we do not require that equivalent propositions are equal, therefore the subobject classifier is not strong.

5.2 The other axioms

Phoa's Axioms are an equivalent formulation of the "Phoa Principle" which states that $\Sigma^\Sigma \cong \{(p,q) \in \Sigma \times \Sigma \mid p \Rightarrow q\}$ [Tay91]. They imply that on Σ the observational order leq defined in the next section coincides with implication.

The continuity axiom states that the canonical limit process, i.e. the ascending chain of natural numbers $(1, 2, 3, \ldots)$ in the domain $\overline{\omega}$ – i.e. ω with a maximal element ∞ – has a supremum, which is important for characterizing suprema. The axiom ensures continuity on the model level (it is a kind of Rice-Shapiro-Theorem) to prove characterization theorems for Σ-cpos. Scott-continuity in our approach follows directly from the definition of supremum which is inspired by the ExPERs rather than Phoa's Σ-spaces.

For the axiomatization of the "ExPER-approach" we need another axiom not used in [Tay91] stating that Σ-propositions are $\neg\neg$-closed (stable). It is a kind of Markov's Principle [4] as Σ corresponds to the Σ_1^0-sentences. It also allows one to use "classical case analysis" for proving Σ-propositions and later for proving equality on domains. In fact, one can show that this axiom is equivalent to the statement that equality on domains (cpo-s) is $\neg\neg$-closed. One more axiom would be needed for dealing with partial map classifiers and lifting, the *Dominance Axiom* (cf. [Ros86]) but we don't go into the details here and refer to [Ros86, Reu95] instead.

6 Σ-posets, Σ-cpos and Σ-domains

6.1 Preorders and suprema

We define the observational preorder $\sqsubseteq$ (`leq`) as introduced in Section 4.

DEFINITION 6.1 (Phoa)

```
leq  ==  [X|Type][x,y:X] {p:X->Sig} (def (p x)) -> def (p y) ;
eq   ==  [X|Type][x,y:X] and (leq x y)(leq y x);
```

PROPOSITION 6.1 The following proposition (in LEGO syntax) can be proved:

```
{X|Type}{f,g:X->Sig} iff (leq f g) ({x:X} (def (f x)) -> def (g x));
```

stating that the the `leq` and the inclusion order are equivalent on powers of Σ. This property also implies that the order of products of our domains is pointwise. As we argued in Section 4 just by definition of `leq` we get:

PROPOSITION 6.2 (*monotonicity*) Any function is monotonic:

```
{X,Y|Type} {f:X->Y} {x,y:X} (leq x y) -> leq (f x)(f y);
```

In LCF a poset must be introduced by a carrier and an ordering (there is no "natural order"). Consequently, in LCF there exist also non-monotonic functions.

In order to achieve that Σ^X are the Scott-open sets on X (cf. Sect. 4) one simply defines the supremum implicitly by

$$\forall P{:}\Sigma^X. \bigsqcup_n x_n \in P \Longleftrightarrow \exists n{:}\mathbb{N}.\,(x_n \in P)$$

(cf. Definition 6.2). This is a difference w.r.t. [Pho90] where the order-theoretic suprema are used. Without further requirements this supremum is not necessarily unique. It will be unique if every object x in X is determined by the set of predicates which hold for x. This will be ensured by Definition 6.3 of Σ-posets. So the considerations above lead to the following definitions:

---

[4] In different axiomatizations, where Σ does not correspond to the Σ_1^0-sentences, it might be better to call this axiom "Σ-propositions are $\neg\neg$-closed".

DEFINITION 6.2 Define the type of ascending chains AC and the binary predicate supr as follows:

```
AC == [X:Type] <f: N->X> {n:N} leq (f n) (f (succ n));

supr == [X|Type][a:AC(X)][x:X]
        {P:X->Sig} iff (def(P x)) ( Ex [n:N] def( P(a.1 n) ) );
```

It is easy to see that this notion of supremum – provided it exists – is also the usual order-theoretic supremum w.r.t. leq. From the definition of suprema it follows immediately that all functions preserve existing suprema.

THEOREM 6.1 (*Scott-continuity*) Any function is Scott-continuous:

```
{X,Y|Type}{f:X->Y} {a:AC(X)}{x:X} (supr a x) -> supr f_o_a (f x) ;
```

where f_o_a = ((compose f a.1), P) represents the chain $f \circ a$ which is ascending in Y as f is monotone (stated by a term P we do not look into).

PROOF: Suppose $x \in A$ and x is the supremum of a, i.e. for any $P \in \Sigma^A$ it holds that $P(x) \Leftrightarrow \exists n{:}\mathbb{N}.\, P(a\,n)$ (*). So for any $Q \in \Sigma^B$ by substituting $Q \circ f$ for P in (*) we conclude that $Q(f(x)) \Leftrightarrow \exists n{:}\mathbb{N}.\, Q(f(a\,n))$, i.e. $f(x)$ is the supremum of $f \circ a$. □

Note that in LCF the type AC cannot be formed (but in HOLCF). We are ready now to proceed to the definition of Σ-posets, where objects are characterized uniquely by their Σ-properties.

6.2 Σ-posets

DEFINITION 6.3 A set $X \in$ Set is called a Σ-poset iff the map $\eta_X{:}X \longrightarrow \Sigma^{\Sigma^X}$ with $\eta_X(x) = \lambda p{:}\Sigma^X.\,p\,x$ is a $\neg\neg$-closed mono. In LEGO we can define this as a predicate on sets:

```
  eta == [X:Type] [x:X][p:X->Sig] p x
poset == [X:Set] and (mono (eta X)) (mapDnclo (eta X));
```

In contrast to [Pho90] the mono is required to be $\neg\neg$-closed. This has the advantage that the observational order on any Σ-poset is pointwise automatically. Such a definition has been already mentioned in [Pho90, p.196]: "*we could call a Σ-space X* **extensional** *if $X \rightarrowtail \Sigma^{\Sigma^X}$ were $\neg\neg$-closed; ... The idea doesn't seem to have been further developed in print.*" and is tributed to Hyland[5]. The name "extensional" indicates the relationship with ExPERs [FMRS92].

As a consequence the observational preorder for Σ-posets is indeed an order and observational equality eq coincides with Leibniz equality Q. In consequence, the equality for Σ-posets is $\neg\neg$-closed. Moreover, for Σ-posets the supremum is unique, so by the Axiom of Unique Choice we get a supremum operator sup : { C | CPO } (AC C.1) -> C.1.

---

[5] There is also a short remark in [Hyl91].

6.3 Σ-cpos

DEFINITION 6.4 A set X is a Σ-cpo (or an *extensional predomain*) iff X is a chain complete Σ-poset or, more formally, iff X is a Σ-poset and it holds that $\forall a{:}AC(X).\, \exists x{:}X.\, \bigsqcup(a,x)$. In LEGO:

```
cpo == [A:Set] and (poset A) ({a:AC X} Ex [x:A] supr a x);
```

The Σ-cpo-s, which will turn out to be a good class of predomains, can be represented by a type in our logic, i.e. we can define the type of Σ-cpo-s.

DEFINITION 6.5 The type of all $X \in \text{Set}$ that are Σ-cpo-s, i.e. $\{X{:}\text{Set}\,|\,\text{cpo}(X)\}$, is called CPO. In LEGO we write: `CPO == <X:Set> cpo X;`

6.4 Admissibility

Admissibility is a concept needed for induction.

DEFINITION 6.6 For any Σ-cpo C a predicate $P \in C \longrightarrow \text{Prop}$ is called *admissible* iff for any ascending chain $a \in AC(C)$ the implication $(\forall n{:}\mathbb{N}.\, P(a\,n)) \Rightarrow P(\bigsqcup a)$ holds. In LEGO we write:

```
admissible == [D|CPO] [P: D.1 -> Prop] {f:AC D.1}
                      ({n:N} P (f.1 n)) -> P (sup_C D f);
```

In LCF admissibility can only be proved syntactically by propagating admissibility accordingly to the construction of a formula and applying the appropriate closure properties. Admissibility is not expressible internally and therefore remains an external concept. Contrary to LCF, the notion of admissibility is expressible in our setting, as we can define the type of ascending chains (like in [Reg94]).

For the admissibility of predicates with negative occurrences of the argument (implications), we need an additional notion. Classically, one uses the following sufficient condition to prove admissibility of implication: if $\neg P$ and Q are admissible, then $\neg P \vee Q$, that is $P \Rightarrow Q$, is admissible too.

This is indeed true in LCF as admissible predicates are closed under disjunction. Unfortunately, in our intuitionistic setting closure under disjunction is not derivable even for classical disjunction $\vee_c$ (i.e. $A \vee_c B \Leftrightarrow \neg\neg(A \vee B)$) as it seems that the proof requires non-constructive choice principles. So we have to use some other sufficient conditions for proving admissibility of implications:

DEFINITION 6.7 For any Σ-cpo C a predicate $P \in C \longrightarrow \text{Prop}$ is called *sufficiently co-admissible* iff for any ascending chain $a \in AC(C)$ the implication $P(\bigsqcup a) \Rightarrow \exists m{:}\mathbb{N}.\, \forall n \geq m.\, P(a\,n)$ holds.

For simplicity we omit the corresponding LEGO code due to lack of space.

PROPOSITION 6.3 Let C be a Σ-cpo and $P, R \in C \to \text{Prop}$. Then the following propositions hold:

(i) If P is sufficiently co-admissible then the predicate $\lambda x{:}X.\,\neg P(x)$ is admissible.

(ii) If P is sufficiently co-admissible and R is admissible then $\lambda x{:}C.\,P(x) \Rightarrow R(x)$ is admissible.

Yet, it might be better to look for other definitions of "admissible" that are more adequate in the intuitionistic setting. The notion of "co-admissibility", however, was sufficient for the correctness proof of Section 9.

6.5 Σ-domains

The Σ-cpos with least element are the natural choice for Σ-domains.

DEFINITION 6.8 A Σ-domain is a Σ-cpo with a least element w.r.t. the leq order. In LEGO:

```
least == [A:Set][m:A] {a:A} leq m a;
dom == [A:Set] and (cpo A)(Ex [bottom: A] least A bottom);
```

For any Σ-domain D we denote the projection on the carrier set by D.c and for the least element $\bot_D$ one can define a function bot_D of type : { D:Dom } D.c using the Axiom of Unique Choice.

Of course, Σ-domains can also be internalized into a type Dom as in Def. 6.5, i.e. Dom == <X:Set> dom X. Note that in LCF there is no type of domains (or posets or cpo-s).

We get the following closure properties for Σ-domains.

THEOREM 6.2 Closure properties:

(1) Let D be a Σ-domain. If P is a $\neg\neg$-closed admissible predicate such that $P(\bot_D)$, then $\{d \in D \mid P(d)\}$ is a Σ-domain with $\bot_D$ as the least element.

(2) If X is a Σ-cpo, E a Σ-domain and $f, g \in D \longrightarrow X$ are such that $f(\bot) = g(\bot)$ then the equalizer of f and g is a Σ-domain with $\bot_D$ as the least element.

(3) Let X be a type and $A : X \longrightarrow$ Set such that for any $x \in X$ we have that $A(x)$ is a Σ-domain. Then $\Pi x{:}X.\,A(x)$ is a Σ-domain, and the least element is $\bot_{\Pi x:X.\,A(x)} = \lambda x{:}X.\,\bot_{A(x)}$.

(4) Let D and E be Σ-domains then $D \times E$ is a Σ-domain where $\bot$ is given by $(\bot_D, \bot_E)$.

(5) Let D and E be Σ-domains then $D \longrightarrow E$ is a Σ-domain where $\bot$ is given by $\lambda x{:}D.\,\bot_E$.

(6) Let D and E be Σ-domains then the strict functions from D to E, short $D \longrightarrow_\bot E$, form a Σ-domain where $\bot$ is given by $\lambda x{:}D.\,\bot_E$ which is strict.

(7) Σ-domains are closed under isomorphism.

The proof uses a Representation Theorem that can be found in [Reu95].

Note that, in order to form the dependent product $\{x:X\}$ (A x) one must know that Set is closed under dependent products i.e. Set is an impredicative universe. This means that – compared to LCF – polymorphic domains as $\{X:\text{Dom}\}$ X belong to Dom again.

For any Σ-domain D one certainly expects to have fixpoints of *arbitrary* endofunctions $D \longrightarrow D$. By the properties we have proved so far about Σ-domains, we are able to perform the "classical" Kleene-construction to get fixpoints.

PROPOSITION 6.4 Let D be a Σ-domain. Any endofunction $f \in D \longrightarrow D$ has a least fixpoint.

By virtue of the Axiom of Unique Choice – analogously to the bottom-case – we get a least fixpoint operator: `fix: { D:Dom } (D.c -> D.c) -> D.c`. Now we can also prove fixpoint induction as usual:

THEOREM 6.3 *(Fixpoint Induction)*
Let D be a Σ-domain, $P \in D \longrightarrow$ Prop an admissible predicate on D, and $f \in D \longrightarrow D$ an endofunction on D. If $P(\bot_D)$ and $\forall d{:}D.\, P(d) \Rightarrow P(f(d))$ then also $P(\text{fix}\, f)$. In LEGO:

```
{D|Dom} {P:D.c->Prop} (admissible P) -> {f:D.c->D.c}
  (P (bot_D D)) -> ( {d:D.c} (P d) -> P (f d) ) -> P( fix D f );
```

In LCF the fixpoint operator is introduced axiomatically. Fixpoint induction is an axiom and not a theorem.

7 Recursive domains

Domain equations are elegantly expressed by mixed-variant functors. To build useful equations, the interesting domain constructors like $\longrightarrow$, $\times$, $\otimes$, $\oplus$, $(\_)_\bot$ etc. must be defined *on Σ-domains*.

7.1 Domain constructors

Most of the constructions follow already from the closure properties of Theorem 6.2. The strict constructors like $\otimes$ and $\oplus$ are more difficult. They must be defined by their universal properties, i.e. as left adjoints to the strict function space and to the diagonal functor in the category of Σ-domains with strict maps, respectively. Otherwise the tupling function and the injection functions, respectively, would not be definable.

Note that for the smash product the projection maps still cannot be defined in general unlike in classical domain theory because they cannot be derived from the universal property. As Hyland puts it *"The smash product is there, but the projections are not in general - they just are not part of the universal structure. ...In classical domain theory how do you tell what is uniformly there from what is accidental??"* [SDT-mailing-list, Wed, 13 Jul 1994]

For the strict constructors defined by the universal properties (which resembles the second order encoding of logical connectives), however, it is not possible to prove the so-called *exhaustion axiom* [Pau87] stating e.g. for $\oplus$ that any $x \in A \oplus B$ is obtained by a left or right injection[6]. It is not clear at the moment how severe this drawback is. For proving equality of functions defined on $A \oplus B$ one can use the universal property of the strict sum. General case analysis, however, seems to be impossible.

7.2 Categories and functors

DEFINITION 7.1 The type of categories can be written in LEGO as follows:

```
Cat ==  <X:Type(0)> <Hom: X->X->Set>
        <o: {A,B,C|X} (Hom B C)->(Hom A B)->Hom A C> <id: {A|X} Hom A A>
        and3 ( {A,B|X} {f:Hom A B} Q (o (id|B) f) f )
             ( {A,B|X} {f:Hom A B} Q (o f (id|A)) f )
             ( {A,B,C,D|X} {h:Hom A B} {g:Hom B C} {f: Hom C D}
                         Q (o (o f g) h) (o f (o g h))             );

ob == [C:Cat] C.1;       hom == [C:Cat] C.2.1;
o  == [C:Cat] C.2.2.1;   id  == [C:Cat] C.2.2.2.1;
```

Note that these are locally small categories as homsets live in Set. In the same line one can define the type of covariant and mixed variant functors. Note that any functor is automatically locally continuous as all functions between Σ-domains are continuous in our setting. It can be easily shown that Σ-domains with strict functions form a category (called DomS) in this (internal) sense.

In LCF (and HOLCF) one cannot define categories or functors without dependent types, so one separates the morphism part from the object part. By contrast to LCF, in the SDT-approach all functors are automatically locally continuous.

THEOREM 7.1 In the category DomS of Σ-domains with strict maps all the following domain constructors are definable as functors: $\longrightarrow$, $\longrightarrow_\bot$, $\times$, $\otimes$, $+$, $\oplus$, $(\_)_\bot$. Moreover, the lifted natural numbers $\mathbb{N}_\bot$ and the Booleans $\mathbb{B}_\bot$ are Σ-domains with flat ordering.

7.3 Minimal solutions of domain equations

As we have dependent products, it is possible to do the Smyth & Plotkin inverse limit construction for solving domain equations. In fact, Σ-domains are closed under equalizers of strict maps and arbitrary products, so one can define the inverse limit. Moreover, one can show that the solutions are minimal in Freyd's sense [Fre91] (i.e. that the fixpoint of the copy functional equals the identity function). Mixed variance functors are coded as functors with two arguments (bifunctors).

---

[6] With "or" we mean classical disjunction.

THEOREM 7.2 Let F be a mixed variant endofunctor in DomS. Then there exists a Σ-domain A and a morphism α from $F A A$ to A which is an iso, such that

$$\text{fix}\,(\lambda h{:}(\text{Hom}\,A\,A).\,\alpha \circ F\,h\,h \circ \alpha^{-1}) = \text{id}_A \;,$$

i.e. A is the so-called *minimal* solution of F. In LEGO:

```
{F:Functor DomS DomS}
  <D:Dom> <alpha: DomS.hom (F.1 D D) D> <alpha_1: DomS.hom D (F.1 D D)>
  and (isopair alpha alpha_1)
    (lfix ([h: DomS.hom D D] o_strict alpha (o_strict (F.2.1 h h) alpha_1))
       (id_strict|D) ) ;
```

where lfix denotes the predicate stating that its second argument is the least fixpoint of the first and o_strict, id_strict denote the composition and identity in DomS.

The minimality condition is in fact important for deriving induction principles. Structural induction can be derived from the minimality condition via the inital F-algebra characterization.

Note that sums are used instead of existential quantifiers. This is convenient since one can extract the corresponding objects and does not have to treat them indirectly via elimination/introduction rules for the existential quantifier. It should be also mentioned that we need (Martin-Löf) identity types for the inverse limit construction that uses dependent families. This is a well-known problem of intensional type theory (see also [Reu95, RS93b]).

Observe that any recursive domain can be derived uniformly just by instantiating the right functor coding the intended domain equation. This is an advantage w.r.t. (HO)LCF where for any recursive type a special theory must be designed with adequate axioms: any recursive type A must be introduced together with a so-called representation type, i.e. $F A A$, a pair ($\alpha : F A A \to A, \alpha^{-1} : A \to F A A$), and two axioms: one to assure that this is a pair of isomorphisms and one stating that the fixpoint of the copy functional is the identity. Structural induction must be derived from fixpoint induction for every type, whereas in our approach it can be obtained by instantiating a general theorem.

8 The Sieve of Eratosthenes – An Example

SDT is appropriate for program verification. This is demonstrated by an example. We prove that the Sieve of Eratosthenes formulated in our setting is correct. The proof follows a rather traditional LCF-style.

The domain of streams over natural numbers (Stream) is obtained by instantiating Theorem 7.2 with the domain equation $S = (\mathbb{N} \times S)_\perp$ and taking the first projection. With the help of the isomorphism (obtained by some more projections) one can define all the basic familiar stream operations, e.g. hd, tl, append, $(\_)_n$ (nth-element). Using the isomorphims one can also *derive* the usual characterization of the stream-order, i.e.

PROPOSITION 8.1 $\forall s,t$:Stream. $s \sqsubseteq t$ iff $\neg\neg((s = \bot) \vee$
($\exists n$:N. $\exists s',t'$:Stream. $(s = \text{append}\, n\, s') \wedge (t = \text{append}\, n\, t') \wedge s' \sqsubseteq t')$.

For the correctness proof structural induction on streams is not sufficient, we also need induction on the length of the streams[7]:

THEOREM 8.1 *(Induction on length)* Let $P \in$ Stream $\longrightarrow$ Prop be an admissible and $\neg\neg$-closed predicate. Then the following induction principle is valid:

$$\forall n\text{:N.}\ \forall s\text{:Stream. (length}\, s\, n) \Rightarrow P(s) \quad \text{implies} \quad \forall s\text{:Stream.}\, P(s).$$

PROOF: First, note that the compact streams are generated by the Kleene chain associated to the copy functional (for streams). By the minimality of the domain Stream one gets immediately its "algebraicity", i.e. – given that compact $n\, s$ yields the prefix of s of length n – $\sup_n$ compact n is the identity on streams. Thus, in order to prove $P(s)$ for an arbitrary stream s one just has to prove $\forall n$:N. $P(\text{compact}\, n\, s)$ which follows by assumption and the fact that compact elements have a length, i.e. $\forall n$:N. $\forall s$:Stream. $\neg\neg\exists k$:N. length (compact $n\, s$) k[8]. □

The functions filter (of type N $\to$ Stream $\to$ Stream) and sieve (of type Stream $\to$ Stream) are defined recursively using the fix operator. Note that no proof of continuity is required, as fixpoints exist for arbitrary maps in SDT. The definitions are standard such that the following crucial properties hold:

1. $(\text{div}\, n\, a) = \text{true} \Rightarrow \text{filter}\, n\, (\text{append}\, a\, s) = \text{filter}\, n\, s$.
2. $(\text{div}\, n\, a) = \text{false} \Rightarrow \text{filter}\, n\, (\text{append}\, a\, s) = \text{append}\, a\, (\text{filter}\, n\, s)$.
3. $(\text{length}\, s\, n) \Rightarrow \exists k$:N. $(\text{length}\, (\text{filter}\, a\, s)\, k) \wedge k \leq n$.
4. sieve(append $n\, s$) = append n (sieve(filter $n\, s$)).

DEFINITION 8.1 Define (recursively) enum n as the stream of natural numbers in ascending order starting with n and define

$$(n\ \varepsilon\ s)\ :\Leftrightarrow\ (\exists k\text{:N.}\ (s)_k = n) \wedge (n \neq \bot_{\text{N}_\bot}).$$

Then the correctness theorem looks as follows:

THEOREM 8.2 For all $x \in \text{N}_\bot$ it holds that

$$x\ \varepsilon\ \text{sieve}(\text{enum}\, 2)\ \text{iff is\_prime}(x).$$

PROOF: The proposition is a consequence of the following lemma[9]:

$\forall s$:Stream. $s \neq \bot_{\text{Stream}} \wedge \text{repitition\_free}(s) \Rightarrow$
$(\forall n$:N. $(s)_n\ \varepsilon\ \text{sieve}(s) \Leftrightarrow \forall k < n.\ \neg\text{div}\, (s)_k\, (s)_n)$

---

[7] The predicate length states whether the stream in its first argument has the length given by the second argument.
[8] The double negation is necessary as we are working in intuitionistic logic; this is why P must be $\neg\neg$-closed.
[9] Note that we are sloppy and sometimes confuse elements of N and $\text{N}_\bot$ omitting the unit up. It is also clear, how to extend operations on N in a strict fashion to $\text{N}_\bot$.

where repitition_free(s) states whether a stream is injective. The lemma is proved by induction on the length of s. In the induction step we need the following additional lemma: let $s \in$ Stream, $n, a \in \mathbb{N}$.

$$n \; \varepsilon \; \text{sieve}(\text{filter} \, a \, s) \; \text{iff} \; \neg(\text{div} \, a \, n) \; \wedge \; n \; \varepsilon \; \text{sieve}(s).$$

Note that the two previous lemmas hold for any binary boolean predicate div as long as it is transitive. □

Whenever we do induction on streams we must prove admissibility of the predicate under investigation. The lemmas above, however, do contain positive existential quantifiers (see the definition of ε), so the syntactic requirements of LCF are useless[10]. By contrast, in our setting one can even *prove* admissibility in the logic. Note that the implication in the first lemma causes problems in the way indicated in Sect. 6.4.

A detailed presentation of the proof (using also LEGO syntax) may appear elsewhere.

9 Conclusions

We have presented a Synthetic Domain Theory, based on a few axioms, that has been completely formalized in type theory. The theory has been shown to be consistent by verifying that the axioms of Sect. 3 and 5 hold in the realizability model of Sect. 2. (cf. [Reu95]) Our setting can be considered a step towards LCF+, i.e. an enhancement of LCF, which is more expressive and permits the treatmeant of domains as sets. Moreover, many principles that are introduced axiomatically in LCF are *theorems* in LCF+ such that one obtains (more) information not only about the "how-s" but also about the "why-s". Of course, if one is not interested in the core theory, one could simply forget it and work with the main theorems as if they were axioms.

Working in a type theoretical setting has another advantage. One can express modules by $\sum$-types. On top of the presented core theory, one could imagine a theory of program modules and modular specifications. Also co-induction principles are still to be implemented. More case studies should be carried out to test how far one can get doing denotational semantics in SDT.

Unfortunately, LEGO is only a proof checker so it does not provide the user comfort of Isabelle or LCF. A theorem prover for ECC (ECC*) could be a future goal. Due to our experiences with this rather big SDT-theory (487 kB) we consider it an important task to develop tools that support modular theories.

Finally, many theoretical questions are still open. Generalizing SDT from Scott domain theory to stable domain theory seems to be a major research topic. But also investigations about admissibility seem to be appropriate.

---

[10] In this special case, however, one could rewrite the definition of ε as a complicated equalizer on Σ since Σ is closed under countable joins.

Acknowledgements

I wish to thank Thomas Streicher for his collaboration on the right axiomatization of Σ-cpos and for his comments on a draft. Thanks to Randy Pollack for hints about the SML-code of the LEGO-system and to everyone on the SDT-mailing-list for discussions and comments. This work was partially sponsored by the DAAD-program VIGONI. I'm grateful to our partners Eugenio Moggi and Pino Rosolini from Genoa for stimulating discussions and suggestions.

References

[Age94] S. Agerholm. *A HOL Basis for Reasoning about Functional Programs*. PhD thesis, BRICS, University of Aarhus, 1994. Also available as BRICS report RS-94-44.

[BM92] R. Burstall and J. McKinna. Deliverables: a categorical approach to program development in type theory. Technical Report ECS-LFCS-92-242, Edinburgh University, 1992.

[Coq86] Th. Coquand. An analysis of Girard's paradox. In *Proc. 1st Symp. on Logic in Computer Science*, pages 227–236. IEEE Computer Soc. Press, 1986.

[FMRS92] P. Freyd, P. Mulry, G. Rosolini, and D. Scott. Extensional PERs. *Information and Computation*, 98:211–227, 1992.

[Fre91] P. Freyd. Algebraically complete categories. In A. Carboni, M.C. Pedicchio, and G. Rosolini, editors, *Proceedings of the 1990 Como Category Theory Conference*, volume 1488 of *Lecture Notes in Mathematics*, pages 95–104, Berlin, 1991. Springer.

[Hyl91] J.M.E. Hyland. First steps in synthetic domain theory. In A. Carboni, M.C. Pedicchio, and G. Rosolini, editors, *Proceedings of the 1990 Como Category Theory Conference*, volume 1488 of *Lecture Notes in Mathematics*, pages 131–156, Berlin, 1991. Springer.

[Koc81] A. Kock. *Synthetic Differential Geometry*. Cambridge University Press, 1981.

[LP92] Z. Luo and R. Pollack. Lego proof development system: User's manual. Technical Report ECS-LFCS-92-211, Edinburgh University, 1992.

[LS95] J.R. Longley and A.K. Simpson. A uniform account of domain theory in realizability models. To be submitted to special edition of MSCS for the Workshop on Logic, Domains and Programming Languages, Darmstadt, Germany, 1995.

[Luo94] Z. Luo. *Computation and Reasoning – A Type Theory for Computer Science*, volume 11 of *Monographs on Computer Science*. Oxford University Press, 1994.

[Pau87] L.C. Paulson. *Logic and Computation*, volume 2 of *Cambridge Tracts in Theoretical Computer Science*. Cambridge University Press, 1987.

[Pho90] W.K. Phoa. *Domain Theory in Realizability Toposes*. PhD thesis, University of Cambridge, 1990. Also available as report ECS-LFCS-91-171, University of Edinburgh.

[Reg94] F. Regensburger. *HOLCF: Eine konservative Erweiterung von HOL um LCF*. PhD thesis, Technische Universität München, November 1994.

[Reu95] B. Reus. *Program Verification in Synthetic Domain Theory*. PhD thesis, Ludwig-Maximilians-Universität München, 1995.

[Ros86] G. Rosolini. *Continuity and effectiveness in topoi*. PhD thesis, University of Oxford, 1986.

[RS93a] B. Reus and T. Streicher. Naive Synthetic Domain Theory – a logical approach. Draft, September 1993.

[RS93b] B. Reus and T. Streicher. Verifying properties of module construction in type theory. In A.M. Borzyszkowski and S. Sokołowski, editors, *MFCS'93*, volume 711 of *Lecture Notes in Computer Science*, pages 660–670. Springer, 1993.

[Str91] T. Streicher. *Semantics of Type Theory, Correctness, Completeness and Independence Results*. Birkhäuser, 1991.

[Tay91] P. Taylor. The fixed point property in synthetic domain theory. In *6th Symp. on Logic in Computer Science*, pages 152–160, Washington, 1991. IEEE Computer Soc. Press.

Function Definition in Higher-Order Logic

Konrad Slind[*]
slind@informatik.tu-muenchen.de

Technische Universität München,
Institut für Informatik, 80290 München, Germany.

Abstract. We use a formally proven wellfounded recursion theorem as the basis upon which to build a function definition facility for Higher Order Logic. This approach offers flexibility in the choice of wellfounded relations used, the deferral of termination arguments, and automatic isolation of termination conditions. Building on this platform, we provide the ability to define recursive functions via pattern matching. The system is parameterized and has been instantiated to quite different theorem provers.

1 Introduction

One use of higher-order logic theorem provers is to verify pure functional programs. By far the easiest way to model such programs is with the built-in functions of the logic. In a logic of total functions, such a naive modelling can capture only a proper subset of all programs, but this subset, although it excludes important examples such as interpreters, is still quite interesting. A genuine advantage is that no extra reasoning infrastructure need be built: the logic already provides it and hence there is no need for a preparatory, and often arduous, phase of "verification theory" construction. Therefore, a verification approach that represents functional programs by functions of higher order logic is attractive in spite of its limitations.

The first step toward making the verification of such programs commonplace is to support their definition in the logic. Here we confront a shortcoming of many higher-order logic theorem-proving systems: often only primitive recursive definitions are supported. The functions definable by higher-order primitive recursion model a large class of programs, but the syntactic restrictions of primitive recursion are bothersome. To avoid this bother, we choose instead to define functions by appeal to the wellfounded recursion theorem. The freedom this gives us in writing right-hand sides of function definitions is complemented by extending the left-hand sides to allow ML-style pattern matching. Thus we arrive at a system wherein functions syntactically similar to ML or Haskell programs can be directly defined and reasoned about in the logic.

Our approach is wholly definitional: we define the basic concepts in the logic, and from these definitions we prove general theorems that our machinery manipulates via inference. In this manner, we can reduce the principle of definition for

---

[*] Research supported by DFG grant Br 887/4-2, *Deduktive Programmentwicklung*

recursive functions to a far simpler principle of abbreviation which we assume is provided by the theorem prover, thus achieving modularity and portability. A consequence of this approach is that no meta-theoretical argument need be given to show that recursive function definitions preserve consistency.

An innovation of this work is that its implementation is usable by more than one theorem proving system. We have instantiated it to both HOL and Isabelle; however, lack of space prevents a description of the system architecture and the instantiations. The details will appear in the author's forthcoming PhD thesis.

In the following sections, we first go into the technical details of our development of wellfoundedness, induction, and recursion. In section 3 we describe the steps that our tool makes when defining a function. Section 4 discusses the implementation of pattern-matching. After that, some examples are given in section 5. In section 6, we discuss some loose ends. In 7 and 8 we survey related work and conclude.

Notation and basic definitions

We use $\equiv$ to show that a definition is being made and $\overline{x}$ denotes a finite sequence of distinct syntactic objects. In parsing logical expressions, earlier members of the following list of infixes have stronger binding power than later members: $=, \wedge, \vee, \supset, \equiv$. All infixes associate to the right. The transitive closure of a relation $R : \alpha \to \alpha \to bool$ is defined as

$$TC\ R\ a\ b \equiv \forall P.\ (\forall xy.Rxy \supset Pxy) \wedge (\forall xyz.Pxy \wedge Pyz \supset Pxz) \supset Pab.$$

Suc denotes the successor function on the natural numbers. The list-processing functions ::, *mem*, *filter*, *length*, and @ (an infix version of *append*) are used; we assume the reader is familiar with their definitions. Function composition is written infix : $f \circ g \equiv \lambda x.f(g\ x)$. $\forall(M)$ denotes the universal quantification of all free variables in M.

Datatypes

For a logical datatype *ty* with constructors $C_1, \ldots, C_n$ our system makes use of the following:

Exhaustion. $\vdash \forall x : ty.\ (\exists \overline{y}.x = C_1 \overline{y}) \vee \ldots \vee (\exists \overline{y}.x = C_n \overline{y})$
Case definition. $(\forall \overline{x}.\ case_{ty}\ f_1 \ldots f_n\ (C_1 \overline{x}) \equiv f_1 \overline{x})$
$\wedge \ldots \wedge$
$(\forall \overline{x}.\ case_{ty}\ f_1 \ldots f_n\ (C_n \overline{x}) \equiv f_n \overline{x})$
Case congruence. $\vdash (O = O') \wedge$
$(\forall \overline{x}.\ (O' = C_1 \overline{x}) \supset f_1 \overline{x} = f'_1 \overline{x})$
$\wedge \ldots \wedge$
$(\forall \overline{x}.\ (O' = C_n \overline{x}) \supset f_n \overline{x} = f'_n \overline{x})$
$\supset case_{ty}\ f_1 \ldots f_n\ O = case_{ty}\ f'_1 \ldots f'_n\ O'.$

2 Wellfoundedness, induction, and recursion

We begin with the notion of wellfoundedness. Roughly speaking, a relation R is *wellfounded* when it has no infinite decreasing chains. In the context of programs, this means that, *if* the arguments to recursive calls can be placed in a wellfounded relation, then the function will terminate and hence is total. There are several equivalent definitions of wellfoundedness; we use the following: $WF(R) \equiv \forall P.\ (\exists w.\ P\ w) \supset \exists min.\ P\ min \land \forall b.\ R\ b\ min \supset \neg P\ b$. From this definition, one can quickly prove a general induction theorem:

$$\vdash WF(R) \supset (\forall x.\ (\forall y.\ R\ y\ x \supset P\ y) \supset P\ x) \supset \forall x.\ P\ x. \qquad (1)$$

For the natural numbers, so-called *mathematical* and *complete* induction are easily derivable from this theorem, by instantiating R to the *predecessor* and $<$ relations, respectively. This is an instance of a more general mechanism which can be applied to derive recursion induction for functions. Unfortunately, for lack of space, we will not be able to explain how this is realized in our work.

Recursion

The statement of the wellfounded recursion theorem uses a ternary operator that restricts a function to a certain set of values. In this definition, the expression $\varepsilon z.\ True$ uses the Hilbert choice operator to denote an arbitrary element of the range of f.

$$(f \mid R, y) \equiv \lambda x.\ \text{if } R\ x\ y \text{ then } f\ x \text{ else } \varepsilon z.\ True. \qquad (2)$$

Now we can formulate the wellfounded recursion theorem, in a form close to how it is stated in the literature[10]:

$$\vdash WF(R) \supset \forall M.\ \exists! f.\ \forall x.\ f(x) = M(f \mid R, x)\ x. \qquad (3)$$

However, we don't use (3) in our development, since we found the following *explicit witness* version to be more useful:

$$\vdash (f = WFREC\ R\ M) \supset WF(R) \supset \forall x.\ f(x) = M\ (f \mid R, x)\ x. \qquad (4)$$

Proof of (4). The statement uses *WFREC*, a recursion operator, which we will define in the course of the proof. We start by defining the set of approximants for a functional M at argument x in the presence of relation R:

$$approx\ R\ M\ x\ f \equiv (f = ((\lambda y.\ M\ (f \mid R, y)\ y) \mid R, x)).$$

By wellfounded induction, any two approximants agree on their common domain:

$$\vdash WF(R) \land transitive(R) \land \\ approx\ R\ M\ u\ f \land approx\ R\ M\ v\ g \\ \supset \forall x. R\ x\ u \land R\ x\ v \supset f(x) = g(x).$$

The following definition chooses an approximant:

$$the\_fun\ R\ M\ x \equiv \varepsilon f. approx\ R\ M\ x\ f.$$

A restricted unfolding theorem can be proven for a chosen approximant, again by wellfounded induction:

$$\vdash WF(R) \wedge \mathit{transitive}(R) \supset$$
$$\mathit{the\_fun}\ R\ M\ x = ((\lambda y.\ M\ (\mathit{the\_fun}\ R\ M\ x\,|\,R, y)\ y)\,|\,R, x).$$

Now we introduce the "bounded fixpoint" operator *WFREC*. It has a somewhat opaque definition, which lifts unwindings to be in the transitive closure of R.

$$WFREC\ R\ M \equiv \lambda x.\ M\ (\mathit{the\_fun}(TC\ R)(\lambda f\ v.\ M\ (f\,|\,R, v)\ v)\ x\,|\,R, x)\ x.$$

The type of WFREC is $(\alpha \to \alpha \to \mathsf{bool}) \to ((\alpha \to \beta) \to (\alpha \to \beta)) \to \alpha \to \beta$. By instantiating the unfolding theorem to $TC\ R$ it is now easy to prove

$$\vdash WF(R) \supset \forall x.\, WFREC\ R\ M\ x = M\ (WFREC\ R\ M\,|\,R, x)\ x,$$

whence we easily get (4) and then (3) follows, using another wellfounded induction to prove uniqueness. ∎

Tobias Nipkow initially proved this theorem in the Isabelle system. The author subsequently proved it in the HOL system. Schwichtenberg and Wainer seem to have independently arrived at a similar formulation [23].

2.1 Primitive and derived relations

Our machinery requires a stock of wellfounded relations. In practice, this stock will be augmented whenever a new datatype is defined. One particularly simple thing to do is to add the predecessor relation for each datatype. (We discuss more sophisticated approaches in section 6.) Here is a small selection:

$$\mathit{pred}\ x\ y \equiv (y = \mathsf{Suc}\ x)$$
$$\mathit{list\_pred}\ l_1\ l_2 \equiv \exists h. l_2 = h :: l_1.$$

We also define means of propagating wellfoundedness. The current collection of these handles lexicographic combinations, inverse images, subsets, and transitive closure. Here we show a few important definitions: those for lexicographic combinations and inverse images. *Measure* functions are also easy to define.

$$(R_1\ \mathsf{X}\ R_2)(u, v)\ (x, w) \equiv R_1\ u\ x \vee (u = x \wedge R_2\ v\ w)$$
$$\mathit{inv\_image}\ R\ f \equiv \lambda x\ y.\ R\ (f\ x)\ (f\ y)$$
$$\mathit{measure} \equiv \mathit{inv\_image}\ <$$

The following theorems are simple exercises:

$$\vdash WF(R) \wedge WF(Q) \supset WF(R\ \mathsf{X}\ Q)$$
$$\vdash WF(R) \supset WF(\mathit{inv\_image}\ R\ f)$$
$$\vdash WF(R) \supset WF(TC\ R)$$

Now, for example, $\vdash \forall f.\ WF(\mathit{measure}\ f)$ is easy to prove using the above theorems, since $<$ is definable as $TC\ \mathit{pred}$.

3 Defining functions

In this section, we will describe the steps taken to define recursive functions. An important part of this process is the extraction of termination conditions; this is detailed in section 3.1. The derived principle of definition takes a description of a recursive function (assume the name of the function is f) and a relation R and performs the following steps:

1. Translates the function description into a functional F;
2. Uses the basic principle of definition to define $f \equiv WFREC\ R\ F$;
3. Applies the recursion theorem (4) to the definition and then performs some basic simplifications; and
4. Extracts termination conditions.

Our design allows further steps to be included as "post-processors" to step 4. In particular, the design allows for a prover that tries to establish the well-foundedness of R as well as a prover that attempts to eliminate the termination conditions.

Example. Consider the following program:

$$variant(x, L) = \text{if } mem\ x\ L \text{ then } variant(x+1, L) \text{ else } x.$$

This function, or close relatives of it, is often found in symbolic systems since it is helpful for renaming bound variables in the course of substitution, or in "renaming variables apart" prior to unification. Trying to define *variant* formally has apparently given some authors difficulty, *e.g.*, see section 4 of [9]. There are at least a couple of relations that explain why *variant* terminates (the second is from Matt Kaufmann).

$$measure\ \lambda(x, L).\ length(filter(\lambda y.\ x \leq y)L)$$
$$measure\ \lambda(x, L).\ (max(L) + 1) - x$$

Using the first as our termination relation, the steps in the definition process elaborate as follows:

Translation. In this simple case, we just λ-abstract the arguments and the recursively occurring function variable to get

$$\lambda variant\ (x, L).\ \text{if } mem\ x\ L \text{ then } variant(x+1, L) \text{ else } x.$$

The translation of definitions with more complex patterns on the left-hand side is described in section 4.

Definition. Make the definition

$$variant \equiv$$
$$WFREC\ (measure\ \lambda(x, L).\ length(filter(\lambda y.\ x \leq y)L))$$
$$(\lambda variant\ (x, L).\ \text{if } mem\ x\ L \text{ then } variant(x+1, L) \text{ else } x).$$

Since the right hand side of the definition has no free variables, this is a mere abbreviation, allowable as a definition by any reasonable higher-order logic implementation.

Apply recursion theorem. By application of *modus ponens* with (4), we get the following theorem (we have shunted the wellfoundedness condition to the assumptions):

$WF(measure \ \lambda(x, L). \ length(filter(\lambda y. \ x \leq y)L))$
$\vdash variant(x, L) =$
   if $(mem \ x \ L)$ then
      $(variant \,|\, measure(\lambda(x, L). \ length(filter(\lambda y. \ x \leq y)L)), (x, L))$
      $(x + 1, L)$
   else x.

Extract termination conditions. The original recursive call $variant(x+1, L)$ has now been converted into a constrained form:

$(variant \,|\, measure(\lambda(x, L). \ length(filter(\lambda y. \ x \leq y)L)), (x, L))(x + 1, L).$

The process of extraction trades such constrained occurrences for the original occurrences, but this trade can only happen when the *termination condition* can be proved. Our machinery enforces this requirement by logic; the replacement is controlled by the following theorem, a trivial consequence of (2) the definition of function restriction:

$$\vdash R \ x \ y \supset (f \,|\, R, y)x = f \ x. \tag{5}$$

Termination conditions are captured and brought out to the top-level of the definition, so that automatic methods can have easy access in their attempts to prove them. If these methods fail, the termination conditions are then also easily accessible for the user to attempt. An important fact is that extraction happens via inference, with the result being a theorem of the form *termination conditions $\supset$ recursion equations*:

$WF(measure \ \lambda(x, L). \ length(filter(\lambda y. \ x \leq y)L))$
$\vdash (mem \ x \ L \supset length(filter(\lambda y. \ x + 1 \leq y)L) < length(filter(\lambda y. \ x \leq y)L))$
   $\supset variant(x, L) = $ if $(mem \ x \ L)$ then $variant(x + 1, L)$ else x.

Section 3.1 explains in detail how termination conditions are captured and brought to the top level.

Now the definition process is over, and it is left to eliminate the assumptions and the antecedent of this theorem. Elimination of the assumption is trivial and will not be discussed. The antecedent can be proved by induction on L, using the following lemma:

$$\vdash (\forall x.P(x) \supset Q(x)) \supset \forall L. length(filter \ P \ L) \leq length(filter \ Q \ L).$$

Finally, the desired unconditional recursion equation is achieved, arrived at through an unbroken chain of inference. As can be seen, the work is mostly automatic. The only creativity required is to find the right relation R, to prove the wellfoundedness of R, and to eliminate any remaining termination conditions. We now turn to the details of extraction.

3.1 Extracting termination conditions

Suppose we are defining a function $f(z) = M$ with recursive calls $f(z_1)\ldots f(z_n)$ occurring in M. Let R be the termination relation. In the instantiated recursion theorem, this gives rise to the constrained occurrences $(f \mid R, z)z_1 \ldots (f \mid R, z)z_n$ and the termination conditions $(R\ z_1\ z)\ldots(R\ z_n\ z)$. Each of these will, in general, only be provable in the context existing at the recursive call site. In general, the *context* $\Gamma(N)$ of a subterm N is a collection of facts that are true because of where N occurs in M. The *full termination condition* of recursive call $f(z_i)$ in context $\Gamma(z_i) = [h_1, \ldots, h_m]$ will be the formula $h_1 \wedge \ldots \wedge h_m \supset R\ z_i\ z$. Our machinery uses *contextual rewriting* to gather the full termination conditions and bring them to the top level. The process has two parts: the accumulation of context, and the capture of termination conditions at a recursive call.

Contextual rewriting Contextual rewriting can be built on top of a *congruence-based* rewrite engine [19]. The congruence-based approach uses the theorems

$$\vdash (M = M') \supset \lambda x.\ M = \lambda x.\ M'$$
$$\vdash (M = M') \wedge (N = N') \supset MN = M'N'$$

as justification for replacement. Contextual rewriting [16] can be implemented by adding extra congruence rules such as those arising from datatype definitions, as well as ones for the conditional and *let*:

$$\vdash (P = P') \wedge (P' \supset x = x') \wedge (\neg P' \supset y = y')$$
$$\supset \text{if } P \text{ then } x \text{ else } y = \text{if } P' \text{ then } x' \text{ else } y'$$
$$\vdash (M = M') \wedge (\forall x.x = M' \supset f\ x = f'\ x) \supset \text{let } f\ M = \text{let } f'\ M'$$

These extra congruence rules are used to track context, as can be seen by the way the congruence rule for the conditional is interpreted by the rewriting engine:

> When a match for 'if P then x else y' is found, first rewrite P to P', then rewrite x under the extra assumption P' to x'; likewise, rewrite y under the assumption $\neg P'$ to y'. Finally, replace 'if P then x else y' by 'if P' then x' else y''.

Replacement of the 3 subexpressions of the conditional is justified by the 3 theorems: $(\Gamma \cup TC_1 \vdash P = P')$, $(\Gamma, P' \cup TC_2 \vdash x = x')$ and, $\Gamma, \neg P' \cup TC_3 \vdash y = y'$. The *new* assumptions TC_1, TC_2, and TC_3 are the captured termination conditions found in the 3 subexpressions of the conditional. As the rewriting process continues, these termination conditions persist as assumptions.

Now we discuss how termination conditions are captured at a recursive call. At a constrained recursive call $(f \mid R, z)z_i$ in context $\Gamma(z_i)$ (having elements $h_1, \ldots, h_m$), the following little proof is performed to allow the replacement of $(f \mid R, z)z_i$ by $f\ z_i$ (MP stands for *modus ponens* and $\wedge *$ for iterated conjunction introduction):

A. $h_1, \ldots, h_m \vdash h_1 \wedge \ldots \wedge h_m$ \hfill $\wedge * \Gamma(z_i)$
B. $\forall (h_1 \wedge \ldots \wedge h_m \supset R\ z_i\ z) \vdash h_1 \wedge \ldots \wedge h_m \supset R\ z_i\ z$ \hfill Assume
C. $h_1, \ldots, h_m, \forall (h_1 \wedge \ldots \wedge h_m \supset R\ z_i\ z) \vdash R\ z_i\ z$ \hfill MP B,A
D. $h_1, \ldots, h_m, \forall (h_1 \wedge \ldots \wedge h_m \supset R\ z_i\ z) \vdash (f\,|\,R, z)z_i = f(z_i)$ \hfill MP (5), C

Now the replacement can be performed. In effect, we have stored the termination condition $\forall (h_1 \wedge \ldots \wedge h_m \supset R\ z_i\ z)$ on the assumption list. As rewriting 'unwinds', each of the extra assumptions $h_1 \ldots h_m$ will be removed at the point it was added to the context. The end result, after rewriting finishes, is the theorem

$$WF(R), \forall (\Gamma(z_1) \supset R\ z_1 z), \ldots, \forall (\Gamma(z_n) \supset R\ z_n z) \vdash f(z) = M$$

in which the termination conditions have been separated from the recursion equations. It is now simple for tools to either automatically attempt to solve them or to present them as goals for the user.

3.2 Definitions without termination relations

The principal obstacle in using our principle of definition for wellfounded recursion is the requirement that a wellfounded relation be given at definition time. After implementing termination condition extraction we realized that, if we leave R variable, we can perform extraction *before* making the definition, provided that in step B above we do not quantify R. This allows one to omit termination relations for a large class of definitions. We now show how. Extraction leaves us with, as before,

$$WF(R), \forall (\Gamma(z_1) \supset R\ z_1 z), \ldots, \forall (\Gamma(z_n) \supset R\ z_n z) \vdash f(z) = M,$$

with the difference that both R and f are free variables. We write these termination conditions as $WF(R), TC_1(R), \ldots, TC_n(R)$, highlighting the fact that they have a single free variable R. Now we may define f by choosing a wellfounded relation meeting the termination conditions:

$$f \equiv WFREC\ (\varepsilon R.\ WF(R) \wedge TC_1(R) \wedge \ldots \wedge TC_n(R))\ M.$$

(For the rest of this discussion, we take R' to abbreviate $(\varepsilon R.\ WF(R) \wedge TC_1(R) \wedge \ldots \wedge TC_n(R))$.) Having made the definition, we can now instantiate the variable f by the constant **f** and eliminate the definition from the recursion theorem, giving the derived definition $WF(R'), TC_1(R'), \ldots, TC_n(R') \vdash \mathbf{f}\ z = M$. We now assume and conjoin the termination conditions, giving

$$WF(R), TC_1(R), \ldots, TC_n(R) \vdash WF(R) \wedge TC_1(R) \wedge \ldots \wedge TC_n(R),$$

to which we apply the Axiom of Choice ($\forall P\ x.\ P\ x \supset P(\varepsilon P)$), after which we can eliminate each R'-hypothesis from the derived definition. This leaves us with

$$WF(R), TC_1(R), \ldots, TC_n(R) \vdash \mathbf{f}\ z = M,$$

a derived definition in which the computed termination conditions and the wellfoundedness requirement can be eliminated at the user's convenience. For nested functions, this technique fails because the termination conditions have not only R free, but also f, and so a definition can not be made.

4 Incorporating pattern-matching

In order to nicely describe functional programs, we want to allow ML-style pattern-matching, *e.g.*, as used in the following version of Euclid's algorithm:

$$gcd(0, y) = y$$
$$gcd(\mathsf{Suc}\ x, 0) = \mathsf{Suc}\ x$$
$$gcd(\mathsf{Suc}\ x, \mathsf{Suc}\ y) = \text{if } (y \leq x) \text{ then } gcd(x - y, \mathsf{Suc}\ y) \text{ else } gcd(\mathsf{Suc}\ x, y - x).$$

This description is intended to represent a function, but which one? A straightforward interpretation of such a description is that to find the value of, *e.g.*, $gcd(m, n)$, one must match the patterns in some order until a match θ is found, then apply θ to the corresponding righthand side, and continue by evaluating the instantiated righthand side. There are a couple of problems with this. First, the meaning of patterns is given operationally. Second, the algorithm is inefficient, since it can happen that a lot of redundant matching occurs in the search for a successful match. Augustsson [3], motivated by the latter consideration, invented a method of matching against all patterns at once. Furthermore, his algorithm also deals with the first problem: the program gets translated into a nested case expression.[2]

Our pattern language differs from that of ML in the following respects:

- We have chosen to exclude overlapping patterns, since overlapping patterns impose an order on rewriting and we want to use each recursion equation as an independent rewrite rule. Definitions with overlapping patterns also lead to difficulties in producing custom induction theorems.
- The cases must be complete. This restriction can probably be eased by using the Hilbert choice operator to choose an arbitrary element of the range in unspecified cases.
- So-called *wildcard* patterns are not yet supported, but this should be possible by translating each wildcard occurrence to a new variable.
- Record patterns, and *as* patterns are not supported.

4.1 Translation of pattern-matching

How is a program written in the pattern-matching style translated into a nested case expression? The following algorithm `trans` (expressed in pseudo-ML), is a simple version of that of Augustsson. `trans` takes two arguments: a stack and a list of *rows*. A row is a list of patterns paired with a right-hand side. In contrast to the interpretation given above, which proceeded row by row, `trans` goes from left to right, column by column. One of the invariants of the algorithm is that, in

---

[2] An alternative is to translate into nested *if*-statements, or decision trees, but then destructors are needed on the right hand sides, a *first order* solution that our higher order setting allows us to avoid.

every call, all rows have the same number of patterns. Assume we are given the program description $(f(pat_1) = rhs_1) \ldots (f(pat_n) = rhs_n)$. The algorithm starts by translating the program description into rows. Then the **trans** function is called. The resulting term is abstracted with respect to f and a variable z (which must not be free in any of $rhs_1, \ldots, rhs_n$):

$$\lambda f\ z.\ \mathtt{trans}([z], [([pat], rhs)_1, \ldots, ([pat], rhs)_n])$$

Variable.

As mentioned, the algorithm proceeds by examining patterns in parallel. This rule covers that case in which the current pattern being examined (in all rows) is a variable. The notation $[v \mapsto z]M$ denotes the substitution of z for v throughout the term M. This translation-time substitution essentially performs α-renaming, which itself sets up parallel substitution at "runtime".

$$\mathtt{trans}(z :: stack, [(v :: pats, rhs)_1 \quad \ldots, (v :: pats, rhs)_n])$$
$$= \mathtt{trans}(stack, \quad [(pats_1, [v_1 \mapsto z]rhs_1), \ldots, (pats_n, [v_n \mapsto z]rhs_n)])$$

Constructor.

In this rule, the current sub-pattern being examined in all patterns is a constructor for a type ty. The problem now splits into n subproblems, one for each constructor $C_1, \ldots, C_n$ of ty. Since each constructor can have repeated occurrences, there is a stage of partitioning the rows into n groups of size $k_1 \ldots k_n$. A row expression $(C_i \overline{p} :: pats, rhs)_j$ should be regarded as the jth element of C_i's partition. After partitioning, C_i is discarded, giving a row expression $(\overline{p}@pats, rhs)_{ij}$. In subproblem i, supposing the constructor C_i has type $\tau_1 \to \ldots \to \tau_j \to \beta$, j new variables $v_1 : \tau_1, \ldots, v_k : \tau_j$ are pushed onto the stack. This vector of variables is denoted $\overline{v : \tau_i}$.

$$\mathtt{trans}(z :: stack, [(C_1\overline{p} :: pats, rhs)_1, \ldots, (C_1\overline{p} :: pats, rhs)_{k_1}, \ldots$$
$$(C_n\overline{p} :: pats, rhs)_1, \ldots, (C_n\overline{p} :: pats, rhs)_{k_n}])$$
$$=$$
$$\text{let } M_1 = \mathtt{trans}(\overline{v : \tau_1}@stack, [(\overline{p}@pats, rhs)_{11}, \ldots, (\overline{p}@pats, rhs)_{1k_1}])$$
$$\ldots$$
$$M_n = \mathtt{trans}(\overline{v : \tau_n}@stack, [(\overline{p}@pats, rhs)_{n1}, \ldots, (\overline{p}@pats, rhs)_{nk_n}])$$
in
$$case_{ty}\ (\lambda \overline{v : \tau_1}.\ M_1) \ldots (\lambda \overline{v : \tau_n}.\ M_n)\ z$$

End.

Conceptually, the search for a match has come to an end: the patterns have been exhausted and the stack is empty, leaving a single right hand side.

$$\mathtt{trans}([\,], ([\,], rhs)) = rhs.$$

Example. The following is the translation of the *gcd* program:

$\lambda gcd\ z.\ case_{prod}$
$\quad (\lambda v\ v_1.\ case_{nat}\ v_1$
$\quad\quad (\lambda v_2.\ case_{nat}(\text{Suc } v_2)$
$\quad\quad\quad (\lambda v_3.\ \text{if } (v_3 \leq v_2)$
$\quad\quad\quad\quad \text{then } gcd(v_2 - v_3, \text{Suc } v_3)$
$\quad\quad\quad\quad \text{else } gcd(\text{Suc } v_2, v_3 - v_2))\ v_1)\ v)\ z.$

5 Examples

The following examples show some simple applications of the package. We have instantiated the parameterized definition algorithm with a wellfoundedness prover, a termination prover (a decision procedure for unquantified statements of linear arithmetic), and some simplification theorems asserting the injectivity of constructors for the datatypes the programs are defined over. The resulting ML function, named `Rfunction`, takes the relation and the description of the program, and returns a conjunction of (perhaps conditional) rewrite rules. In all the examples, the wellfoundedness condition is automatically proved.

Example 1. *Ackermann's function*

This celebrated example grows faster than any first order primitive recursive function. Its termination relation is a lexicographic combination of the predecessor relation for natural numbers.

$\quad$ `Rfunction` $(pred\ \text{X}\ pred)$
$\quad\quad (ack(0, n) = n + 1)\ \wedge$
$\quad\quad (ack(\text{Suc } m, 0) = ack(m, 1))\ \wedge$
$\quad\quad (ack(\text{Suc } m, \text{Suc } n) = ack(m, ack(\text{Suc } m, n)))$

The automatically extracted termination conditions

$(\underline{\text{Suc } m = \text{Suc } m} \vee m = \text{Suc } m \wedge 0 = \text{Suc } 1)\ \wedge$
$(\text{Suc } m = \text{Suc}(\text{Suc } m) \vee \underline{\text{Suc } m = \text{Suc } m \wedge \text{Suc } n = \text{Suc } n})\ \wedge$
$(\underline{\text{Suc } m = \text{Suc } m}$
$\quad \vee (m = \text{Suc } m\ \wedge$
$\quad\quad (\text{Suc } n = \text{Suc } ((ack\,|\,(\lambda(s,t)(u,v).u = \text{Suc } s \vee (s = u \wedge v = \text{Suc } t)),$
$\quad\quad\quad (\text{Suc } m, \text{Suc } n))(\text{Suc } m, n))))$

are trivial (as the underlined subterms show) and automatically proved by rewriting. The theorem

$\quad \vdash (ack(0, n) = n + 1)\ \wedge$
$\quad\quad (ack(\text{Suc } m, 0) = ack(m, 1))\ \wedge$
$\quad\quad (ack(\text{Suc } m, \text{Suc } n) = ack(m, ack(\text{Suc } m, n)))$

is returned.

Example 2. *Normalization of conditional expressions*

In this example, due to Boyer and Moore, assume that we have already declared a logical datatype (named *cond*) of conditional expressions with the following constructors:

$$\text{A} : individual \to cond$$
$$\text{IF} : cond \to cond \to cond \to cond$$

The following then defines a normalization function for such expressions.

Rfunction (*measure M*)
  ($norm(\text{A } i) = \text{A } i$) $\wedge$
  ($norm(\text{IF}(\text{A } x) \; y \; z) = \text{IF}(\text{A } x) \; (norm \; y) \; (norm \; z)$) $\wedge$
  ($norm(\text{IF}(\text{IF } u \; v \; w) \; y \; z) = norm(\text{IF } u \; (\text{IF } v \; y \; z) \; (\text{IF } w \; y \; z))$))

The measure function M, attributed to Robert Shostak, is defined by primitive recursion:

$$M(\text{A } i) \equiv 1$$
$$M(\text{IF } x \; y \; z) \equiv Mx + (Mx * My) + (Mx * Mz)$$

The system returns 3 termination conditions for *norm*:

$$Mz < M(\text{IF}(\text{A } x) \; y \; z) \; \wedge$$
$$My < M(\text{IF}(\text{A } x) \; y \; z) \; \wedge$$
$$M(\text{IF } u \; (\text{IF } v \; y \; z) \; (\text{IF } w \; y \; z)) < M(\text{IF}(\text{IF } u \; v \; w) \; y \; z)$$

which when expanded with the definition of M, give the goal

$$Mz < 1 + My + Mz \; \wedge$$
$$My < 1 + My + Mz \; \wedge$$
$$Mu + Mu * (Mv + Mv * My + Mv * Mz) +$$
$$Mu * (Mw + Mw * My + Mw * Mz)$$
$$<$$
$$(Mu + Mu * Mv + Mu * Mw) +$$
$$(Mu + Mu * Mv + Mu * Mw) * My +$$
$$(Mu + Mu * Mv + Mu * Mw) * Mz$$

Our postprocessor cannot prove this automatically, although some highly automated systems may be able to. With the tools available to us, however, the simple induction lemma $\vdash \forall x. \; 0 < Mx$ needs to be given, and arithmetic laws for distribution of products over sums must also be manually applied before the termination condition yields to manual application of our postprocessor.

Example 3. *Nested recursion*

A possible difficulty of directly using the wellfounded recursion theorem is raised by nested recursion.[3] For example, consider McCarthy's 91 function:

Rfunction $(measure \; \lambda x. \; 101 - x)$
   $(ninety1 \; x = \text{if} \; (x > 100) \; \text{then} \; x - 10 \; \text{else} \; ninety1(ninety1(x + 11)))$.

There are two termination conditions: the first is automatically disposed of, and the second is presented to the user as

$$\neg x > 100 \supset 101 - ninety1(x + 11) < 101 - x.$$

By use of the fact $\neg(x > 100) \supset ((101 - y < 101 - x) = (x < y))$, we get

$$x \leq 100 \supset x < ninety1(x + 11). \qquad (6)$$

This property involves *ninety1*. In approaches where the termination conditions need to be proved before defining the function, one would have to prove a property of *ninety1* before it had been defined: evidently, there is a kind of bootstrapping problem! Fortunately, our approach, which might be characterized as *early* definition, will have already defined the constant *ninety1*, and (6) is directly provable. Clearly, (6) can be established by an exhaustive examination of the value of $ninety1(x + 11)$ for all x less than 101. Instead we proved (6) by wellfounded induction along *measure* $\lambda x. \; 101 - x$, i.e., the given relation. This, plus the ability to expand the definition of *ninety1*, plus a few properties of $<$, sufficed to prove termination. That this is possible is at variance with the observation of Paulson that partial correctness and termination need to be proved together for nested recursions [21]. Perhaps this discrepancy can be explained by our early definition approach; if not, which nested recursion termination proofs actually require partial correctness properties may be an interesting theoretical question.

Example 4. Quicksort

This example displays the ability to define functions without giving a termination relation (the entrypoint for this is named function). It also show that higher-order constructs used to get efficiency and generality in programs are easily supported. To start, we define a function that partitions a list around a predicate.

function
   $(part(P, [\,], l_1, l_2) = (l_1, l_2)) \land$
   $(part(P, h :: t, l_1, l_2) =$
      if $P \; h$ then $part(P, t, h :: l_1, l_2)$ else $part(P, t, l_1, h :: l_2))$

---

[3] Ackermann's function is also a nested recursion, but of an extremely simple form.

And now Quicksort:

> function
> $(qsort(ord,[\,]) = [\,]) \wedge$
> $(qsort(ord, h :: t) =$
>   let $(l_1, l_2) = part(\lambda y.\ ord\ y\ h, t, [\,], [\,])$
>   in
> $qsort(ord, l_1)@[h]@qsort(ord, l_2))$.

Examining the following theorem returned from the definition of *qsort*, we see that 3 termination conditions have been placed on the assumptions. The conclusion is a conjunction. The first conjunct is the definition and the second is a principle of recursion induction for *qsort*, which our package also derives (for clarity and space, we omitted the induction theorems from the other examples).

$[WF\ R,$
$\forall l_1\ l_2\ ord\ h\ t.\ ((l_1, l_2) = part((\lambda y.\ ord\ y\ h), t, [\,], [\,])) \supset R(ord, l_2)(ord, h :: t),$
$\forall l_1\ l_2\ ord\ h\ t.\ ((l_1, l_2) = part((\lambda y.\ ord\ y\ h), t, [\,], [\,])) \supset R(ord, l_1)(ord, h :: t)]$
$\vdash$
$((qsort(ord, [\,]) = [\,]) \wedge$
$(qsort(ord, h :: t) =$
    let $(l_1, l_2) = part((\lambda y.\ ord\ y\ h), t, [\,], [\,])$
    in
    $qsort(ord, l_1)@[h]@qsort(ord, l_2)))$
$\wedge$
$(\forall P.$
 $(\forall ord.\ P(ord, [\,])) \wedge$
 $(\forall ord\ h\ t.$
    $(\forall l_1 l_2.\ ((l_1, l_2) = part((\lambda y.\ ord\ y\ h), t, [\,], [\,])) \supset P(ord, l_2)) \wedge$
    $(\forall l_1 l_2.\ ((l_1, l_2) = part((\lambda y.\ ord\ y\ h), t, [\,], [\,])) \supset P(ord, l_1)) \supset P(ord, h :: t))$
 $\supset \forall v\ v_1.\ P(v, v_1))$

The user is now free to decide when termination is to be proved. This may allow a clean division of labour in large-scale formalizations: blocks of functions can be parsed, typechecked, and defined very quickly with function, leaving termination obligations to be proved in a separate phase. Caution must be exercised since nonsensical descriptions like $f(x) = f(x+1)$ will be accepted by function; however, the termination conditions that arise will be very difficult!

6 Further Topics

We have now covered the basics of function definition, but this is only the beginning of the story. There are several areas where more support would improve the usefulness of the system:

- When a datatype is defined, various wellfounded relations can be automatically defined, *e.g.*, the immediate subterm, proper subterm (as the transitive closure of the immediate subterm relation), and the measure of the size of the datatype. The latter elements of this trio properly include the former, hence some systems [18, 4] automatically use just the *size* measure, which has the added bonus of being trivial to prove wellfounded. An alternative approach might be to find a single master definition of size for all datatypes and use that as a measure. This is possible in systems that represent datatypes in a larger type having its own notion of size [15]. However, it is overly constraining to assume that all implementations will define dataypes in this manner.

- It would be a useful exercise to integrate an automatic component that would take the termination conditions arising from an invocation of `function` and attempt to find a wellfounded relation and prove the termination conditions. Some interesting work in this area can be found in [7].
- Heuristics automating inductive proof, *e.g.*, the rippling technique [5] have received much attention recently, and we hope to somehow marry this work to ours.

7 Related work

The landmark work of Boyer and Moore has been deeply influential in the field of theorem proving and, in the case of function definition, provides an interesting contrast to our approach.

- They operate in the context of an unquantified first order logic; therefore the pattern matching translations that we make use of are not available to them.
- When given a function to define, *Nqthm* attempts to prove the termination lemma automatically, using a hard-wired termination relation. If that proof fails, the user must state and prove the lemma before the function can be defined. Our approach relieves the user of the burden of formulating termination lemmas, and does not have a fixed termination relation.
- In the case of nested recursions, the *Nqthm* user must prove that the nested description is equivalent to an already defined function, while we provide the full power of higher-order logic to directly attack the problem of proving termination.
- The *Nqthm* system provides a great deal of automated support for proof of program properties while we, as yet, provide only very simple support.

Other systems providing definition facilities akin to ours are LAMBDA [6] and PVS [18]. The research of [6] is quite similar to ours, except that they take a meta-theoretic approach and also avoid termination issues. Some past work on wellfounded recursion in the HOL system is [24]. Wellfounded induction and recursion have also been treated in Constructive Type theory [20, 17]. ALF [12], LAMBDA and Coq [13] support ML-style pattern matching in definitions. A lambda calculus with first class patterns is treated in [11].

In contrast to the difficulties attendant on proving termination in logics of total functions, LCF and its latter-day implementations in higher-order logic theorem provers (HOLCF[22], HOL-ST[1]) defines functions via a domain theoretic fixpoint operator. This has the nice property that termination proofs need never be done; however, the user pays the price of having two somewhat incompatible notions of function space in such a system. Agerholm has gone on to do some interesting work which combines a simplified domain-theory with termination proofs, resulting in a flexible system for total function definition[2].

In [21] Paulson used LCF to explore termination proofs for various normalizers for conditional expressions. Walther has investigated a class of terminating functions, focusing on a *size* measure on datatypes[25]. Automation of termination proofs is also investigated in [14]. Holger Busch has made an extensive investigation into wellfounded induction in the LAMBDA system [8].

8 Conclusions

We have presented a parameterized system for making recursive definitions of terminating functions. To our knowlege, this work provides the first definition package that is directly based on a formally proven wellfounded recursion theorem. As a result of our design, it is also the first such package to run on top of more than one theorem prover. We allow functions to be described in the pattern-matching style popular in ML-style languages. We have also presented a method that cleanly and soundly allows termination arguments to be deferred for the class of non-nested equations.

Acknowledgments. This work has benefitted greatly from frequent discussions with Tobias Nipkow. I also thank Matt Kaufmann for taking the time to explain how the ACL2 prototype handles the *variant* example.

References

1. S. Agerholm. LCF examples in HOL. *The Computer Journal*, 38(2):121–130, July 1995.
2. S. Agerholm. Non-primitive recursive function definition. In E. T. Schubert, P. J. Windley, and J. Alves-Foss, editors, *Proceedings of the 8th International Workshop on Higher Order Logic Theorem Proving and Its Applications (LNCS 971)*, pages 17–31, Aspen Grove, Utah, September 1995. Springer Verlag.
3. Lennart Augustsson. Compiling pattern matching. In J.P. Jouannnaud, editor, *Conference on Functional Programming Languages and Computer Architecture (LNCS 201)*, pages 368–381, Nancy, France, 1985.
4. Robert S. Boyer and J Strother Moore. *A Computational Logic*. Academic Press, 1979.
5. A. Bundy, A. Stevens, F. van Harmelen, A. Ireland, and A. Smaill. Rippling: A heuristic for guiding inductive proofs. *Artificial Intelligence*, 62:185–253, 1993.
6. Simon Finn, Mike Fourman, and John Longley. Partial functions in a total setting. To appear in Journal of Automated Reasoning, 1996.

7. Juergen Giesl. Termination analysis for functional programs using term orderings. In *Proceedings of the 2nd International Static Analysis Symposium*, Glasgow, Scotland, 1995. Springer-Verlag.
8. H. Busch. Unification based induction. In L.J.M. Claesen and M.J.C. Gordon, editors, *International Workshop on Higher Order Logic Theorem Proving and its Applications*, pages 97–116, Leuven, Belgium, September 1992. IFIP TC10/WG10.2, North-Holland. IFIP Transactions.
9. P. V. Homeier and D. F. Martin. A verified verification condition generator. *The Computer Journal*, 38(2):131–141, July 1995.
10. Peter Johnstone. *Notes on logic and set theory*. Cambridge University Press, 1987.
11. Delia Kesner, Laurence Puel, and Val Tannen. A typed pattern calculus. *Information and Computation*, 124(4):32–61, 1996.
12. Lena Magnusson and Bengt Nordstrom. The ALF proof editor and its proof engine. In *Types for Proofs and Programs (LNCS 806)*, pages 213–237, Nijmegen, Netherlands, 1994. Springer-Verlag.
13. Pascal Manoury. A user's friendly syntax to define recursive functions as typed λ-terms. In *Types for Proofs and Programs: International Workshop TYPES'94*, number 996 in Lecture Notes in Computer Science, Baastad, Sweden, June 1995. Springer Verlag.
14. Pascal Manoury and Marianne Simonot. Automatizing termination proofs of recursively defined functions. *Theoretical Computer Science*, (135):319–343, 1994.
15. Tom Melham. Automating recursive type definitions in higher order logic. In Graham Birtwistle and P.A. Subrahmanyam, editors, *Current Trends in Hardware Verification and Automated Theorem Proving*, pages 341–386. Springer-Verlag, 1989.
16. Tobias Nipkow. Term rewriting and beyond—theorem proving in Isabelle. *Formal Aspects of Computing*, 1:320–338, 1989.
17. Bengt Nordstrom. Terminating general recursion. *BIT*, 28:605–619, 1988.
18. S. Owre, J. M. Rushby, and N. Shankar. PVS: A prototype verification system. In Deepak Kapur, editor, *11th International Conference on Automated Deduction*, LNAI 607, pages 748–752, Saratoga Springs, New York, USA, June 15–18, 1992. Springer-Verlag.
19. Lawrence Paulson. A higher order implementation of rewriting. *Science of Computer Programming*, 3:119–149, 1983.
20. Lawrence Paulson. Constructing recursion operators in intuitionistic type theory. *Journal of Symoblic Computation*, 2:325–355, 1986.
21. Lawrence Paulson. Proving termination of normalization functions for conditional expressions. *Journal of Automated Reasoning*, 2:63–74, 1986.
22. Franz Regensburger. *HOLCF: Eine konservative Einbettung von LCF in HOL*. PhD thesis, Institut für Informatik, Technische Universität München, 1994.
23. H. Schwichtenberg and S. Wainer. Ordinal bounds for programs. In Jeff Remmel, editor, *Feasible Mathematics II*, pages 387–406. Birkhäuser, 1994.
24. M. van der Voort. Introducing well-founded function definitions in HOL. Leuven, Belgium, September 1992. IFIP TC10/WG10.2, Elsevier Science Publishers.
25. Christoph Walther. On proving the termination of algorithms by machine. *Artificial Intelligence*, 71(1):101–157, 1994.

Higher-Order Annotated Terms for Proof Search

Alan Smaill and Ian Green

Department of Artificial Intelligence, University of Edinburgh,
Edinburgh EH1 1HN, Scotland.
Email: A.Smaill, I.Green@ed.ac.uk

Abstract. A notion of *embedding* appropriate to higher-order syntax is described. This provides a representation of annotated formulae in terms of the difference between pairs of formulae. We define substitution and unification for such annotated terms. Using this representation of annotated terms, the proof search guidance technique of *rippling* can be extended to higher-order theorems. We illustrate this with two selected examples using our implementation of these ideas in λProlog.

1 Introduction

There is an increasing need to automate theorem proving in higher-order (HO) logics. Theories for hardware and software synthesis and verification are typically expressed in HO logics. The encoding of propositional and first-order (FO) logics in logic frameworks means that proof-search is lifted away from the object logic into a HO setting. In this paper we address the problem of controlling proof search in such a setting; in particular our approach is a development of techniques currently used to control first-order term rewriting.

Here the heuristic of *difference reduction* to problem solving is central: find the differences in term structure between the goal G and the assumptions H_n; then rewrite G in an effort to remove those differences in term structure, in order that an appeal to one or more of the assumptions can be made. Differences are called *wave-fronts* and shared term structure is called the *skeleton*.

Rippling is difference reduction by the application of certain, well-constrained rewrite rules. The key idea of rippling is that terms are explicitly decorated with annotations that describe explicitly wave-fronts and skeletons; similarly, rewrite rules are annotated and the rewrite relation restricted by requiring that annotation in the goal matches annotation in the rule. In this way the differences are explicitly manipulated by the annotated rewrite rules. By insisting that annotated rewrite rules are *skeleton-preserving* rippling is further restricted. Skeleton preservation ensures that if all the differences are removed, the assumption(s) can be used.

Basin and Walsh [3] and Hutter [16] have developed first-order rippling calculi: formal systems for manipulating annotated terms. However, there are some difficulties in extending such calculi to the λ-calculus, as is required when, for example:

- rippling first-order terms containing meta-variables that stand for functions, and, more generally,

– rippling in logics containing the λ-calculus.

The crux of the problem is that annotations on λ terms behave badly under reduction. Our solution is a new formalisation of rippling in which annotated terms are captured via *term embeddings*. We show how term embeddings can be defined uniformly for both first- and higher-order syntax, then how term embeddings capture rippling. We believe this presentation provides a clear account of rippling that generalises to the higher-order case.

Overview. In §2 we present standard FO embedding; in §3 we present our extension to HO embedding.

In §4 we show how embeddings are related to rippling. With HO embeddings, we are able to give a description of HO rippling. §5 gives some examples carried out with our prototype λProlog implementation. §6 discusses other work and draws some conclusions from our experiences.

2 Term Embeddings: the first-order case

The notion of an *embedding* (or homeomorphic embedding) between trees is known from its use within rewriting (see [18, p. 31]). For our purposes, following [7], we use a slightly different notion, where the order of the subtrees at a node is taken to be significant, and where the labels of a node and its image under an embedding are required to coincide.

More precisely, an embedding between terms t_1, t_2 is defined follows. Consider terms t_1, t_2 in a standard first-order syntax as labelled trees in the usual way. Then an embedding is an injective map e from the nodes of t_1 to the nodes of t_2 which maps labels to identical labels, which preserves the tree order, and which also preserves "horizontal order". Such an embedding, when it exists, can be represented by a tree of the shape of t_1 labelled with the addresses of the corresponding image nodes in t_2; where e is a tree representing an embedding, we write $e : t_1 \hookrightarrow t_2$, or simply $t_1 \hookrightarrow t_2$ if the identity of e is not important.

This notion can be captured inductively. For terms $t_1 \equiv f(u_1, \ldots, u_n)$ and $t_2 \equiv g(v_1, \ldots, v_m)$,

$$t_1 \hookrightarrow t_2 \text{ iff } \exists i\ t_1 \hookrightarrow v_i$$
$$\text{or } f \equiv g \text{ and } \forall i\ u_i \hookrightarrow v_i$$

This definition forms the basis for the computation of embeddings, by returning an appropriately labelled tree; this tree has the same shape as the term t_1, and the labels give the address of the image of the node under the embedding map. For an example, consider the terms $plus(x, y)$ and $plus(s(x), y)$ shown in figure 1.

The pair of terms together with the embedding together describe an *annotated term*, namely t_2 decorated so as to indicate which parts of the term tree are in the image of the embedding, and which are not; in this way we are able to capture the "differences" between terms (more on this in §4).

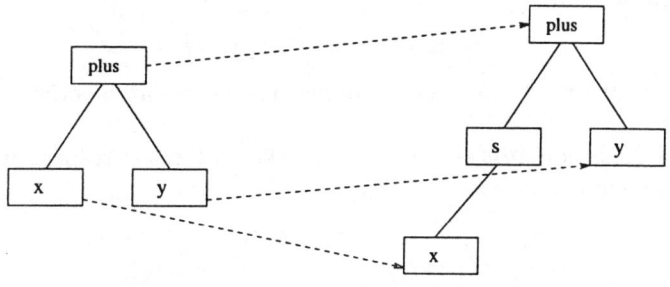

Fig. 1. An embedding

In earlier descriptions of rippling (e.g., [9]), annotated terms are depicted by boxing the wave-front and underlining the skeleton. For example, the embedding shown in figure 1 would be $plus(\boxed{s(\underline{x})}, y)$. We shall adopt this notation where is it convenient to do so.

In this earlier work, operations on annotated terms were defined on an enlarged term algebra, with extra functors wrapped around subterms to indicate which parts of a term belong to a skeleton. For example, the term above corresponds to $plus(wf(s(wh(x))), y)$. This notation necessitates the use of non-standard matching and substitution, for cases where the annotational syntax has to be distinguished from the underlying term structure.

2.1 Substitution and Unification

When reasoning with annotated terms, we will use the annotations for search guidance; the annotations should not however interfere with the standard logical operations on the underlying unannotated term. Thus we want to define substitution on embeddings so that it is sound with respect to the substitution on standard terms.

We define two functions *erase* and *skel* from annotated terms to plain terms by

$$erase(e: t_1 \hookrightarrow t_2) =_{def} t_2$$

and

$$skel(e: t_1 \hookrightarrow t_2) =_{def} t_1.$$

Now suppose that σ is a substitution on terms. Say that an embedding e' *extends* e if the tree e' is obtained from e by replacing some of the leaves of e with new trees, while preserving all the labels within the original tree e. It is easy to show the following in the first-order case.

Proposition 1. *If $e: t_1 \hookrightarrow t_2$ is an annotated term, and σ a substitution, then there is a unique $e': t_1\sigma \hookrightarrow t_2\sigma$ such that e' extends e.*

We therefore define
$$(e : t_1 \hookrightarrow t_2)\sigma =_{def} e' : t_1\sigma \hookrightarrow t_2\sigma$$
where e' is this unique extension of e. In general, there can be other embeddings of $t_1\sigma$ in $t_2\sigma$.

From this definition, it is easy to see that *skel* and *erase* respect substitution. That is, for all embeddings $e : t_1 \hookrightarrow t_2$,
$$erase((e : t_1 \hookrightarrow t_2)\sigma) = (erase(e : t_1 \hookrightarrow t_2))\sigma$$
$$skel((e : t_1 \hookrightarrow t_2)\sigma) = (skel(e : t_1 \hookrightarrow t_2))\sigma$$

A *unifier* for two annotated terms, $e : t_1 \hookrightarrow t_2$, $f : u_1 \hookrightarrow u_2$, is thus in the usual way a substitution σ such that
$$(e : t_1 \hookrightarrow t_2)\sigma = (f : u_1 \hookrightarrow u_2)\sigma,$$
where the equality entails equality of domain terms, and of range terms. From our previous remark, a unifier in this sense is clearly also a unifier for the erasures of the embeddings, that is for t_2, u_2.

A most general unifier for $e : t_1 \hookrightarrow t_2$, $f : u_1 \hookrightarrow u_2$, when it exists, is simply a most general unifier for the unification problem
$$\langle t_1 = u_1, t_2 = u_2 \rangle,$$
provided that the corresponding embeddings coincide. Thus annotated unification is a *restriction* of standard unification, which can be captured for example by a generate-and-test procedure.

3 Embeddings: a higher-order version

When reasoning about languages with variable-binding constructs, it has become normal to deal with binding and substitution once and for all in such a way that the operations can be applied uniformly over a range of situations; see for example [15, 13].

When we come to extend the idea of term embedding to higher-order syntax, we need to solve some new problems. If we treat terms of the λ-calculus like first order terms, then we will not get the behaviour we want. In particular, an operation such as β-reduction should not affect the existence of an embedding, as we do not want to distinguish between inter-convertible terms. Since there is no logical significance to the difference between such terms, it would seem unnatural for a proof guidance technique to distinguish between them.

For example, taking a and b to be constant terms, a embeds in $(\lambda x.b)\ a$ according to our initial definition, yet a is not embedded in the reduct b. The embedding should be insensitive to such reductions. For abstraction terms having the form $\lambda x.A$ for some λ term A, there must be at least one occurrence of x in A to ensure that a β-reduction of $\lambda x.A\ B$ preserves any embedding into it. This is precisely the restriction in Church's λI-calculus [2, §2.2.2].

The following definition of HO embedding exploits the type structure and the syntax of λI-calculus:

3.1 Extending the syntax

We suppose that the syntax at hand can be described as follows.

Take simply-typed λI-calculus with products, over a number of base types which represent the syntactic categories of the language. Now suppose that a number of typed constants are given within this system. The exposition of [12] is an example of this approach, though products are not used there, nor is there a restriction to λI.

The terms of interest are thus the typable terms of this calculus, formed from the constants via typed λI-abstraction, composition, and pairing. The projection terms normally associated with the product are excluded for the same reason that we have restricted attention to λI.

We now define an appropriate notion of embedding for this situation.

Definition of HO-embeddings Here t_i, u_i range over terms, and $A, B, \ldots$ over types. The notation $(t, u) : A \times B$ is used for products. In the following definition, inter-convertible terms are identified.

There are four cases:

1. **base**
$$t \hookrightarrow t \text{ for atomic } t$$

2. **product**
$$t \hookrightarrow (u_1, u_2) \text{ iff } t \hookrightarrow u_1$$
$$\text{or } t \hookrightarrow u_2$$
$$\text{or } t = (t_1, t_2) \text{ and } t_i \hookrightarrow u_i \quad (i = 1, 2)$$

3. **abstraction**
$$t \hookrightarrow \lambda x.u \text{ iff for all } z, t \hookrightarrow (\lambda x.u)z$$
$$\text{or } t = \lambda y.t_1 \text{ and for all } z, (\lambda y.t_1)z \hookrightarrow (\lambda x.u)z$$

4. **application**
$$t \hookrightarrow u_1 u_2 \text{ iff } t \hookrightarrow u_1$$
$$\text{or } t \hookrightarrow u_2$$
$$\text{or } t = t_1 t_2 \text{ and } t_i \hookrightarrow u_i \quad (i = 1, 2)$$

The abstraction case here reduces the problem at functional types to that of finding embeddings for atomic types, at the cost of requiring that they exist uniformly. This is the method that is used to deal with the case of standard variable binding constructs such as quantifiers.

The use of product types here is for efficiency reasons. It is normal to use curried syntax in a higher-order system, and our definition allows this; however this means that a given term has typically many more subterms than in a standard first-order representation, and it is convenient to ignore such terms by representing tuples via products.

This definition also allows us to compute embeddings as labelled trees. The notions of erasure and skeleton are as before. We have also the analogous result.

Proposition 2. *Suppose that $e : t \hookrightarrow u$, and σ is a higher-order substitution. Then there is an $e' : t\sigma \hookrightarrow u\sigma$.*

This uses restriction to λI-calculus and the result for atomic types extended by virtue of the universal quantification in case (3) of the definition of HO embeddings.

As in the first-order case, this notion of substitution provides a definition of unification between two annotated terms. A *unifier* for two annotated terms $e : F_1 \hookrightarrow F_2$, $f : G_1 \hookrightarrow G_2$ is a substitution σ such that

$$(e : F_1 \hookrightarrow F_2)\sigma = (f : G_1 \hookrightarrow G_2)\sigma,$$

where higher-order terms may now be substituted for variables.

Example. In §5.1 we will see an example of rippling involving an introduced constant lim, and some additional symbols:

$$\begin{aligned}
\text{lim} &: (term \to term) \to term \to term \to form \\
f &: term \to term \\
g &: term \to term \\
+ &: (term \times term) \to term \\
a &: term \\
l &: term \\
m &: term
\end{aligned}$$

Then then the following HO embedding exists

$$e : (\text{lim } f\ a\ l) \hookrightarrow (\text{lim } (\lambda u.(f\ u) + (g\ u))\ a\ (l + m))$$

by virtue of the fact that for all terms z,

$$(\lambda y.(f\ y))\ z \hookrightarrow (\lambda u.(f\ u) + (g\ u))\ z.$$

Remark. While the above proposition establishes that our definition enjoys useful properties with respect to λI terms, when we come to look at synthesis examples in section 5 we need to consider a slightly wider class of terms. We are currently investigating how the definition above can be generalised.

4 Rippling: a proof search technique

We now illustrate the use of terms annotated by means of embeddings to guide proof search. This approach to the search for proofs involving induction was proposed in [8] and has since been developed and also applied to non-inductive proofs, where a logical connection is sought between two syntactically similar formulas [9, 25, 3].

We first introduce a well-founded *measure* on embeddings. Recall that embeddings are represented as trees labelled by addresses, whose lengths increase towards the leaves of the tree. We define an order on such trees. For each node, we associate a weight: the difference between the length of the associated address, and the depth in the tree. This is an indication of the difference between the erasure and the skeleton, in this branch of the skeleton. We now form a list of the sum of the weights at successive depths of the embedding tree: call this the *measure list*.

One difference reduction heuristic is to move differences *outward* and for this we must compare two measure lists in *reverse* lexicographic order, if they are of the same length; a longer list is larger in the order than a shorter one. A smaller embedding in this order is thus one where syntactic difference has been reduced, deep in the term structure.

In proofs by induction, using a constructor formulation, this is used as follows. Given an induction hypothesis $\Phi(x)$ and goal $\Phi(s(x))$, there is an embedding $e : \Phi(x) \hookrightarrow \Phi(s(x))$. Recursion equations and lemmas are available as rewrite rules to be used on the goal — rewriting here is extended to take account of quantification. Typically, these rewrite rules are not confluent, and include potentially looping rules, such as associativity. Among the possible rewrites, the following heuristic can be used: use rewrites which give a new goal G' such that there is an embedding $e' : \Phi(x) \hookrightarrow G'$, and such that $e' < e$, ie the embedding is smaller in the measure defined above. This gives at each step the following picture (where e, e' are embeddings and the vertical arrow indicates rewriting). We call such a rewriting step a *wave rewrite*, following the terminology of [8]; the successive application of such waves is known as *rippling*.

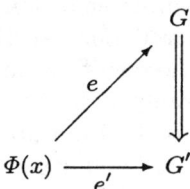

There are three conditions on wave rewriting:

1. *soundness* – the rewriting should correspond to logically valid inference;

2. *skeleton invariance* — the appropriate skeleton embeds in the goal, before and after rewriting;
3. *measure decreasing* — the embedding in the rewritten formula is smaller in the order.

Wave-rules are derived from implications and equalities (thus respecting the first requirement), and embeddings constructed so as to respect the second and third requirements.

We can depict the possible ways a wave rewrite may be made on the goal by considering the rewrite rules available, and annotating them with embeddings: these annotated rules are called *wave-rules*. In general there is more than one way in which this annotation can be added, and it may be added so that the rule can be used in either direction.

Since we are reasoning backwards, there is a polarity inversion to the direction in which rules based on implication may be used. A logical implication $A \to B$ may be used as a rewrite rule $A \Rightarrow B$ (at positions of negative polarity) and $B \Rightarrow A$ (positive polarity). Equalities are used in either direction. In all cases the direction of the wave-rule is given by the double arrow $\Rightarrow$.

Example wave-rule We will see in §5 some example proofs using rippling. There use is made of the associativity of disjunction:

$$A \vee (B \vee C) \equiv (A \vee B) \vee C$$

which can be used as a rewrite in either direction, depending upon the skeleton to be preserved, and the direction in which the wave-front is to be moved. For example, the two wave-rules below are derived from the associativity property:

$$A \vee \boxed{(\underline{B} \vee C)} \Rightarrow \boxed{(A \vee B) \vee C}$$

$$\boxed{(A \vee \underline{B})} \vee C \Rightarrow \boxed{A \vee (\underline{B} \vee C)}$$

In the first rule the skeleton is $A \vee B$ (which embeds into the LHS and RHS of the wave-rule), and in the second it is $B \vee C$ (also embedded in both sides). Thus one is able to use rewrites in *both* directions during rippling.

Given the definition of higher-order embeddings from above, these ideas can be generalised. Rewriting in this situation can be done using the ideas of [12]. The use of annotations to guide the rewriting can then follow the same strategy as in the first-order case; some example are given in §5.

When these annotations are used within proof search, under some circumstances it is useful to use skeleton-preserving equations as rewrites in a direction which increases the measure defined above. This corresponds to moving differences away from the root of the term (examples of this are given in §5).

We incorporate this possibility by adding to the embedding an indication of *inward* or *outward* direction. We use the notation $\boxed{s(\underline{x})}^{\downarrow}, \boxed{s(\underline{x})}^{\uparrow}$ respectively. In harmony with this, we define an extended well-founded measure, which is a lexicographic combination of a lexicographic ordering on measure lists (for inbound differences) and a reverse lexicographic ordering on measure lists (for outbound differences).

4.1 Middle Out Reasoning

The use of meta-variables in implemented systems to delay commitment to certain choices in proof search is by now commonplace (this is used in Lego and Isabelle, for example [20, 23]). This is a form of *middle out reasoning*. We use such meta-variables in dealing with existentially quantified goals; the interaction between the instantiation of meta-variables and the wave annotations has been a stumbling block in earlier attempts to extend the scope of rippling.

The use of embeddings enables us to deal with this situation. Here we allow the terms and formulas we manipulate to contain meta-variables. An annotated term is simply an embedding as before, in which either the erasure, or both the skeleton and erasure, contain meta-variables. Now instantiation of such meta-variables is defined for formulas in the standard way for the simply-typed λI-calculus; for annotated formulas, the propositions above assure us that embeddings will be instantiated in a sensible way.

The treatment of termination has to be different in the case where we deal with meta-variables. The proofs of the termination of rippling given for earlier versions ([9, 5]) are necessarily restricted to the ground case: some additional mechanism is needed to deal with the middle-out case. The problem is that if one of the rewriting steps causes the instantiation of a meta-variable, as we want, then the resultant embedding will not be smaller in our given measure. The heuristic in this case must provide some way of controlling the instantiation; recomputation of the embedding can be used as part of this process.[1]

We note here that we want annotations to *guide* the instantiation through the application of rewrites; we do not allow the computation of embeddings to cause any instantiation on its own. This means that we use the weaker notion of embeddability rather than embedding in practice here.

5 Examples, Results

In this section we show some examples to illustrate higher-order embeddings. The examples were carried out on a prototype λProlog implementation of the Clam proof-planning system [10].

5.1 LIM+ theorem

This theorem was proposed by Bledsoe as a benchmark theorem for automated theorem provers [6]. LIM+ has two hypotheses, and the difference reduction

---

[1] The heuristic used in the implementation is very simple: an upper bound is placed on the number of ripple steps which *increase* the measure. (See [24] for a discussion of this heuristic.)

proof will exploit both of these. It is possible to ripple simultaneously towards both hypotheses (such a proof of LIM+ is shown in [25]), but our λProlog implementation currently does not support this improvement. Instead, we ripple towards a single hypothesis only.

The definition of lim : $(term \to term) \to term \to term \to form$ is as follows:

$\lim(f, a, l) \equiv$
$\quad \forall \epsilon.(0 < \epsilon \to \exists \delta.(0 < \delta \land \forall x.(x \neq a \land |x - a| < \delta \to |(f\ x) - l| < \epsilon))),$

and the $\lim^+$ theorem is:

$$\lim(f_1, a, l) \land \lim(f_2, a, m) \to \lim(\lambda x.(f_1\ x) + (f_2\ x), a, l + m).$$

Here f and g are functions on the reals, a is a point at which we are taking the limits, and l and m are the limiting values of f and g at a.

We shall use the following HO rewrite rules in the proof:

$(\boxed{U_1 + U_2}^\uparrow) - (\boxed{V_1 + V_2}^\uparrow) \Rightarrow \boxed{(U_1 - V_1) + (U_2 - V_2)}^\uparrow$

$|\boxed{U + V}^\uparrow| < E \Rightarrow \boxed{|U| + |V|}^\uparrow < E$

$\boxed{U + V}^\uparrow < W \Rightarrow U < \boxed{half(W)}^\downarrow \land V < half(W)\ ^\uparrow$

$\forall x. \boxed{P_1\ x \land P_2\ x}^\uparrow \Rightarrow \boxed{\forall x.P_1\ x \land \forall x.P_2\ x}^\uparrow$

$\exists \delta. P\ \delta \land \forall x.(C\ x \land Q\ x < \delta) \to \boxed{(R_1\ x) \land (R_2\ x)}^\uparrow \Rightarrow \begin{array}{l} \exists \delta.P\ \delta \land \forall x.(C\ x \land Q\ x < \delta) \to R_1\ x\ \land \\ \exists \delta.P\ \delta \land \forall x.(C\ x \land Q\ x < \delta) \to R_2\ x \end{array}^\uparrow$

$0 < E \to \boxed{P \land Q}^\uparrow \Rightarrow 0 < \boxed{half(E)}^\downarrow \to P \land 0 < half(E) \to Q\ ^\uparrow$

We have based our rewrite system on the axiomatisation given by Bledsoe [6] for first-order resolution theorem provers. There are some differences: Bledsoe uses unary minus and commutativity and associativity of +, rather than the distributivity lemma that we use here. Most deviant from his presentation however is the rule concerning existential quantification. This rule expresses the fact that we can distribute $\exists \delta$ through $\land$ providing the context is appropriate (obviously there is no such distributivity rule in general). The primary justification of this rule is $U < X \land U < Y \to U < \min(X, Y)$ (which Bledsoe admits as an axiom). We intend to investigate how these existential variants can be constructed "on the fly".

The ripple proof. We start the proof by replacing each occurrence of lim with its definition, then embedding the left conjunct of the antecedent into the consequent. The resulting annotated goal is

$$\forall \epsilon.(0 < \epsilon \rightarrow \exists \delta.(0 < \delta \wedge$$
$$\forall x.(x \neq a \wedge |x - a| < \delta \rightarrow ||\boxed{(f\,x) + (g\,x)}^{\uparrow} - \boxed{\underline{l} + m}^{\uparrow}| < \epsilon))).$$

Exhaustive rippling with the rules above put the goal into a normal form thus

$$\forall \epsilon.(0 < \boxed{half(\underline{\epsilon})}^{\downarrow} \rightarrow$$
$$\exists \delta.(0 < \delta \wedge \forall x.(x \neq a \wedge |x - a| < \delta \rightarrow |(f\,x) - l| < \boxed{half(\underline{\epsilon})}^{\downarrow}))) \wedge$$
$$\forall \epsilon.(0 < half(\epsilon) \rightarrow$$
$$\boxed{\exists \delta.(0 < \delta \wedge \forall x.(x \neq a \wedge |x - a| < \delta \rightarrow |(g\,x) - m| < half(\epsilon))))}^{\uparrow}$$

at which point the hypothesis may be used. Notice that the presence of the remaining difference (two occurrences of $\boxed{half(\underline{\epsilon})}^{\downarrow}$) does not prevent this appeal to the hypothesis because the ϵ in the assumption is universally quantified and so we may use any instance of it. The ability to use universal positions in this way is what motivates the inwards rippling introduced in §4.

5.2 Program synthesis

This example illustrates the synthesis of functional programs in type theory. We work in a constructive type theory called *Oyster* [10] (which is a close relation to Nuprl [11]). In this setting the problem of (recursive) program synthesis from specifications is the problem of (inductive) theorem proving: from a proof of

$$\forall input.\exists output.(spec\ input\ output) \tag{1}$$

a program satisfying the specification *spec* can be extracted.

On a conjecture of the above form, our synthesis proof-plan (see [24]) chooses an appropriate induction scheme and then searches for an existential witness using middle-out reasoning. A meta-variable N (here of type *term* $\rightarrow$ *term*) ranging over object-level terms is introduced to stand for this witness.

$$\forall input.(spec\ input\ (N\ input)). \tag{2}$$

As an example we will give a proof of

$$\forall l : list(nat), m : list(nat).\exists n : list(nat).$$
$$\forall x : nat.(x \in l \vee x \in m) \rightarrow x \in n,$$

where at least the following rewrites are available

$$X \in nil \Rightarrow false \qquad (3)$$

$$X \in \boxed{H :: \underline{T}}^\uparrow \Rightarrow \boxed{X = H \lor \underline{X \in T}}^\uparrow \qquad (4)$$

$$\boxed{P \lor \underline{Q}}^\uparrow \lor R \Rightarrow \boxed{P \lor \underline{Q \lor R}}^\uparrow \qquad (5)$$

$$\boxed{P \lor \underline{Q}}^\uparrow \to \boxed{P \lor \underline{R}}^\uparrow \Rightarrow Q \to R \qquad (6)$$

The proof-plan chooses an $h :: t$ induction on the input list l. This results in the annotated step-case goal

$$\begin{array}{l} l, m, t, n_0 : list(nat), h : nat \\ \forall x. x \in t \lor x \in m \to x \in n_0 \\ \vdash \\ \forall x. x \in \boxed{h :: \underline{t}}^\uparrow \lor x \in m \to x \in (N\, h\, t\, m\, n_0) \end{array} \qquad (7)$$

An embedding captures the difference between the induction hypothesis and the goal.

Rippling with (4) then (5) gives

$$\begin{array}{l} l, m, t, n_0 : list(nat), h : nat \\ \forall x. x \in t \lor x \in m \to x \in n_0 \\ \vdash \\ \forall x. \boxed{x = h \lor \underline{x \in t \lor x \in m}}^\uparrow \to x \in (N\, h\, t\, m\, n_0) \end{array} \qquad (8)$$

At this point no more ripples are possible: for example, rippling prevents the repeated use of the associativity of $\lor$. Rule (4) is applicable if we allow the measure to increase, as we do according to §4.1. Applying this measure-increasing rule instantiates N by higher-order unification to $\lambda uvwx.(N_1\, u\, v\, w\, x) :: (N_2\, u\, v\, w\, x)$. These fresh variables N_1 and N_2 come from the higher-order unification procedure of λProlog: they capture a most general instantiation of N. It is the generality of the substitution which allows subsequent proof to proceed middle-out. (The rule (3) is also applicable at this point, instantiating N to be the constant function returning nil. However, this instantiation is rejected since it does not preserve the embedding.)

$$\begin{array}{l} l, m, t, n_0 : list(nat), h : nat \\ \forall x. x \in t \lor x \in m \to x \in n_0 \\ \vdash \\ \forall x. \boxed{x = h \lor \underline{x \in t \lor x \in m}}^\uparrow \\ \qquad \to \boxed{x = (N_1\, h\, t\, m\, n_0) \lor \underline{x \in (N_2\, h\, t\, m\, n_0)}}^\uparrow \end{array} \qquad (9)$$

N_1 is instantiated to a projection onto its first argument by the application of rule (6). The resulting goal is

$$l, m, t, n_0 : list(nat), h : nat \\ \forall x. x \in t \vee x \in m \rightarrow x \in n_0 \\ \vdash \\ \forall x. x \in t \vee x \in m \rightarrow x \in (N_2\ h\ t\ m\ n_0) \tag{10}$$

and this can be finished by fertilisation, with the instantiation of N_2 to a projection onto its last argument. Composing the instantiations gives $(N\ h\ t\ m\ n_0) = h :: n_0$. From this proof a functional program for list append is automatically extracted; different proofs yield different functions still satisfying the specification.

6 Related work & conclusions

The idea of rippling dates back to Aubin [1]; Alan Bundy is responsible for rippling as we have presented it here, and for the conception of its termination measure and proof-planning in general.

Only recently has there been an interest in higher-order rippling: initially brought about by the need to treat binding operators correctly, and latterly by the need to ripple essentially higher-order syntax (for example in Bledsoe's LIM family of theorems which were reported in [25]).

Chuck Liang was perhaps the first to integrate rippling into a higher-order language, although the related term rewriting was essentially first-order ([19]). His representation used λ abstraction to represent context (the wave-front), the (position of the) wave-hole being marked by the bound variable; the wave-hole term is carried separately. Thus annotated terms have type $(i \rightarrow i) \times i$. (There is a wave-operator to map terms of this type into i.) Left- and right-projection deliver the wave-front and hole; applying the former to the latter gives the erasure. This representation suffers from a certain lack of expressiveness and combinatorial complexity when one attempts to generalise it to the coloured case (see [14] for details, extensions and generalisations). Liang hints at the difficulty in making sensible HO annotations.

In parallel with the development of rippling in Edinburgh, Dieter Hutter has formulated a sound calculus of annotated FO terms in which one can express rippling [16]. In much the same vein, Basin and Walsh describe a slightly less expressive FO calculus for rippling, and provide a termination proof. Hutter does not consider termination; Basin and Walsh have a simplification ordering for rippling ([4]).

In each of these calculi the underlying term algebra of the logic is extended in order to represent annotated terms. It is our experience that such an approach makes it very difficult to demonstrate the necessary properties of rippling. Furthermore, one is obliged to devise and implement new "extended" calculi on a logic-by-logic basis. Recently Hutter and Kohlhase have extended Hutter's first-order calculus to the λ-calculus ([17]), and we are comparing our approach with theirs. We do not yet understand the representative strength of their calculus, although we expect that some restriction of it will be equivalent to embeddings.

Higher-order embeddings treat annotated terms in a more abstract way. Correctness of unification and rewriting are immediate from the given logic. None of the basic logical machinery needs modification.

It would be interesting to see if embeddings and the associated operations can be implemented without calling on full higher-order unification, for example using the restriction to β_0-unification possible with *higher-order patterns* [21, 22].

The use of embeddings to represent annotated terms allows concise and implementable characterisation of wave annotations that is closer to their "intended meaning". It thus allows a systematic approach to the use of annotations to guide proof search in higher-order proof systems.

Acknowledgements

The authors would like to thank members of and visitors to the DReaM group for feedback on this work. Some earlier ideas related to this work are due to Jason Gallagher. We would also like to thank Paul Jackson of Edinburgh University and anonymous referees for their comments on the paper.

References

1. R. Aubin. Some generalization heuristics in proofs by induction. In G. Huet and G. Kahn, editors, *Actes du Colloque Construction: Amélioration et vérification de Programmes*. Institut de recherche d'informatique et d'automatique, 1975.
2. H. P. Barendregt. *The Lambda Calculus*. Elsevier, 1985.
3. David Basin and Toby Walsh. A calculus for and termination of rippling. Technical report, MPI, 1994. To appear in special issue of the Journal of Automated Reasoning.
4. David Basin and Toby Walsh. A calculus for rippling. In *Proceedings of CTRS-94*, 1994.
5. David Basin and Toby Walsh. Termination orders for rippling. In Alan Bundy, editor, *12th Conference on Automated Deduction*, Lecture Notes in Artificial Intelligence, Vol. 814, pages 466–83, Nancy, France, 1994. Springer-Verlag.
6. W.W. Bledsoe. Challenge problems in elementary calculus. *Journal of Automated Reasoning*, 6(3):341–359, 1990.
7. A. Boudet and H. Comon. About the theory of tree embedding. In M.-C. Gaudek and J.-P. Jouannaud, editors, *TAPSOFT '93: Theory and Practice of Software Development*, number 668 in LNCS, pages 376–90. Springer-Verlag, 1993.
8. A. Bundy. The use of explicit plans to guide inductive proofs. In R. Lusk and R. Overbeek, editors, *9th Conference on Automated Deduction*, pages 111–120. Springer-Verlag, 1988. Longer version available from Edinburgh as DAI Research Paper No. 349.
9. A. Bundy, A. Stevens, F. van Harmelen, A. Ireland, and A. Smaill. Rippling: A heuristic for guiding inductive proofs. *Artificial Intelligence*, 62:185–253, 1993. Also available from Edinburgh as DAI Research Paper No. 567.

10. A. Bundy, F. van Harmelen, C. Horn, and A. Smaill. The Oyster-Clam system. In M.E. Stickel, editor, *10th International Conference on Automated Deduction*, pages 647–648. Springer-Verlag, 1990. Lecture Notes in Artificial Intelligence No. 449. Also available from Edinburgh as DAI Research Paper 507.
11. R.L. Constable, S.F. Allen, H.M. Bromley, et al. *Implementing Mathematics with the Nuprl Proof Development System*. Prentice Hall, 1986.
12. A. Felty. A logic programming approach to implementing higher-order term rewriting. In L-H Eriksson et al., editors, *Second International Workshop on Extensions to Logic Programming*, volume 596 of *Lecture Notes in Artificial Intelligence*, pages 135–61. Springer-Verlag, 1992.
13. A. Felty. Implementing tactics and tacticals in a higher-order logic programming language. *Journal of Automated Reasoning*, 11(1):43–81, 1993.
14. J. K. Gallagher. *The Use of Proof Plans in Tactic Synthesis*. PhD thesis, University of Edinburgh, 1993.
15. R. Harper, F. Honsell, and G. Plotkin. A framework for defining logics. *Journal of the ACM*, 40(1):143–84, 1992. Preliminary version in LICS '87.
16. D. Hutter. *Pattern-Direct Guidance of Equational Proofs*. PhD thesis, University of Karlsruhe, 1991.
17. D. Hutter and M. Kohlhase. A colored version of the λ-calculus. SEKI-report sr-95-05, University of Saarland, 1995.
18. J.W. Klop. Term rewriting systems. In S. Abramsky, D. Gabbay, and T.S.E. Maibaum, editors, *Handbook of Logic in Computer Science, vol 2*, volume 2, pages 1–116. Clarendon Press, Oxford, 1992.
19. Chuck Liang. λProlog implementation of ripple-rewriting. In *Proceedings of the 1992 Workshop on the λProlog Programming Language*, University of Pennsylvania, Philadelphia, PA, USA, July-August 1992.
20. Z. Luo and R. Pollack. Lego proof development system: User's manual. Report ECS-LFCS-92-211, Department of Computer Science, University of Edinburgh, May 1992.
21. D. Miller. A logic programming language with lambda abstraction, function variables and simple unification. In *Extensions of Logic Programming*, volume 475 of *Lecture Notes in Artificial Intelligence*. Springer-Verlag, 1991.
22. T. Nipkow. Higher-order critical pairs. In *Proc. 6th IEEE Symp. Logic in Computer Science*, pages 342–349, 1991.
23. L. Paulson. Natural deduction as higher order resolution. *Journal of Logic Programming*, 3:237–258, 1986.
24. Alan Smaill and Ian Green. Automating the synthesis of functional programs. Research paper 777, Dept. of Artificial Intelligence, University of Edinburgh, 1995.
25. Tetsuya Yoshida, Alan Bundy, Ian Green, Toby Walsh, and David Basin. Coloured rippling: An extension of a theorem proving heuristic. In A.G. Cohn, editor, *In proceedings of ECAI-94*, pages 85–89. John Wiley, 1994.

A Comparison of MDG and HOL for Hardware Verification

Sofiène Tahar[§] and Paul Curzon[‡]

[§]University of Montreal, IRO Department, Canada.
[‡] University of Cambridge, Computer Laboratory, UK.

Abstract. Interactive formal proof and automated verification based on decision graphs are two contrasting formal hardware verification techniques. In this paper, we compare these two approaches. In particular we consider HOL and MDG. The former is an interactive theorem proving system based on higher-order logic, while the latter is an automatic system based on Multiway Decision Graphs. As the basis for our comparison we have used both systems to independently verify a fabricated ATM communications chip: the Fairisle 4 by 4 switch fabric.

1 Introduction

Formal hardware verification techniques are starting to attract widespread interest due to their potential to give very strong results about the correctness of designs. Two very different forms of formal verification have arisen: interactive proof and automated decision graph techniques. The aim of this paper is to compare and contrast these two approaches.

In the interactive proof approach, the circuit and its behavioral specification are represented in the underlying logic of a general purpose theorem prover. The user interactively constructs a formal proof which proves a theorem stating the correctness of the circuit. Many different proof systems with various forms of interaction have been used for this purpose. In this paper we consider one such system: HOL [7]. It is an LCF style proof system based on higher-order logic.

In the automated decision graph approach the circuit is represented as a decision diagram, and techniques such as reachability analysis are used to automatically verify given properties of the circuit or verify machine equivalence. We consider the MDG system. It uses a new class of decision graphs called Multiway Decision Graphs [3]. They subsume the class of Bryant's Reduced Ordered Binary Decision Diagrams [1] while accommodating abstract sorts and uninterpreted function symbols.

As the basis of our comparison of HOL and MDG, we have used both to independently verify the Fairisle [10] 4 by 4 switch fabric[1]. This is a fabricated chip which forms the heart of an ATM communication switch. It does the actual switching of data cells from input ports to output ports within the switch,

---

[1] See URL http://www.cl.cam.ac.uk/Research/HVG/atmproof/ for more details of Fairisle, the 4 by 4 fabric design and both the MDG and HOL verification projects.

arbitrating clashes and sending acknowledgments. It was not designed for the verification case study. Indeed it was fabricated and in use, carrying real user data, prior to any formal verification attempt.

There has been a vast amount of work on formal hardware verification. We mention here only that which is directly related to our study on verifying network hardware components.

J. Herbert [8] used HOL to formally verify the ECL chip: a local area network interface which formed part of the Cambridge Fast Ring. This is of roughly similar complexity to the circuit we considered, though our HOL proof took less time, demonstrating the increased maturity of the system.

B. Chen *et. al* at Fujitsu Digital Technology Ltd. [2] verified an ATM circuit that makes high-speed switching operations at 156 MHz and consists of about 111K gates. When the circuit was manufactured it showed an abnormal behavior under certain circumstances. Using the SMV tool [11], the authors identified the design error by checking some properties expressed in Computational Tree Logic [11]. Due to the restriction of the Boolean computation used by SMV and in order to avoid a state space explosion, they had to abstract the data width of addresses from 8 bits to 1 bit, and the number of addresses in the Write Address FIFO from 168 to 5. Although the design error was diagnosed, there is no proof showing that the abstracted circuit was itself correct.

K. Schneider *et. al* [12] formally verified the Fairisle 4 by 4 switch fabric using a verification system based on the HOL theorem prover: MEPHISTO. They described the structure of each of the modules used in the hardware design hierarchically down to the gate level and provided their behavioral specifications using hardware formulas. Although they automated the verification of lower-level hardware modules which implement the top-level block units, they have not accomplished the complete verification of the intended overall behavior of the switch fabric against its implementation.

The outline of the paper is as follows. In Section 2 we give a brief overview of the particular hardware considered: the Fairisle 4 by 4 switch fabric. We describe its verification using HOL in Section 3 and using MDG in Section 4. For each we overview the verification method, tools and our experiences on this case study. Finally, in Section 5 we draw conclusions. Since we have considered only a single case study it should be noted that such conclusions cannot be definitive.

2 The Fairisle 4 by 4 Switch Fabric

The Fairisle switch forms the heart of the Fairisle network. It consists of three types of components: input port controllers, output port controllers and a switch fabric. Each port controller is connected to a transmission line and to the switch fabric. The port controllers synchronize incoming and outgoing data cells, appending control information to the front of the cells in a routing byte (header). This byte is stripped off before the cell reaches the output stage of the fabric. A cell consists of a fixed number of data bytes which arrive one at a time. The fabric switches cells from the input ports to the output ports according to the routing

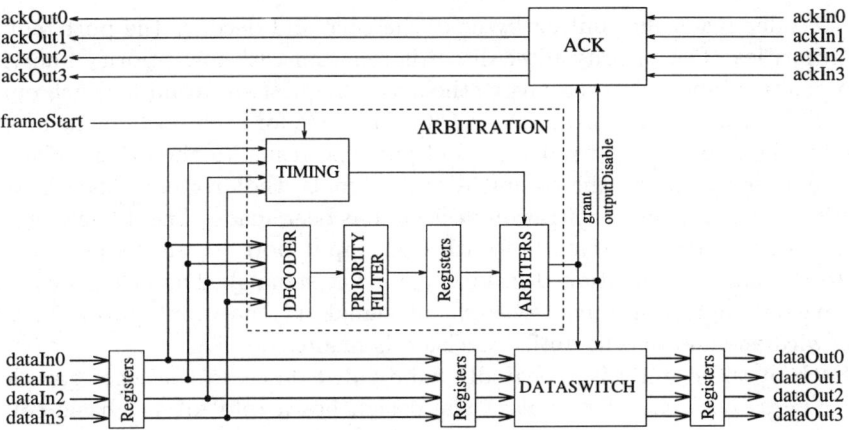

Fig. 1. The Fairisle Switch Fabric

byte. If different port controllers inject cells destined for the same output port controller (indicated by route bits in the routing byte) into the fabric at the same time, then only one will succeed. The others must retry later. The routing byte also includes a priority bit that is used by the fabric during arbitration. It takes place in two stages. First, high priority cells are given precedence, and for the remaining cells the choice is made on a round-robin basis. The input controllers are informed of whether their cell was successful using acknowledgment lines. The fabric sends a negative acknowledgment to the unsuccessful input ports, but passes the acknowledgment from the requested output port to the successful input port. The port controllers and switch fabric all use the same clock, hence bytes are received synchronously on all links. They also use a higher-level cell frame clock–the *frame start* signal. It ensures that the port controllers inject data cells into the fabric synchronously so that the routing bytes arrive at the same time. In this paper, we are concerned with the verification of the switch fabric which is the core of the Fairisle ATM switch.

The behavior of the switch fabric is cyclic. In each cycle or frame, it waits for cells to arrive, reads them in, processes them, sends successful ones to the appropriate output ports, and sends acknowledgments. It then waits for the arrival of the next round of cells. The cells from all the input ports start when the *active* bit of any one of their routing bytes goes high. The fabric does not know when this will happen. However, all the input port controllers must start sending cells at the same time within the frame. If no input port raises the active bit throughout the frame then the frame is inactive–no cells are processed. Otherwise it is active.

Figure 1 shows a block diagram of a 4 by 4 switch fabric. It is composed of an arbitration unit (timing, decode, priority filter and arbiters), an acknowledgment unit and a dataswitch unit. The timing block controls the timing of the decision with respect to the frame start signal and the time the routing byte arrives.

The decoder reads the routing bytes of the cells and decodes the port requests and priorities. The priority filter discards requests with low priority and those from inactive inputs. It then passes the actual request situation for each output port to the arbiters. The arbiters (in total four—one for each port) make arbitration decisions for each output port and pass the result to the other units with the grant signal. Using the output disable signals, the arbiters indicate to the other units when a new arbitration decision has been made. The dataswitch unit performs the actual switching of data from input port to output port according to the latest arbitration decision (the grant signals). The acknowledgment unit passes appropriate acknowledgment signals to the input ports. Negative acknowledgments are sent until a decision is made.

Each of these units is repeatedly subdivided down to the logic gate level, providing a hierarchy of modules. The design has a total of 441 basic components (a multiple input logic gate or single bit flip flop). It is built on a 4200 gate equivalent Xilinx programmable gate array. The switching element can be clocked at 20 MHz and currently frame start pulses occur every 64 clock cycles. The hardware was originally described in the Qudos HDL hardware description language which was used for generating the Xilinx netlist. The Qudos simulator was used to perform the original (non-formal) validation.

3 The HOL Verification

The HOL90 theorem proving system is an LCF style theorem prover for higher-order logic [7]. The original HOL system was intended as a tool for hardware verification. However, it is actually a general purpose proof system that has subsequently been used in a wide variety of application areas. Proofs are input to the system as calls to Standard ML functions. Because of the use of an abstract type to represent theorems, the user can have a great deal of confidence in the results of the system. Programming errors cannot cause a non-theorem to be erroneously proved unless they are in a few simple functions corresponding to the primitive inference rules of the system.

The verification of the 4 by 4 switch fabric used standard techniques [6]. We give only a brief overview. Structural and behavioral specifications of each module were given in higher-order logic. A correctness theorem was then independently proved for each module that its implementation satisfied (implied) the specification. Finally, the correctness theorems for the separate modules were used to prove a correctness theorem for the whole design. The verification was conducted down to the level of the basic logic gates used by the simulator. As in the simulator they were described behaviorally rather than structurally. The modular nature of the proof facilitates the management of the complexity of large designs.

In conducting the proof, the verifier needs a very clear understanding of why the design is correct, since a proof is essentially a statement of this. Thus performing a formal proof involves a deep investigation of the design. It also provides a means to help achieve that understanding. Having to write formal specifications for each module helps in this way, but having to formulate the reasons why the implementation has that behavior gives much greater insight. In addition to uncovering errors, this can serve to highlight anomalies in the design and suggest improvements, simplifications or alternatives [5].

The Structural Specifications No simplification was made to the implementation to facilitate the verification. While some simplification was made to the surface description (such as grouping components into extra modules), the netlists of the structural specifications used corresponded to that actually implemented. The basic building blocks used were logic gates and single bit registers. These corresponded to the basic units of the simulator used by the designers. Qudos structural descriptions can be mimicked very closely in HOL up to surface syntax. However, the extra expressibility of HOL was used to simplify and generalize the description. For example, in HOL words of words are supported. Therefore, a signal carrying 4 bytes can be represented as a word of 4 8-bit words, rather than as 4 separate signals or as one 32-bit signal. This allows more flexible indexing of bits, so that the module duplication operator FOR can be used. To illustrate the expressibility of HOL, we consider the Qudos HDL description of the following multiplexing component of the dataswitch–DMUX4T2:

```
DEF DMUX4T2(d[0..3],x:IN;dOut[0..1]:IO); xBar:IO;
BEGIN
      Clb:=XiCLBMAP5i20(d[0..1],x,d[2..3],dOut[0..1]);
      InvX:= XiINV(x,xBar);
      B[0]:= AO(d[0],xBar,d[1],x,dOut[0]);
      B[1]:= AO(d[2],xBar,d[3],x,dOut[1]);
END;
```

The Clb statement is a dummy declaration providing information about the way the component design should be mapped into a Xilinx gate array. XiINV is an inverter and the AO components are AND-OR logic gates. Using HOL, this module can be expressed as follows with only a single occurrence of AO rather than two as in the Qudos version.

DMUX4T2$((d,x),dOut)$ = LOCAL $xBar$.
    XiINV$(x,xBar)$ $\wedge$
    FOR i :: TO 2 .
        AO((SBIT 0 (SBIT i d),$xBar$,SBIT 1 (SBIT i d), x), SBIT i $dOut$)

In HOL, arithmetic can also be used to specify which bit of a word is connected to an input or output of a component. For example, we can specify that for all i, the $2i$-th bit of an output is connected to the i-th bit of a subcomponent. This again meant that a single module could be used instead of needing to write essentially identical pieces of code several times.

The Behavioral Specifications The behavioral specification against which the structural specification was verified describes the actual un-simplified behavior of the switch fabric. It is presented at a similar level of abstraction to that used by the designers, describing the behavior over a frame in terms of timing diagrams represented as interval temporal operators. Within the interval, the values output are functions of the input values and state at earlier times.

As an example, consider the specification for the acknowledgment signal on a frame where cell headers arrive at time t_h. The predicate AFRAME specifies that we are dealing with intervals corresponding to such active frames. The $ackOut$ signal must be zeroed until time $t_h + 3$. Thereafter, its value depends on the arbitration decision made. This depends on the value of the data injected into the fabric at time t_h (the header), the value of the last arbitration decision, and the value of the acknowledgments coming in from the output ports. This behavior is specified by a function argument to the interval operator DURING. We omit the details here for the purposes of exposition.

(AFRAME t_s t_h t_e fs ...) ⊃
    STABLE $(t_s + 1)$ $(t_h + 3)$ $ackOut$ (ZEROW ...) ∧
    DURING $(t_h + 3)$ $(t_e + 1)$ $ackOut$
        (λt. ... $(d\ t_h)$ ... $(last\ (t_h + 2))$ ... $(ackIn\ t)$ ...)

The correct operation of the fabric relies on an assumption that the environment maintains the frame structure of repeated frame start signals and that cells will not arrive at certain times within a few clock cycles of the frame start. The cycles on which the cells cannot arrive was specified and verified precisely.

Time Taken The module specifications (both behavioral and structural) were written prior to any proof. This took between one and two person-months. No breakdown of this time has been kept. Much of the time was spent in understanding the design. The structural specifications were adapted directly from the Qudos HDL. The behavioral specifications were more difficult. The specifier had no previous knowledge of the design. There was a good English overview of the intended function of the switch fabric. This also outlined the function of the major components. While it gave a good introduction, it was not sufficient to construct an unambiguous behavioral specification of all the modules. The behavioral specifications were instead constructed by analyzing the HDL. This was very time-consuming.

Approximately two person-months were spent performing the verification. Of this one week was spent proving theorems of general use. Approximately 3 weeks were spent verifying the upper modules of the arbitration unit, and a further week was spent on the top two modules of the switch. 3-4 days were spent combining the correctness theorems of the 43 modules to give a single correctness theorem for the whole circuit. The remaining time of just over two weeks was spent proving the correctness theorems for the 36 lower level units. The proofs of the upper-level modules were generally more time-consuming for several reasons: there were more intervals to consider; they gave the behavior of several outputs; and those behaviors were defined in terms of more complex notions. They also contained more errors which severely hampered progress. The verifier had not previously performed a hardware verification, though was a competent HOL user. Apart from standard libraries, the work did not build directly on previous theories.

The machine time taken to completely rebuild the proofs from scratch by rerunning the scripts in batch mode is several hours on a Sparc 10. Single theories representing individual modules generally take minutes to rebuild. In the initial development of the proof the machine time is generally not critical, as the human time is so much greater. However, since the proof process consists of a certain amount of replay of old proofs, a speed up would be desirable.

If changes are made to the design, it is important that the new verification can be done quickly. Since proof is very time consuming this is especially important. This is attacked in several ways in the HOL approach: the proofs can be made generic; their modular nature means that only affected modules need to be reverified; and proofs of modules which have changed can often be replayed with only minor changes. While the 4 by 4 switch fabric took several months to specify and verify, modified versions took only a matter of hours or days [4]. Generic proofs were not used to as great an extent as was possible in this study as it was generally easier to reason about specific values than general ones. Furthermore, there were many different ways that the design and its submodules could be made generic. It was not clear which if any of these might be utilized in subsequent designs. It thus seemed sensible in the first instance to stick closely to the actual design. Indeed the limited ways that the proofs were made generic turned out not to cover design changes incorporated into later designs.

One of the biggest disadvantages of the HOL system is that its learning curve is very steep. Furthermore, interactive proof is generally a time-consuming activity even for the expert. Much time is spent dealing with trivial details of a proof. Recent advances in the system such as new simplifiers and decision procedures may alleviate these problems. However, more work is needed to bring the level of interaction with the system closer to that of an informal proof.

Errors No errors were discovered in the fabricated hardware. Errors that had inadvertently been introduced in the structural specifications (and could just as easily have been in the implementation) were discovered. The original versions of the behavioral specifications of many modules contained errors.

A strong indication of the source of detected errors was obtained. Because each module was verified independently, the source of an error was immediately narrowed down to being in the current module, or in the specification of one of its submodules. Furthermore, because performing the proof involves understanding why the design is correct, the exact location of the error was normally obvious from the way the proof failed. For example, in one of the dataswitch modules, two wires were inadvertently swapped. This was discovered because the subgoal ([T, F] = [F, T]) was generated in the proof attempt. One side of this equality originated from the behavioral specification and one from the structural specification. It was clear from the proof attempt that two wires had been swapped and also which signals they were from the context of the subgoal. It was not immediately clear in which specification they had been swapped.

A further example of an error that was discovered concerned the time the grant signal was read by the dataswitch. It was specified that the two bits of the grant signal from each arbiter were read on a single cycle. However, the implementation read them on consecutive cycles. This resulted in a subgoal of the form *grant t* = *grant* (*t*+1). No information was available in the goal to allow this to be proven, suggesting an error. In this case it was in the specification.

Occasionally false alarms occurred: an unprovable goal was obtained, suggesting an error. However, on closer inspection it was found that the problem was that information had been lost in the course of the proof. For example, if $t_1 < t_2$ is turned into $t_1 \leq t_2$ during the proof, the information that the two times are not equal is lost. Such a false alarm could lead to an unnecessary change in the implementation being made.

Many trivial typing errors were caught at an early stage by type-checking. However, many other trivial mistakes were made over the size of words and signals. For example, words of size 4 by 2 were inadvertently specified as 2 by 4 words. These errors were found during the proof process. It would have been much better if they had been picked up earlier. This would have been possible if dependent typing had been available.

Scalability In theory, the HOL proof approach is scalable to large designs. Because the approach is modular and hierarchical, increasing the size of the design does not necessarily increase the complexity of the proof. However, in practice the modules higher in the hierarchy do take longer to verify, partly because there are more cases to consider. This is made worse if the interfaces between modules are left containing lots of low level detail. For example, in the proof of the switch fabric, low level modules required assumptions to be made about their inputs. These assumptions had to be dealt with in the proofs of higher level modules adding extra proof work manipulating and discharging them. If the proof is to be tractable for large designs, it is important that the interfaces between modules are as clean as possible. This is demonstrated by the fact that two of the upper most modules took approximately half of the total verification time–a matter of weeks. However, it should be noted that the very top module which simply added various delays to various inputs and outputs of the main module, only took a day to verify.

4 The MDG Verification

In the second study, the same circuit was verified using a decision graph approach. A new technique called abstract implicit enumeration has been developed where decision graphs are used to represent sets of states as well as the transition and output relations [3]. Based on this technique hardware verification tools have been developed which perform combinational circuit verification, safety property checking and equivalence checking of two state machines.

The formal system underlying MDGs is many-sorted first-order logic augmented with a distinction between abstract and concrete sorts. Concrete sorts have enumerations, while abstract sorts do not. A data value can be represented by a single variable of abstract sort, rather than by concrete Boolean variables, and a data operation can be represented by an uninterpreted function symbol (cross-operator). MDGs permit the description of the output and next state relations of a state machine in a similar way to the way ROBDDs do for FSMs. We call the model an Abstract State Machine (ASM) since it may represent an unbounded class of FSMs, depending on the interpretation of the abstract sorts and operators. For circuits with large datapaths, MDGs are thus much more compact than ROBDDs. As the verification is independent of the width of the datapath, the range of circuits that can be verified is greatly increased.

We described the actual hardware implementation of the switch fabric at two levels of abstraction. We gave a description of the original Qudos gate-level implementation and a more abstract RTL description which holds for an arbitrary word width. Using the MDG tools, we verified the gate-level implementation against the abstract (RTL) hardware model. The n-bit words of abstract sort of the latter were instantiated to 8 bits using uninterpreted functions which encode and decode abstract data to Boolean data and vice-versa [13].

Starting from timing-diagrams describing the expected behavior of the switch fabric, we derived a complete high-level behavioral specification in the form of a state machine. This specification was developed independently of the actual hardware design and includes no restrictions with respect to the frame size, cell length and word width. Using implicit reachability analysis, we checked its equivalence against the RTL hardware model when both seen as abstract state machines. That is, we ensured that the two machines produce the same observable behavior by feeding them with the same inputs and checking that an invariant stating the equivalence of their outputs holds in all reachable states [9].

By combining the above two verification steps, we hierarchically obtain a complete verification of the switch fabric from a high-level behavior down to the gate-level implementation. Prior to the full verification, we also checked both behavioral and RTL structural specifications against several specific safety properties of the switch. Here, we combined an environment state machine with each switch fabric specification yielding a composed machine which represented the required platform for checking if the invariant properties hold in all reachable states of the specification. Although the properties we verified do not represent the complete behavior of the switch fabric, we were able to detect several injected design errors in the structural description.

When an invariant is not satisfied during the verification process, a counterexample is provided to help with identifying the source of the error. Like ROBDDs, the MDGs require a fixed node ordering. Currently, the node ordering has to be given by the user explicitly. Unlike ROBDDs where all variables are Boolean, every variable used in the MDGs needs to be assigned an appropriate sort and type definitions must be provided for all functions. Rewrite rules may need to be provided to partially interpret the otherwise uninterpreted function symbols.

The Structural Specification As with the HOL study, we translated the Qudos HDL gate-level description into a suitable HDL description; here a Prolog-style HDL, called MDG-HDL. As in the HOL study, extra modularity was added over the Qudos descriptions, while leaving the underlying implementation unchanged. A structural description is usually a (hierarchical) network of components (modules) connected by signals. The MDG-HDL comes with a large library of predefined commonly used basic components (such as logic gates, multiplexors, registers, bus drivers, ROMs, etc.). Multiplexors and registers can be modeled at the Boolean or the abstract level using abstract terms as inputs and outputs. A translator from a subset of VHDL into MDG-HDL is under development.

As an example, the following is an MDG-HDL description of the DMUX4T2 module given in Section 3:

module(DMUX4T2
   port(**inputs**(($d0, bool$), ($d1, bool$), ($d2, bool$), ($d3, bool$)), ($x, bool$)),
      outputs(($dOut0, bool$), ($dOut1, bool$))),
   structure(
      signals($xBar, bool$),
      component(InvX, **NOT**(**input**(x), **output**($xBar$))),
      component(AO_0, **AO**(**input**($d0, xBar, d1, x$), **output**($dOut0$))),
      component(AO_1, **AO**(**input**($d2, xBar, d3, x$), **output**($dOut1$))))).

Here, the components **NOT** and **AO** are basic components provided by the MDG-HDL library. Note also that the data sorts of the interface and internal signals must always be specified.

Besides the gate-level description, we also provided a more abstract (RTL) description of the implementation which holds for arbitrary word width. Here, the data-in and data-out lines are modeled using an abstract sort *wordn*. The active, priority and route fields are accessed through corresponding cross-operators (functions). In addition to the generic words and functions, the RTL specification also abstracts the behavior of the dataswitch unit by modeling it using abstract data multiplexors instead of logic gates. We thus obtain a simpler implementation model of the dataswitch which reflects the switching behavior in a more natural way and is implemented with fewer components and signals. For example, a set of four DMUX4T2 modules is modeled using a single multiplexor component. For more details about the abstraction techniques used refer to [13].

The Behavioral Specification MDG-HDL is also used for behavioral descriptions. A behavioral description is given by high-level constructs as ITE (If-Then-Else) formulas, CASE formulas or tabular representations. The tabular constructor is similar to a truth table but allows first-order terms in rows. It can be used to define arbitrary logic relations. In the MDG study, we gave the behavioral specification of the switch fabric in two different forms: 1) as a complete high-level behavioral state machine and 2) as a set of properties which reflect the essential behavior of the switch fabric as it is used in its environment.

The main behavioral description of the switch fabric was as an abstract state machine (ASM) which reflects its complete behavior under the assumption that

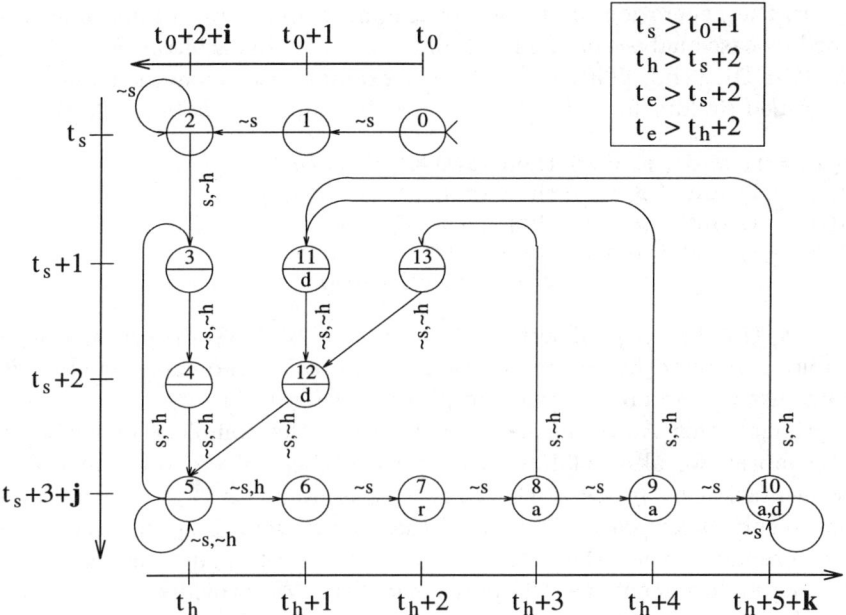

Fig. 2. ASM Behavioral Specification

the environment maintains certain timing constraints on the arrival of the frame start signal and headers. A schematic representation of the ASM specification of the 4 by 4 switch fabric is shown in Figure 2. The symbols t_0, t_s, t_h and t_e in the figure represent the initial time, the time of arrival of the frame start signal, the time of arrival of the routing bytes and the time of the end of a frame, respectively. There are 14 conceptual states: States 0, 1 and 2 along the time axis t_0 describe the initial behavior of the switch fabric. States 2, 3, 4 and 5 along the time axis t_s describe the behavior of the switch on the arrival of a frame start signal. States 6 to 13 along the time axis t_h describe the behavior of the switch fabric after the arrival of the headers. The waiting loops in states 2, 5 and 10 are illustrated in the figure by the non-zero natural numbers i, j and k, respectively. Figure 2 also includes many meta symbols used to keep the presentation simple. For instance, the symbols s and h denote a frame start and the arrival of a routing tag (header), respectively, and the symbol "$\sim$" denotes negation. The symbols a, d and r inside a conceptual state represent the computation of the acknowledgment output, the data output and the round-robin arbitration, respectively. The absence of an acknowledgment or a data symbol means that no computation takes place and the default value is output.

To formally describe this ASM using MDGs, we first introduced some basic sorts, constants and functions (cross-operators), e.g. a concrete sort *port* = $\{0,..,3\}$, an abstract sort *wordn*, a constant *zero* of sort *wordn* and a cross-operator *rou* of type [*wordn* $\rightarrow$ *port*] representing the route field in a header.

Further, the generation of the acknowledgment and data output signals is described by case analysis on the result of the round-robin arbitration. This is done in MDG-HDL using ITE-constructs. For example, the acknowledgment output is described by four formulas determining the value of $ackOut_i$, $i \in \{0,..,3\}$:

if $((co_0 = 1)$ **and** $(ip_0 = i))$ **then** $(ackOut_i = ackIn_0)$
ef $((co_1 = 1)$ **and** $(ip_1 = i))$ **then** $(ackOut_i = ackIn_1)$
ef $((co_2 = 1)$ **and** $(ip_2 = i))$ **then** $(ackOut_i = ackIn_2)$
ef $((co_3 = 1)$ **and** $(ip_3 = i))$ **then** $(ackOut_i = ackIn_3)$
$\qquad\qquad\qquad$ **else** $(ackOut_i = 0)$

where co_i ($i \in \{0,..,3\}$) of sort *bool* and ip_i ($i \in \{0,..,3\}$) of sort *port* are state variables generated by the round-robin computation and corresponding to the output disable and grant signals, respectively (Figure 1).

Although this ASM specification describes the complete behavior of the switch fabric, we also validated (in an early stage of the project) the fabric implementation by property checking. This is useful as it gives a quick verification result at low cost. We verified that the structural specification satisfies its requirements when the ATM switch fabric works under the control of its operating environment, i.e. the port controllers. We provided for this purpose a set of properties which reflect the essential behavior of the switch fabric, e.g. for checking of correct priority computation, circuit reset or data routing. We first simulated the environment as a state machine with one state variable s of enumerated (concrete) sort [1..68]. This allowed us to map the time points t_0, t_s, t_h and t_e to specific states. We then described the properties as invariants which should hold in all reachable states of the specification model. The following is an example of a property which checks for correct routing to port 0. It is expressed in MDG-HDL using an ITE construct.

if $(s \in \{17,..,68\})$ **and** $priority[0..3] = [1,0,0,0]$ **and** $route[0] = 0$
then $dataOut[0] = dataIn'[0]$

Here $priority[0..3]$ indicates the priority bits for all input ports, $route[0]$ represents the routing bits for input port 0 and $dataIn'[0]$ is the data input on port 0 delayed by 4 clock cycles. Further examples of properties are described in [13].

Time Taken The user time required for the specification and verification is hard to determine since it included the improvement of the MDG package, writing documentation, etc. The translation of the Qudos design description to the MDG-HDL gate-level structural model was straightforward and took about one person-week. The description of the RTL structural specification including modeling required about one person-week. The time spent for understanding the expected behavior and writing the behavioral specification was about one person-week. The time taken for the verification of the gate-level description against the RTL model, including the adoption of abstraction mechanisms and correction of description errors, was about two person-weeks. The verification of the RTL structural specification against the behavioral model required about

Verification	CPU Time (s)	Memory (MB)	MDG Nodes Generated
Gate-Level to RTL	183	22	183300
RTL to Beh. Model	2920	150	320556
P1: Data Output Reset	202	15	30295
P2: Ack. Output Reset	183	15	30356
P3: Data Routing	143	14	27995
P4: Ack. Output	201	15	33001
Error (i)	20	1	2462
Error (ii)	1300	120	150904
Error (iii)	1000	105	147339

Table 1. Experimental Results for the MDG Verification

one person-week of work. The user time required to set up four properties, build the environment state machine, conduct the property checking on the structural specification and interpret the results was about one person-week. Checking of these same properties on the behavioral specification took about one hour. The average time for the injection and verification of an introduced design error was less than one hour. The experimental results in machine time are shown in Table 1 including CPU time (on a SPARC station 10), memory usage and number of MDG nodes generated.

A disadvantage of MDGs is that much verification time is spent finding an optimal variable ordering. This is crucial since a bad ordering easily leads to a state space explosion. This occurred after an early ordering attempt. For more information about the variable ordering problem, which is common to all ROBDD-based systems, see [1].

Because the verification is essentially automatic, the amount of work re-running a verification for a new design is minimal compared to the initial effort since the latter includes all the modeling aspects. Much of the effort is spent on determining a suitable variable ordering. Depending on the kind of design changes adopted, it is not obvious if the original variable ordering could still be used on a modified design without major changes.

The MDG gate-level specification is a concrete description of the fabricated implementation. In contrast, the RTL structural and ASM behavioral specifications are generic. They abstract away from frame, cell and word sizes, provided the environment timing assumptions are kept. Design implementation changes at the gate-level that still satisfy the RTL model behavior would hence not affect the verification against the ASM specification. For property checking, specific assumptions about the operating environment were made, (e.g. that the frame interval is 64 cycles). This is sound since the switch fabric will in fact be used under the behest of its operating environment, i.e. the port controllers. However, while this reduces the verification cost, it has the disadvantage that the verification must be completely redone if the operating environment changes. Still, the work required is minor as only a few parameters have to be changed in the description of the environment state machine (which is a simple machine [13]).

Errors As with the HOL study, no errors were discovered in the implementation. For experimental purposes, however, we injected several errors into the implementation and checked them using either the set of properties or the behavioral model. Errors were automatically detected and identified using the counterexample facility. The injected errors included the main errors introduced in the HOL study, discussed in Section 3. We summarize here three further examples. (i) We exchanged the inputs to the JK Flip-Flop that produces the output disable signal. This prevented the circuit from resetting. (ii) We used, at one point, the priority information of input port 0 instead of input port 2. (iii) We used an AND gate instead of an OR gate within the acknowledgment unit, thus producing a faulty $ackOut[0]$ signal. Experimental results for these three errors, which have been checked by verifying the RTL model against the behavioral specification, are reported in Table 1.

While checking properties on the hardware structural description, we also discovered some errors that we mistakenly introduced in the structural specifications. However, we were able to easily identify and correct these errors using the counterexample facility of the MDG tools. Also during the verification of the gate-level model, we found a few errors in the description that were introduced during the translation from Qudos HDL to MDG-HDL. These were easily removed by comparing both descriptions, since they included the same collection of gates. Finally, many trivial typing errors were highlighted at an early stage of the description process by the error messages output after each compilation of the specification's components.

Scalability Like any FSM-based verification system, the MDG proof approach is not directly scalable to large designs. This is due to the possible state space explosion that results from large designs. Unlike other ROBDD-based approaches, however, MDGs do not need to cope with the datapath complexity since they use data of abstract sort and uninterpreted functions. Still, a direct verification of the gate-level model against the behavioral model or even against the set of properties is practically impossible. We overcame this problem by providing an abstract RTL structural specification which we instantiated for the verification of the gate-level model. In order to handle large designs, major efforts are in general required to set up the appropriate model abstraction levels.

5 Conclusions

The MDG and HOL structural descriptions are very similar, both to each other and to the original designer's description. HOL provides significantly more expressibility allowing more natural specifications. Some generic features were included in the MDG description that were not in the HOL description. This could have been done with only minimal additional effort, however.

The behavioral descriptions of the two approaches are totally different. The MDG specification is based on a state machine model while the HOL one is based on interval temporal logic operators, explicitly describing the timing behavior using a set of formulas that include different scenarios of the switch fabric behavior, e.g. active or inactive frames. Both describe the behavior in a clear and comprehensive form. Which of these is preferred is perhaps a matter of taste.

An advantage of MDG is that a property specification is easy to set up and verify. Expected operating conditions can be used to simplify this, even if the full specification is more general. This is useful for verifying that a specification satisfies its requirements. It can greatly reduce the full verification cost by catching errors at an early stage.

Writing the behavioral specifications was far slower in HOL, as separate specifications were needed for each module. In MDG this was not necessary because the whole design was verified in one go, rather than a module at a time. This also reduced the MDG verification time because fewer mistakes were made.

Both approaches successfully highlight errors, and help determine their location. However, the way this information manifests itself differs. MDG is more straightforward, outputting a trace of the input sequence that leads to the erroneous behavior. In HOL, errors manifest themselves as unprovable goals. The form of the goal, the context of the proof and the verifier's understanding of the proof are combined to track down the location, and understand its cause.

The HOL verification was much slower, taking a matter of months. This time includes the verification of each of the modules and the verification of their combination. Using HOL, a large number of lemmas had to be proved and much effort was required to interactively create the proof scripts. For example, the time spent for the verification of the dataswitch unit was about 3 days. Here the proof script was about 530 lines long (17 KB). The MDG verification was achieved automatically without the need of a proof script. All that was required was the careful management of the MDG node ordering (as with ROBDDs). However, this is a matter of hours or at most a few days of work.

In both the HOL and MDG approaches, the amount of work necessary to verify a modified design, once the original has been verified, is greatly reduced. Both allow generic verification to be performed, though HOL has the potential to be more flexible. Because MDG is automated and fast, the re-verification times would largely be just the time taken to modify the specifications and to find a new variables ordering. In the HOL approach, the behavioral specifications of many modules and the proof scripts themselves may need to be modified.

An advantage of the HOL approach in contrast to the MDG method is the confidence in the tool the LCF approach offers. Although the MDG software package has been successfully tested on several benchmarks and has been considerably improved, it is not yet a mature tool. It cannot guarantee the same level of proof security as HOL. The main advantage of the MDG approach is that it is much quicker and is automatic. On the other hand the theorem proving approach is potentially scalable and involves a comprehensive investigation of why the design works correctly. However, these advantages are only likely to be realized in practice if the level of proofs which must be provided to the system can be raised closer to the level of informal proofs.

Acknowledgements We are grateful to Zijian Zhou, Xiaoyu Song and Eduard Cerny at the University of Montreal, Canada and Michel Langevin at GMD-SET, Germany for initiating and advocating this study. Ian Leslie and Mike Gordon at Cambridge University were also of great help. This work was partially funded by EPSRC research agreements GR/J11133 and GR/K10294.

References

1. R. Bryant. Graph-based Algorithms for Boolean Function Manipulation. *IEEE Transactions on Computers*, C-35(8):677–691, August 1986.
2. B. Chen, M. Yamazaki and M. Fujita. Bug Identification of a Real Chip Design by Symbolic Model Checking. In *Proc. of the Int. Conf. on Circuits And Systems*, pages 132–136, June 1994.
3. F. Corella, Z. Zhou, X. Song, M. Langevin and E. Cerny. Multiway Decision Graphs for Automated Hardware Verification. *Formal Methods in System Design*, To appear. Available as IBM research report RC19676(87224), July 1994.
4. P. Curzon. Tracking Design Changes with Formal Machine-checked Proof. *The Computer Journal*, 38(2):91–100, July 1995.
5. P. Curzon and I.M. Leslie. A Case Study on Design for Provability. In *Proc. of the Int. Conf. on Engineering of Complex Computer Systems*, pages 59–62, IEEE Computer Society Press, November 1995.
6. M.J.C. Gordon. HOL: A Proof Generating System for Higher-order Logic. In G. Birtwistle and P.A. Subrahmanyam, editors, *VLSI Specification, Verification and Synthesis*, pages 73–128. Kluwer Academic Publishers, 1988.
7. M.J.C. Gordon and T.F. Melham. *Introduction to HOL: A Theorem Proving Environment for Higher-order Logic*. Cambridge University Press, 1993.
8. J.M.J. Herbert. Case Study of the Cambridge Fast Ring ECL Chip using HOL. Technical Report 123, University of Cambridge, Computer Laboratory, February 1988.
9. M. Langevin, S. Tahar, Z. Zhou, X. Song and E. Cerny. Behavioral Verification of an ATM Switch Fabric using Implicit Abstract State Enumeration. In *Proc. of the Int. Conf. on Computer Design*, IEEE Computer Society Press, October 1996.
10. I.M. Leslie and D.R. McAuley. Fairisle: An ATM Network for the Local Area. *ACM Communication Review*, 19(4):327–336, September 1991.
11. K.L. McMillan. *Symbolic Model Checking*. Kluwer Academic Publishers, 1993.
12. K. Schneider and T. Kropf. Verifying Hardware Correctness by Combining Theorem Proving and Model Checking. In J. Alves-Foss, editor, *International Workshop on Higher Order Logic Theorem Proving and Its Applications: B-Track: Short Presentations*, pages 89–104, August 1995.
13. S. Tahar, Z. Zhou, X. Song, E. Cerny and M. Langevin. Formal Verification of an ATM Switch Fabric using Multiway Decision Graphs. In *Proc. of the Great Lakes Symp. on VLSI*, pages 106–111, IEEE Computer Society Press, March 1996.

A Mechanisation of Computability Theory in HOL

Vincent Zammit

Computer Laboratory, University of Kent, United Kingdom

Abstract. This paper describes a mechanisation of computability theory in HOL using the Unlimited Register Machine (URM) model of computation. The URM model is first specified as a rudimentary machine language and then the notion of a computable function is derived. This is followed by an illustration of the proof of a number of basic results of computability which include various closure properties of computable functions. These are used in the implementation of a mechanism which partly automates the proof of the computability of functions and a number of functions are then proved to be computable. This work forms part of a comparative study of different theorem proving approaches and a brief discussion regarding theorem proving in HOL follows the description of the mechanisation.

1 Introduction

The theory of computation is a field which has been widely explored in mathematical and computer science literature [4, 12, 13] and several approaches to a standard model of computation have been attempted. However, each exposition of the theory centres on the basic notion of a computable function, and as such, one of the main objectives of a mechanisation of computability in a theorem prover is the formal definition of such functions. The mechanisation illustrated in this paper includes also the proof of a number of basic results of the theory and the implementation of conversions and other verification tools which simplify further development of the mechanisation.

This work is part of a comparative study of an LCF [9] style theorem proof assistant (namely the HOL system [5, 6]), and a non-LCF style theorem proving environment based on constructive type theory (such as the ALF system [1, 7] and the Coq proof assistant [3]). The definitions and proofs of even the most trivial results of computability tend to be of a very technical nature much similar to the proofs of theorems one finds in mathematical texts, and thus this theory offers an extensive case study for the analysis of the two approaches of mechanical verification. The verification styles are also compared to the way proofs of mathematical results are represented in texts. It is expected that this comparative study will contribute to the identification of possible enhancements to the theorem proving styles. Although it is not the scope of this paper to give full details of this comparative study, a brief discussion regarding theorem proving in HOL is given after the description of the implementation.

This particular mechanisation of the theory is based on the URM model [11] of computation and much of the implementation is based on the definitions and results described in [4]. The next section gives a brief discussion on URM computability, however it is strongly suggested that the interested reader consults the literature ([4, 12, 13]). The rest of the paper illustrates the actual mechanisation and includes the specification of the notion of a computable function in HOL, the proof of a number of results of computability and a mechanism for constructing and proving the computability of functions.

2 The URM Model of Computation

An Unlimited Register Machine (URM) consists of a countably infinite set of registers, usually referred to as the memory or store, each containing a natural number. The registers are numbered R_0, R_1, ..., R_n, ..., and the value stored in R_n is specified by r_n. A URM executes a finite program constructed from the following four different types of instructions:

Zero: ZR n sets r_n to 0.
Successor: SC n increments r_n.
Transfer: TF n m copies r_n to R_m.
Jump: JP n m p jumps to the pth instruction (starting from 0) of the program if $r_n = r_m$.

A program counter keeps track of the current point in program execution, and the configuration of a URM is given by a pair (p, r) consisting of the program counter and the current store. A configuration is said to be *initial* if the program counter is set to the index of the first instruction, and it is said to be *final* if the program counter exceeds the index of the last instruction.

Given a program P and an initial configuration c_0, a *computation* is achieved by executing the instructions of the program one by one altering the URM configuration at each step. An execution step of a URM with a final configuration has no effect on the current configuration. A computation is thus an infinite sequence of configurations $\langle c_0, c_1, c_2, \ldots \rangle$ and is denoted by $P\langle c_0 \rangle$, or simply by $P(r)$ where $c_0 = (0, r)$. The store r is usually represented by a sequence of register values $(r_0, r_1, \ldots)$ and a finite sequence $(r_0, r_1, \ldots, r_n)$ represents the store where the first $n+1$ registers are given by the sequence and the rest contain the value 0, which is the initial value held in each register. We use the notation $P\langle c \rangle \rightarrow_n c'$ to express that P alters the URM state from c to c' in n steps.

A computation is said to *converge* if it reaches a final configuration, otherwise it is said to *diverge*. The *value* of a convergent computation is given by the contents of the first register R_0 of the final configuration.

The computation of a program can be used to define an n-ary partial function by placing the parameters in the first n registers of a cleared[1] URM store and then executing the program returning the contents of the first register as the function value. Formally, a program P is said to compute an n-ary function f

---
[1] all registers containing 0.

if, for every $a_0, \ldots, a_{n-1}$ and v, $P(a_0, \ldots, a_{n-1})$ converges to v if and only if $f(a_0, \ldots, a_{n-1}) = v$. This definition implies that $P(a_0, \ldots, a_{n-1})$ diverges if and only if $f(a_0, \ldots, a_{n-1})$ is undefined. A function is said to be *URM-computable* if there is a program which computes it.

The URM model of computation is proved to be equivalent to the numerous alternative models such as the Turing machine model, the Gödel-Kleene partial recursive functions model, and Church's lambda calculus [4, 11] in the sense that the set of URM computable functions is identical to the set of the functions computed by any other model.

3 Mechanisation of URM Computability

An Unlimited Register Machine can be regarded as a simple machine language and as such its formal specification in HOL is similar to that of real world architectures [14].

3.1 The URM Instruction Set

A URM store can be represented as a function from natural numbers to natural numbers and configurations as pairs consisting of a natural number signifying the program counter and a store,

```
store   == :num → num
config  == :num × store
```

The syntax of the URM instruction set is specified through the definition of the type :instruction using the type definition package [8] of HOL

```
instruction ::= ZR num
              | SC num
              | TF num → num
              | JP num → num → num
```

and programs are defined as lists of instructions.

The semantics of the instruction set is then specified through the definition of a function exec_instruction: instruction -> config -> config such that given an instruction i and a configuration c, exec_instruction i c returns the configuration achieved by executing i in configuration c.

```
⊢def (∀n c. exec_instruction (ZR n) c
      = (SUC (FST c),(λx. (x = n) → 0 | (SND c x)))) ∧
     (∀n c. exec_instruction (SC n) c
      = (SUC (FST c),
          (λx. (x = n) → (SUC (SND c n)) | (SND c x)))) ∧
     (∀n m c. exec_instruction (TF n m) c
      = (SUC (FST c),(λx. (x = m) → (SND c n) | (SND c x)))) ∧
     (∀n m p' c. exec_instruction (JP n m p') c
      = (((SND c n = SND c m) → p' | (SUC (FST c))),SND c))
```

The execution of a number of steps of a URM program is then given by the primitive recursive function EXEC_STEPS: num -> program -> config -> config, such that EXEC_STEPS n P $c_0 = c_1$ if and only if $P\langle c_0 \rangle \to_n c_1$

⊢$_{def}$ (∀P c. EXEC_STEPS 0 P c = c) ∧
  (∀n P c. EXEC_STEPS (SUC n) P c
  = EXEC_STEPS n P (EXEC_STEP P c))

where EXEC_STEP: program -> config -> config represents one step execution of a given program,

⊢$_{def}$ ∀P c. EXEC_STEP P c
  = ((Final P c) → c | (exec_instruction (EL (FST c) P) c))

and the predicate Final: program -> config -> bool holds for final configurations.

3.2 Computations

A finite list of natural numbers is transformed into an initial URM configuration by the function set_init_conf: (num list) -> config, and CONVERGES: program -> (num list) -> num -> bool and DIVERGES: program -> (num list) -> bool represent converging and diverging computations respectively. It is shown that a program converges to a unique value unless it diverges.

⊢ ∀P l. (∃!v. CONVERGES P l v) ∨ DIVERGES P l

3.3 Computable Functions

Since the functions which are considered are not necessarily total, a polymorphic type of *possibly undefined values* is defined. Elements of this type are either undefined or have a single value:

'a PP ::= Undef
       | Value 'a

The domain of functions is then chosen to be the type of possibly partial numbers. Since the functions have different arities, the codomain is chosen to be the type of lists of numbers, where the length of the list represents the function's arity:

pfunc == :num list → num PP

A program computes a function if and only if it converges to the value of the application of the function whenever this is defined,

⊢$_{def}$ ∀n P f. COMPUTES n P f
  = (∀l v. (LENGTH l = n) ⇒
    (CONVERGES P l v = (f l = Value v)))

such that COMPUTES n P f holds if P computes the n-ary function f. Finally, a function is computable if there is a program which computes it.

⊢$_{def}$ ∀n f. COMPUTABLE n f = (∃P. COMPUTES n P f)

3.4 Manipulating URM Programs

The proof that a particular function is computable usually involves the construction of a URM program which computes it. The URM instruction set is rudimentary and it would be impractical as a general purpose programming language without a mechanism for concatenating program segments, and without a number of program modules performing simple but often used tasks.

An operator $\widehat{++}$ can be defined such that, given two programs P_1 and P_2, the computation of $P_1\widehat{++}P_2$ is given by the individual computation of the two programs. This is achieved by first adding the length of P_1 to the destination of the jumps in P_2 and then appending the two programs together using the normal list concatenation function. The destination of the jumps in P_2 need to be altered since URM jump instructions are absolute, rather than relative. However, in order that the required property is achieved, the programs must be in *standard form*, in the sense that the destinations of all their jumps are less than or equal to the length of the program. This is required so that the program counter of any final configuration is equal to the length of the program; in particular the program counter of a final configuration of P_1 is equal to its length and thus the next instruction executed in the combined computation of $P_1\widehat{++}P_2$ is the first one in P_2. This does not constitute any restrictions since it is proved that any program can be transformed into standard form by setting the destination of out of range jumps to the length of the program. Moreover, it can be proved that $P_1\widehat{++}P_2$ diverges if one of the component programs diverges.

The transformation of programs into standard form is given by the function SF: program -> program and the proof that for any program P, its behaviour is equivalent to SF P is done by first showing that after a single step of the execution of both programs the resulting configurations are equivalent. Two configurations are equivalent either if they are the same, or both are final and have the same store. This result is then extended for any number of execution steps and finally it is proved that

⊢ ∀P l v. CONVERGE (SF P) l v = CONVERGE P l v

by showing that if one program converges in a number of steps then the other converges in the same number of steps.

The concatenation operator $\widehat{++}$ is defined in HOL as the function SAPP: program -> program -> program, and since it is often required to concatenate more than two programs, a function SAPPL: program list -> program which concatenates a given list of programs is defined as well.

The following three simple program modules which are used quite often in the construction of general URM programs:

- SET_FST_ZERO n stores the value 0 in the registers $(R_0, R_1, \ldots, R_n)$.
- TRANSFER_FROM $p\ n$ stores $(r_p, r_{p+1}, \ldots, r_{p+n-1})$ into $(R_0, R_1, \ldots, R_{n-1})$.
- TRANSFER_TO $p\ n$ store $(r_0, r_1, \ldots, r_{n-1})$ into $(R_p, R_{p+1}, \ldots R_{p+n-1})$.

are defined as follows:

$\vdash_{def}$ ∀n. SET_FST_ZERO n = GENLIST ZR (SUC n)

$\vdash_{def}$ ∀p n. TRANSFER_FROM p n = GENLIST (λx. TF (p + x) x) n

$\vdash_{def}$ ∀p n. TRANSFER_TO p n
         = REVERSE (GENLIST (λx. TF x (p + x)) n)

where GENLIST and REVERSE are defined in the List theory of HOL. The programs yielded by these functions are proved to converge and to convey their expected behaviour by induction on the number of steps of execution of the programs.

Another program module, which is given by $[P\ p_s \xrightarrow{n} p_v]$, or by the term PSHIFT $P\ p_s\ n\ p_v$, is defined. This program module executes P, taking its n parameters from the memory segment at offset p_s rather than from the first n registers. Also, this program stores the value of the computation in R_{p_v} rather than the first register. This is defined by:

$\vdash_{def}$ ∀P ps n pv. PSHIFT P ps n pv
         = SAPPL [SET_FST_ZERO (MAXREG P);
                  TRANSFER_FROM p n;
                  P;
                  [TF 0 pv]]

where MAXREG P is the maximum register used by P and is denoted by $\rho(P)$. The proofs that $[P\ p_s \xrightarrow{n} p_v]$ diverges if and only if P diverges, and that if the former converges it yields the expected configuration, are done by applying the result that the computation of programs constructed by $\widehat{++}$ is made up from the computations of the constructing programs.

4 Constructing Computable Functions

In this section we show that a number of basic functions are computable and that the family of computable functions is closed under the operations of substitution, recursion and minimalisation. These results yield a mechanism for constructing computable functions and automatically proving their computability. Moreover, the set of functions which are constructed by the above operations, which is called the set of partial recursive functions, is equal to the set of computable functions [4, 10]. Thus particular functions can be proved to be computable by proving their equality to some partial recursive function. However this process is not decidable; nevertheless a number of symbolic animation tactics [2] have been implemented which simplify the proof of theorems stating such an equality.

The verification of the closure properties involves the construction of a URM program which is proved to compute the function constructed by the particular operation being considered. Due to space limitations, only the proof of the closure under recursion is illustrated in detail.

4.1 The Basic Functions

The following three basic functions are considered:

1. The zero functions (each of different arity) which return 0 for any input:
   $\forall n, x_0, \ldots, x_{n-1}.\mathcal{Z}(x_0, \ldots, x_{n-1}) = 0$,
2. the successor function which increments its input by one: $\forall x_0.\mathcal{S}(x_0) = x_0 + 1$,
3. and projections, which return a particular component from a given vector:
   $\forall n, i < n, x_0, \ldots, x_{n-1}.\mathcal{U}_n^i(x_0, \ldots, x_{n-1}) = x_i$.

These functions are defined in HOL as follows:

$\vdash_{def}$ ∀l. ZERO l = Value 0

$\vdash_{def}$ ∀l. SUCC l
   = ((LENGTH l = 1) → (Value (SUC (HD l))) | Undef)

$\vdash_{def}$ ∀i n l. PROJ i n l = ((i < n) → (Value (ZEL i l)) | Undef)

and are proved to be computable by showing that the programs [ZR 0] and [SC 0] compute $\mathcal{Z}$ and $\mathcal{S}$ respectively; and that the projection $\mathcal{U}_n^i$ is computed by [TF i 0] for $i < n$ and since it is undefined for $i \geq n$ it is computed by [JP 0 0 0]. The function ZEL i l returns the $(i+1)$th element of l if i is less than the length of l; otherwise it returns 0.

4.2 Substitution

The substitution of k n-ary functions $\underline{g} = (g_0, \ldots, g_{k-1})$ into a k-ary function f gives the n-ary function produced by applying f on the results of the applications of $\underline{g}$. That is,

$$f \hat{\circ} \underline{g}(x_0, \ldots, x_{n-1}) = \\ f(g_0(x_0, \ldots, x_{n-1}), \ldots, g_{k-1}(x_0, \ldots, x_{n-1})).$$

This is defined in HOL as the function FSUBS: pfunc -> pfunc list -> pfunc.

$\vdash_{def}$ ∀f l. APPLY f l
   = ((ALL_EL DEFINED l) → (f (MAP VALUE l)) | Undef)

$\vdash_{def}$ ∀f gl l. FSUBS f gl l = APPLY f (MAP (λg. g l) gl)

In order to prove that computable functions are closed under substitution, it is required to show that given the programs $P_f, P_{g_0}, \ldots, P_{g_{k-1}}$ which compute the functions $f, g_0, \ldots, g_{k-1}$ respectively, a program $P_{f\hat{\circ}\underline{g}}$ can be constructed which computes $f \hat{\circ} \underline{g}$. Such a program is shown in Fig. 1. The program parameters are first transferred to some memory location at offset p_s. The programs P_{g_i} for $i < k$ are then executed one at a time storing their results into another memory segment starting at p_v, and finally P_f is executed on the results. The value of p_s is chosen to be $\max(n, k, \rho(P_f) + 1, \max(\rho(P_{g_0}), \ldots, \rho(P_{g_{k-1}})) + 1)$ so that the contents of this memory segment is not altered during the program execution. Similarly, p_v is set to $p_s + n$.

```
start: TRANSFER_TO p_s n              store parameters in (r_{p_s},...,r_{p_s+n-1})
inner: [P_{g_0} p_s  →ⁿ  p_v]
         ⋮
        [P_{g_i} p_s  →ⁿ  (p_v + i)]    for each i < k execute P_{g_i}
                                         storing its result in R_{p_v+i}
         ⋮
        [P_{g_{k-1}} p_s  →ⁿ  (p_v + k - 1)]
outer: [P_f p_v  →ⁿ  0]                execute P_f on the values returned by the P_{g_i}'s
```

Fig. 1. The program $P_{f \hat{\circ} \underline{g}}$

4.3 Recursion

Given an n-ary base case function β and an $(n+2)$-ary recursion step function σ, the $(n+1)$-ary recursive function $\mathcal{R}(\beta;\sigma)$ is defined as follows:

$$\mathcal{R}(\beta;\sigma)(0, x_0, \ldots, x_{n-1}) = \beta(x_0, \ldots, x_{n-1})$$
$$\mathcal{R}(\beta;\sigma)(x+1, x_0, \ldots, x_{n-1}) = \sigma(x, \mathcal{R}(\beta;\sigma)(x, x_0, \ldots, x_{n-1}), x_0, \ldots, x_{n-1})$$

and is specified in HOL as the function FREC: pfunc -> pfunc -> pfunc.

$\vdash_{def}$ (∀basis step l. RECURSION basis step 0 l = basis l) ∧
       (∀basis step n l.
        RECURSION basis step (SUC n) l =
        (let r = RECURSION basis step n l in
         (DEFINED r) → (step (CONS n (CONS (VALUE r) l)))
                     | Undef))

$\vdash_{def}$ ∀basis step l.
       FREC basis step l =
       ((l = []) → (basis [])
                 | (RECURSION basis step (HD l) (TL l)))

Given the programs P_β and P_σ which compute the functions β and σ respectively, the program $P_{\mathcal{R}(\beta;\sigma)}$ shown in Fig. 2 computes the recursive function $\mathcal{R}(\beta;\sigma)$. The value of p_c is chosen to be $\max(\rho(P_\beta)+1, \rho(P_\sigma)+1, n+2)$ so that the registers starting at p_c are not used by P_β and P_σ; and the values of p_v, p_s and p_x are chosen to be $p_c + 1$, $p_v + 1$ and $p_s + n$ respectively. The register R_{p_c} is used to store a counter for the number of times the inner loop

```
start: TRANSFER_TO p_v (n + 1)      Store (x, x_0, ..., x_{n-1}) in (R_{p_v}, R_{p_s}, ..., R_{p_s+n-1})
       TF 0 p_x                     Set r_{p_x} to the value x
       [P_β p_s  ⁿ⟶ p_v]            Execute P_β
loop:  JP p_c p_x final             While r_{p_c} < x
       [P_σ p_c  ⁿ⁺²⟶ p_v]          Execute P_σ
       SC p_c                       Increment the counter r_{p_c}
       JP 0 0 loop
final: TF p_v 0                     Return the final value r_{p_v}
```

Fig. 2. The program $P_{\mathcal{R}(\beta;\sigma)}$

is executed. The value of each recursion step is stored at R_{p_v} and the memory segment $(R_{p_s}, \ldots, R_{p_s+n-1})$ is used to store the function's parameters, which are transferred by the first step of the program. The register R_{p_x} stores the depth of the recursion such that the inner loop is repeated r_{p_x} times after the code computing the base case function, $[P_\beta\ p_s \xrightarrow{n} p_v]$, stores $\beta(x_0, \ldots, x_{n-1})$ into r_{p_v}. The final value of p_v is then transferred into the first register R_0.

This program can be divided into three parts, which we call P_{start}, P_{loop} and P_{final}. These are represented in HOL by the terms

```
Pstart = SAPPL [TRANSFER_TO p_v (n + 1);
                [TF 0 p_x];
                PSHIFT P_β p_s n p_v]

Ploop  = let Ps = PSHIFT P p_c (n + 2) p_v in
         APPEND (SAPP [JP p_c p_x (3 + LENGTH Ps)]
                      Ps)
                [SC p_c;
                 JP 0 0 0]

Pfinal = [TF p_v 0]
```

and $P_{\mathcal{R}(\beta;\sigma)}$ is then given by

```
SAPPL [^Pstart; ^Ploop; ^Pfinal]
```

This program is then proved to compute the recursive function by considering whether the base case function β and the step function σ are defined:

- If $\beta(x_0, \ldots, x_{n-1})$ is defined then

- P_β converges, and so does $[P_\beta\ p_s \xrightarrow{n} p_v]$. As a result P_{start} converges to a final configuration containing the value of $\beta(x_0,\ldots,x_{n-1})$ (which is equal to $\mathcal{R}(\beta;\sigma)(0,x_0,\ldots,x_{n-1})$) in the first register, the parameters $(x_0,\ldots,x_{n-1})$ stored in $(R_{p_s},\ldots,R_{p_s+n-1})$ and r_{p_x} set to the depth x;
- If, also the step function σ is defined for all values of $i \leq x$ then
    * all programs $[P_\sigma\ p_c \xrightarrow{n+2} p_v]$ converge for each value of the recursion counter r_{p_c}, and hence P_{loop} converges, placing the final value of the application of σ (which is equal to $\mathcal{R}(\beta;\sigma)(x,x_0,\ldots,x_{n-1})$) in (R_{p_v});
    * and finally P_{final} stores the value of $\mathcal{R}(\beta;\sigma)(x,x_0,\ldots,x_{n-1})$ into the first register. Thus, whenever $\mathcal{R}(\beta;\sigma)$ is defined, $P_{\mathcal{R}(\beta;\sigma)}$ converges to the required value.
- On the other hand, if σ is undefined for some non-zero value $i \leq x$ then
    * the program $[P_\sigma\ p_c \xrightarrow{n+2} p_v]$ diverges when $r_{p_c} = i-1$, thus P_{loop} diverges and so does $P_{\mathcal{R}(\beta;\sigma)}$. However, if the step function is undefined then $\mathcal{R}(\beta;\sigma)$ is undefined as well. Hence, in this particular case, the function is undefined and the program diverges (as expected).
- Now, if the base case function β is undefined then
    - P_β diverges. As a result, all the programs constructed from it using the $\widehat{++}$ operator diverge. In particular the programs $[P_\beta\ p_s \xrightarrow{n} p_v]$, P_{start} and $P_{\mathcal{R}(\beta;\sigma)}$. Also, given that β is undefined, then so is $\mathcal{R}(\beta;\sigma)$, and even in this final case the recursive function is undefined and the program diverges.
- Thus
    1. $P_{\mathcal{R}(\beta;\sigma)}$ converges to $\mathcal{R}(\beta;\sigma)(0,x_0,\ldots,x_{n-1})$ whenever the latter is defined; and
    2. $P_{\mathcal{R}(\beta;\sigma)}$ diverges whenever $\mathcal{R}(\beta;\sigma)(0,x_0,\ldots,x_{n-1})$ is undefined.
- So, $P_{\mathcal{R}(\beta;\sigma)}$ computes $\mathcal{R}(\beta;\sigma)$ proving that the latter is computable.

The same method of considering whether the constituting functions of an operation are defined or not, is used in the proofs that substitution and minimalisation of computable functions yield functions which are also computable.

4.4 Minimalisation

The unbounded minimalisation of an $(n+1)$-ary function f is the n-ary function given by:

$$\mu_x(f(x,x_0,\ldots,x_{n-1})=0) = \begin{cases} \text{the least } x \text{ s.t. } f(x,x_0,\ldots,x_{n-1})=0, \text{ and for} \\ \quad \text{all } x' \leq x\ f(x',x_0,\ldots,x_{n-1}) \\ \quad \text{is defined} \\ \text{undefined} \quad \text{if no such } x \text{ exists.} \end{cases}$$

This is formalised in HOL by the following definition:

$\vdash_{def} \forall f \; l.$
```
    FMIN f l =
    (let Z x = f (CONS x l) = Value 0 in
     let n = FIRST_THAT Z in
     (Z n ∧
       (∀m. m ≤ n ⇒
         DEFINED (f (CONS m l)))) → (Value n) | Undef)
```

where FIRST_THAT R returns the first natural number n such that $R(n)$ holds, if such an n exists, or any particular value otherwise.

$\vdash_{def} \forall P. \; \text{FIRST\_THAT } P = (\varepsilon n. \; P \; n \land (\forall m. \; P \; m \Rightarrow n \leq m))$

Given that P_f computes f, the program $P_{\mu_x(f(x) \mapsto 0)}$ shown in Fig. 3 computes the minimalisation function $\mu_x(f(x) \mapsto 0)$. This is done by executing P_f until it returns the value 0. The register at position p_c is used as a counter which is incremented each time P_f is executed and is returned by $P_{\mu_x(f(x) \mapsto 0)}$ if it terminates. The value of p_c is set to $\max(\rho(P)+1, n+1)$, so that it will not be used by P_f, also the parameters are stored at the memory segment starting at p_s which is set to $p_c + 1$. The value of p_0 is chosen to be $p_s + n$ and is not altered during program execution, so that $r_{p_0} = 0$.

---

start: TRANSFER_TO p_s n Store parameters in $(R_{p_s}, \ldots, R_{p_s+n-1})$
fetch: $[P_f \; p_c \xrightarrow{n+1} 0]$ Execute P_f
       JP 0 p_0 final   Until it returns 0
       SC p_c Otherwise, increment r_{p_c}
       JP 0 0 fetch   Jump back to fetch
final: TF p_c 0   Return r_{p_c}

Fig. 3. The program $P_{\mu_x(f(x) \mapsto 0)}$

4.5 Proving the Computability of Particular Functions

The operations mentioned above make up the language of partial recursive functions and are sufficient to build up the family of computable functions. A HOL conversion simulating the application of function terms constructed using these constructs is implemented. The three basic functions are automated by simply

Function Description	Notation	Partially recursive equivalent
Parameter rearrangement	$f \circ v_{i_0,\ldots,i_{n-1}}$	$f \hat{\circ} (\lambda j.\mathcal{U}_n^{i_j})$
Identity	ι	$\mathcal{U}_1^0$
One	1_c	$\mathcal{S}\hat{\circ}(\mathcal{Z})$
Addition	$+_c$	$\mathcal{R}(\iota; \mathcal{S} \circ v_1)$
Multiplication	$\times_c$	$\mathcal{R}(\mathcal{Z}; +_c \circ v_{1,2})$
Factorial	fact_c	$\mathcal{R}(1_c; \times_c \hat{\circ}(\mathcal{S}, \iota))$
Predecessor	pred_c	$\mathcal{R}(\mathcal{Z}; \mathcal{U}_2^0)$
Subtraction	$-'_c$	$\mathcal{R}(\iota; \text{pred}_c \circ v_1)$
	$-_c$	$-'_c \circ v_{1,0}$
Power	e'_c	$\mathcal{R}(1_c; \times_c \circ v_{1,2})$
	e_c	$e_c \circ v_{1,0}$
Conditional	if_c	$\mathcal{R}(\mathcal{U}_4^3; \mathcal{U}_2^0)$
Check if 0^a	is0	$\text{if}_c \hat{\circ}(\iota, \mathcal{Z}, 1_c) \circ v_{0,0,0}$
Check if non 0	non0	$\text{if}_c \hat{\circ}(\iota, 1_c, \mathcal{Z}) \circ v_{0,0,0}$
Difference	$\vert -_c (x_0, x_1) \vert$	$+_c \hat{\circ}(-'_c, -_c)$
Equality	$=_c$	$\text{is0} \hat{\circ}(\lambda(x_0,x_1). \vert -_c (x_0, x_1) \vert)$
Inequality	$\neq_c$	$\text{is0} \hat{\circ}(=_c)$
Conjunction	$\wedge_c$	$\text{non0} \hat{\circ}(\times_c)$
Disjunction	$\vee_c$	$\text{non0} \hat{\circ}(+_c)$
Minimal inverse	f^{-1}	$\mu_y((\neq_c \hat{\circ}(f, \iota) \circ v_{0,0})(y) \mapsto 0)$

Fig. 4. A list of computable functions

---

[a] can also be used as a negation operator

rewriting with the appropriate definitions; substitution is automated by first evaluating each of the substituting functions and then evaluating the function into which these functions are substituted. A function defined by recursion is animated by evaluating either the base case function, or the step function recursively. A function constructed by minimalisation, $\mu_x(f(x) \mapsto 0)$ is animated by first proving that if for some i, $f(i) = 0$ and for all $j < i$, $f(j)$ is defined and is greater than 0, then $\mu_x(f(x) \mapsto 0) = i$. By evaluating $f(j)$, for $j = 0, 1, \ldots$ one can construct a thoerem which states that $\forall j'.j' < j \Rightarrow f(j) > 0$ until $j = i$ and thus $f(j) = 0$. This theorem and the evaluation of $f(i)$ are then used to prove that $\mu_x(f(x) \mapsto 0) = i$, thus evaluating the minimalisation function.

This conversion is used to prove that particular functions are equal to some specified partial recursive function. In general an equality to a function constructed by substitution is proved by applying this conversion on the functional application term and then proving the equality of the resulting terms (often by simple rewriting of the definitions); and equality to a function defined by recursion is proved by mathematical induction and then applying this conversion on the base case and induction step subgoals. Since the simulation of minimalisation involves a possibly non-terminating fetching process (when the function is total and never returns 0) the execution of this conversion may diverge, in such case this conversion cannot be used in the required proof; although if the fetching process terminates the proof of the equality of functions defined by minimalisation is otherwise relatively straightforward.

Moreover, since partial recursive functions are constructed from three basic functions which are proved to be computable and by three operators which are proved to preserve computability, the process of proving that such functions are computable can be automated. Given a conversion which automates this mechanism, the proof that a function is computable simply involves showing that it is equal to some partial recursive function. Figure 4 lists a number of functions which are proved to be computable.

5 Theorem Proving in HOL

The proofs of most of the results in this mechanisation tend to be quite elaborate and involve the consideration of details which are often omitted in the proofs given in mathematical texts. This is often the case with mechanical verification since most of the definitions and proofs done by hand and represented in texts omit a number of steps which are considered to be trivial or not interesting to the reader. For example the proof that for every URM program one can construct a program in standard form which has an equivalent behaviour (Sect. 3.4) is considered to be trivial in [4], although it requires a considerable number of lemmas concerning the behaviour of executing URM programs in general and programs transformed into standard form by the function SF. Also, in the proofs that computable functions are closed under the operations of substitution, recursion and minimalisation, little or no attention is given in showing that the program constructed to compute the required function diverges whenever the function is undefined, probably because this part of the proof is considered uninteresting. It is to note, however, that such proofs usually offer an interesting challenge in a mechanisation.

However, an advantage of theorem proving in HOL and other LCF style theorem provers is the availability of a flexible general purpose meta-language which can be used in the implementation of program modules which simulate the behaviour of the formal definitions as well as intelligent algorithms which automate parts of the verification process. In this particular implementation, such a mechanism is found to be quite useful in the verification of general computable functions. On the other hand, theorem proving in Coq is usually done by applying tactics through the specification language Gallina. This offers the advantage that unless a number of specialised tactics need to be implemented, the user does not need any knowledge on how the actual terms and theorem are represented in the implementation of the theorem prover and is thus easier to learn than HOL. Another advantage of having a specification language which bridges the user from the meta-language is that proofs can be easily represented as lists, or trees of tactics in a format which can be read by the user. However such an approach has the disadvantage that it reduces the flexibility of the system and discourages the user from implementing his or her own tactics.

Type theories allow the definition of dependent types where a type can be parametrised by other types, and thus offer a more powerful and flexible type definition mechanism than the one available in HOL. For example, the type of functions which are considered for computability are defined as mappings from lists of numbers to possibly partial numbers (Sect. 3.3) and the type itself contains no information regarding the function's arity. In an implementation in Coq, an n-ary partial function is defined as a single valued relation between vectors of n elements and natural numbers. Vectors are defined as the following inductive dependent type:

```
Inductive vector [A: Set]: nat → Set
   := Vnil:  (vector A 0)
    | Vcons: (n: nat)A → (vector A n) → (vector A (S n)).
```

and partial functions as a record with two fields:

```
Record pfunc [arity: nat] : Type := mk_pfunc
  { reln:       (Rel (vector nat arity) nat);
    One_valued: (one_valued (vector nat arity) nat reln) }.
```

The first field `reln` represents the function as a relation and the second field `One_valued` is a theorem which states that `reln` is single valued. Relations are defined by:

```
Definition Rel := [A,B: Set]A → B → Prop.
```

and the predicate `one_valued` by:

```
Definition one_valued
   := [A,B: Set][R: (Rel A B)](a: A)(b1, b2: B)
      (R a b1) → (R a b2) → (b1 = b2).
```

Finally the type of all partial functions is defined as the dependent product

```
Inductive pfuncs: Type
   := Pfuncs: (n: nat)(pfunc n) → pfuncs.
```

Possible enhancements to theorem proving in the HOL system include mechanisms for naming or numbering assumptions in a goal-directed proof, for allowing constant redefinition and the declaration of local definitions. Proofs found in mathematical texts often contain definitions which are only used in showing a number of particular results. Such definitions can be made local to the results which require them. For example, in the proof that computable functions are closed under recursion (Sect. 4.3) the constants `Pstart`, `Ploop` and `Pfinal` are used only in obtaining this particular results. Such constants are defined as local meta-language variable definitions, however a more elegant approach would involve a theory structure mechanism where constants can be defined local to a theory module and made invisible outside their scope.

6 Conclusion

The mechanisation of computability theory discussed above includes the definition of a computable function according to the URM model and the proof of various results of the theory of which we have given particular attention to the closure property of computable functions under the operations of substitution, recursion and minimalisation. This result is used in the implementation of a process which automates the proof of the computability of functions constructed from these operations and a number of basic computable functions. In the future, it is expected that the theory will be extended through the proof of a number of theorems including the denumerability of computable functions, the S_n^m theorem and the universal program theorem.

A mechanisation of computability theory is also being implemented in the Coq proof assistant. This implementation uses the partial recursive function model of computation and will include the proof of the results mentioned in the previous paragraph. These two implementations and other work in Alf are expected to yield a comparative study of the different theorem proving approaches.

7 Acknowledgements

I thank my supervisor, Simon Thompson, for his continuous support and for his comments on the material presented in this paper, as well as the anonymous referees for their constructive comments on the first draft of this paper.

References

1. Thorsten Altenkirch, Veronica Gaspes, Bengt Nordström, and Björn von Sydow. *A User's Guide to ALF*. Chalmers University of Technology, Sweden, May 1994.
2. J. Camilleri and V. Zammit. Symbolic animation as a proof tool. In T.F. Melham and J. Camilleri, editors, *International Workshop on Higher Order Logic Theorem Proving and its Applications*, volume 859 of *Lecture Notes in Computer Science*, pages 113–127, Malta, September 1994. Springer-Verlag.
3. C. Cornes et al. The Coq Proof Assistant Reference Manual, Version 5.10. Rapport technique RT-0177, INRIA, 1995.
4. N.J. Cutland. *Computability: An introduction to recursive function theory*. Cambridge University Press, 1980.
5. M. Gordon. HOL a machine oriented formulation of higher order logic. Technical Report TR-68, Computer Laboratory, Cambridge University, July 1985.

6. M.J.C. Gordon and T.F. Melham. *Introduction to HOL: a theorem proving environment for higher order logic.* Cambridge University Press, 1993.
7. Lena Magnusson and Bengt Nordström. The ALF proof editor and its proof engine. In Henk Barendregt and Tobias Nipkow, editors, *Types for Proofs and Programs*, pages 213–237. Springer-Verlag LNCS 806, 1994.
8. T.F. Melham. Using recursive types to reason about hardware and higher order logic. In G.J. Milne, editor, *International Workshop on Higher Order Logic Theorem Proving and its Applications*, pages 27–50, Glasgow, Scotland, July 1988. IFIP WG 10.2, North-Holland.
9. L.C. Paulson. *Logic and computation : interactive proof with Cambridge LCF.* Cambridge tracts in theoretical computer science, 1987.
10. H. Rogers. *Theory of recursive functions and effective computability.* McGraw-Hill, 1967.
11. J.C. Shepherdson and H.E. Sturgis. Computability of recursive functions. Technical Report 10, J. Assoc. Computing Machinery, 1967.
12. R. Sommerhalder and S.C. van Westrhenen. *The theory of computability: programs, machines, effectiveness and feasibility.* Addison-Wesley publishing company, 1988.
13. G.J. Tourlakis. *Computability.* Reston Publishing Company, 1984.
14. P.J. Windley. Specifying instruction-set architectures in HOL: A primer. In T.F. Melham and J. Camilleri, editors, *International Workshop on Higher Order Logic Theorem Proving and its Applications*, volume 859 of *Lecture Notes in Computer Science*, pages 440–456, Malta, September 1994. Springer-Verlag.

Author Index

Agerholm, S. 1, 17
Basin, D. 33
Becker, B. R. 235
Beylin, I. 17
Black, P. E. 51
Blumenröhr, C. 157
Brackin, S. H. 61
Busch, H. 77
Butler, M. 93
Collins, G. 109
Coupet-Grimal, S. 125
Crégut, P. 251
Curzon, P. 415
Dutertre, B. 141
Dybjer, P. 17
Eisenbiegler, D. 157
Friedrich, S. 33
Gordon, A. D. 173
Gordon, M. 191
Green, I. 399
Harrison, J. 203, 221
Heckman, M. R. 235
Heyd, B. 251

Howe, D. J. 267
Jakubiec, L. 125
Kolyang 283
Kumar, R. 157
Laibinis, L. 315
Larsson, M. 299
Levitt, K. N. 235
Långbacka, T. 93
Melham, T. 173
Nazareth, D. 331
Nipkow, T. 331
Olsson, R. A. 235
Peticolas, D. 235
Pusch, C. 347
Reus, B. 363
Santen, T. 283
Slind, K. 381
Smaill, A. 399
Tahar, S. 415
Windley, P. J. 51
Wolff, B. 283
Zammit, V. 431
Zhang, C. 235

Springer-Verlag and the Environment

We at Springer-Verlag firmly believe that an international science publisher has a special obligation to the environment, and our corporate policies consistently reflect this conviction.

We also expect our business partners – paper mills, printers, packaging manufacturers, etc. – to commit themselves to using environmentally friendly materials and production processes.

The paper in this book is made from low- or no-chlorine pulp and is acid free, in conformance with international standards for paper permanency.

Lecture Notes in Computer Science
For information about Vols. 1–1045

please contact your bookseller or Springer-Verlag

Vol. 1046: C. Puech, R. Reischuk (Eds.), STACS 96. Proceedings, 1996. XII, 690 pages. 1996.

Vol. 1047: E. Hajnicz, Time Structures. IX, 244 pages. 1996. (Subseries LNAI).

Vol. 1048: M. Proietti (Ed.), Logic Program Syynthesis and Transformation. Proceedings, 1995. X, 267 pages. 1996.

Vol. 1049: K. Futatsugi, S. Matsuoka (Eds.), Object Technologies for Advanced Software. Proceedings, 1996. X, 309 pages. 1996.

Vol. 1050: R. Dyckhoff, H. Herre, P. Schroeder-Heister (Eds.), Extensions of Logic Programming. Proceedings, 1996. VII, 318 pages. 1996. (Subseries LNAI).

Vol. 1051: M.-C. Gaudel, J. Woodcock (Eds.), FME'96: Industrial Benefit and Advances in Formal Methods. Proceedings, 1996. XII, 704 pages. 1996.

Vol. 1052: D. Hutchison, H. Christiansen, G. Coulson, A. Danthine (Eds.), Teleservices and Multimedia Communications. Proceedings, 1995. XII, 277 pages. 1996.

Vol. 1053: P. Graf, Term Indexing. XVI, 284 pages. 1996. (Subseries LNAI).

Vol. 1054: A. Ferreira, P. Pardalos (Eds.), Solving Combinatorial Optimization Problems in Parallel. VII, 274 pages. 1996.

Vol. 1055: T. Margaria, B. Steffen (Eds.), Tools and Algorithms for the Construction and Analysis of Systems. Proceedings, 1996. XI, 435 pages. 1996.

Vol. 1056: A. Haddadi, Communication and Cooperation in Agent Systems. XIII, 148 pages. 1996. (Subseries LNAI).

Vol. 1057: P. Apers, M. Bouzeghoub, G. Gardarin (Eds.), Advances in Database Technology — EDBT '96. Proceedings, 1996. XII, 636 pages. 1996.

Vol. 1058: H. R. Nielson (Ed.), Programming Languages and Systems – ESOP '96. Proceedings, 1996. X, 405 pages. 1996.

Vol. 1059: H. Kirchner (Ed.), Trees in Algebra and Programming – CAAP '96. Proceedings, 1996. VIII, 331 pages. 1996.

Vol. 1060: T. Gyimóthy (Ed.), Compiler Construction. Proceedings, 1996. X, 355 pages. 1996.

Vol. 1061: P. Ciancarini, C. Hankin (Eds.), Coordination Languages and Models. Proceedings, 1996. XI, 443 pages. 1996.

Vol. 1062: E. Sanchez, M. Tomassini (Eds.), Towards Evolvable Hardware. IX, 265 pages. 1996.

Vol. 1063: J.-M. Alliot, E. Lutton, E. Ronald, M. Schoenauer, D. Snyers (Eds.), Artificial Evolution. Proceedings, 1995. XIII, 396 pages. 1996.

Vol. 1064: B. Buxton, R. Cipolla (Eds.), Computer Vision – ECCV '96. Volume I. Proceedings, 1996. XXI, 725 pages. 1996.

Vol. 1065: B. Buxton, R. Cipolla (Eds.), Computer Vision – ECCV '96. Volume II. Proceedings, 1996. XXI, 723 pages. 1996.

Vol. 1066: R. Alur, T.A. Henzinger, E.D. Sontag (Eds.), Hybrid Systems III. IX, 618 pages. 1996.

Vol. 1067: H. Liddell, A. Colbrook, B. Hertzberger, P. Sloot (Eds.), High-Performance Computing and Networking. Proceedings, 1996. XXV, 1040 pages. 1996.

Vol. 1068: T. Ito, R.H. Halstead, Jr., C. Queinnec (Eds.), Parallel Symbolic Languages and Systems. Proceedings, 1995. X, 363 pages. 1996.

Vol. 1069: J.W. Perram, J.-P. Müller (Eds.), Distributed Software Agents and Applications. Proceedings, 1994. VIII, 219 pages. 1996. (Subseries LNAI).

Vol. 1070: U. Maurer (Ed.), Advances in Cryptology – EUROCRYPT '96. Proceedings, 1996. XII, 417 pages. 1996.

Vol. 1071: P. Miglioli, U. Moscato, D. Mundici, M. Ornaghi (Eds.), Theorem Proving with Analytic Tableaux and Related Methods. Proceedings, 1996. X, 330 pages. 1996. (Subseries LNAI).

Vol. 1072: R. Kasturi, K. Tombre (Eds.), Graphics Recognition. Proceedings, 1995. X, 308 pages. 1996.

Vol. 1073: J. Cuny, H. Ehrig, G. Engels, G. Rozenberg (Eds.), Graph Grammars and Their Application to Computer Science. Proceedings, 1994. X, 565 pages. 1996.

Vol. 1074: G. Dowek, J. Heering, K. Meinke, B. Möller (Eds.), Higher-Order Algebra, Logic, and Term Rewriting. Proceedings, 1995. VII, 287 pages. 1996.

Vol. 1075: D. Hirschberg, G. Myers (Eds.), Combinatorial Pattern Matching. Proceedings, 1996. VIII, 392 pages. 1996.

Vol. 1076: N. Shadbolt, K. O'Hara, G. Schreiber (Eds.), Advances in Knowledge Acquisition. Proceedings, 1996. XII, 371 pages. 1996. (Subseries LNAI).

Vol. 1077: P. Brusilovsky, P. Kommers, N. Streitz (Eds.), Mulimedia, Hypermedia, and Virtual Reality. Proceedings, 1994. IX, 311 pages. 1996.

Vol. 1078: D.A. Lamb (Ed.), Studies of Software Design. Proceedings, 1993. VI, 188 pages. 1996.

Vol. 1079: Z.W. Raś, M. Michalewicz (Eds.), Foundations of Intelligent Systems. Proceedings, 1996. XI, 664 pages. 1996. (Subseries LNAI).

Vol. 1080: P. Constantopoulos, J. Mylopoulos, Y. Vassiliou (Eds.), Advanced Information Systems Engineering. Proceedings, 1996. XI, 582 pages. 1996.

Vol. 1081: G. McCalla (Ed.), Advances in Artificial Intelligence. Proceedings, 1996. XII, 459 pages. 1996. (Subseries LNAI).

Vol. 1082: N.R. Adam, B.K. Bhargava, M. Halem, Y. Yesha (Eds.), Digital Libraries. Proceedings, 1995. Approx. 310 pages. 1996.

Vol. 1083: K. Sparck Jones, J.R. Galliers, Evaluating Natural Language Processing Systems. XV, 228 pages. 1996. (Subseries LNAI).

Vol. 1084: W.H. Cunningham, S.T. McCormick, M. Queyranne (Eds.), Integer Programming and Combinatorial Optimization. Proceedings, 1996. X, 505 pages. 1996.

Vol. 1085: D.M. Gabbay, H.J. Ohlbach (Eds.), Practical Reasoning. Proceedings, 1996. XV, 721 pages. 1996. (Subseries LNAI).

Vol. 1086: C. Frasson, G. Gauthier, A. Lesgold (Eds.), Intelligent Tutoring Systems. Proceedings, 1996. XVII, 688 pages. 1996.

Vol. 1087: C. Zhang, D. Lukose (Eds.), Distributed Artificial Intelliegence. Proceedings, 1995. VIII, 232 pages. 1996. (Subseries LNAI).

Vol. 1088: A. Strohmeier (Ed.), Reliable Software Technologies – Ada-Europe '96. Proceedings, 1996. XI, 513 pages. 1996.

Vol. 1089: G. Ramalingam, Bounded Incremental Computation. XI, 190 pages. 1996.

Vol. 1090: J.-Y. Cai, C.K. Wong (Eds.), Computing and Combinatorics. Proceedings, 1996. X, 421 pages. 1996.

Vol. 1091: J. Billington, W. Reisig (Eds.), Application and Theory of Petri Nets 1996. Proceedings, 1996. VIII, 549 pages. 1996.

Vol. 1092: H. Kleine Büning (Ed.), Computer Science Logic. Proceedings, 1995. VIII, 487 pages. 1996.

Vol. 1093: L. Dorst, M. van Lambalgen, F. Voorbraak (Eds.), Reasoning with Uncertainty in Robotics. Proceedings, 1995. VIII, 387 pages. 1996. (Subseries LNAI).

Vol. 1094: R. Morrison, J. Kennedy (Eds.), Advances in Databases. Proceedings, 1996. XI, 234 pages. 1996.

Vol. 1095: W. McCune, R. Padmanabhan, Automated Deduction in Equational Logic and Cubic Curves. X, 231 pages. 1996. (Subseries LNAI).

Vol. 1096: T. Schäl, Workflow Management Systems for Process Organisations. XII, 200 pages. 1996.

Vol. 1097: R. Karlsson, A. Lingas (Eds.), Algorithm Theory – SWAT '96. Proceedings, 1996. IX, 453 pages. 1996.

Vol. 1098: P. Cointe (Ed.), ECOOP '96 – Object-Oriented Programming. Proceedings, 1996. XI, 502 pages. 1996.

Vol. 1099: F. Meyer auf der Heide, B. Monien (Eds.), Automata, Languages and Programming. Proceedings, 1996. XII, 681 pages. 1996.

Vol. 1100: B. Pfitzmann, Digital Signature Schemes. XVI, 396 pages. 1996.

Vol. 1101: M. Wirsing, M. Nivat (Eds.), Algebraic Methodology and Software Technology. Proceedings, 1996. XII, 641 pages. 1996.

Vol. 1102: R. Alur, T.A. Henzinger (Eds.), Computer Aided Verification. Proceedings, 1996. XII, 472 pages. 1996.

Vol. 1103: H. Ganzinger (Ed.), Rewriting Techniques and Applications. Proceedings, 1996. XI, 437 pages. 1996.

Vol. 1104: M.A. McRobbie, J.K. Slaney (Eds.), Automated Deduction – CADE-13. Proceedings, 1996. XV, 764 pages. 1996. (Subseries LNAI).

Vol. 1105: T.I. Ören, G.J. Klir (Eds.), Computer Aided Systems Theory – CAST '94. Proceedings, 1994. IX, 439 pages. 1996.

Vol. 1106: M. Jampel, E. Freuder, M. Maher (Eds.), Over-Constrained Systems. X, 309 pages. 1996.

Vol. 1107: J.-P. Briot, J.-M. Geib, A. Yonezawa (Eds.), Object-Based Parallel and Distributed Computation. Proceedings, 1995. X, 349 pages. 1996.

Vol. 1108: A. Díaz de Ilarraza Sánchez, I. Fernández de Castro (Eds.), Computer Aided Learning and Instruction in Science and Engineering. Proceedings, 1996. XIV, 480 pages. 1996.

Vol. 1109: N. Koblitz (Ed.), Advances in Cryptology – Crypto '96. Proceedings, 1996. XII, 417 pages. 1996.

Vol. 1111: J.J. Alferes, L. Moniz Pereira, Reasoning with Logic Programming. XXI, 326 pages. 1996. (Subseries LNAI).

Vol. 1112: C. von der Malsburg, W. von Seelen, J.C. Vorbrüggen, B. Sendhoff (Eds.), Artificial Neural Networks – ICANN 96. Proceedings, 1996. XXV, 922 pages. 1996.

Vol. 1113: W. Penczek, A. Szałas (Eds.), Mathematical Foundations of Computer Science 1996. Proceedings, 1996. X, 592 pages. 1996.

Vol. 1114: N. Foo, R. Goebel (Eds.), PRICAI'96: Topics in Artificial Intelligence. Proceedings, 1996. XXI, 658 pages. 1996. (Subseries LNAI).

Vol. 1115: P.W. Eklund, G. Ellis, G. Mann (Eds.), Conceptual Structures: Knowledge Representation as Interlingua. Proceedings, 1996. XIII, 321 pages. 1996. (Subseries LNAI).

Vol. 1117: A. Ferreira, J. Rolim, Y. Saad, T. Yang (Eds.), Parallel Algorithms for Irregularly Structured Problems. Proceedings, 1996. IX, 358 pages. 1996.

Vol. 1120: M. Deza. R. Euler, I. Manoussakis (Eds.), Combinatorics and Computer Science. Proceedings, 1995. IX, 415 pages. 1996.

Vol. 1121: P. Perner, P. Wang, A. Rosenfeld (Eds.), Advances in Structural and Syntactical Pattern Recognition. Proceedings, 1996. X, 393 pages. 1996.

Vol. 1122: H. Cohen (Ed.), Algorithmic Number Theory. Proceedings, 1996. IX, 405 pages. 1996.

Vol. 1125: J. von Wright, J. Grundy, J. Harrison (Eds.), Theorem Proving in Higher Order Logics. Proceedings, 1996. VIII, 447 pages. 1996.